DIRECTORY OF COMPUTING DIAGRAMS

Descriptive and Inferential Statistics

An Introduction

Fourth Edition

HERMAN J. LOETHER
California State University, Dominguez Hills

DONALD G. McTAVISH
University of Minnesota

ALLYN AND BACON
Boston   London   Toronto   Sydney   Tokyo   Singapore

Senior Editor: Karen Hanson
Executive Editor: Susan Badger
Series Editorial Assistant: Susan Hutchinson
Production Administrator: Marjorie Payne
Editorial-Production Service: Raeia Maes
Cover Administrator: Linda Dickinson
Cover Designer: Suzanne Harbison
Composition Buyer: Linda Cox
Manufacturing Buyer: Louise Richardson

This book is printed on
recycled, acid-free paper.

Copyright © 1993, 1988, 1980, 1974, by Allyn and Bacon
A Division of Simon & Schuster, Inc.
160 Gould Street
Needham Heights, MA 02194

Library of Congress Cataloging-in-Publication Data

Loether, Herman J.
    Descriptive and inferential statistics : an introduction / Herman
J. Loether, Donald G. McTavish. — 4th ed.
        p.   cm.
    Includes index.
    ISBN 0-205-14019-X
    1. Sociology—Statistical methods.   2. Sampling (Statistics)
3. Statistical hypothesis testing.   I. McTavish, Donald G.
II. Title.
HN25.L62   1993
001.4′22—dc20   92-10348
                                                            CIP

Printed in the United States of America

10 9 8 7 6 5 4 3 2 1   98 97 96 95 94 93

To our wives, Louise Loether and Janet McTavish, and our children,
Chris Loether, Kathleen, Karen, and Steve McTavish,
with love and appreciation

Contents

PART V   Inferential Statistics: Sampling and Probability

PART VI   Inferential Statistics: Hypothesis Testing

14   The Logic of Hypothesis Testing   475

15   Testing Hypotheses about Central Tendency, Dispersion, and Form   507

16   Hypothesis Tests of Association   574

Preface

In the social sciences generally and in sociology in particular, statistical tools are increasingly used in substantive inquiry. Whether the inquiry is descriptive, exploratory, or hypothesis testing in orientation, statistical tools are used because they help the investigator understand the relationships in (often) large data sets. Large-scale survey research is endemic in social science research and in the popular press. Special-purpose data sets exist on individuals questioned at different points of their lives, on the clients and workers in organizations such as health care, social service, or criminal justice agencies (*e.g.*, people diagnosed as having AIDS, subscribers of cable TV, applicants for jobs), and, of course, the massive society-wide censuses of the United States and most other modern nations. In most of these instances, the prudent investigator will turn to statistical tools to help in the analysis of his or her substantive problem.

Rather than present applied statistics as a field that stands by itself, divorced from any particular subject matter, we have chosen to focus on inquiry and have portrayed statistics as a set of tools that sociologists use in the process of making decisions about their data. It is not the only tool, of course, but it is an increasingly important tool. We have tapped the literature of sociology, government sources, and our own research for studies and data to serve as points of departure for the discussion of particular statistical techniques. When the subject matter of statistics is integrated into inquiry in the subject matter of sociology, students realize its value for sociology and for them.

By focusing on the uses of statistics in sociology it is possible to show how statistical techniques are actually applied, how these applications relate the ideal forms of many techniques to the actual data available, and what the consequences of this adjustment may be. In this book we have made a point of calling to the reader's attention the meanings and quality of the data to be analyzed. For example, we introduce measurement and the relationship of conceptual and operational definitions of concepts, which have consequences for statistical analysis. Various scales and indices are also illustrated. The text assumes data are fallible, not idealized. Thus, consequences of decisions about treatment of nonresponse are incorporated. Again, distributions are real instead of idealized. In discussions of sampling, implications of discrepancies between a sampling frame and a population are introduced as are practical alternatives to theoretically desired sampling procedures. We hope this will lead the reader to become a more informed user of statistics and to be more cautious about the unqualified acceptance of statistical findings.

In a first course in statistics, we are convinced that students need to gain experience by actually conducting some inquiry in which they are interested and by using statistical techniques. Thus, we recommend giving students access to an important, interesting data set(s) and the computer means for exploring their ideas. By using standard statistical software with a computer on a major, real data set permits the student to see many instances of a statistical measure and gain some insight into the value of statistics for making comparisons in the context of an important research problem. This contrasts sharply with traditional approaches that focus on hand computation of a given statistic (*e.g.*, standard deviation or correlation coefficient) on one or a few unrelated bits of artificial data.

Although statistics is based on mathematical considerations, we do not feel it necessary at this level to present rigorous, mathematical definitions of concepts or formal mathematical proofs of theorems. The underlying logic of the field of statistics is relatively simple and can be understood, to a large extent, from a verbal presentation. For the student who is pursuing statistical reasoning further, mathematically oriented course work in statistics is essential, but for a beginning course, the point of statistics appears to be better presented if it focuses on the reasons for research grounded in important issues of a particular field.

Unlike many traditional treatments of statistics, this text begins with the use of statistical reasoning to *describe* data. We find that this foundation is essential for understanding the role of statistics in inquiry. This is developed from univariate to bivariate to multivariate description. In each case, alternative approaches to description (often based on characteristics of the data) are grouped together and the rationale for a choice between them is given. Thus the student is better prepared to diagnose an analysis problem and consider a range of options with different strengths. Graphic analysis of data is also underscored. The latter half of the text is devoted to the problem of *statistical inference*. Here issues of sampling, sampling error, and the logic of inference using confidence limits and tests of significance are presented. We have presented this section in the context of real problems of data collection and sources of bias and error. Related statistical tests are again grouped to provide an organized way to consider options for univariate, bivariate, and multivariate analysis. The separation of descriptive and inferential statistics helps clarify the various ways in which statistics are useful in inquiry.

Most chapters in this edition carry the same titles as those in the previous edition, but substantial changes have been made in the contents of several of them. Overall, we have tried to tighten up the presentation by editing for clarity and by eliminating some techniques that are little used (*e.g.*, Chapters 2 and 4 were shortened, the triangular plot was eliminated from Chapter 4, some of the hand computation routines were taken out of some computing diagrams, etc.). Reluctantly, we have omitted the chapter on a collection of techniques we refer to as "space analysis" (*e.g.*, cluster analysis, factor analysis, multidimensional scaling), due to space limitations. In many places, the references and data have been significantly extended and updated. Unfortunately, however, much of the detailed 1990 Census information was not available at the time revisions were made. More substantial changes in individual chapters are noted below.

Chapter 1 was extended to underscore the role of computers and statistical packages a student might encounter and to provide sources for them. We have also tried to underscore the importance of a healthy suspicion about the accuracy of new statistical software and of checking computer results to be sure that errors do not go undetected.

Chapters 2 and 4 were shortened. The triangular plot was omitted from Chapter 4.

Chapter 5 begins with a new, more relevant and useful example. The grouped computation for the median has been dropped from the text. Further linkages have been made between approaches to ordinal measures of association and univariate measures of variation. The coefficient of variation was dropped as well. Diagrams in Chapter 5 omit computation of moments since skewness and kurtosis measures are more likely to be used from SPSS and SAS computer output rather than computed by hand.

Chapter 7 was significantly revised to incorporate the uncertainty coefficient, not tau-y, as the nominal measure of association, in line with its availability in SPSS.

Chapter 8 has an added section on using the Consumer Price Index to change current dollars to standard dollars, an example of standardization. This is an adjustment that is likely to be useful to students and in studies of economic welfare and trends.

Chapter 9 incorporates new data. The hand computation diagrams for regression and partial correlation have been dropped in deference to the likely use of computers for these procedures.

The new Chapter 10, introducing inferential statistics, has been shortened and examples have been revised substantially. Instead of pages of random numbers in the appendix, a short basic program to compute pseudorandom numbers has been used.

Chapter 11 includes new boxed material on dealing with nonrandom factors and weighting for disproportionate samples. The concept of a sampling frame has been updated as well. The section on sampling distribution of proportions has been moved from Chapter 12 to Chapter 11 to complete the discussion of sampling distributions in a more compact and complete way.

Chapter 12 includes a box on the finite population correction of the standard error for samples that are nearly enumerations.

Chapter 13 has a shortened and sharpened discussion of confidence limits.

Chapter 14 includes a new section on the analysis of power of a test.

Chapter 15 has been extended significantly to include nonparametric statistical tests (randomization test, sign test, direct difference test for matched samples, and the runs test). These not only introduce a subarea of statistical tools but also illustrate statistical creativity in re-thinking a difficult problem in order to solve it effectively.

Finally, Chapter 16 includes more discussion of the use of computer statistical packages.

In this and the previous edition, we have moved further on the assumption that many colleges and universities will provide students with ready access to a microcomputer or mainframe computer. Step-by-step *computing diagrams* are at the ends of chapters and many have been simplified to focus on raw data formulas and to exclude most of the so-called "short-cut" computational procedures needed for hand computation. Some small, basic-language programs are provided for useful procedures not currently incorporated in standard statistical packages. We have also provided more detail on available statistical software and where it can be acquired.

A set of computer disks is available from the publisher and includes data and computer instructions that may be of use to an instructor using this book. These disks can be modified and printed for use as student handouts. The data disk includes a variety of modest-sized data bases that may be used with homework, laboratory problems, or for examinations. These data are provided in a form that can be printed out in student handouts or transferred to mainframe computers or microcomputers for student use with statistical program packages such as SPSS, SAS, or MINITAB. In addition, disks are available that include examples of how each of several different statistics packages may be used to apply the techniques covered in this book. These text files can be modified to include local conventions and used as handouts to explain the appropriate statistical package to students. Included are instructions for MINITAB, SPSS/PC, SPSSX, SAS, BMDP/PC, and SYSTAT.

Universities and colleges in the United States appear to be in a transitional phase and have a great variety of types of mainframe and microcomputer equipment and software and very different policies of student access. Thus, we have attempted to support a variety of ways in which examples and problems can be approached, expecting that very soon there will be realistic support for greater educational access to microcomputer technology.

We have retained all of the teaching aids that appeared in the third edition and, where necessary, we have revised or expanded them. These aids are the following:

1. *Diagrams* that follow the discussion of each technique and that summarize the essential steps in its use. These diagrams are placed at the end of the chapter and may serve as a review of the techniques and as useful step-by-step guides to computation.
2. *Loop-back boxes* placed at strategic points throughout the text to flag important points in preceding discussions and to refer the reader to other sections of the text that may help clarify the subjects covered.
3. *Key concept* lists are provided at the end of each chapter. These help flag key ideas covered in the text and serve as a focus for reviewing.
4. *Discussion questions* at the end of each chapter require the reader to integrate materials from the chapter or to apply what has been learned. These questions are supplemented by computational problems.
5. *References*, annotated and listed at the end of each chapter, provide a starting point for the reader who wishes to explore some of the topics covered in more depth.

It is our experience that students who take advantage of these learning aids will benefit from the effort they devote and will gain a more thorough understanding of the subject matter of the book.

We solicit your reactions, suggestions, and reference to examples you have found to be useful in learning and teaching statistical approaches to sociological inquiry.

All cross-reference citations within the text use section numbers. The first digit of a section number is the chapter; two-part section numbers indicate major section divisions (*e.g.*, 3.2); three-part section numbers (*e.g.*, 3.1.2) indicate minor section divisions; the numbering sequence runs from major to minor divisions (if any) and back to the next major division (thus, in Chapter 3 we have Sections 3.1; 3.1.1; 3.1.2; 3.1.3; 3.2; 3.3; 3.3.1; etc.). In addition, major or minor numbered sections may have subsections designated by an added lower-case letter (thus: Sections 3.1a; 3.1b; 3.1.1a; 3.1.1b; etc.), usually indicating a paragraph heading for a particular topic.

For material in the Appendix, we are indebted to the literary executor of the late Sir Ronald A. Fisher, F. R. S., and to Oliver and Boyd, Edinburgh, for permission to reprint Tables III, IV, and V from their book, *Statistical Tables for Biological, Agricultural, and Medical Research.*

We are indebted to Edmund D. Meyers, Jr., and to the late Richard J. Hill, who read the entire original manuscript, and to Marian Deininger who read the original descriptive sections. Kirby McMaster of California Polytechnical Institute, Pomona, and Charles Y. Glock of the University of California gave us detailed critiques of the first edition of the book. James Wood, San Diego State University, provided helpful comments and suggestions on improving our presentation of ordinal measures of association. We wish to acknowledge the helpful assistance of Deming Wang, University of Minnesota, who provided an extensive analysis of the uncertainty coefficient and its relationship to other measures of association as well as other suggestions to improve the manuscript. Comments and suggestions by Robert Leik, University of Minnesota, and Lance Egley, were also very helpful. Their suggestions for improvements in the manuscript had a major influence on the changes made in succeeding editions. Other colleagues, too numerous to name, also contributed valuable suggestions and pointed out errors in the earlier editions. The suggestions of all of these have added significantly to the quality of the current book.

Finally, we are indebted to our students at California State University, Los Angeles, California State University, Dominguez Hills, and the University of Minnesota who contributed by their questions and reactions to the development of this book.

This has been a joint product of the authors. We have both been so deeply involved in the writing that we are no longer able to sort out individual contributions. The order in which our names appear on the title page could easily have been reversed. They are merely listed in alphabetical order.

H. J. L.
D. G. M.

CHAPTER 1

An Introduction to Statistics in Sociology

How many people do you know in your community? Most of them? Only a few in your social circles? Do you know your grocer (or banker, member of the school board, neighbor, or gas station operator) and speak to them by name on a conversational basis? Sociologists find that people in small towns tend to be acquainted, at least informally, with a high proportion of the people in their town. This is not true of larger urban areas. The larger the place, the smaller the proportion of people with whom one is acquainted. Freudenburg (1986), a sociologist, calls this community characteristic "density of acquaintanceship." He studied the phenomenon in four small western towns. One of these was an energy "boom town." It grew very rapidly over the 1970s decade, swollen by workers in a new energy-production industry. He found that 92% of the people he sampled in the three stable small towns said they knew and spoke with their grocer, for example. Only 63% of the people in the energy "boom town" were able to say this.

Why should this be so? Freudenburg argues that it is simply physically impossible to know most of the people in a large community. Furthermore, people in larger communities are not as likely to be (nor do they plan to be) long-term residents. Larger populations also tend to include people from a greater variety of backgrounds and sometimes this leads to residential segregation, which further reduces the opportunities for acquaintanceship.

What difference does that make? Some have argued that in communities with a low density of acquaintanceship people are more likely to feel isolated and outside the mainstream of community life. But research does not bear this out. Even in communities characterized by low density of acquaintanceship, most people seem to have a circle of acquaintances and close friends. Freudenburg found, however, that communities with lower density of acquaintanceship tend to have more problems (and more serious problems) in three areas: (a) *control of deviance* (higher density of acquaintanceship apparently involves more neighborly watchfulness and informal social control); (b) *socialization* of the young (higher density of acquaintanceship makes socialization a community as well as a family affair); and (c) *caring for those in need* of help (higher density of acquaintanceship

1

seems to involve more favorable conditions for shared and informal community "caring" functions). And it wasn't just the new residents who accounted for these problems!

Freudenburg bases his conclusions on a variety of kinds of information. He studied the four towns first hand, visiting them over a period of years and conducting interviews with people who knew and lived in those communities. He also carefully selected a sample of households from the towns at one point in time, and conducted a survey in which nearly 600 interviews were completed. This helped guard against biases that might have crept into more informal visits (*e.g.*, talking only with a select and nonrepresentative group of informants) and provide information on the same questions asked of everyone in the sample. Finally, Freudenburg used census data and town records on such things as crime rates, mental health services, and the size of the police force. He was able to compare his data with statistics from other community studies in the United States and Canada.

What implications does this study have for solving problems of crime and welfare in communities, for introducing major new industries into smaller towns, or for developing strategies for community organization? Do the findings still apply, for example, when larger rural towns rapidly lose population? Can our legislators rely on this information in designing governmental policies? How can we, as citizens, assure ourselves that conclusions from studies such as this one are true?

Clearly, we would want legislators to have sufficient understanding of statistics to make a responsible judgment.* Furthermore, we would want sufficient insight into these matters to assure ourselves both that good information is properly used by those in a position to influence important decisions, and that unworthy information is recognized and set aside.

One of the most impressive developments in contemporary society has been the vast growth of research into all aspects of our physical environment, our society, and ourselves. Increasingly, statistically analyzed research results and statistical models are used as a basis for gaining insight and making decisions. To understand contemporary society, to understand the basis of many decisions, to follow intelligently and evaluate the output of others' studies, or to handle important lines of thought and reasoning, knowledge of elementary statistics is essential. The demand for research is growing, and research is being conducted by neighborhoods, political parties, and public and private agencies, as well as university investigators. Chances are rather high that you will become involved in the conduct and evaluation of some kind of research yourself. Certainly, we cannot read critically and interpret most of the literature in sociology, much less make original contributions to knowledge in this area, without a relatively detailed knowledge of how to interpret statistical arguments and descriptions and apply them to our own ideas and research. For leader or legislator or citizen, for poet or social scientist, the need for statistical insight seems to be a fact of contemporary life and a powerful tool for understanding and decision making.

---

* The Freudenburg study used statistical techniques such as percent change, rates, arithmetic means, and *F*-tests for statistical significance, all of which are discussed later in this text.

1.1   SOME BASIC CONCEPTS IN STATISTICS

The Freudenburg study involved four communities from one western U.S. area during the 1970s. For each community, Freudenburg uses degree of acquaintance-ship as well as other properties or characteristics such as size, crime rates, and the average number of years people had been residents of the community. Properties of communities such as degree of acquaintanceship and level of crime are called **variables**, or ways in which communities could differ from each other that are of interest to investigators. In other words, a *variable* is a measurable characteristic of a "case" or "unit of analysis" (such as a community) that can differ in value from one unit to another. Sometimes it is helpful to refer to the scores, or measured values, on variables (such as those Freudenburg collected) as **statistical observations** or simply **observations**. Typically, statistical observations are expressed as numbers (e.g., population size, crime rate) although numbers are not essential to the use of statistics.

It is important to know what one's statistical observations are, of course. The number 1,000 has little meaning until one learns that it is the size in number of residents of a community in the western United States in 1979. Notice that the statistical observation was described in terms of three things:

1. A **variable**—what the characteristic is that was measured (e.g., size in terms of number of people).
2. A **case**—what the variable was measured upon, the "unit of analysis" (e.g., a community).
3. The **location** of cases in *time* and *place*—where and when this measured community is located (e.g., western United States, 1979).

The second point deserves further comment. Part of Freudenburg's study dealt not with communities but with individual people—the survey, for example, and the people with whom he talked when he visited these towns. As he points out, acquaintanceship could be studied as a property of individuals or as a property of communities. The kind of "object" being studied makes considerable difference. The consequences for individual people of being socially isolated is one thing (e.g., mental health and suicide). Consequences for individuals living in communities with a high (or low) density of acquaintanceship are much different (e.g., problems with social control, socialization, and welfare). Furthermore, what it means for old residents (or newcomers) to live in a place with high (or low) degree of acquaintanceship (or the rapid change in this regard, as in "boom towns") is also a different phenomenon. The kind of "object" being measured is called the **case** or **unit of analysis**, and in the Freudenburg study cited earlier, the unit of analysis was the community, not the individual resident. Sociology often deals with different units of analysis, and this makes clarity on this point very important.

Freudenburg's study involved four cases that were selected largely because they were convenient representatives of a class of communities he was interested in studying and had an opportunity to study. One part of his study, how-ever, illustrates how a more representative sampling can occur. He sampled households in the communities at one point in time, asking adults about the people in town they knew, their length of residence, and other impressions about changes in their community. Scores (measurements or "observations") on each of the variables he measured for each of the 597 adults included in this survey constitute what is known as a **statistical sample**. Scores on each of the variables he measured, but for all adults in these communities (several thousand people) in that same year, would constitute what is called a **statistical population** or **universe**. It is composed of scores on all the variables the investigator wants to measure on all of the "objects" at the time and place indicated; in short, a statistical population is *all relevant statistical observations*. In this study, the statistical population was **finite**, since, in principle, the statistical observations in such a population could be counted. Often this is not the case, particularly in scientific pursuits that hope to develop principles applying to certain kinds of phenomena wherever and whenever they may occur. An **infinite** statistical population, sometimes called a **conceptual** or **general universe**, would be one in which relevant statistical observations are *not limited to any particular time and/or location*. All cities of the past, present, and future; all students any time or place they may exist; or all families that ever have or ever will exist would be examples of infinite statistical populations or conceptual universes.

BOX 1.1   FOR THOSE WHO LIKE OVERVIEWS

There are certain basic *themes* and ideas that unite the various specific techniques in different chapters of this book. These themes are reviewed in Chapter 17, and it may be helpful to you to read through that chapter now. Chapter 17 should alert you to possible ways of organizing your thinking about statistics.*

Whether the universe being studied is infinite or finite, investigators rarely enumerate an entire statistical population (*i.e.*, they seldom examine *all* statistical observations); rather, they select a subset of the units of analysis called a **statistical sample**. Freudenburg's study was based, in part, on a sample from a defined statistical population. Although any subset of a statistical population would constitute a sample, in practice some kinds of samples may not represent very well the population from which they were drawn. Usually investigators are interested in selecting an appropriate sample that will give them accurate information about the population from which it was drawn. They make inferences about the population, but they want to do this in a way that is sound, practical, and economical. Samples are, of course, always finite, although they may also be large, as in the samples used by the Census Bureau to describe U.S. citizens. The field of inferential statistics, the second part of this book, is concerned with the ways in which reliable samples may be drawn and how inferences can be based on them (see Chapter 11 for a thorough survey of sampling procedures in sociology).

---

* "Loop-back" boxes such as this will be used here and there in the text to flag important ideas you should be sure to get at that point, or to suggest points where you should pull yourself back and reflect on your grasp of the argument being made. We will refer and cross-refer you to points where necessary ideas can be found and reviewed.

1.2   THE FIELD OF STATISTICS

Like most words, the word *statistics* is used in several ways in everyday speech. Some use it to refer to virtually any number that describes something, such as the height of Mont Blanc, the number of babies born in Vermont, or the turning radius of a Ford. *Statistics* is also used in two more technical ways: first, to refer to a certain subject matter or field of intellectual pursuit, and, second, in a specific and narrow sense (which will be defined later) related to the description of statistical samples.

The field of statistics, broadly viewed, encompasses a range of techniques and human activities that border on mathematics on one side, and on the other, on the problems that investigators and people in everyday life want solved. Basically, statistics involves logical thinking about statistical samples or populations, either for the purpose of arriving at a compact and meaningful description, or for the purpose of assessing some of the risks involved in extending conclusions based on samples to conclusions about populations. Logical thinking about such collections of scores involves knowing which comparisons to make as well as how to make the comparisons in an informative way. The field of statistics includes a growing "kit" of techniques particularly useful for making these descriptions and inferences.

The subject matter of statistics may be divided into two parts. One part, called **descriptive statistics**, consists of tools and issues involved in describing collections of statistical observations, whether they are samples or total populations. The other part, called **inferential** or **inductive statistics**, deals with the logic and procedures for evaluating risks of generalizing from descriptions of samples to descriptions of populations.

Roughly the first half of this book will focus on descriptive statistics. An analysis of how statisticians make inferences by drawing reliable samples is pursued in the latter half of the book. One last terminological distinction may be worth making at this point. A second technical meaning of the word *statistics* refers to any summary measure of a characteristic or property of a *sample* of statistical observations. However, if the collection of statistical observations that is described is a statistical *population*, then a summary measure of a characteristic or property of these observations is called a **parameter**. For instance, the average degree of acquaintanceship of all U.S. communities in 1979 could be treated as a population *parameter*, and the average degree of acquaintanceship in the four communities sampled by Freudenburg could be called a *statistic*. A statistic and a parameter may be the same kind of summary measure (*i.e.*, a percentage or average or ratio). It is *what* is described—population or sample—that makes the difference between these terms. The focal problem of inferential statistics is the relationship between a statistic and a parameter. The problem of descriptive statistics is the development of ways to analyze samples *or* populations, that is, to develop different kinds of statistics and parameters that could be used on any given collection of scores both to describe and to analyze them.

The field of statistics has grown enormously in the past 50 years or so. Doctorates are offered in this specialty, and there are professional associations and professional journals that report new techniques and raise and resolve new issues. Student memberships are offered for these, and it would be worthwhile for you as a student to look at some of the ideas and publications in this field. Its history makes fascinating reading.

1.2a   A Bit of History

The development of the field of statistics is one of the more interesting segments of our intellectual history. One theme in statistics (a word coined in a 1770 British publication) has been the interest of politicians in characteristics of their citizenry, usually for taxation purposes. In early "political arithmetic" was the work of Englishman John Graunt (1620–1674), who was one of the first to describe gross population data in his book *Natural and Political Observations Made upon the Bills of Mortality*, which described 1604–1661 London death records. This was shortly followed by an analysis by English astronomer Edmund Halley to "ascertain the prices of annuities upon lives," the first life tables. Quetelet (1796–1874), Belgian supervisor of statistics and "father of the quantitative method in sociology," and Galton (1822–1911) extended work in statistical description and analysis, the latter developing percentiles, the median, correlation, and regression—all ideas discussed later in this book. Karl Pearson (1857–1936), English biologist, extended statistics in the areas of regression, correlation, and chi-square.

A *second theme* is that of probability. In China as early as 300 B.C. there is reported discussion of the probability of an unborn child being a boy or girl. Early pondering on outcomes of games of chance (*e.g.*, dice rolling, card dealing) led to the development of formal rules of probability and their application to the fledgling field of astronomy by such men as Galileo (1564–1642), Pascal (1623–1662), and Fermat (1601–1665). The French astronomer Demoivre (1667–1754) was first to describe the *normal curve* as a description of errors in his observations. James Bernoulli (1654–1705) is credited with what was later called the "Law of Large Numbers," a principle central to statistical inference that states what the expected sample value of proportions will be, and how differences between sample and true proportions decrease as size of sample increases. Gauss (1777–1855) extended thinking on probability and worked with the concepts of standard deviation and standard error and, under the pseudonym "Student," Gossett (1876–1937) extended further our knowledge of sampling distributions. In the early 20th century, Ronald A. Fisher developed the analysis of variance and other statistical tests, and contributed in important ways to knowledge about the design of experiments. You will encounter the names of other pioneers in statistics as you read through this book—names such as Cochran, Yule, Yates, Spearman, Markoff, and Poisson. An interesting presentation of the work of some of these pioneers appears in James R. Newman's *The World of Mathematics* (Simon and Schuster, New York, 1956). The development of statistics is far from finished, and one of the exciting current topics in sociology is the tailoring and creation of new and more useful statistical techniques.

BOX 1.2 TERMS TO CONSIDER FURTHER

Some of the terms defined here will be used in later chapters. To familiarize yourself with these, it might be helpful at this point to flip through a sociology book or journal to see if you know the meanings of these terms and can find examples of each of them. Especially important ideas are the following in Sections 1.1 and 1.2:

> *Statistical observation*
> *Variable*
> *Unit of analysis, case*
> *Statistical population*
> *Statistical sample*

1.3 THE RELATIONSHIP BETWEEN STATISTICS AND SOCIOLOGY

Sociologists are interested in studying individual social actors, groups, or organizational processes, interaction patterns, norms, and social behavior of various types. Typically, sociologists concerned with, say, the age structure of societies and its consequences for kinship structures might begin by simply gathering information on age structure from each of a series of societies with existing records. To highlight the age structure rather than simply the size of the society, they will use statistical procedures for creating, say, averages or proportions of those over 65, or perhaps "dependency ratios" (*i.e.*, the ratio of the number of young and old people to the number of middle aged). Since the data are quite likely based on a careful sample of existing societies, they may well want to know how much confidence they should place in these figures, and again they will turn to statistics, especially inferential statistics, to help reason out the risks involved in using a sample. They may also examine the record-creating procedures in different societies for evidence of bias, and perhaps compute an index number of some sort to show the way two or more variables are related—a correlation coefficient, for example.

Simple statistical description frequently leads to more detailed theoretical work and, perhaps, the creation of some type of model or theory about the way age structure would be expected to influence kinship. Usually this process will lead sociologists to take account of other variables, such as the level of technology in a society, or sex ratios, and generally they will then have to set out to measure these variables on a sample of societies. If they are fortunate, information may already exist in some archive, but usually they will have to organize records for their own purposes or gather new data. These processes pose issues of *validity* (*i.e.*, questions of whether you are measuring what you want to measure) and of *reliability* (*i.e.*, questions of whether the measurement process itself is stable), and these again may be examined with statistical tools. In fact, they will probably express their hypotheses or theories in terms of statistical descriptions and will compare their revised theories with new data in terms of statistical tools.

A brief look at some of the main journals in the field, the *American Sociological Review*, the *American Journal of Sociology*, or *Social Forces*, for example, should convince you of the centrality of statistical expression and analysis in our field. An increasingly interesting area of sociology, in fact, is the development of more useful statistical tools for handling some interesting new questions about social behavior. One of the purposes of this book is to help you move to the forefront of sociology where interesting phenomena are described in new and different ways.

1.4   WHAT YOU CAN EXPECT IN STUDYING STATISTICS

You will do much better, we believe, if you have some knowledge of what to expect in general from your encounter with statistics. Most students seem to find the area somewhat different from others they know, requiring something of a new approach.

First and most basic, the field of statistics is oriented toward drawing logical conclusions. This implies that there is some type of *question* or *problem* that commands an interest in finding an answer. Freudenburg's study addressed several different questions: Does the density of acquaintanceship drop when small towns greatly increase their size? What seems to account for lowered density of acquaintanceship? What consequences are there for individuals? For communities? Procedures for supporting the claim that an answer has indeed been found are basically those of logic. Statisticians develop various measures, examine their logical implications, and then critique and develop statistical models that may be of use in finding answers in certain restricted situations. Although this characterizes many of the interests scholars and citizens have about the world, it by no means exhausts the things they are interested in doing to discover or experience what their world is like. It is, nevertheless, a surprisingly useful approach.

Second, the field of statistics is quite similar to a language. There are several key concepts and technical terms, some of which we have already introduced, and there is a kind of grammar involved in using them. The language is relatively carefully worked out and consistent, although sometimes the same idea may be symbolized by somewhat different symbols in different contexts. Once you learn how to express yourself in the language, you will find that you have a very powerful way to make important distinctions and express logical relationships. In fact, it is often true that difficult problems disappear once expressed in a clear and consistent fashion. The early part of the book will emphasize the important "language of statistics."

Third, although it is true that many of the summary measures and lines of reasoning are expressed in numerical form, there is a basic underlying logic of statistical reasoning that can be discussed and handled apart from much of mathematics. Mathematical-logical procedures, however, are such powerful and logically clear tools for handling statistical reasoning that they are nearly always of interest to the professional statistician. This means, for example, that it is often helpful to look for ways to express ideas and relationships in numerical form, although this is not always true nor formally necessary. It also means that one could neatly express the logical relationships between different ideas by mathematical notation and derivation. In this beginning treatment of statistics we will highlight the underlying logic without resorting to more than elementary algebra, but it is clear that we are imposing a limitation on expression that could not be conveniently imposed on general treatments of statistical reasoning. Other books use more mathematical expression to explain the reasoning involved, and you may want to consult one of these references from time to time.

Finally, the field of statistics focuses upon situations where there are repeated or repeatable operations—such as in taking a measurement of the organizational involvement of each of 137 people who are members of the local YMCA, or as in tossing an imaginary coin an infinite number of times and recording whether the outcome each time is "heads" or "tails," or as in drawing samples of organizations in order to measure the relationship between the height of their stratification system and productivity. In principle, these repeated or repeatable operations (*i.e.*, samplings, tossings, measurings, etc.) must be such that there could be a range of different outcomes. For example, in principle, people might have different degrees of organizational involvement, coins might turn up "heads" or "tails," and an organization might have a "tall" or "flat" stratification hierarchy, or have high or low productivity.

If one's approach does not involve *logical* progress toward solution of an interesting problem, if there is no interest in *supportable* findings, if *repeated or repeatable* operations and the possibility of *differences* in outcomes are not involved, then the field of statistics probably does not have much to offer. The surprising thing is that statistical reasoning *is* so useful in handling many of the truly important problems confronting scholars in sociology and many of the issues confronting those who govern and are governed. Basically, we are interested in drawing logically supportable conclusions in situations where we have carried out some kind of measurement procedure on a number of cases (often on samples of cases). We need to summarize our findings clearly, assess risks of inference, and draw valid conclusions. Most of all we need to get to the point where we can use useful tools to promote important inquiry in our field without losing sight of the original purpose of our inquiry. We hope the organization of this book will help.

1.5   THE ROLE OF COMPUTERS

The rapid advances in computer technology have had a dramatic impact on the field of statistics. Computers are widely used now to analyze quantitative data. Several highly sophisticated and comprehensive packages of computer software are available not only for large (mainframe) and medium-sized (mini-) computers, but also for the small but powerful microcomputers (also known as personal computers) whose use is so widespread.* The most popular of these software packages

* By one estimate there were more than 70 million personal computers in use throughout the world in 1991 (Magid, 1991).

are SPSS (Statistical Package for the Social Sciences), SAS, and BMDP (Biomedical Programs). Although less complete than these three, MINITAB is another package available for mainframes, minicomputers, and microcomputers. There are also numerous statistical packages designed particularly for use on microcomputers.* One of the most elaborate of these is one called SYSTAT (references to these statistical packages are provided at the end of this chapter).

Some or all of these packages are available at universities and colleges throughout the United States (and many other countries), as well as the hardware (computers) to run them. In addition, government and private agencies that engage in research make extensive use of such packages. These days the lone researcher, too, is almost certainly equipped with a microcomputer and statistical software.

The speed and efficiency with which statistical software handles the analysis of data are so great that a computer **should** be used whenever quantitative research is undertaken. The speed and efficiency with which analysis proceeds is roughly proportional to the size and capacity of the computer used. Mainframe computers can generally handle larger and more complex data bases than can mini- or microcomputers. Even microcomputers that are "state of the art" can now handle relatively large data bases with comparative ease and tolerable rates of speed.

It is more than likely that computers of one sort or another will be available to you in connection with your statistics course and most assuredly will be available to you should you become involved in research. Accordingly, it is assumed, for purposes of this book, that any necessary computations will take place on a computer. Consequently, emphasis is put on applications and interpretations of statistical techniques, rather than on the mechanics of their computation.

Prior to the availability of computers, conceptualization in social science research tended to be oversimplified and testing of conceptual schemes was piecemeal. It was not at all unusual to devote each of a number of research projects to small segments of a single conceptual scheme in hopes of later fitting their separate findings together in some meaningful way. Now, however, much more elaborate conceptual schemes are being constructed and more comprehensive tests of such schemes are being conducted. While this book deals with techniques normally executed by computers, Chapters 9 and 16, in particular, provide examples of some of the more sophisticated analytical and inferential techniques that lend themselves to computer analysis. Because it is now practical to work with these more sophisticated models, the social science disciplines are developing increasingly realistic explanations of social phenomena and progressively more efficient predictions of their occurrence.

Unfortunately, the influence of computers on statistical analysis has not all been positive. Because powerful statistical software is so readily available, its use has often been abused. Once data are reduced to numerical codes and are input to a computer in an acceptable format, the statistical software will perform any type of analysis of which it is capable, without regard for the quality of the data. Consequently, persons who are naive about the constraints on legitimate

---

* When statistical packages for microcomputers were evaluated in 1989 for *PC Magazine*, over 200 such packages were available (Goldstein, 1989:99).

statistical analysis can use computers and statistical packages with relative ease to perform analyses that are completely without merit. Often, in their ignorance, these people assume that, because the analysis was successfully carried out by the computer, "it must be right." As a result, contemporary research literature is full of flagrant examples of misuses of statistics. "User-friendly" software is not entirely positive in its effects. As a matter of fact, it is counterproductive to the extent that it facilitates the misuse of the statistical techniques programmed.

It is very importat to realize that a computer is only a tool for the rapid processing of data in accordance with specific and detailed instructions programmed into it by a human agent. The computer is incapable of making any judgments about the appropriateness of the instructions it is commanded to carry out. Consequently, the person controlling the computer **must be qualified** to make informed judgments about what analysis is appropriate to do in the first place and about how the results of the analysis should be evaluated. One lesson that you can ill afford to neglect in reading this book is that, to be an intelligent user of statistics, you must understand (1) what assumptions you are making about your data in using a particular statistical technique, (2) what type of analysis of the data the technique performs, and (3) how the results of the analysis may be interpreted.

1.6   PURPOSES AND OBJECTIVES OF THIS BOOK'S APPROACH

The aim of this beginning course in sociological statistics is to introduce you to those aspects of statistical description and inference that will be most useful to you as a consumer and potential creator of statistical information. The emphasis will be on applications of statistical reasoning, description, and inference and on some of the theory behind it, as well as on interpretation of these procedures in the context of sociological research. At the end of the course you should be able to read intelligently and evaluate statistical arguments made in social science research literature.

To accomplish this, we are presenting an elementary text with a coverage of current techniques somewhat broader than usual; we incorporate in a subordinate role the methodological issues that you will encounter in more depth in other social science research methods courses; we attempt to emphasize a user's approach and understanding in the context of social science problems and current usage; and we minimize the more formal mathematical mode of expression as well as minimizing most aspects of clerical mathematics. We have assumed that the more clerical or computational aspect of statistics is less important at an elementary level, and in any event will most likely be handled by a computer when the need to compute arises. It is our view that it is much more important to gain a comprehension of statistical measures through *several* encounters with summary measures in real research contexts, rather than to emphasize computational routine or shortcuts on a few fictitious examples of "error-free" data.

Ideas are fun, particularly when one has the tools to rub them together in interesting ways. It is hoped that the pursuit of statistics in sociology will be interesting, challenging, and fun, too. Some of the ideas behind *measurement* in sociology are introduced in the next chapter.

CONCEPTS TO KNOW AND UNDERSTAND

statistical observation   unit of analysis, case
statistical population   statistic
statistical sample   parameter
descriptive statistics   variable
inferential statistics

QUESTIONS AND PROBLEMS

1. Some sociologists study the phenomenon of power balance in families by
   having a family (parents and children) play a game of some type in a laboratory
   setting. As the game progresses the investigators record verbal and nonverbal
   interaction (*e.g.*, who asks questions of whom, who provides information or
   gives direction or commands). What are the unit of analysis and the main variable of interest to the investigator? Give an example of a statistical observation
   that could be made. Define a statistical population such an investigator could
   study.

2. Select any sociological journal or research monograph that reports the results
   of research and then select an interesting article. Identify the statistical
   observation, unit(s) of analysis, and variables the investigator uses. Then
   describe the sample and population they are dealing with. Point out situations
   where descriptive and inferential statistics are used, and give an example of a
   statistic and a parameter. See your instructor for a list of journals.

3. Usually, early in a course, it is helpful to gather some actual statistical observations yourself. To do so you might do one of the following:
   a. Go to the cafeteria and observe people sitting at tables. How many people are
      sitting together at each table? (Note that you must define what you mean
      by "sitting together" in order to count numbers of people.) Record numbers
      of males and females in each "sitting" group.
   b. Observe shoppers in a shopping mall or district. Observe the sizes of groups
      of shoppers and the distribution of shoppers by gender in each group
      observed. Classify the shoppers as children, young adults, middle-aged
      adults, and older adults. Observe the distribution of these age categories in
      each group of shoppers. What problems do you have in deciding on age?
   c. Check the vital statistics section of the newspaper for seven consecutive
      days. Note the age difference of each couple applying for a marriage license.
      Make note of whether the male is older or younger than the female and the
      number of years' difference.

GENERAL REFERENCES

Blalock, Hubert M., Jr., *Social Statistics*, Revised Edition (New York, McGraw-Hill), 1979.
    Often it helps to read the discussion of some statistical procedure in more than
    one text, and the intermediate level text *Social Statistics* would be especially
    recommended for this purpose. It is detailed, relevant to our field, and covers most of
    the topics included in this book.

Brent, Edward, and Ronald E. Anderson, *Computing Applications in the Social Sciences* (New York, McGraw-Hill), 1990.

Hacking, Ian, *The Emergence of Probability: A Philosophical Study of Early Ideas About Probability, Induction and Statistical Inference* (New York, Cambridge University Press), 1975.

An informative account of the history of statistical inference.

Norusis, Marija J., *The SPSS Guide to Data Analysis for SPSSX* (Chicago, SPSS, Inc.), 1991.

Rowland, David, Daniel Arkkelin, and Larry Crisler, *Computer-based Data Analysis Using SPSSX in the Social and Behavioral Sciences* (Chicago, Nelson-Hall), 1991.

Tufte, Edward R. (ed.), *The Quantitative Analysis of Social Problems* (Reading, Mass., Addison-Wesley), 1970.

This book is a collection of articles on social problems and studies about them that use statistics or a quantitative approach. Although some of the issues are beyond the scope of this text, the articles are all interesting and help point out some of the difficulties and successes in studies for which statistical analysis may be relevant.

Wallis, W. Allen, and Harry V. Roberts, *Statistics: A New Approach* (New York, The Free Press of Glencoe), 1956.

Chapters 1 and 2 present an interesting introduction to statistics and some of the uses to which it is put. A large number of brief illustrations are presented.

LITERATURE CITED

Freudenburg, William R., "The Density of Acquaintanceship: An Overlooked Variable in Community Research?", *American Journal of Sociology*, 92:1 (July 1986), pp. 27–63.

Goldstein, Richard, "Understanding Numerical Analysis: Working with Odds," *PC Magazine* (March 14, 1989), pp. 94–99.

Magid, Lawrence J., "Computer File," *Los Angeles Times*, August 8, 1991, p. D3.

Newman, James R., *The World of Mathematics* (New York Simon & Schuster), 1956.

STATISTICAL SOFTWARE PACKAGES

BMDP   For mainframes and minicomputers, and BMDP/PC for microcomputers. BMDP Statistical Software, 1440 Supulveda Boulevard, Los Angeles, CA 90025.

MINITAB   For mainframes, minicomputers, and microcomputers. Minitab, Inc., 3081 Enterprise Drive, State College, PA 16801.

SAS   For mainframes, minicomputers, and microcomputers. SAS Institute, SAS Circle, Box 8000, Cary, NC 27512-8000.

SPSS   For mainframes and minicomputers, and SPSS/PC+ for microcomputers. SPSS, Inc., 444 N. Michigan Drive, Chicago, IL 60611.

SYSTAT   For microcomputers. Systat, Inc., 1800 Sherman Avenue, Evanston, IL 60201.

StatXact   For microcomputers, (nonparametric tests). Cytel Software Corporation, 137 Erie Street,Cambridge, MA 02139.

CHAPTER 2

Measuring Variables

Sociologists have consistently found a relationship between autonomy in work and self-esteem. When there is freedom from continuous surveillance and frequent supervisory intervention, workers tend to have a greater sense of self-esteem. Schwalbe (1985) argues that there are several different ways that autonomy influences self-esteem; it is a reward for reliability and competent performance, an indicator of status in the workplace, and freedom to act and take responsibility for success. This relationship is of interest to sociologists because it concerns consequences of social structure for individuals in organizations and helps explain why some workers have high self-esteem and others have lower self-esteem.

Schwalbe's research involved on-the-job interviews with 103 people in five job groups: first-line supervisors, engineers, maintenance mechanics, secretaries, and production/assembly workers. Since the job groups have different degrees of autonomy, he could ask about their self-esteem and make comparisons between the different job groups. Research in this area requires that sociologists be able to define and measure a number of different variables, among them "self-esteem."

Thus, in addition to the problems of locating and selecting research subjects, securing their cooperation, and recording their responses, there are two basic problems to be solved that lead directly to a consideration of what statistical reasoning is all about. These problems are (a) classifying subjects by degree of self-esteem, and (b) comparing self-esteem scores for job groups providing different levels of work autonomy. The first problem will be discussed in the remainder of this chapter. The second problem will be dealt with in the next chapter where some of the difficulties and possibilities for making valid comparisons will be described.

2.1 MEANINGS AND MEASUREMENT

The first problem, noted above, is measuring self-esteem. **Measurement**, most simply stated, is a procedure for carefully classifying "cases" [called research subjects, respondents, or, more generally, "objects," such as a person, an interacting pair of people (a dyad), a company, or a whole society] into previously defined categories of some variable. A **variable** is simply any characteristic or property of a case that has a series of two or more possible categories into which a "case" could potentially be classified. Self-esteem is a variable characteristic of individuals. Classifying a person as having "high" or "low" self-esteem is an example of a simple measurement process.

To develop a useful measurement of "self-esteem," for example, sociologists need to define the concept. Schwalbe defines self-esteem as positive or negative feelings toward oneself, or an affective response to what one thinks about oneself—for example, feeling good about oneself because others seem to see one as competent and effective. This definition is called a **conceptual definition**. In measurement, a necessary place to start is with a conceptual definition of each variable. A good definition is clearly stated and contains sufficient information so that researchers can develop ways to classify cases on the basis of that concept.

The next step is to develop practical procedures that permit classifications to be made. How can one tell whether a worker has "high" or "low" self-esteem? One way might be to simply ask a worker: "Do you feel positively or negatively about yourself?" and record + or − or "no answer" according to the response. Alternatively, one could ask a person: "Please rate yourself from 1 to 10 in terms of how you feel about yourself. One is feeling very negatively about yourself and 10 is feeling very positively about yourself." The result would be a number from 1 to 10 (or "no response"). These specific procedures to measure a concept are called **operational definitions**, operations by which measurements of the variable are actually carried out.

A widely used procedure for measuring self-esteem is a series of questions rather than just one question, as shown in Figure 2.1. This particular measurement procedure is called a **Likert scale** because it is composed of several statements all aimed at measuring the same concept and respondents are asked to indicate how strongly they agree or disagree with each statement. Notice that some of the statements have the 1–4 responses reversed. This is because some statements refer to positive feelings about oneself and other statements refer to negative feelings. An overall "self-esteem" score is created by adding up the numbers of the responses a person circled for each question. Scores could range from 10 (all 1's, indicating a strongly positive self-esteem) to 40 (all 4's, indicating a strongly negative self-esteem).

Usually there are many different operational definitions of a given concept like self-esteem.* Some of these measurement procedures are better than others. Two general questions can be asked about measurement procedures: (a) Does the procedure actually measure the concept we want it to measure? and (b) Can we depend upon the procedure to give the same result if the same cases were measured under the same conditions (i.e., assuming no real change on the variable being measured)? The first question refers to **validity** and cannot be answered without a good conceptual definition of the concept to use as a reference; the second question refers to the **reliability** of a measurement procedure. Scales are often preferred because they can be more reliable than single-item measures. Statistical analysis can help evaluate both reliability and validity of measures.

---

* It is interesting to note that the problem of definition is sufficiently serious for whole books to be written about it. One interesting and useful book is by Robinson (1954).

Circle the answer for each question that most closely expresses your feelings.	Strongly agree	Agree	Disagree	Strongly disagree
1. I feel that I'm a person of worth, at least on an equal basis with others.	1	2	3	4
2. I feel that I have a number of good qualities.	1	2	3	4
3. All in all, I am inclined to feel that I am a failure.	4	3	2	1
4. I am able to do things as well as most other people.	1	2	3	4
5. I feel I do not have much to be proud of.	4	3	2	1
6. I take a positive attitude toward myself.	1	2	3	4
7. On the whole, I am satisfied with myself.	1	2	3	4
8. I wish I could have more respect for myself.	4	3	2	1
9. I certainly feel useless at times.	4	3	2	1
10. At times I think I am no good at all.	4	3	2	1

Source: Rosenberg (1965).

FIGURE 2.1   Rosenberg Self-Esteem Scale

2.1a Data

The result of a measurement process is classification of each case into a category of a variable. The result of this process is a **statistical observation** or, simply, an **observation** that is usually labeled with an appropriate name or number. Other words for a statistical observation are "datum" or "score." A collection or group of several of these scores is called **data.*** Descriptive statistics are aimed at describing and analyzing data.

The meaning of a score resulting from a measurement process depends upon the way variables are defined and the adequacy of the measurement process itself. The way these scores are statistically described depends on the information they contain, and upon the specific use of the variable in an overall research project.

Three differences between variables will be pointed out in the following discussion: (a) level of measurement (Section 2.1.1), (b) scale continuity (Section 2.1.2), and (c) the role the variable plays in a research project (Section 2.1.4). Information on the first and second of these comes ultimately from the conceptual definition of a variable. Information on the third comes from theory and the research problem.

---

* Note that the word *data* is plural. One way an investigator's data *are* described is by using the statistical techniques discussed in this book.

2.1.1   Level of Measurement

To illustrate some of the different types of information a defined variable is intended to convey, it is convenient to distinguish among four **levels of measurement**. These four types of measurement are traditionally listed in statistics texts primarily as a way to reinforce the importance of knowing what information is contained in scores that are assigned.

This distinction is one of several that have an important influence on the kinds of statistical summary one chooses, and it certainly influences the kinds of interpretations warranted. It is worth noting that some scholars distinguish more levels of measurement than these four, and they may include different mixtures of these that feature some of the properties of the levels mentioned below (Stevens, 1946).

2.1.1a   Establishing Categories of Measurement

Any adequately specified measurement procedure must have a category into which each and every case can be classified; it must be **inclusive**. If only the categories 0, 1, and 2 were available to classify individuals on the number of groups to which they belong, there would be no category for those belonging to more than two groups. Such a situation could be corrected by making the last category "2 or more" or by providing a category labeled "other," if additional categories could not be defined in a more positive way. The "other" category is sometimes used simply to create a logically **exhaustive** or inclusive classification system. In most actual research it turns out that categories for "don't know" (often symbolized in printed tables as DK) and "no answer" (often symbolized as NA) are often needed as well. These categories pose problems for the thinking statistician.

Another property of a good classification system is that its categories are **mutually exclusive** as well as exhaustive; that is, it is possible to classify an individual case into one and *only one* category of the classification system. If a variable has two categories "married" and "has children," for example, it would not be adequately defined in this sense, because married people with children could be classified into both categories. The categories are not mutually exclusive. "Ever married" and "never married," on the other hand, are mutually exclusive. It is true, of course, that individuals change over time, but at any given point when the measurement is made it should be possible for a case to fall in only one category.

It is preferable, generally, to have a measurement procedure that is as **precise** as possible, that is, one that makes more rather than fewer distinctions. The measurement scheme "never married," "currently married," and "other" would be less adequate in this regard than "never married," "currently married," "divorced," "widowed," and "separated," simply because there are more categories in the second scheme. Knowledge that a subject is "divorced" is less ambiguous and conveys more information than knowledge that they are an "other." Adequately defined variables are generally classified into the following four levels of measurement.

2.1.1b   Nominal Variables

Nominal variables are those that are properly defined with logically exhaustive and mutually exclusive categories, so that equivalences or differences are clearly established. Good nominal variable measures are *precise* and have *positively defined* categories. There is no implication, however, of any ordering of categories; there is no implication that distance exists between one category and the next; the categories are simply logically *different* or distinct from each other. Many scientific variables are of this kind. In sociology, nominal variables include marital status, gender, group membership, religious affiliation, and kind of role relationship.

The idea of level of measurement refers to the relationship between *categories* of a variable. For nominal variables the categories are simply different, whether or not they are labeled with numbers. It would not be appropriate to add together, for example, the category labels of a nominal variable. Sometimes students wonder why formulas are used to compute statistics on nominal variables if this is so. The reason is that the formulas make use of category frequencies, the number of cases in each category, and not the numerical label that may be used to identify any particular category of the variable. It *does* make logical sense to say that there are twice as many "ever married" people as "never married" people, but it would not make sense to add up numbers that may be used to label marital status categories (although it would make logical sense to do this for a variable such as age).

2.1.1c   Ordinal Variables

Ordinal variables are those defined to include the features of nominal variables, plus the feature that categories are *ordered* or *ranked* in a described way. For example, "friendliness," as we usually think of it, is defined as an ordinal variable. Categories generally distinguished (low, medium, and high, for example) not only imply categorical differences but also imply an order from low to high on that variable. We are not able to define *amount* of difference between these categories however; it is only sequence that we can define in an ordinal variable. Traditional class grading schemes are often thought of as ordinal properties of students. Social class is ordinal (*e.g.*, lower-lower, upper-lower, lower-middle, upper-middle, lower-upper, upper-upper class), no matter how precisely it may be measured. In principle, it is not defined in such a way that a "degree" or "unit" of social class is meaningful or distinguishable, although we feel we can distinguish rankings by social class. Rank order is defined in variables such as self-esteem, desirability, segregation, and many other attitudinal variables used in sociology.

The values assigned to the categories of an ordinal scale may be expressed either in words or in numbers. Friendliness, for example, as defined above could take the values low, medium, and high. On the other hand, the prestige rankings of universities might be expressed as numbers (e.g., 1st, 2nd, 3rd, etc.). Note, however, that these numbers are ordinal rather than cardinal. The usual arithmetic operations of addition, subtraction, multiplication, and division are not permissible with ordinal numbers. It makes no sense whatsoever to add two numbers such as 1st and 3rd. Consequently, ordinal level measurement does not lend itself to computations that require the use of arithmetic. It is permissible, however, to count the numbers of observations that fall into each category of the scale.

2.1.1d   Interval Variables

Interval variables, a third level of measurement which is usually distinguished, include the logical and ranking features of ordinal and nominal variables, but in addition, the categories are defined in terms of a standard *unit of measurement*, such as the individual "social actor," the year of age, or degree. Definition of a unit that can be unambiguously detected and counted is a difficult undertaking. You may recall accounts of the very elaborate definitional procedures to establish the foot or meter, the degree Fahrenheit, the ton, or the U.S. dollar. The U.S. Bureau of Standards, a special governmental agency, is concerned with the definition of certain standard measurement units such as these. Given a recognizable unit such as "one social actor," however, one is able to compare the differences between pairs of values on an interval scale. Therefore, although we cannot say that a value of 40 is twice a value of 20, we can say that a difference in values of 40 points (*e.g.*, the difference between scores of 20 and 60) is twice as large as a difference of 20 points (*e.g.*, the difference between scores of 20 and 40). An interval scale, then, allows us to compare the difference between families with one or two members and those with three or four members. The *differences* are the *same*. We can distinguish this sameness by virtue of having defined "one social actor." The scale of measurement of an interval variable is defined in terms of a standard unit size.

Determining the interval for a variable is an unusually important step up from nominal and ordinal variables, in that it greatly increases the amount of information that is contained in numerical scores. Given interval variables, a large amount of mathematical manipulation is possible, and conclusions can be drawn that are often not at all obvious from the raw scores themselves. Much of statistics implies the use of interval scores, and because of the added power of these procedures, there is a strong motivation to use these techniques whenever they can be justified.

2.1.1e   Ratio Variables

Finally, *ratio* variables include the foregoing characteristics and, in addition, the scale of measurement has a defined origin, or zero point. A zero point is a theoretically defined point on a scale representing the absence of any of the property being measured. Not all properties can easily be measured from zero. Zero social class, for example, is not a meaningful concept thus far in sociology. On the other hand, zero inches of length or families of size zero is meaningful in this sense. Prior to the development of theoretical ideas of an absolute zero point in measurement of temperature, other temperature scales used the concept of zero degrees (as in Centrigrade or Fahrenheit) as a relative measurement that did not correspond to a theoretical absence of any positive features of the property "temperature." In the Fahrenheit and Centigrade scales, since degree-units were defined, but zero did not represent any absolute in temperature, their degree-unit would be classified as an interval, not a ratio variable. With the development of the Kelvin scale, temperature became a ratio variable.

If it is possible to define a variable as a ratio variable, then it is also possible to talk about comparable score ratios (*e.g.*, 2 years of education is half of 4 years and this ratio is the same as the ratio of 12 years to 24 years). These statements would not be possible if one could not count units (*e.g.*, year units) from a meaningful zero starting point. For example, if one is interested in the variable "amount of learned knowledge," the number of years of education might be used as an indicator of this concept. In this case, however, because of the way "amount of learned knowledge" is usually defined, we may not have available a specified unit of "learned knowledge" nor a meaningful zero point on the "learned knowledge" scale. Notice that the way a variable is defined governs its level of measurement. We probably would not want to say, for example, that a person with four years of education has exactly twice the amount of learned knowledge that a person with only two years of education has.

In many cases, once one has defined a standard unit of measurement in an interval scale, a definite zero point may also be established. For this reason, variables in sociology that we are able to measure with this degree of sophistication may qualify as ratio scales rather than interval scales. Examples of such variables would include counts like "family size," "number of levels in an organizational hierarchy," "group size," "number of years of formal education," and perhaps "social status."*

In statistics, the techniques appropriate for the analysis of ratio-scale data may also be used for the analysis of interval-scale data; consequently, we will not make a distinction between interval and ratio levels of measurement, but will refer to either type as "interval/ratio" or simply "interval" variables.

Figure 2.2 illustrates the cumulative character of levels of measurement. The important point is that the way one defines a variable has consequences for the categories developed, the relationships between them, and the meaning of an assigned score.†

2.1.1f   Using Level of Measurement Information

Labels for categories can be of a wide variety of different types. Names ("married," "high self-esteem," etc.) are often convenient to use, although sometimes letters or, more frequently, numbers are used. The utility of numbers lies in the possibility of an unlimited set of symbols and in the fact that there is a set of powerful logical rules (mathematics) for handling numbers. The symbols one uses and the available means for operating on the symbols has far greater consequences than is immediately apparent. It leads most scientists to the use of numbers and to various branches of mathematics for handling these number codes, although in principle the logical distinctions could be handled (usually ineffi-

---

* An example of work toward establishing a ratio scale of measurement for social status is Jones and Shorter (1972). See also Hamblin (1971).
† A very helpful chapter on measurement in sociology, and on some of the meanings that may be available in scores, is Clyde Coombs' Chapter 11 in Festinger and Katz (1953). Also see Blalock (1982).

Defined Characteristics
of Category Systems

Level of Measurement	Exhaustive, Mutually Exclusive Categories	Categories Defined to Have an Ordering	Defined, Standard Unit of Measurement	Meaningful Zero Point on Scale
Nominal	Yes	No	No	No
Ordinal	Yes	Yes	No	No
Interval	Yes	Yes	Yes	No
Ratio	Yes	Yes	Yes	Yes

FIGURE 2.2   PROPERTIES OF DIFFERENT LEVELS OF MEASUREMENT

ciently and with great effort and risk of confusion) by other code schemes such as words.

A word of caution is necessary at this point. It is clear that number symbols could be used to label the categories of a variable like marital status, as well as the categories of the variable age. Marital status, however, is a nominal variable, since its categories have no defined ordering, and age is usually defined as an interval/ratio variable. For marital status of a group of individuals, one could not use the mathematical operation of adding up the number labels, for example, because such a manipulation of the scores would rely upon information about ordering and relative spacing of categories, and these are not part of the (usual) definition of marital status. The adding operation is, of course, appropriate when the variable is years of age, an interval/ratio variable. To take another example, if one were to score 10 individuals on gender, assigning the number code 0 for males and 1 for females, a collection of ten scores as the following might be available.

$$0 \quad 0 \quad 1 \quad 0 \quad 0 \quad 1 \quad 1 \quad 1 \quad 0 \quad 1$$

If these 10 numbers were added and then divided by 10 to create a simple average score for the group, the average gender would be 0.5. Adding and dividing scores in this manner clearly involves the assumption that females are higher—one unit higher—than males. A statement that the average gender of this group of scores is 0.5 is nonsense, but, on the other hand, if this were reinterpreted as a proportion or percentage (*e.g.*, 50% of the group are female), the results can be meaningfully interpreted.

There are statistical procedures that are created in such a way that only nominal or only ordinal or only interval/ratio meanings in scores are relied upon. It is most appropriate logically to use the statistical procedure that features the

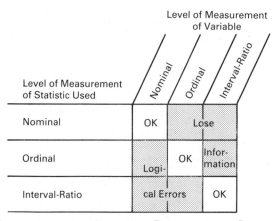

FIGURE 2.3   CORRESPONDENCE OF STATISTICAL PROCEDURES AND LEVEL OF MEASUREMENT

kind of information contained in the scores. As indicated in Figure 2.3, if statistics appropriate for lower levels of measurement were to be used on scores defined at a higher level of measurement, no *technical* error would be made, since the properties of levels of measurement are cumulative. An interval measure contains all the properties of a nominal variable plus added properties—added information. However, although no technical error is made, a serious research problem would arise in that this procedure does not use all of the information readily available in the higher level score.

A technical "error" *is* made, however, if we use statistical procedures appropriate to higher levels of measurement when the data are defined at a lower level. In this case, we would be acting by our statistical operations as if the scores contained more information than they were defined to have—that is, we would have assumed a property of order where there was none, or of defined distances where there were none. Consequences for the interpretation of research findings when this technical error is made have been the subject of detailed discussion in sociology (see Burke, 1971; Labovitz, 1970; replies by Vargo *et al.*, 1971). The strong preference of most careful investigators is to select appropriate statistics for the data at hand.

Statistical procedures appropriate to higher levels of measurement generally have the advantage of permitting more compact descriptions of data, as we shall see. In the case of our dichotomous (two-category) nominal variable of gender in the example above, the "average gender" could be readily reinterpreted as a percentage: a meaningful interpretation. On the other hand, if we discovered that the average "marital status" scores were 2.563, we would be essentially at a loss to provide a meaningful interpretation of this figure. Some research on the issue indicates that for many purposes, a procedure that moves from ordinal to interval statistics while using ordinal data does not result in great errors in interpretation of the statistical results. The important point is to know what differences between scores constitute meaningful variation, and then to use statistical procedures that highlight meaningful variation in scores, ignoring undefined or meaningless differences.

An associated problem is that of not collecting sufficient information in the process of measurement to classify a case adequately. For example, if one wanted

---

BOX 2.1 LEVELS OF MEASUREMENT—SOME REVIEW EXAMPLES

It would be well to pause at this point and think of some examples of variables that fit each *level of measurement* we have been discussing. If you are clear about this first classification of variables, then you are ready to go on to the other two ways to classify variables—by scale continuity, and by role in a research project.

As a review starter, try defining "age" first as a nominal variable, then as an ordinal variable, and finally as an interval or ratio variable.

Note: Lack of information does not change the level of measurement of a variable. To decide, use the conceptual definition. A potential source of invalidity is an operational definition that doesn't allow one to use the variable as it is conceptually defined. A source of lack of precision may be grouping categories together.

---

to measure age (an interval/ratio characteristic) but found that employment forms only listed the person as "old," "middle-aged" or "young," one would have to treat the variable as ordinal because information on number of units (years) is not available in this classification for use in computation of interval/ratio statistics. If the information is not contained in the data one gathers, no amount of statistical manipulation will create it.

2.1.2 Scale Continuity

Another distinction in addition to level of measurement involves whether or not variables are defined to have a continuous or discrete scale of measurement. A **continuous** variable, as Figure 2.4 suggests, is defined in such a way that in principle individuals could have infinitely varied fractional scores—that is, scores located at any point on an unbroken scale. The same concept can be stated in at least three other ways. First, given any two scores, however close together, in

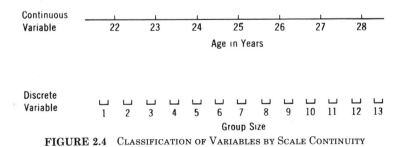

FIGURE 2.4 CLASSIFICATION OF VARIABLES BY SCALE CONTINUITY

principle a third score could always fall between them; or second, the probability that any two scores will be identical is zero; or third, an infinite number of different scores is possible between any two points on the scale. In any case, this concept implies an unbroken scale of possible score values. In sociology such variables as age, alienation, segregation, and social class are defined (usually) as continuous variables. There is a well-known debate in the social stratification field over whether the phenomenon of social class is most usefully defined as discrete or continuous (Duncan and Artis, 1951).

A contrasting type of variable is a **discrete variable**. This type is defined in such a manner that, again in principle, only a *limited set* of distinct scores can be achieved. Stated differently: (a) the scale of measurement is broken by "gaps" in the numbering scale that cannot contain measured cases; only integers are needed to label scale categories appropriately, not decimal fractions; or (b) within a finite range there are only a finite number of possible score categories. Thus a variable such as family size (as usually defined) is a discrete variable, because families can only be of size 1 or 2 or 3 . . . etc.; there can be no example of fractional size such as 2.5. This is true of most "count"-type variables; nominal variables are always discrete. It should be noted that "average family size" is a different variable. It is measured on "cases" that are *groups*, such as states or nations or cities; it is possible for that kind of variable to take on fractional values, since it is usually defined as a continuous variable characteristic of *groups* of observations.

2.1.3   Grouping Scores Into Classes

From the start, most measurement processes in sociology have tended to group scores into defined classes. In the case of nominal and ordinal variables, it is usually quite clear that categories of social class, anomie, marital status, or productivity have to be defined before one can begin to measure or ask into which of these categories the individual falls. Actually, this is the case with interval and ratio variables as well.

2.1.3a   Units of Measurement and Nearest Whole Units

Typically, investigators confronted with the problem of measuring an interval variable decide upon a **unit of measurement** such as the inch, foot, year, or thousands of people. They then count up the number of these "units" that can be said to characterize the object being measured rather than attempt to measure continuously the exact amount of these variables.* The interpretation of continuous

* A thermometer measures temperature in a continuous fashion, since the physical height of a column of fluid varies with temperature. We usually record temperature, however, as if the variable were discrete, in terms of degrees or tenths of degrees. An analog computer works on the principle of continuous measurement by varying electrical current. A digital computer, on the other hand, operates in terms of discrete states where a switch is either on or off, or a magnet is polarized in one direction or not, etc. Most measurement of continuous variables involves counting units of measurement to yield discrete-looking scores. However, the differences, built in by definition, between discrete and continuous variables mean that scores can be manipulated and interpreted differently.

variable scores is different from the interpretation of discrete scores, and this difference is shown in the way continuous variables are measured.

For continuous variables, the problem is one of deciding where any given case should fall in a classification scheme, and then characterizing the category so that its meaning can be interpreted. Take the age category, for example, that we usually refer to as "65." Assuming that age has been measured by asking individuals what age they were, at their *nearest* birthday, the category could be said to include those who are past their 64th birthday by just over a half a year (half a unit of measurement) and that category would still include persons who are almost half a year over their 65th birthday. To be precise, the category would extend from 64.5 to 65.5 years, as illustrated in the following diagram:

Measuring Age to the Nearest Whole Year:

| 62.5 | 63.5 | 64.5 | 65.5 | 66.5 | "real limits" or class boundaries |

| 63.0 | 64.0 | 65.0 | 66.0 | class midpoints |

A category of an interval variable is generally referred to by its **midpoint**. In this case, using 1 year as a *unit* of measurement, the category midpoints would be 63.0, 64.0, 65.0, 66.0, etc. Scores, or measured values, would be classified into the nearest category, which means that scores in the category 65.0 may be as small as 64.5 and as large as 65.5. These limits, extending one-half unit on either side of the midpoint, are called **real limits** or **class boundaries**.*

2.1.3b Significant Digits and Rounding

How far one should round off an answer, to how many digit places, is determined basically by the degree of precision in the original measurement process. If a measurement is exact (*e.g.,* counting the number of people in a group, a discrete variable), then it is appropriate to retain all significant digits in a score and to carry out computations involving these scores to as many decimal places as desired, treating the results as meaningful and significant (although this could easily be carried to the point of absurdity, *e.g.,* average family size = 2.5632567). However, if measurement is inexact, as it is in the case of continuous variables, then the answer should not be expressed to more digits than are significant in the original scores (sometimes the rule is to admit one more digit than this). Thus, if one were measuring age in years, answers should be rounded to whole years (or, perhaps, tenths of years) even though the computational process (as in computing an average) may result in a fractional number that could be carried out to many decimal places. If births are recorded to the nearest 100, then the answers from computations using these scores should be expressed to the nearest 100 births. To express the result to more digits gives readers a distorted impression of accuracy. It is appropriate, however, to retain as many digits as possible throughout a com-

---

* Students are usually concerned about classification of scores near class boundaries—a relatively rare and minor problem in most research. What if "super-precise" subjects reported their ages as 65.51 years or 65.49 or 65.50; how shall one classify these age figures? The rules of *rounding* solve the problem. These rules are given in Figure 2.5, and they would result in classifying the first age above as 66, the next as 65, and the third as 66.

If a number such as 34.6 were to be rounded to a whole number, the digit to the right of the place to which it is to be rounded should be examined. In this case, that digit is "6." Then:

(a) If that digit is greater than 5, round up by increasing the digit to the left by one. In the example above, this would result in a rounded number of 35.

(b) If that digit is less than 5, round down by simply dropping digits to the right of the last place in the desired answer.

(c) If the digit is exactly 5 (followed by zeros), then the rule is to round up if the last digit in the desired answer is odd, and round down if the last digit in the desired answer is even. This is the so-called "engineer's rule" and its logic is simply that odd and even numbers are likely to occur equally frequently so that no bias in rounding up or down will occur if this is always followed.

Some examples should help.

65.60	could be rounded to	66
64.50	could be rounded to	64
65.50	could be rounded to	66
64.51	could be rounded to	65
65.49	could be rounded to	65

FIGURE 2.5   RULES FOR ROUNDING OFF

putational procedure rather than rounding at each step, since this practice avoids accumulation of small rounding errors at each computational step. Only the final answer should be rounded. On the other hand, at any point where digits are to be dropped (*i.e.*, the number is too big for your calculator) they should be rounded off, rather than simply dropped, so that systematic distortions are not built into the result.

These principles can be quickly illustrated if we think of an example where two ages are to be added together, say the age 65 and the age 69. A score called 65 (the midpoint of its category) could be anything between 64.5 and 65.5 in actual value. Likewise a score called 69 could be between 68.5 and 69.5. Adding the class midpoints of these two scores gives a total of 134 (65 + 69 = 134), but the actual sum of the two scores could be as small as 133 or as large as 135 (64.5 + 68.5 = 133.0 and 65.5 + 69.5 = 135). Perhaps the sum should be expressed with only two significant digits as about 130 or as 134, but certainly not 134.0, which gives the impression that measurement was to the nearest tenth of a year. With more scores, of course, the problem becomes potentially more serious.

When a classification scheme is published in a table or chart, it is common practice to express the category boundaries in a fashion that clearly indicates that the categories do not, in fact, overlap, but are mutually exclusive. Thus the series of age categories

Years Age
31–35
36–40
41–45

would be preferred over an expressed system such as:

Years Age
30–35
35–40
40–45

However expressed, the meaning of a score on a continuous variable extends from real limit to real limit.

2.1.3c   Measurement to the Last Whole Unit

Measurement to the **last whole unit** is sometimes used rather than measurement to the *nearest* whole unit. This introduces a complication that is worth a word of caution. A common example of this is in the measurement of age where individuals typically are asked for or give information on their ages at their *last* birthday even though their next birthday may be nearer. The meaning, then, of the age expressed as "65" would change, and the data would have to be treated slightly differently as a result. The real limits of the age 65 measured to the *last full year* would be 65.0 and 66.0, as shown in the diagram below.

Age Measurement to the Last Whole Year:

Although the difference between measurement to the "nearest" or "last" whole unit may seem to be slight and may not lead to gross errors in interpretation if caution and consistency are adhered to, failure to use the appropriate midpoint will have the effect of adding (or subtracting) a constant to each score. It is important that information measured by a measurement procedure be correctly represented.

The relationship between scale continuity and levels of measurement is shown in Figure 2.6A, and it is usually helpful at this point if you read a sociology text or journal article and attempt to classify the variables used in terms of *both* scale continuity and level of measurement. An abstract from an article by Marsh is reproduced in Figure 2.6B. Can you identify the different variables, units of analysis, and types of information probably contained in scores on the different variables?

For most statistical purposes, the distinction between continuous and discrete variables, where ordinal and interval/ratio variables are concerned, is of less importance.* In some treatments of statistics and in other parts of statistics than the descriptive, the distinction would be more consistently carried through. Here we will make the distinction at points (such as in graphic presentation of

---

* Note that we have used the term *variable* in a broad sense in this chapter to refer to properties we may be interested in measuring, regardless of their level of measurement or scale continuity. In many statistics texts the term *variable* is reserved for only those characteristics that are defined as interval or ratio in level of measurement and, perhaps, are also continuous. Other characteristics, in this usage, are then referred to as "attributes" or "categorical" data. Whichever usage you and your instructor prefer, the important point is that you should carefully inquire into the meaning of the scores that are assigned.

Level of Measurement	Scale Continuity	
	Discrete	Continuous
Nominal	sex a	~~Logically Impossible~~
Ordinal	b	d
Interval-Ratio	family size c	e

To illustrate the use of the figure above, sex (a nominal, discrete variable the way it is usually defined) is an example in cell *a*. Meanwhile, family size (a ratio variable, and a discrete variable, if it is defined in terms of number of related people) is an example of a variable appropriately classified in cell *c* above.

FIGURE 2.6A CLASSIFICATION OF VARIABLES

The following is an abstract from an article by Robert M. Marsh (of Brown University) in *Social Forces*, 50 (December, 1971), p 214–222. Most journals include these brief abstracts or "roadmaps" to a research report, and in most cases they indicate the main variables, the ideas to be examined, and some of the main conclusions. Can you classify the variables Marsh talks about in terms of Figure 2.6A above?

THE EXPLANATION OF OCCUPATIONAL
PRESTIGE HIERARCHIES

Abstract
The ranking of occupational prestige in Taiwan is highly similar to that in the United States and the numerous other societies—both modernized and relatively non-modernized—in which such studies have been conducted. Previous explanations of this important cross-societal invariant, occupational prestige evaluation, have emphasized *common structural features* of any complex society, whether industrialized or not. It is not clear how this proposition can be empirically falsified. A more satisfactory explanation is sought in terms of *specific properties of occupational roles* in any society, viz., education, responsibility (or authority), and income. The relationship of each of these variables to prestige is analyzed with data from Taiwan, Denmark, and the United States.

Source: Marsh (1971).

FIGURE 2.6B CLASSIFICATION OF VARIABLES IN RESEARCH

data) where it is necessary to describe data accurately. Our main point here is that the scores one chooses to summarize contain certain kinds of information, and the manipulations and conclusions one is able to draw from them depend upon a clear understanding of their meaning. The results of statistical description do not contain any more information or meaning than the scores that were originally produced by measurement.

2.1.4   Roles Variables Play in Research

A third and crucial way in which variables differ from each other is in the use made of them in research. Usually, two main categories of variables are distinguished: **independent** and **dependent**.

Frequently, an investigator begins his research with a curiosity about the ways in which a certain variable fluctuates from case to case in a population. It is noted, for example, that some marriages last a long time and others last but a brief time, some persons have more power than others, some have more income than others, some are more respected, some organizations are more effective than others, and so on. The next question is usually: Why? Why does the variable vary as it seems to do? This then leads to explanations and predictions of the variation in one variable based on variations in other variables, and this eventually becomes the general substance of theory or scientific knowledge.

2.1.4a   Dependent Variables

The variable of prime interest, the one whose variation is to be explained in the research, is called a **dependent** variable. Differences between scores on this variable are thought to depend upon certain other variables. Dependent variables are usually defined as those that are influenced by some other variables, but sometimes they are defined merely as the crucial variable whose variation an investigator is interested in examining.

2.1.4b   Independent Variables

Independent variables are those that are thought to influence dependent variables; that is, they are explanatory variables that may help account for why a dependent variable fluctuates as it does in some population. If one thinks in cause/effect or influencer/influenced terms, as most people do who are interested in theory development, the relationship between independent and dependent variables is frequently expressed in terms of a directional arrow or path pointing from the independent variable at the left to the dependent variable at the right, as follows:

An investigation of how to look at data to see whether this type of argument is reasonable will be one of the central purposes of this book. Some statistical procedures are meaningful only if we can clearly settle the role the variable plays in the research.

2.1.4c   Intervening Variables

Referring back to the example of work autonomy and self-esteem, cited at the beginning of this chapter, since self-esteem was the variable of central concern, and was thought to be influenced by other variables, it would clearly be the dependent variable. Whether or not an individual has autonomy in work is then the independent variable, because it is seen as having an influence on the dependent variable. One way autonomy affects self-esteem is by first influencing perceived status of a worker in the workplace. This three-variable relationship could be shown as follows with work status **intervening** between the other two variables:

The role a variable plays in one piece of research may not be the same as the role it plays in another. For example, it may be reasonable to think about the influence of an individual's level of self-esteem at one point in time on the likelihood of being given autonomy at a later point in time. In that research, the role of the variables would be just the opposite from the role they played in Schwalbe's study.

Research can be more complex in the sense of involving relationships between more than two or three variables. Often there are several independent (or dependent) variables, and not infrequently some variables intervene (and are called intervening variables for that reason), as illustrated in the three-variable example above and the five-variable example below.

Elder, in a study of marriage mobility of women from middle and working class families, hypothesized that the following five variables would all be related (Elder, 1969).

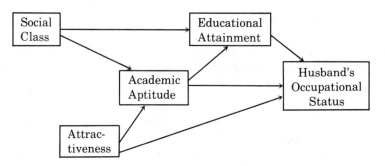

Women in his sample were first rated on "attractiveness" and some other personal variables; "husband's occupational status" was measured in a re-interview several years later. In this study, "husband's occupational status" is the primary dependent variable, and the other four variables are independent variables that are thought to help explain or account for score differences on the dependent variable. One could focus on other pairs of variables, however. For example, "social class" is an independent variable with respect to the dependent variable, "educational attainment"; one could go further and speak of "educational attainment" as intervening between "social class" and "husband's occupational status." The diagram shown here in effect illustrates theoretical relationships between variables that Elder believes exist, and his statistical analysis will be focused on the question of whether the data he examines behave as his model predicts.

Usually, a great deal of clarity can be gained in reading a research report if one first draws out the relationships hypothesized in the article in a fashion similar to that shown by Elder. The abstract at the beginning of most journal articles usually presents the model the investigator is considering. The important point here is that variables do play different roles in research; their role will generally determine the kinds of questions that are raised and, in turn, the kinds of statistical operations that are most appropriate in answering the questions.

2.2   MEASUREMENT IN SOCIOLOGY

The crucial role of measurement in sociology is becoming recognized, and there are a number of sociologists working on problems of how to measure significant variables adequately, how to move toward interval and ratio scales without leaving all sense behind, and how to actually carry out measurement of rapid-paced interaction on subjects who often are influenced by the measurement process itself.

Statistical procedures, as we will show later on, aid a great deal both in focusing attention on the need for better measurement and in providing ways to assess the quality of measurement scales. With increased sophistication in statistics and with more investigators aware of and concerned about the problem, it appears to us that some exciting and significant strides will be made in sociological measurement. We already have a number of measurement devices of impressive and demonstrated quality, and many, such as social class measures, have received repeated examination and improvement.* In a few sentences we would like to point out some of the issues and developments.

Because much of what sociology is interested in predicting and explaining turns out to be variable characteristics of people, interacting pairs, and the various groups people create and engage in, it is natural that individual people are among the most helpful sources of information. Much measurement in the past and present has involved **questioning** people. Particularly in the area of attitude research, this questioning has taken the form of batteries of questions that are then examined for consistency and combined into attitude scales such as Rosenberg's self-esteem scale (Rosenberg, 1965). One of the problems with this approach is that people sometimes become aware of the purpose of the questions and specifically are influenced by this knowledge. A very interesting book about possible ways to solve the problem of "reactive" measurements, those which also influence the measurements taken, is *Nonreactive Measures in the Social Sciences*, by Webb and others (1981). Such measurement problems, of course, are evident in every area of inquiry. In other cases, respondents are not particularly knowledgeable and simply do not know the answers to the questions asked. Where this is the case, alternative measurement procedures are pursued.

* For an inventory of sociological measures, see Bonjean *et al.* (1967); for more specialized collections of measurement scales, see Shaw and Wright (1967), Price (1972), and Miller (1991).

A number of electronic devices have been invented that greatly aid the measurement process. One device, for example, involves a simple 10-key keyboard much like that on a calculator with computer memory. By setting up numerical codes for categories of interaction, an investigator can make records of the type of interaction in group settings. Times are recorded, which then permits an examination of the shuttle of interaction through time, in terms of meaningful variables. This permits a kind of observation of interaction that pencil and paper methods cannot achieve, of course, and with the improvement in equipment comes improvement in the quality of measurement.*

Personal **observation** by a trained social scientist is another approach to measurement of sociological variables. In laboratory settings, of course, more control and more options for measurement are available, but many situations of interest either cannot be readily captured in a laboratory or cannot be artificially created. War might be an extreme example. On the other hand, some aspects of competition and conflict can be simulated, and sometimes this takes the form of a respondent playing a carefully devised game where measurements have been built into the process of playing or scoring outcomes.

Records created in the normal process of agency operation, or especially created records from the U.S. Census or from vital statistics, for example, form the basis for much other measurement and research work. Unfortunately, many people think only of questionnaire research as the basic source of information for sociology, and one of the reasons this brief section is included now is to get you to think seriously about measurement possibilities and be sensitive to the issues and possibilities implied by different types of data.

2.2a Creating Variables

An integral part of the conceptualization process, so vital for conducting sound research, is the process of constructing variables to represent concepts to be studied. Given a clear-cut definition of a concept it should be possible to identify an empirical indicator or indicators of that concept. Although a single indicator may be employed, it is generally the case that multiple indicators will give better results, as in the case of the self-esteem scale. Accordingly, many of the variables utilized in sociological research are composite variables consisting of two or more indicators of the concept of interest (e.g., socioeconomic status is a composite index that is usually constructed from the three indicators income, occupation, and education).

In a study of the "academic mind" Lazarsfeld and Thielens (1958) looked at

---

* Early instruments to do this in laboratory settings were developed by Bales at Harvard. The device described here is called "MIDCARS," and it was developed and refined by Richard Sykes at the University of Minnesota. The cassette tapes are converted to computer tape, and the computer is able to carry out the needed statistical analysis using programs written by the researchers. This is only one example of a number of technological developments that have impressive consequences for the quality of sociological measurement.

the eminence of faculty members. In the course of their study they developed two indices of eminence, one based upon the honors a faculty member had received and the other upon the faculty member's productivity. The honors index consisted of four indicators—possession of a Ph.D., publication of three or more papers, holding office in a professional association, and experience as a professional consultant. The productivity index also consisted of four related, but somewhat different, indicators—having written a dissertation, having published at least one paper, having read three or more papers at meetings, and having published at least one book (Lazarsfeld and Thielens, 1958). The researchers found that, although their honors and productivity indices were not the same, they were closely interrelated and thus could be used interchangeably as a measure of eminence in relating that concept to variables such as age and rate of promotion in academic rank.

According to Lazarsfeld (1972), the process of constructing a variable involves the following four steps:

1. The concept of interest is put into words and is communicated by examples. The concept is defined.

2. As the concept is being analyzed and defined a number of potential indicators are identified. With expanded discussion of the concept the number of potential indicators is increased. The total number of potential indicators is known as the *universe of indicators*.

3. The universe of indicators is likely to be large; therefore, it is necessary, from a practical standpoint, to select a subset of the indicators to form the composite variable to be used for the empirical work at hand.

4. The subset of indicators is combined in some fashion to form the composite variable.*

This process is the one used by Lazarsfeld and Thielens in constructing their honors and productivity indices. As a matter of fact, they extracted two subsets of indicators and developed two separate indices of eminence.

Ideally, the process of variable construction should be carried out in the conceptualization stage of a research project—before any data are collected. When this is, in fact, done, the resulting variables are more likely to be satisfactory measures of the concepts of interest than if it is not done prior to the collection of data. In point of fact, in many studies variable construction takes place after the data have been collected. When this is the case, it is often difficult to identify available indicators that adequately measure that which the researcher wishes to measure. Thus, indicators such as assessed value of a house may have to be substituted for the more relevant indicator "income" in the construction of socioeconomic status as a composite variable. Even in this kind of *ex post facto* situation, however, the innovative researcher is often able to construct quite satisfactory composite variables.

---

* Formation of the composite variable can be done systematically through the use of such techniques as factor analysis. See Harman (1967) for a comprehensive treatment of factor analysis and Heise (1974), Armor (1974), and Allen (1974) for a discussion of the use of factor analytic techniques for the construction of composite variables.

2.2b   Validity and Reliability

Regardless of the procedure by which a measurement is taken, the issue of quality of the procedure has to be resolved before statistical manipulation is useful or interpretable. The quality of measures is usually phrased in terms of two questions: (a) Is the procedure valid? and (b) Is the procedure reliable? **Validity** is usually taken to mean the extent to which a measurement process is able to make distinctions based (only) on the variable one intends to measure. For example, measurement procedures, that measured family size by recording the number of people in a dwelling unit would be less valid to the extent that boarders, visitors, and domestic help are not identified and eliminated from the count.

There are several ways by which validity of a procedure can be checked. Generally, these involve the use of some criterion measure and comparison of scores on this and the new measurement procedure. Statistical tools help in making these contrasts and comparisons.

The problem of **reliability** refers to the stability of the measurement process itself when applied under standard conditions. If there is a good deal of sloppiness in the measurement, then to that extent the measurement is unreliable. A classic example of an unreliable measurement instrument would be an elastic measuring tape, which would be likely not to give the same reading upon repeated measurement of the same object (where the object itself is known not to have changed). One would have to separate simple change in the phenomenon under study from unreliability of a measurement instrument, of course, and again statistical procedures (often correlation coefficients, discussed in Chapter 7, or path coefficients, discussed in Chapter 9) are available to help in the evaluation of reliability.

Validity and reliability are linked. To the extent that a measurement is unreliable, it cannot be said to be a valid measure of an intended variable. Again this relationship can be expressed in terms of statistical descriptions that will be developed in this volume. We will not develop the topic of validity and reliability at any great length here (that will be a main topic in a research methods course), but we will develop some of the elementary tools of thought that are helpful in handling these and other questions of inquiry.

2.3   SUMMARY

This chapter has dealt with the central problem of measurement. It is central because the scores we will be describing statistically are the result of a measurement process, and their meaning and quality depend upon what happens there. It is central, too, because the information contained in scores helps determine what sorts of statistical summary are appropriate and how they can be interpreted. Finally, as sociologists, we are interested in making significant statements about phenomena we find interesting, and that means we have a central concern for the ability of scores to capture variations we intend to examine and believe are important.

The next chapter begins the more organized pursuit of statistics itself by showing the importance of comparison and the various ways comparisons may be made statistically. The next chapter will also establish some of the basic themes that serve to organize the whole field of sociological statistics.

CONCEPTS TO KNOW AND UNDERSTAND

measurement
   conceptual definition
   operational definition
level of measurement
   nominal
   ordinal
   interval
   ratio
scale continuity
   continuous variables
   discrete variables
role of variables in research
   dependent
   independent
   intervening
reliability

validity
inclusive
mutually exclusive
precision
grouping into classes
   class midpoint
   real limits
   measurement
      to nearest whole unit
      to last whole unit
significant digits
rounding
concept
indicator
variable construction
scale

QUESTIONS AND PROBLEMS

1. Pick an interesting variable and define it (you may want to look up other definitions and critique them). Is it adequately defined and, if not, what difference does poor definition make, specifically? How can it actually be measured? What other variables might be used as independent or intervening variables with this variable? (Examples of variables might be alienation, urban segregation, age, social class, marital status, deviance, level of industrialization, social integration, or morale.)

2. By reviewing research articles in journals, you can find challenging ways to examine problems with variables. First, you might pick out and classify variables in abstracts of research reports in the sociological literature. It usually helps to read the entire article, too. Second, you might at this point pick out variables and begin to specify how you would prefer to measure and relate them to hypotheses in a study of your own.

3. Identify the most likely scale of measurement (nominal, ordinal, interval, or ratio) and the continuity or discreteness of the following variables:
  a. Years of service on a job
  b. Occupational prestige
  c. Number of organizations to which a person belongs

    d. Population density of a city

    e. Home ownership (of a person)

4. Round the following numbers to whole numbers:

    a. 45.73   d. 25.5

    b. 38.1   e. 26.501

    c. 27.49   f. 22.5

GENERAL REFERENCES

Carmines, Edward G., and Richard A. Zeller, *Reliability and Validity Assessment* (Beverly Hills, Calif., Sage Publications), 1979.

Lieberman, Bernhardt, *Contemporary Problems in Statistics: A Book of Readings for the Behavioral Sciences* (New York, Oxford University Press), 1971.

    A reader, this book includes in Section 1 a number of articles on measurement. The article by S. S. Stevens is included, as well as more detailed treatments of problems and possibilities in measuring.

Miller, Delbert C., *Handbook of Research Design and Social Measurement*, Fifth edition (Newbury Park, Calif., Sage Publications), 1991.

    This book provides compact summaries of various facets of research and statistics. Part I includes an outline of steps in typical research projects and Part III includes comments on statistical analysis. The book also includes a sampling of scales used to measure sociological variables.

Mueller, John H., Karl F. Schuessler, and Herbert L. Costner, *Statistical Reasoning in Sociology*, third edition (Boston, Houghton Mifflin), 1977.

LITERATURE CITED

Allen, Michael Patrick, "Construction of Composite Measures by the Canonical-Factor-Regression Method," in Herbert L. Costner, editor, *Sociological Methodology 1973–1974* (San Francisco, Jossey-Bass Publishers), 1974, pp. 51–78.

Armor, David J., "Theta Reliability and Factor Scaling," in Herbert L. Costner, editor, *Sociological Methodology 1973–1974* (San Francisco, Jossey-Bass Publishers), 1974, pp. 17–50.

Blalock, Hubert M., *Conceptualization and Measurement in the Social Sciences* (Beverly Hills, Calif., Sage), 1982.

Bonjean, Charles M., Richard J. Hill, and S. Dale McLemore, *Sociological Measurement: An Inventory of Scales and Indices* (San Francisco, Chandler Publishing Company), 1967.

Burke, Cletus J., "Measurement Scales and Statistical Models," Chapter 7 in Bernhardt Lieberman, *Contemporary Problems in Statistics: A Book of Readings for the Behavioral Sciences* (New York, Oxford University Press), 1971.

Duncan, Otis Dudley, and Jay W. Artis, "Some Problems of Stratification Research," *Rural Sociology*, 16 (March 1951), pp. 17–29.

Elder, Glen H., Jr., "Appearance and Education In Marriage Mobility," *American Sociological Review*, 34 (August 1969), pp. 519–533.

Festinger, Leon, and Daniel Katz, *Research Methods in the Behavioral Sciences* (New York, Holt, Rinehart and Winston), 1953. Chapter 11 by Clyde H. Coombs, is entitled, "Theory and Methods of Social Measurement."

Hamblin, Robert L., "Ratio Measurement for the Social Sciences," *Social Forces*, 50 (December 1971), pp. 191–206.

Harman, H. H., *Modern Factor Analysis*, revised edition (Chicago, University of Chicago Press), 1967.

Heise, David R., "Some Issues in Sociological Measurement," in Herbert L. Costner, editor, *Sociological Methodology 1973–1974* (San Francisco, Jossey-Bass Publishers), 1974, pp. 1–16.

Jones, Bryan D., and Richard Shorter, "The Ratio Measurement of Social Status: Some Cross-Cultural Comparisons," *Social Forces*, 50 (June 1972), pp. 499–511.

Labovitz, Sanford, "The Assignment of Numbers to Rank Order Categories," *American Sociological Review*, 35 (June 1970), pp. 515–524.

Lazarsfeld, Paul F., "Problems in Methodology," in Paul F. Lazarsfeld, Anna K. Pasanella, and Morris Rosenberg, editors, *Continuities in the Language of Social Research* (New York, The Free Press), 1972, pp. 17–24.

Lazarsfeld, Paul F., and Wagner Thielens, Jr., *The Academic Mind: Social Scientists in a Time of Crisis* (New York, The Free Press), 1958.

Marsh, Robert M., "The Explanation of Occupational Prestige Hierarchies," *Social Forces*, 50 (December 1971), pp. 214–222.

Price, James L., *Handbook of Organizational Measurement* (Lexington, Mass., D. C. Heath and Company), 1972.

Rosenberg, Morris, *Society and the Adolescent Self-Image* (Princeton, N.J., Princeton University Press), 1965.

Robinson, Richard, *Definition* (New York, Oxford University Press), 1954.

Schwalbe, Michael L., "Autonomy in Work and Self-Esteem," *The Sociological Quarterly*, 26 (4, 1985), pp. 519–535.

Shaw, Marvin E., and Jack M. Wright, *Scales for the Measurement of Attitudes* (New York, McGraw-Hill), 1967.

Stevens, S. S., "On the Theory of Scales of Measurement," *Science*, 103 (1946), pp. 677–680.

Vargo, Louis G., et al., "Replies and Comments," *American Sociological Review*, 36 (June 1971), pp. 517–522.

Webb, Eugene J., et al., *Nonreactive Measures in the Social Sciences*, second edition (Boston, Houghton Mifflin), 1981.

CHAPTER 3

The Basic Logic of
Valid Comparison

Not long ago, an advertisement proclaimed that a major brand of cigarettes had 16% less tar and nicotine. While we probably could agree that less is better than more, the ad clearly goes on to suggest that because of this, the specific brand advertised is better than other cigarettes in this respect. Such an extension of the argument is, of course, inappropriate given only this evidence. The reason? Comparison information is not given. The 16% figure cannot be interpreted. Does this brand have less tar and nicotine than it did formerly? Is this 16% less than the cigarette with the next least tar? Is it merely 16% less than the average of all tobacco products, or 16% less than the cigarette with the most tar? Even with the 16% drop, we do not know whether the levels of tar and nicotine are at a dangerous level or not, and there is no information that would permit one to judge the accuracy of the figure itself.

The problem is that a comparison is implied, but only half of the comparative information is given. Without some standard or something specified with which to compare the 16%, little can be concluded from this information.

This kind of problem confronts nearly everyone, whether sociologist or lawyer or layperson. We are nearly always faced with comparative problems that call for some way to make a clear and valid contrast between things we want to compare. Is a figure of 11.3% of the population in the 65-and-older age bracket (as it was in the U.S. in 1980) a high or a low figure? A major university has a staff with 6.6% who are from minority groups. Does this indicate discrimination or not? In each case, comparative information is needed to reach a conclusion.

Appropriate comparisons are central to any inquiry, as the following illustrations suggest.

Do radio and television reports of National election results from the east coast influence voting behavior on the west coast where the polls close later? A sample of California voters was interviewed the day before and immediately after the 1964 Presidential elections. Among those voting after the eastern polls closed, [and] who were exposed to election broadcasts of eastern poll outcomes, 97 percent reported voting for Johnson, the eventual landslide winner. Among late voters who did not hear the broadcasts, 96 percent reported voting for Johnson; [this is] a comparison suggesting that eastern election victory reports had little overall effect on the later vote outcome. (Mendelsohn, 1966).

Where does the after-tax, personal income go? In 1973 23.5% of the total went to the tenth of the population with the highest incomes and 2.5% of the total went to the tenth of the population with the lowest incomes, an outcome generally predictable from social stratification theory (U.S. Department of Commerce, 1976:477).

Presently the birth rate in Africa is 45 births per 1,000 population, compared to 31 for Latin America, 28 for Asia, 16 for North America, and 13 for Europe. It is estimated that it will take 24 years to double the population of Africa, compared to 30 years for Latin America, 39 for Asia, 98 for North America, and 248 for Europe (Population Reference Bureau, 1986).

"Do you think everyone should have the right to criticize the government, even if the criticism is damaging to our national interests?" In a 1970 national survey, 61 percent of the college graduates said "yes," 44 percent with a high school degree said "yes" and 28 percent with a grade school education said "yes." The higher the education, overall, the higher the percentage of "yes" responses. (Erskine, 1970:491).

According to the 1983 National Health Interview Survey, older people are less likely to eat snacks everyday. Of those 75 or older 26.6% report snacking every day, and this percentage steadily increases across age categories to 40.2% of those 20–34 reporting snacking every day. Also, 61.1% of those 75 and older, but only 26.3% of the 20–34 year olds, report being lifetime abstainers from drinking alcohol (National Center for Health Statistics, 1986).

The percent illiterate in the U.S. dropped from 20% in 1870 to 11% in 1900, to 4% in 1930, to 1% in 1969, and to 0.5% in 1981 (U.S. Bureau of the Census, 1984:385).

Ross et al. created indicators of well-being for all 3,097 counties in the United States. One indicator was "social status," including measures of family income, poverty, years of education, and housing with complete plumbing. For 1970, the social status index for the United States as a whole was set at 100. The index was highest for metropolitan areas (123.5 for central city areas) and lowest for totally rural areas adjacent to large cities (87.3). Northeastern and western regions were highest and north central and southern counties were lowest on this indicator of well-being (Ross *et al.*, 1979).

Does ordinal position in the family have an influence on potential scientific creativity? No, according to Lois-Ellin Datta, who studied a group of male high school seniors who competed in the 1963 Westinghouse Science Talent Search. Average creativity ratings for oldest boys was 6.1 on her scale and it was 6.0 for youngest boys. Other factors, however, influenced average rankings of both oldest and youngest boys (Datta, 1968).

Throughout these examples you will notice that there is a heavy emphasis upon comparison as the basis on which questions are posed and answered. These examples also indicate several ways in which comparisons can be made. At this point, two questions arise: first, for a comparison to be valid, *what* should be compared, and second, *how* should a comparison be made?

3.1   KINDS OF COMPARISONS

The answer to the question, "What should be compared?" depends largely on what one is interested in studying. A clearly stated question or research problem is basic to all comparisons. Without a clearly formulated problem, it is difficult to decide which of the myriad possible comparisons one should make and how it should be interpreted. Comparisons are aided if variables are carefully formulated and measured, and if objects to be measured are clearly and consistently identified so that comparable groups can be contrasted. Problems of theory, conceptualization, measurement, and research design cannot be overemphasized in statistics, although they are usually reserved as topics for a course in research methods. Here we will repeatedly call attention to these issues; you may want to read about some of them in a more systematic research methods text as well.*

In general, comparisons are made (a) *between groups* (group–group comparisons), either within the same study or in different studies, (b) *between a group and an individual* from that group (group–individual comparisons), or (c) *between study outcomes and standards* that have been established by previous research or are predictions derived from a formal model or theory the investigator is interested in testing.

3.1.1   Group–Group Comparisons

Group–group comparisons might be made between an experimental group that has been given some type of special treatment, such as a new instructional program, and a control group that has not had special treatment.[†] Comparisons could also be made between groups that have different characteristics but are from the same population, such as old versus young people or males versus females. Different contrasted groups might also represent different populations, such as a cross-cultural study where different societies are compared or an urban study where regions of a city (central city and suburbs) are contrasted. Which groups one compares depends to some degree, of course, on what problem one is studying. If the problem is whether there is a difference in reactions to stress by men and women, then comparisons of stress scores between gender groupings would clearly be the relevant contrast. The problem then becomes the statistical one of summarizing the stress scores for each gender group so that the two summaries can be compared. In each instance, the groups that are compared are themselves composed of a number of scores measured on a series of definable cases or objects. Groupings could consist of measurements made on cases other than individual people, of course, such as groups of different kinds of organizations, or different kinds of dyads, or different types of encounters between people and organizations.

---

* An excellent research methods text that examines these issues is Orenstein and Phillips (1978). Also, see Riley (1963); Babbie (1989); Bailey (1987); Nachmias and Nachmias (1987).
[†] Contrasts that are set up so that some subjects are assigned to an experimental group and some to control groups are usually discussed under the topic "experimental design." See Stouffer (1950) for a good reference to study design that introduces some alternatives.

3.1.2   Group–Individual Comparisons

Another kind of comparison is between a group and an individual who is a part of the group itself. Examples of individual–group comparisons would be a contrast of one person's grade in a course with the average grade for that class, or the increase in cost of living in one city compared to an overall, national average cost-of-living rise. Ranking states by the percentage of their population in institutions provides a somewhat different way to express where an individual unit (a state) falls in comparison to others in a group. The Ross study cited above contrasts individual counties as well as regions of the United States in terms of their socioeconomic index. The southern region, with an index of 87.8, was below the national average of 100, whereas the West had an index of 111.6.

3.1.3   Study Outcome versus Standard Comparisons

Finally, there are comparisons in which data one has collected in a study are contrasted with some outside standard. Outside standards come from a number of sources. For example, one might have past data with which to compare some new information, such as comparisons between the percentage of males in a sample one has just drawn versus the percentage of males found in the latest census figures for that sampled area. Comparisons may be made between a study in a new setting versus a previous study in a similar area. Theory provides a powerful source of expectations or standards for comparison. Social stratification theory suggests that lower class individuals will be relatively deprived (have poorer "life chances") on many of the characteristics society values, as compared to upper class individuals. Population models of the transition of societies from nonindustrial to industrial predict a change from high birth and death rates to low birth and death rates, and a resulting change in the age distribution and rate of growth of the society. One could compare changes in a society over time with such expectations to see if the theory of population transition works out as expected, or whether some modification or qualification in the theory needs to be made. Clearly, the more carefully thought out the standard and the more precisely it permits predictions to be made, the more interesting and useful it is to compare study data with the standard. This, in fact, is one of the main reasons why sociological theory is so important as a context for learning statistical reasoning.

Some time ago, Samuel Stouffer suggested the theory of "intervening opportunity," according to which the number of persons who make a move of a given distance is directly proportional to the number of opportunities at that distance, and inversely proportional to the number of intervening opportunities. Does this theory apply to the number of people who travel a given distance in the selection of marriage mates? A study could be designed in which the premarital residence of married partners is plotted on a map, and then the number of available partners of equivalent social class, ethnicity, education, and religion at that point and in intervening regions between the two partners' premarital residences is calculated. If the theory holds, the number of partners whose eventual spouses lived greater distances apart should decrease as intervening opportunities for marital partners are increased, and there should be an increase in longer moves as the number of potential partners at a given distance increases relative to intervening marital opportunities. This is an example of comparison between a study outcome and outcomes that would be predicted from a theoretical statement.

3.1.4   An Example of Comparison Operations

Suppose we start with a small study in which we are interested in the institutionalized **population** of certain states. The institutional solution to problems of care and custody might be expected to be related to a more urbanized style of life with greater density of population, greater mobility of families, more specialization of jobs, and bureaucratization. Rural states might be expected to make less use of institutions as a solution to care and custody problems of the handicapped, homeless children, old people, and the various categories of deviants, because families and neighbors are more readily available alternatives to institutional care. Given this formulation it is clear that rural–urban states are to be *compared* on number of inmates in institutions.

Table 3.1 provides a list of 1980 data on the number of inmates of custodial and care institutions for all 50 states and the District of Columbia. Rural states are defined here as those states with less than 73.7% (the overall national percentage) of their people living in urban areas. The variable, "number of *inmates of institutions*" is defined as a discrete, ratio variable measured to the nearest one thousand persons for each state on a scale that begins with zero.

First, before we proceed with our comparison example, we have to be alert to the possibility of error in the data. The number of inmates in the third score in the urban column is for Connecticut, which has a total population of 3,108,000, so clearly it could not have over 3.6 million inmates of institutions! Upon checking, we would find the last two zeros were a clerical error. The number should have been 36 (thousands) rather than 3,600 (thousands). We do not know if there was bias in the original data-gathering operation. The data came from a source that is usually reliable and careful. One should keep in mind that the data are for 1980 and that this may have been an unusual year. Since the results of a comparison depend upon the quality of the original data, it is always worthwhile to check carefully for accuracy.

If you inspect the 51 scores in Table 3.1 you will probably get some idea of whether there is a rural–urban difference in these scores. That impression is likely to reflect more accurately true rural–urban differences than, say, visits to the 50 states and the District of Columbia. Nevertheless, it is clear that without some type of ordering of even this small amount of data one could not say, in general, how much difference there is between rural and urban states. Certainly if more variables were measured or if there were more cases (*i.e.*, if we measured counties rather than states), as there usually are in sociological studies, the problem of drawing a conclusion would be greatly compounded. Our first task, then, is to develop some of the basic ways to go about making valid comparisons.

TABLE 3.1   NUMBER OF INMATES OF CARE AND CUSTODIAL INSTITUTIONS FOR STATES AND THE DISTRICT OF COLUMBIA, CLASSIFIED AS RURAL OR URBAN,[a] U.S. CENSUS, 1980, NUMBERS IN 1000's

Rural States	Urban States
12.	62.
8.	10.
5.	~~3600~~ 36.
100.	172.
96.	49.
52.	102.
80.	39.
57.	10.
51.	72.
41.	135.
48.	21.
9.	18.
9.	7.
20.	6.
29.	201.
6.	6.
45.	
9.	
51.	
26.	
49.	
30.	
36.	
32.	
22.	
24.	
41.	
35.	
7.	
7.	
4.	
5.	
38.	
22.	
3.	

Total Inmates:

Rural States	1109.
Urban States	946.
All States	2055.

Source: U.S. Bureau of the Census, 106 ed. (1986).

[a] An "urban" state is defined as one with 73.7% or more of its population residing in urban areas. This figure is the percentage of people in the United States who reside in urban areas. Other states are classified as "rural". Inmates include people in prisons, mental institutions, and nursing homes, according to the U.S. Census.

3.2　BASIC COMPARISON OPERATIONS

One of the central problems of descriptive statistics is that of developing techniques for making valid comparisons among collections of quantitative data. In this chapter we will develop some of the basic ways data can be organized and explore some of the summary indices that can be used in making comparisons. Chapter 4 deals with graphic ways to analyze contrasts of interest to us, and later chapters will move toward more general and searching procedures for describing quantitative information in ways that facilitate valid comparisons.

There are two general operations that are usually involved in comparisons. One is the organization of scores into some convenient form or distribution, and the other is the arithmetic manipulation of scores, usually by means of subtraction and/or division. You will find that the division idea, the creation of a ratio of one number (numerator) to another (denominator), is one of the basic organizing themes of descriptive and inferential statistics. The problem is, of course, what to divide into what! The remaining part of this chapter is a discussion of these basic comparison operations, and you will find that they underlie most of the concepts in this book.

BOX 3.1　Main Points in the Argument

The **first chapter** introduced some of the important concepts of the language of statistics such as "variable," "statistical observation," and "sample." **Chapter 2** emphasized the importance of knowing something about the scores one has generated from a measurement process; also, the distinctions between "level of measurement," "scale continuity," and "independent" and "dependent" variables were discussed. At that point we were ready to make use of these ideas in sociological research.

The first part of **this chapter** emphasizes the point that almost all research involves some kind of comparison (three general kinds of comparison were noted). Now, the problem is how to set up a valid comparison. We have just introduced the first (relatively simple-minded) approach to comparison—make a list of scores. Next, we will talk about the very *important idea of* **creating distributions of scores** (three general kinds of distributions will be discussed plus some of the problems associated with their creation and meaning). Finally, we will discuss several kinds of *ratios* that are very valuable statistical tools for summarizing data and making valid comparisons.

The next two chapters will continue the theme started here. **Chapter 4** will discuss graphic means for making comparisons, an all too frequently underrated approach, and **Chapter 5** will discuss ways to summarize whole distributions of scores by using some selected index numbers. Chapter 5 depends rather heavily upon your understanding of the notion of "distributions," which we are about to begin here.

3.2.1 Organization of Scores

An unordered listing or accumulation of scores, as in Table 3.1, is relatively useless in itself for comparison purposes simply because it does not communicate the information that is there in a way that permits a focused contrast. Somewhat more helpful is an ordered set of observations (or scores), called an **array**, shown in Table 3.2. An array is created by sorting the scores into order from lowest number

TABLE 3.2   SCORES FROM TABLE 3.1 IN AN ARRAY, SEPARATELY FOR RURAL AND URBAN STATES

Rural States (N = 35)		Urban States (N = 16)
3.	Lowest Score	6.
4.		6.
5.		7.
5.		10.
6.		10.
7.		18.
7.		21.
8.		36.
9.		39.
9.		49.
9.		62.
12.		72.
20.		102.
22.		135.
22.		172.
24.		201.
26.		
29.		
30.		
32.		
35.		
36.		
38.		
41.		
41.		
45.		
48.		
49.		
51.		
51.		
52.		
57.		
80.		
96.		
100.	Highest Score	

Source: Data in Table 3.1.

BOX 3.2 Symbol Notation in This Book

At this point a number of useful symbols are being introduced to simplify some often used statistical expressions. Your attention is called to the following frequently used symbols.

N refers to sample size (the number of cases). An example of its use is in Table 3.3C.

X_i refers to a single score, the ith score. In grouped distributions it refers to the midpoint of a class, and these two uses will be clear from the context in which they are used. Sometimes, for simplicity, the subscript is dropped and only X is used.

f_i refers to the frequency in the ith category of a frequency distribution.

$\sum\limits_{i=1}^{N}$ this is the *summation* operator. It instructs one to add up whatever is listed after the symbol. *Limits* of the summation operator are given below and above Σ (the capital Greek letter sigma). In the instance shown above, the lower limit is where i equals 1 and the upper limit where i equals N.

$\sum\limits_{i=1}^{k} f_i = N$ merely instructs one to add up frequency of cases in each of the k categories of a distribution, starting with the first category $(i = 1)$ and stopping after all k category frequencies have been summed. This is a more elaborate way of symbolizing what is meant by N.

$\sum\limits_{i=1}^{N} X_i$ symbolizes the operation of adding up raw scores (X_i), starting at the first score in a list $(i = 1)$ and ending when all N scores had been summed. Sometimes, in statistical notation, the limits of the summation operator are omitted because they almost always require addition over *all* the N scores. We will generally follow this practice, but we will include limits when needed for clarity.

---

of inmates to the highest number of inmates. This ordering makes evident where one group starts and stops as compared to other groups, and by close examination one might be able to discern the general concentration of scores in the array. One could, for example, locate the score in the middle (later we will refer to this score as the 50th percentile or median) and gain some idea of the distance between the lowest and highest scores (later we will refer to this as the crude range). If the scores are for a nominal variable, simple clustering of like scores together would be analogous to an array of ordinal or interval/ratio scores.

At this point the process of organizing data for the purpose of comparison has to stop, *unless* we are prepared to lose or deemphasize some detail in order to highlight or present some interesting characteristic. This is, in fact, usually a desirable step to take, since we want to compare scores in some particular respect rather than compare all of the raw scores with each other in various groups. In fact, one of the keys to valid comparison is to be selective in eliminating from the comparison those things that are distracting or contaminating as far as the comparison of interest is concerned. One of the important points beginning investigators sometimes forget is that irrelevant information, no matter how precise and accurate, is not only *not* positively worthwhile, but is often a distinct handicap to making clear comparisons and drawing correct conclusions.

3.2.2   Grouping Scores into Classes

One of the procedures by which raw scores are further organized is that of creating a series of categories and then grouping the scores into these categories. Sometimes very little information is lost in the process. For example, if marital status were measured (Table 3.3A lists scores on marital status for 16 respondents) and a category system such as that in Table 3.3B were created, no information is lost by classifying the raw data in 3.3A into these categories. Clearly the "4" in the

---

TABLE 3.3   CREATING A DISTRIBUTION OF DATA: AN EXAMPLE

A. Raw Data on a Marital Status Question Asked of 16 Adults

Married	Single	No response
Single	Separated	Separated
Single	Divorced	Married
Divorced	Widowed	Married
Single	Single	Separated
Married		

B. Frequency Distribution of the Scores from 3.3A Above

Score Categories	Frequency of Cases in Each Score Category
Single	5
Married	4
Widowed	1
Separated	3
Divorced	2
No Response	1
Total cases	16

C. Grouped Frequency Distribution (Widowed, Separated, and Divorced from Table 3.3B Above Have Been Grouped into a Single Category Called "Other")

X_i	f_i
Single	5
Married	4
Other	6
No Response	1
	$N = 16$

"Married" category indicates that there were four subjects with the same "score"; that is, they were each married. If we group the categories of marital status further, such as in Table 3.3C, we have lost some information about whether some of the six respondents in the "other" category are widowed, divorced, or separated. All we know from Table 3.3C is that there were six who were either divorced or widowed or separated. The organization of scores in B and C are called *frequency distributions*, a topic discussed in the next section. First, a note on "nonresponse."

The **No Response** category poses some problems. Clearly, it is not a marital status, but just as clearly it is a type of result of the measurement process. Usually, nonresponse is considered to reflect on the quality of the measurement procedures (and thus, most investigators take great pains to keep the rate of nonresponse very low and double-check possible biases that nonresponses may create in the analysis of data), and nonresponses would be reported in a footnote to a table rather than being included as a separate category of the variable being statistically summarized. On the other hand, one could be interested in studying nonresponses in which case this category would be included in statistical analysis. The fact that some groups (*e.g.*, older people) are more likely than others to respond with "no response" is one of the factors a thoughtful investigator considers carefully in making statistical summaries.

3.2.3   Distributions

Three types of distributions are commonly used to organize data further: (a) frequency distributions, (b) percentage distributions, and (c) cumulative frequency (or percentage) distributions.

3.2.3a   Frequency Distributions

In creating **frequency distributions**, we first list score categories and then tally up the number of cases that fall in each category.* Diagram 3.1 (p. 69) lists some of the main rules for creating a frequency distribution. Any one of the frequencies in the f column is called a **cell frequency**, or category frequency, and is symbolized by the small letter f or f_i where the subscript i refers to the ith class of the variable. The total number of cases is simply the sum of the frequency column and is symbolized by the letter N (or Σf_i).

Whereas category labels for nominal variables are quite explicit (as in Table 3.3B, for example), those for continuous variables need further explanation. Categories for the age distribution below, for example, are expressed in terms of a **unit of measurement**, in this case, "1-year" units.

* A "tally" is created by putting a mark beside the category into which a score is to be classified, for each score, as one moves through a list of raw scores. It is usually a good idea to move systematically through a list of scores rather than jump around in an attempt to count all scores of a certain value, since this usually leads to fewer clerical errors. The following would be a tally of scores from which Table 3.3B was created.

JHH Single	‖‖ Separated
‖‖‖ Married	‖ Divorced
‖ Widowed	‖ No Response

AGE OF STUDENTS IN A CLASS

Age (Nearest Year)	Real Limits	f_i
11	10.5–11.5	3
12	11.5–12.5	6
13	12.5–13.5	9
14	13.5–14.5	4
		$N = 22$

If age is measured to the *nearest* whole year, however, the three people in the 11-year-old class could have had ages between 10.5 and 11.5. We do not know what their exact age is but we know, if this has been accurately done, that they are nearer to 11 than they are to 12 or 10. The **real limits** (also called class boundaries and symbolized by L_i for lower real limit and U_i for upper real limit) of category 11 are 10.5 and 11.5, as discussed in Chapter 2. Eleven is thus the **midpoint** (symbolized by X_i) of a category, which potentially covers a range of scores. For purposes of further summary work, we use the value of the class midpoint to stand for the scores in that class. The **class width** (also called the **interval width** and symbolized by w_i) is 1-year, the difference between the real limits of that class. Each category has real limits that extend half of one unit (here the unit is 1-year) above and half a unit below the expressed limits of the class.

The problem of classification of age scores is handled for the investigator by rounding rules discussed in Chapter 2.

Sometimes a frequency distribution is *grouped* into fewer categories for ease of presentation. The above distribution of age might be grouped as follows:

Age in Years (Expressed Limits)	Real Limits	Class Midpoint X_i	f_i
11–12	10.5–12.5	11.5	9
13–14	12.5–14.5	13.5	13
			$N = 22$

Here again, since the unit of measurement is 1 year of age, and measurement is to the *nearest* whole unit (year), the real limits extend one-half of one unit (half a year in this case) on either side of the *expressed* category limits. The interval width is again the difference between real limits for a class, and in this example both classes have a width of 2 years

(3.1)
$$w_i = U_i - L_i$$

Thus, $12.5 - 10.5 = 2$ and $14.5 - 12.5 = 2$

Notice that the wider the categories become, the greater the loss of detailed information about specific scores. Again, the scores in each category are known by their midpoints, which can be found by adding half the interval width to the lower real limit of a class, as follows:

(3.2)
$$X_i = L_i + \frac{w_i}{2}$$

The midpoint of a class is referred to by the symbol X_i where i refers to the specific category being considered. This is the same symbol used for a raw score; but notice that, when data are grouped, the most reasonable single value to use for that raw score is the midpoint of the class into which it is classified. The X_i stands for "score"—raw scores where these are available; class midpoints where they are not. The usage is always clear in context. For the first class in the grouped frequency distribution above, the midpoint is $(10.5 + 2/2 = 10.5 + 1.0 = 11.5)$.

BOX 3.3   AN EXCEPTION IN CLASS LIMITS

The important exception to this class interval procedure is where measurement is to the *last* whole unit (rather than the *nearest* whole unit), a frequently used way of measuring age. The difference this makes in category midpoints and limits is discussed in Section 2.1.3c.

While grouping is frequently helpful to summarize information, it also may pose problems, depending upon how the scores are distributed within categories and upon the number of categories that are distinguished. The frequency distribution in Table 3.4A includes a set of 17 scores classified in a six-category distribution, each category width being one unit. The third column in Table 3.4A (headed fX_i, or f_i times X_i) shows the total of the scores within each class. In the "1" category, for example, there are 10 scores, each with the value of 1 (the midpoint of that class). The total or sum of these scores would be 1 times 10, or 10. There are two scores with the value 2 in the distribution, so the total of these scores would be $2 \times 2 = 4$, and so forth. If we add up the third column we would get the total of all of the 17 scores, and this value is 40 (also symbolized as ΣfX or the sum of the f times X column). Notice that the interval width is 1 for each category in this distribution.

Now, let us group these same scores into three categories instead of six. Each category will have a width of 2 rather than 1. The cell frequencies would be combined from cell frequencies in Table 3.4A, and they would still total to 17 cases, because we are not adding or dropping any cases. The total value of the 17 scores (ΣfX) in Table 3.4B appears to differ from the same data classified into six categories in 3.4A.

We are again using the class midpoint as the value that stands for each of the scores in a category. The product of the midpoint and the category frequency (fX) yields the total of the scores that fall in that class. Twelve scores with an assumed value of 1.5 is a total of 18.0, for example. The sum of the fX column is the total of all scores. Notice that this figure is not equal to 40, the sum of the fX column in the distribution in Table 3.4A. Finally, let us group the original six categories into only two, with a width of 3 years of age, as shown in Table 3.4C. Here the total of the scores appears to be 46.0 rather than 43.5 or 40.0. The difference is called

TABLE 3.4   AN EXAMPLE OF GROUPING ERROR

A

X_i	f_i	fX_i	$w = 1$
1	10	10	
2	2	4	
3	1	3	
4	0	0	
5	1	5	
6	3	18	
	$N = 17$	$40 = \Sigma fX_i$	

B

Class	X_i	f_i	fX_i	$w = 2$
1–2	1.5	12	18.0	
3–4	3.5	1	3.5	
5–6	5.5	4	22.0	
		$N = 17$	$43.5 = \Sigma fX_i$	

C

Class	X_i	f_i	fX_i	$w = 3$
1–3	2	13	26	
4–6	5	4	20	
		$N = 17$	$46 = \Sigma fX_i$	

grouping error, and it occurs because the midpoints of classes in this example do not adequately represent the value of the cases that fall in each class. The midpoint 2, for the 1–3 class in Table 3.4C, includes 10 scores of 1, 2 scores of 2, and 1 score of 3, and simply does not reflect the value of the scores in that class very well. Ideally, the cases in a class would be distributed rather uniformly across the interval. For this reason, especially where further computation is to occur, statisticians prefer a larger number of categories (such as 10, 15, or 20, depending on their purposes) or better yet, the raw scores themselves. Where grouping occurs, categories should be set up with some care and sensitivity to the way class midpoints reflect the value of cases in the class.* Other rules for creating frequency distributions are more a matter of convenience. Some useful guidelines are listed in Diagram 3.1 (p. 69).

The frequency distribution is exceedingly helpful in summarizing scores, partly because one can gain a sense of the way scores are spread over a scale, and

---

* It probably goes without saying that statistical tools that have power to summarize data to facilitate comparison also are capable of summarizing errors and distorting comparisons if they are not appropriately used. The grouping error shown in Table 3.4, for example, comes about in spite of the fact that there are no technical errors in creating those frequency distributions. If one wanted to overemphasize the total scores one could, of course, present only Table 3.4C, thereby misrepresenting the data and misleading the reader. Improper presentation of data can come out of inexperience as well as intention. A very nice presentation of some of the ways people inadvertently or intentionally give a distorted picture is given in the book *How to Lie with Statistics* by Huff (1954).

partly because this distribution is often used as a basis for computing other summary measures. We will avoid using grouped distributions as a basis for presenting summary statistics later in this book. Instead we will give only the raw score form for formulas to compute summary measures. It should be clear from the discussion above that we could readily convert raw score formulas to formulas for grouped data if we remember that class midpoints stand for each score in a class, and so instead of summing up the scores themselves we have to multiply class midpoints by class frequencies before summing. The reason we are going to omit grouped score formulas for the most part is that the logic behind statistical reasoning can be presented quite nicely with raw scores, and most investigators and students have access to computers where programs typically accept raw score rather than grouped data.

Table 3.5 presents frequency distributions of the data in our previous example in Table 3.1 on number of institutional inmates in various states. Notice that the frequency distributions for rural and urban states shown in Table 3.5 allow us to get a much better picture of the distribution of scores than did the raw data in Table 3.1 or the array in Table 3.2. Here it is apparent that the urban states are less concentrated in the low-number-of-inmates categories, for example.*

3.2.3b   Percentage Distributions

In creating **percentage distributions**, we simply divide each cell frequency by the total number of cases and multiply by 100.[†] Thus, the 0–24 category in Table 3.5A, with a frequency of 16 out of 35, corresponds to a percentage:

(3.3)
$$\text{Percent} = \frac{f_i}{N}(100)$$

$$= \frac{16}{35}(100) = 45.7\%$$

The total of the percentage distribution should equal 100.0% except for very minor differences due to rounding.[‡]

---

* While the number of inmates is, as defined, a discrete variable, we will not treat these data differently from a continuous variable for two reasons. First, the difference in the kinds of descriptions we will be making is often of less importance than other distinctions we have to make, and, second, the results are generally identical. For example, if we treated the real limits of the first class in the distribution in Table 3.5 as extending from -0.5 to $+24.5$, it would be apparent that the category width is indeed 25. This is the same result one would get if one counted the interval widths in thousands. The class midpoint is $-0.5 + \frac{1}{2}(25) = 12.0$, and so forth. There are points in statistics where different computations would be implied by differences between discrete and continuous variables. Here it is important because it is one aspect of the meaning of a score, and it will be sufficient to keep its meaning in mind in interpreting computations.

[†] A *proportion* is simply a cell frequency divided by the total number of cases in a distribution. It is sometimes used instead of a percentage (which is the same ratio multiplied by 100). Both proportions and percentages convey the same information about a distribution, however.

[‡] Some investigators make slight adjustments in the distribution so that the total actually comes out to 100.0%, and others let the total equal what it will, with a footnote indicating that the small difference (if any) is due to rounding errors. Rounding errors should be very small; otherwise, one would suspect some computational error has been made.

TABLE 3.5 Distributions of States on Number of Inmates of Care and Custodial Institutions by Rural and Urban States, U.S. Census, 1980, Numbers in 1000's[a]

A. *Rural States*

| Number of Inmates in Thousands | Distributions | | | |
	Frequency	Percent	Cumulative Frequency	Cumulative Percent
0–24	16	45.7	16	45.7
25–49	12	34.3	28	80.0
50–74	4	11.4	32	91.4
75–99	2	5.7	34	97.1
100–124	1	2.9	35	100.0
125–149	0	0	35	100.0
150–174	0	0	35	100.0
175–199	0	0	35	100.0
200–224	0	0	35	100.0
Totals	35	100.0 (35)		

B. *Urban States*

| Number of Inmates in Thousands | Distributions | | | |
	Frequency	Percent	Cumulative Frequency	Cumulative Percent
0–24	7	43.8	7	43.8
25–49	3	18.8	10	62.6
50–74	2	12.5	12	75.1
75–99	0	0	12	75.1
100–124	1	6.2	13	81.3
125–149	1	6.2	14	87.5
150–174	1	6.2	15	93.7
175–199	0	0	15	93.7
200–224	1	6.2	16	99.9
Totals	16	99.9 (16)		

Source: Data in Table 3.1.
[a] Where sum of percentages does not equal 100.0 it is due to rounding.

While the frequency distribution permitted a somewhat clearer summary and contrast between urban and rural states than the original raw scores or arrays, the percentage distribution contributes a great deal more to the ease and validity of comparison because it removes an important source of possible error. Notice that the percentage is a simple ratio, one that, incidentally, is easily understood in our culture since we tend to think in terms of parts of 100. Significantly, it takes account of the comparison problem of different totals, converting each total to 100. While 16 and 7 in Table 3.5A and B were not directly comparable because there were, after all, different numbers of rural and

TABLE 3.6 PERCENTAGE DISTRIBUTION OF REASONS GIVEN BY 6,292,000 MEN AGED 18–64 FOR A RECENT CHANGE IN RESIDENCE WITHIN THE SAME COUNTY, 1963

Reasons	Percentage
To take a job	2.7
To look for work	1.0
Job transfer	0.4
Commuting or Armed Forces	7.5
Housing	60.4
Change in Marital Status	11.0
Join or move with family	8.1
Health	1.1
Other reasons	7.9
	100.1
	(6,839,000)

Source: U.S. Bureau of the Census (1966).

urban states, 45.7% and 43.8% are comparable because these are expressed as parts of the same total, namely 100%. One could, then, compare percentage distributions even though there were differing total frequencies, because that distorting feature of a distribution has been eliminated by computing percentages. Many comparisons that you will be making will be in terms of percentages, and you will get the feeling that they convey a good deal of useful information for making valid comparisons.

Table 3.6 presents data on reasons given by men aged 18 to 64 for a recent change of residence within the same county. It would probably not be hard to find someone who would claim, on the basis of these data, that 2.7% of the 6,292,000 men whose responses are summarized in the table moved to take a job. This is, of course, an error. The base of the percentages in Table 3.6 is not individual *men* who moved, but the 6,839,000 *reasons* that these 6,292,000 men gave for making the move. Some people gave more than one reason. The correct interpretation of the percentage would be that 2.7% of the reasons given for moving within the same county were to take a job. A good caution in reading a percentage table is to find out what 100% (the base or denominator of the percent) refers to. This total is supplied in parentheses below the total of the percentage column.

3.2.3c Cumulative Distributions

Table 3.5 also presents two **cumulative distributions**, one for *frequencies* and one for *percentages*.* A cumulative distribution is formed by indicating for each category the number (or percentage) of cases that fall below the upper real

---

* Just as frequency distributions are symbolized by the small letter f, percent and cumulative distributions have traditional labels. Percent distributions are labeled "%" or "Pct" and cumulative distributions are labeled "cum %" or "cum f." Some authors use capital F to stand for a cumulative frequency distribution, but the more explicit "cum f" will be used in this text.

limit of that class.* It is the sum of the frequency (percentage) in a given class and the frequencies (percentages) in all of the classes below that class. Thus for rural states in Table 3.5A, the cumulative frequency of 28 means that by the upper real limit of the "25–49" class (that is, by the time one reaches the score of 49.5), 28 states have been categorized out of a total of 35 states. By the time one reaches the top of the "112" class (using the midpoint to identify the class), all 35 cases have been classified and all of the cases have also been classified below the upper real limit of the 212 class. The same type of expression applies to the cumulative percentage distribution. By the upper real limit of the "25–49" class, 80.0% of the cases have been classified, or, 80.0% of the cases fall below the upper real limit of that class.

Cumulative distributions are useful for comparison if one wants to compare the way cases are spread across a scale. For example, for rural states, 97.1% have numbers of inmates below 99.5 (thousand) while for urban states, only 75.1% have numbers of inmates below 99.5 (thousand).

3.2.3d   Percentiles

The smallest score below which a given percentage of cases falls is referred to as a **percentile**. The smallest score below which 100% of the cases falls, for example, is known as the 100th percentile. For urban states this value is 224.5 (thousands) of inmates. One could be interested in any percentile, of course: for rural states the 45.7 percentile is the score 24.5 (thousands) of inmates; the 75.0th percentile for urban states is 74.5 (thousands). That means that about three-quarters of the urban states have number-of-inmate scores below 74.5 (thousands) and one-quarter above this score. The 50th percentile is a frequently used percentile in statistics (later to be called the *median*). Frequent reference is also made in research to the 10th, 25th, 33rd, 66th, 75th, 90th, 95th, and 99th percentiles.[†] Results of tests, such as the College Board Exam or the Graduate Record Exam, are reported in terms of percentiles—the percentage of individuals in some defined test-norm group that have scores equal to or less than a particular person's score. Diagram 3.2 (p. 71) illustrates and explains how any percentile can be computed from a frequency distribution.

3.2.3e   The Presentable Table

Some tables are simply working data for the researcher to use (*e.g.*, a tally or output from computer programs) and may contain a variety of information for double-checking and different kinds of preliminary comparisons (*e.g.*, Table 3.5).

* In this book, we will adopt the convention of creating cumulative distributions by starting to accumulate from the low scores and accumulating so that N or 100% is by the high-score category. Here the low-score category is 0–24 (the smallest number of inmates) and the high-score category is 200–224 (the largest number of inmates). It has nothing to do with the geography of the printed page. In some cases the small scores will be printed at the bottom of the table and in other cases (as in Table 3.5) the small scores are printed at the top. This convention merely saves presentation of alternative formulas where accumulation is from high to low scores.
† It should be noted that any "fractile" could be computed using the general logic shown here. Percentiles (100ths), deciles (10ths), and quartiles (4ths) are only the most frequently used.

When it comes to presenting a distribution for others to read and interpret, special care must be exercised to make the table clear and complete. A good example is found in Table 3.6. You should note several important features of tables that are polished for presentation. First, it has a title that is very clear, stating: (a) what the numbers in the distribution are (*e.g.*, frequencies, percentages, cumulative percentages; (b) what the variable is that has been measured; and (c) the nature of the research cases that were measured (*e.g.*, the kind of sample or collection of cases and how many there are). Usually, the time and place location of the data are indicated. Second, the **title** usually makes reference to a table footnote indicating the source of the data. The footnote may also clarify other details of computation or characteristics of the sample. Third, the categories of the variable are listed in a column down the left side of the table (called the **stub** of the table), and the columns of the table are labeled (called the **headings**); for example, the name of the variable is shown above the column of category labels and the heading of the distribution itself shows what kind of distribution it is. Usually, only the relevant distribution is shown. If it is a frequency distribution, the total number of cases is shown. If it is a percent distribution, the sum of the percentages (100%) is shown with the total number of cases (N) in brackets underneath (this gives readers enough information to re-create a frequency distribution by multiplying the percentage in each category by N and dividing by 100). Creating a readable and clear table is a difficult and creative task and it is often useful to get feedback from others on whether the table is clear and useful.

3.2.4 Other Uses of Ratios for Comparison

The percentage is a good example of the use of a ratio to create numbers that can be compared more validly. It takes account of the total number of cases, converting it to 100 for each distribution, and because of that, percentages can be directly compared without the distortion created by differing numbers of cases in the different distributions. There are other situations where ratios can be used and where other sources of distortion are controlled to provide a basis for comparison. A classic example is the sex ratio.

The *sex ratio* is usually computed as a ratio of the number of men in a group divided by the number of women in that same group. Because the resulting number is a decimal, it is traditionally multiplied by 100 so that it reflects the number of males in the population per 100 females (rather than the number of males per 1 female).* Table 3.7 shows the sex ratio for the United States over three census years and 1985 for the whole population and for various age groups. Although slightly more males than females are born, this ratio reduces to nearly an even split at under 45 years of age, and decreases dramatically in the older years as women outlive men. From other information we know that the life expectancy at birth was 70.9 years for men and 78.2 years for women in 1982. By the year 2050

---

* William Petersen, in his book *Population* (1961, Note 4; p. 72), notes that the European tradition is to define the sex ratio as the number of females per 100 males. More recently, the U.S. and European tradition is to express the sex ratio per 1000 rather than per 100, thus eliminating a decimal position. A number of other ratios useful in the field of demography are defined and used in this book.

TABLE 3.7   Sex Ratios by Age Group, U.S. 1950–1970 and 1985

| Age | Males Per 100 Females | | | |
	1950	1960	1970	1985
All ages	98.7	97.1	94.8	95.1
Under 65	99.6	98.6	97.7	99.5
At birth	105.8	105.5	104.2	104.9
Under 45	99.4	99.5	99.5	101.8
45–64	100.2	95.7	91.7	93.0
65 and over	89.7	82.8	72.2	67.8
65–74	93.0	87.0	77.6	78.4
75 and over	82.7	75.1	64.0	54.3

Source: Brotman (1970), U.S. Bureau of the Census (1986).

one estimate by the U.S. Census is a life expectancy at birth of 71.8 years for men and 81 years for women.

Another interesting ratio is the *dependency ratio*, which is designed to show the relationship between the population in the middle years (21–65) and that at the young and older years. It is computed by using a numerator that is an estimate of the number of dependents (*i.e.*, those "culturally defined" as younger than "adult" or "probably retired," namely below 15 and above 64) and a denominator which is the number of people in the middle years (*i.e.*, culturally defined as in their "productive" or "self-supportive" years), and multiplied by 100. Dependency ratios are shown in Table 3.8 for several different countries for the early to middle 1980s, again permitting a comparison across societies. The next question, of course, is why do the ratios differ from country to country as they do?

TABLE 3.8   Dependency Ratios for Selected Countries for Census Dates in the Mid-1980s

Zimbabwe	113
Ethiopia	100
Nigeria	100
Libya	96
Mexico	85
Iran	85
Venezuela	75
South Africa	72
Columbia	70
Chile	61
Taiwan	54
Sweden	54
United States	52
Canada	47
Japan	47

Source: Population Reference Bureau (1986).

Turning to our earlier example of the number of inmates for different states, suppose we wanted to compare New Hampshire's 7,700 inmates and Massachusetts' 61,500 inmates. It seems likely that a direct comparison of these two figures may be misleading if for no other reason than because Massachusetts had 5.7 million people within its borders and New Hampshire had slightly less than 1 million. Other things being equal, one might expect a bigger state to have more inmates. We could create a ratio, then, that is a percentage of the total population of a state who are inmates of care or custodial institutions. Table 3.9

TABLE 3.9 PERCENTAGE OF THE POPULATION OF STATES WHO ARE INMATES OF CARE OR CUSTODIAL INSTITUTIONS FOR RURAL AND URBAN STATES, U.S. CENSUS, 1980

Rural States	Urban States
1.03	1.07
.84	1.11
1.06	1.16
.84	.98
.89	.67
.95	.90
.87	.93
1.21	1.55
1.26	.73
1.42	.95
.97	.74
1.36	.65
1.35	.48
1.27	.78
1.24	.85
.93	.59
.84	
.48	
.87	
.85	
.89	
.83	
.77	
.81	
.85	
1.07	
.97	
1.16	
.91	
.73	
.74	
.41	
.92	
.83	
.75	

Source: Data in Table 3.1, presented in the same order.

shows these ratios in the same order for the rural and urban states as in Table 3.1. By inspection, it is hard to reach a definite conclusion about whether rural or urban states have a higher ratio (percentage) of their population as inmates in institutions, but it would appear that rural states have a higher percentage in institutions than urban states—a result, if true, that would be opposite to our earlier prediction. At this point we need some further means for statistical comparison between these two sets of percentages. These measures will be developed in Chapter 5.*

3.2.5   Time-based Ratios

Some specialized ratios are especially useful for measuring the amount of some variable characteristic that occurs in a given time period or the change from one period to the next. Two of these time-based measures are the *rate* and *percentage change*.

3.2.5a   Rates

One way of defining a **rate** is as a ratio of the number of events that actually occurred in a given time period to the number of such events that might potentially have occurred during the same period. Thus we have a formula that, stated in prose, looks like this:

(3.4)
$$\text{Rate} = \frac{\text{Number of events occurring during a time period}}{\substack{\text{Potential number of events that could have} \\ \text{occurred during the same time period}}}$$

The birth rate is one such use of a rate. The *crude birth rate* is simply the ratio of the number of births in a given year divided by the total number of people in a given society at the middle of that year. The midyear population is used because it is probably a good average of the number of people in the society during the year (in times of increasing population, for example, this number would exceed the actual population early in the year but be below the population achieved by the end of the year). The total population is used as an estimate of the potential number of births in a society.

Unfortunately, the birth rate defined above does not take account of the fact that some societies have an unusually large number of prepuberty females, males, or older people, which makes the total population a poor estimate of the potential number of births. A better measure of the extent to which a society is producing a maximum number of births would be the number of births divided by the number of women in that society, or better yet, divided by the number of women aged 15 to, say, 45. In each instance, the improvement in the rate results from finding a more refined estimate of the number of *potential* births in a society in a given time

* By way of preface to Chapter 5, you could compute the summary ratio called the average (arithmetic mean). The average percentage of inmates for the 35 rural states is 0.93% and for the 16 urban states it is 0.88%. The average (weighted average in this case) is but one of several simple ratios that greatly facilitate comparisons of groups of scores such as these.

TABLE 3.10 RETENTION RATE FOR FIRST-TIME COLLEGE STUDENTS PER 1000 STUDENTS STARTING AT GRADE 5 IN THE UNITED STATES

1924	118
1926	129
1928	137
1930	148
1932	160
1934	129
1936	121
1938	—
1940	—
1942	205
1944	234
1946	283
1948	301
1950	308
1952	328
1954	343
1956	357

Source: U.S. Bureau of the Census (1966:112).

period. Specialists in population use rates such as the *age-specific fertility rate* (defined as the number of births per 1000 women in a specific age group in a year), or the *fertility ratio* (defined as the number of children under 5 per 1000 women in the 15–45 childbearing period) for these reasons.*

Often rates, like proportions, result in small decimal values, and it is traditional to multiply the rate by some power of 10 so that the result is a comfortable whole number. The death rate from cancer in the U.S. general population during 1982 was 187.2 deaths per 100,000 population. The suicide rate for the same period was 12.2 per 100,000 (U.S. Bureau of the Census, 1984).

A somewhat different use of the rate is illustrated in Table 3.10. Is there an increasing chance for people to go to college? Is the educational system retaining more people, longer than before? The *retention rate* shown in Table 3.10 is the ratio of the number of first-time college students in a given year to the number of students in the fifth grade 8 years before. Of those who were in fifth grade 8 years ago, how many entered college?

It should be clear by now that any summary measure is useful only if one is clear about the components that make it up. This is not so much a comment on the obscurity of summary measures as it is on the need for a clear understanding of what one wants to know and what the information permits one to infer, and it is not a problem particularly unique to statistics. Some time ago, an investigator, T. N. Ferdinand, illustrated how a seeming contradiction between statements based on

---

* An interesting example of the problem of defining and refining ratios in the area of crime statistics may be found in an article by Chiricos and Waldo (1970). They develop ratios to measure certainty of punishment and severity of punishment plus measures of percentage change, which are discussed later in this chapter.

rates could arise if one is not rather clear about what the rate is designed to represent (Ferdinand, 1967). Federal reports for some time had been reporting increases in the number and also the rate of crimes in the United States, but in Ferdinand's careful study of criminal patterns in Boston since 1849, he found declines in crime rates. How could this be so? Ferdinand explained it this way. Using a society with a population of some given size, say 1 million, for illustration, in one year, 800,000 people (80%) might live in villages and the rest might live in urban areas. It is known that villages have a lower crime rate than urban areas, so the following crime rates might hold:

RESIDENTIAL DISTRIBUTION AND CRIME RATE, YEAR ONE

Location of the Population	Percent	Crime Rate
Villages	80	40 per 1000 population
Urban areas	20	100 per 1000 population
Overall crime rate for this year = 52 per 1000 population		

In a later year two things could happen. First, there might be a shift of population from low-rate areas (villages) to higher-rate areas (urban areas), but there might also be a decrease in crime rates in, say, the urban areas, as shown below.

RESIDENTIAL DISTRIBUTION AND CRIME RATE, YEAR ONE

Location of the Population	Percent	Crime Rate
Villages	10	40 per 1000 population
Urban areas	90	60 per 1000 population
Overall crime rate = 58 per 1000 population		

The result of these two shifts, a drop in rate in urban areas and a shift of population to the higher rate areas, results in an overall increase in crime rate. Rates of subpopulations have gone down, but the overall rate has increased.

3.2.5b Percentage Change

Another ratio that is sensitive to time and change, and the last of those we will discuss in this book, is called **percentage change**. It is defined as the ratio of the amount of change between two time periods to the amount at the start multiplied by 100.

(3.5) $$\text{Percentage change} = \frac{(\text{Amount at Time 2}) - (\text{Amount at Time 1})}{(\text{Amount at Time 1})}(100)$$

It is used to express the amount of change in a variable relative to the starting value of that variable. For example, the population of the United States changed (increased) by $+11.4\%$ between the 1970 and 1980 censuses. This is computed by using the 1970 population of 203,302,000 and the 1980 population of 226,546,000 as follows:

$$\text{Percentage change} = \frac{226,546,000 - 203,302,000}{203,302,000}(100) = +11.4\%$$

The logic of this ratio is that the amount of change, the difference in the numerator, should take account of the size at the starting point. It is one thing for a society of 203,302,000 to increase in 10 years by 23,244,000, but it would be much more startling if a society starting with only 4 million people made this amount of change in the same 10-year period. If the change is a reduction, then the percentage change will turn out to be negative.

There are situations where it might make more sense to measure change in terms of an ending point rather than in terms of change from a starting point. Hans Zeisel points out some of these situations in his excellent book, *Say It with Figures* (1985). For example, if we were interested in measuring percentage change in a student's grade point average (GPA) from the student's freshman to senior year, we might reason that it is harder to make an increase of 1.0 overall if the student started out with a 3.0 as a freshman than if the student started out with a 1.5 GPA as a freshman. In the latter case, several A's might raise the GPA to 2.5, but no amount of A (where A = 4) work would be sufficient to raise the 3.0 to an overall GPA of 4.0. Students are approaching an upper limit, and the closer one approaches the limit the more difficult it is to make any given percentage change. In such a case, one might use the ending GPA rather than the beginning GPA as the figure in the denominator of the percentage change formula. Of course, such a computation would have to be clearly explained, since the percentage change is usually computed with the beginning amount in the denominator.

Zeisel makes a similar point for change in sales volume from one year to the next of businesses that already control most of the market versus those just starting in business. It would be harder for one of the largest companies to double its sales volume than for a very small business to double its sales.

An important area of study in demography is the study of migration, both between states and within the boundaries of states. Percentage change in the population of states or regions provides the basis for many such studies.

3.3 SUMMARY

In this chapter we have considered the basic problem of making valid comparisons to reach important substantive conclusions. Almost any thoughtful inquiry involves a *comparison* of one group with another, an individual with the rest of his or her group, or the outcome of a study with some previously established standard or prediction. Just *what* should be compared and *how* it might be compared are the major topics introduced here. The following chapters in descriptive statistics have as one basic theme the formulation of procedures for making meaningful and valid comparisons. Often these comparative procedures involve some form of the simple idea of a *ratio*.

Some of the procedures, such as those for creating distributions, involve organizing data in such a way that certain information contained in the data is made more apparent in order to permit comparisons to be made more readily.

Frequency, *percentage*, and *cumulative* distributions were discussed in this light, as well as *tallies*, *arrays*, and *unordered lists* of data. Where *grouped distributions* are used, detailed knowledge of individual scores is lost, and scores within a category are referred to by their *class midpoint*. Among the problems that confront a person making a comparison is the problem of irrelevant information. Loss of irrelevant detail is often helpful in exposing the desired information that the data contain.

BOX 3.4   SOME GUIDELINES FOR MAKING VALID COMPARISONS

1. *Know what you want to compare*, what variable(s) should be compared for which groups. Is the comparison to be made between groups, between an individual and a group or between the study group and some standard (previous study, predicted value, etc.)? A good line of reasoning or theory helps immensely at this point.

2. *Have comparable measures on comparable cases*. Training observers, providing standard interview schedules, questions that have been pretested, or special recording devices, etc. generally help assure comparability. The same scale of measurement should generally be used on all groups to be compared and the rules for "sameness" should be such that this claim can be checked. Furthermore, cases that are measured should be the same across groups. If individuals are measured in one group, individuals should be measured in other comparison groups. In general, the sweetness of apples and the sweetness of sweethearts are not comparable.

3. *Research design* is an important topic here, too. We will not go over design possibilities, but you should consult a research methods text on design *before* you become deeply involved in your own study.

4. Where there is a shifting base, consider taking account of this in terms of a *ratio* of some type.

5. Where the features to be compared are "buried" in accurate but irrelevant information, consider some of the *data summary* operations discussed in this chapter and in the rest of the book.

A second source of problems for those who would make comparisons is the *shifting base of comparison*. The number of inmates in rural states could not be directly compared with that for urban states, (a) because the totals involved more states classified as rural and fewer states classified as urban, and (b) because the urban states had larger populations to start with than did rural states, so that on this basis alone one might expect differences between states in the number of inmates. The solution was to create and examine a number of ratios that took out the effect of different kinds of incomparability. In one instance, we computed percentages of rural (and urban) states with lower or high numbers of inmates. In another, we computed the percentage of a state's population that were inmates of institutions so that we could compare these percentages for rural and urban states. The point is that a ratio was the mechanism by which we arrived at summaries that could be properly compared only when certain distorting irrelevancies were taken into account.

There are a number of different kinds of ratios. Two time-based ratios discussed in this chapter are the *rate* and the *percentage* change.

The next chapter will introduce graphic procedures for making comparisons. You will notice that often ratios, percentages, and rates are used in a graph, and graphic analysis provides some additional analytic insight that is useful in making valid comparisons.

CONCEPTS TO KNOW AND UNDERSTAND

types of comparisons
array
grouped frequency distribution
   class
   midpoint (X_i)
   cell or class frequency (f_i)
   interval width (w_i)
   real limits (L_i, U_i)
   total number of cases (N)
grouping error

distributions
   frequency (f)
   percentage $(\%)$
   cumulative frequency $(\text{cum}f)$
   cumulative percentage $(\text{cum}\%)$
percentile
ratio
rate
percentage change
how to set up a table

BOX 3.5 GENERAL SOCIAL SURVEYS

One of the most widely used collections of social science data is a series of nearly annual surveys conducted since 1972 by the National Opinion Research Center (NORC), a research center affiliated with the University of Chicago.* Called "General Social Surveys" (GSS), they are unusual in that interviewers use standard questionnaires that repeat earlier questions on a wide variety of subjects so that across-time comparisons can be made on many important variables of interest to social scientists. A core set of questions is repeated in each survey. Other items are repeated on a cycle every 2 or 3 years. About 1,500 60-minute interviews are conducted with a cross section of persons 18 years of age or older, living in noninstitutional arrangements within the United States. A large number of variables are measured, such as age, gender, education, alienation, aspects of work and family, and various attitudes and behaviors and experiences. Many of these items can be combined in new measures too. The sample is a probability sample, which means that inferential statistics can be used to make inferences to the overall population. NORC checks the data, prepares them on computer tape, and prepares the all-important codebook that explains each variable and how it is coded. These data tapes are available for purchase, and many departments and colleges have these data available for students and faculty to use in their research. You may be asked by your instructor to use these data for computer-based analysis in which you would use a statistical program such as SAS or SPSS to get descriptive and inferential statistics to answer your own research questions.

---

* Codebooks and information about the General Social Surveys are available from the National Opinion Research Center, University of Chicago, 6030 South Ellis Avenue, Chicago, IL 60637.

QUESTIONS AND PROBLEMS

1. Find an interesting article from one of the sociology journals (*American Sociological Review, American Journal of Sociology, Social Forces, Sociometry, Sociological Quarterly*, etc.) and analyze the kinds of contrasts or comparisons its line of reasoning implies. What kind of comparisons are they? If standards are involved, where do the standards come from? What kinds of values might be computed to aid in the comparison?

2. Using a newspaper or magazine, find illustrations of individual–group, group–group, and study–standard types of comparisons. How are these contrasts made, and how would you make them if you were the investigator on the case?

3. Find illustrations of each of the summary measures we have discussed thus far: ratios, rates, distributions, percentiles, and percentage change.

4. The following set of scores is relative pay levels for unskilled factory workers for selected metropolitan areas of the United States in 1986. The base number for this index is 100; therefore, any score less than 100 would be considered substandard and any score over 100 would be above the standard. Using these scores (a) set up a frequency distribution, (b) compute the corresponding percentage distribution, (c) compute the cumulative frequency distribution, and (d) compute the cumulative percentage distribution. What do these distributions tell you about the variable classified?

112, 91, 81, 79, 101, 97, 81, 120, 102, 107, 92, 68, 88, 130, 118, 88, 132, 94, 79, 100, 95, 85, 74, 87, 72, 90, 108, 75, 74, 100, 97, 100, 74, 71, 94, 112, 98, 89, 69, 133, 81, 95, 89, 84, 114, 114, 92, 102, 114, 81, 85, 105, 100, 91, 71, 140, 116, 130, 126, 109, 78, 94, 118. (*Source:* Bureau of Labor Statistics, 1989)

5. The following data are percentages of persons 65 years old or older in Metropolitan Statistical Areas (larger urban places) of the United States in 1980. Set up a frequency distribution for these data and include the following: (a) frequencies, (b) midpoints of the classes, (c) percentages, (d) cumulative frequencies, and (e) cumulative percentages.

PERCENTAGES OF PERSONS 65 YEARS OLD OR OLDER IN METROPOLITAN
STATISTICAL AREAS, 1980

Case	Percent 65+	Case	Percent 65+	Case	Percent 65+
1	13.1	11	7.0	21	12.3
2	8.1	12	10.3	22	11.9
3	12.9	13	12.2	23	11.1
4	10.7	14	11.7	24	6.9
5	7.8	15	12.2	25	11.5
6	17.2	16	10.6	26	9.6
7	8.4	17	12.0	27	10.9
8	7.8	18	9.2	28	9.9
9	9.7	19	7.5	29	8.9
10	10.1	20	13.0	30	8.4

(continues)

PERCENTAGES OF PERSONS 65 YEARS OLD OR OLDER IN METROPOLITAN
STATISTICAL AREAS, 1980

Case	Percent 65+	Case	Percent 65+	Case	Percent 65+
31	7.7	81	14.3	131	9.7
32	10.8	82	11.7	132	11.6
33	7.2	83	7.6	133	13.4
34	10.6	84	7.6	134	11.9
35	10.9	85	9.2	135	13.5
36	8.9	86	9.9	136	11.0
37	10.4	87	9.9	137	11.3
38	11.7	88	8.3	138	9.0
39	8.5	89	9.9	139	11.1
40	7.3	90	8.3	140	13.4
41	9.0	91	12.1	141	13.3
42	8.4	92	10.4	142	13.3
43	8.2	93	8.9	143	8.5
44	7.9	94	8.4	144	14.1
45	8.9	95	9.2	145	9.9
46	10.7	96	12.7	146	11.0
47	9.8	97	18.1	147	9.6
48	7.8	98	22.0	148	9.1
49	6.8	99	15.7	149	11.7
50	7.9	100	11.0	150	9.2
51	10.5	101	11.1	151	7.3
52	9.3	102	10.2	152	9.0
53	6.4	103	9.6	153	10.3
54	13.6	104	10.2	154	9.9
55	6.6	105	11.0	155	12.6
56	10.9	106	10.1	156	7.5
57	9.6	107	10.4	157	13.2
58	8.0	108	10.5	158	13.5
59	9.8	109	9.1	159	9.9
60	10.1	110	12.2	160	11.3
61	9.8	111	12.1	161	15.6
62	10.2	112	9.1	162	9.8
63	9.6	113	12.6	163	9.9
64	11.8	114	8.9	164	9.4
65	11.4	115	15.5	165	10.4
66	10.3	116	9.8	166	11.5
67	11.4	117	13.2	167	12.8
68	10.7	118	11.1	168	11.3
69	12.4	119	9.9	169	10.7
70	7.3	120	10.9	170	21.5
71	6.4	121	12.4	171	11.7
72	6.1	122	10.0	172	10.1
73	6.2	123	9.6	173	13.5
74	11.7	124	10.9	174	10.7
75	9.8	125	8.6	175	7.6
76	9.4	126	10.9	176	23.3
77	9.7	127	11.5	177	9.9
78	11.4	128	11.7	178	13.3
79	10.5	129	11.4	179	11.5
80	11.2	130	11.7		

Source: U.S. Bureau of the Census (1984: 19–21).

GENERAL REFERENCES

Hagood, Margaret, and Daniel Price, *Statistics for Sociologists* (New York, Holt), 1952.
    Chapter 6 is a useful introduction and organization of the field of descriptive statistics.
    Chapter 7 deals with rates, ratios, proportions, and percentage change.
Hammond, Phillip E., *Sociologists at Work* (New York, Basic Books), 1964.
    This book is a collection of discussions about the background of some significant
    sociological research by a baker's dozen of sociologists. In addition to being a very
    readable and useful insight into research, the book presents the lines of reasoning that
    result in a need for data on specific kinds of contrasts between individuals, groups, and
    standards.
Zeisel, Hans, *Say It with Figures*, 6th ed. (New York, Harper & Row), 1985.
    This book contains some clear discussion of the idea of comparison and some of the
    ways ratios and percentages may be used. Chapter 1 deals with percentages, another
    chapter illustrates how the "don't know" category might be handled, and Chapter 5
    explains how index numbers might be constructed for comparison purposes.

LITERATURE CITED

Babbie, Earl, *The Practice of Social Research*, fifth edition (Belmont, Calif., Wadsworth),
    1989.
Bailey, Kenneth D., *Methods of Social Research*, third edition (New York, The Free Press),
    1987.
Brotman, Herman B., "Facts and Figures on Older Americans," Administration on Aging,
    U.S. Department of Health, Education and Welfare, Publication No. 182, 1970. (Data
    were originally from U.S. Census publications.)
Chiricos, Theodore G., and Gordon P. Waldo, "Punishment and Crime: An Examination of
    Some Empirical Evidence," *Social Problems*, 18 (Fall 1970), pp. 200–217.
Datta, Lois-Ellin, "Birth Order and Potential Scientific Creativity," *Sociometry*, 31 (March
    1968), pp. 76–88.
Erskine, Hazel, "The Polls: Freedom of Speech," *Public Opinion Quarterly*, 34 (Fall, 1970),
    p. 491.
Ferdinand, Theodore N., "The Criminal Patterns of Boston Since 1849," *American Journal of
    Sociology*, 73 (July 1967), pp. 84–99.
Huff, Darrell, *How To Lie with Statistics* (New York, Norton), 1954.
Kolko, Gabriel, *Wealth and Power in America: An Analysis of Social Class and Income
    Distribution* (New York, Praeger), 1962.
Mendelsohn, Harold, "Election-Day Broadcasts and Terminal Voting Decisions," *Public
    Opinion Quarterly*, 30 (Summer 1966), pp. 212–225.
Nachmias, David, and Chava Nachmias, *Research Methods in the Social Sciences*, third
    edition, (New York: St. Martin's Press), 1987.
National Center for Health Statistics, C. A. Schoenborn, and B. H. Cohen, "Trends in
    Smoking, Alcohol Consumption, and Other Health Practices among U.S. Adults, 1977
    and 1983." Advance data from *Vital and Health Statistics*, No. 118, DHHS Pub. No.
    (PHS) 86–1250, Public Health Service, Hyattsville, Md., June 30, 1986.
Orenstein, Alan, and William R. F. Phillips, *Understanding Social Research: An Introduction*
    (Boston, Allyn and Bacon), 1978.
Petersen, William, *Population* (New York, Macmillan), 1961.
Population Reference Bureau, Inc., *1986 World Population Data Sheet*, Washington, D.C.,
    April 1986.

Riley, Matilda White, *Sociological Research: A Case Approach* (New York, Harcourt, Brace), 1963.

Ross, Peggy J., Herman Bluestone, and Fred K. Hines, "Indicators of Social Well-being for U.S. Counties," Economic Development Division, Economics, Statistics, and Cooperatives Service, U.S. Department of Agriculture. Rural Development Research Report No. 10, Washington, D.C., May, 1979.

Stouffer, Samuel A., "Some Observations on Study Design," *American Journal of Sociology,* 40 (January 1950), pp. 355–361.

U.S. Bureau of the Census, *Statistical Abstract of the United States: 1966* (87th Edition) (Washington, D.C.), 1966.

————, *Current Population Reports*, Series P-23, No. 34 (Washington, D.C.), 1971(a).

————, "Population Characteristics," *Current Population Reports*, Series P-20, No. 217 (Washington, D.C., U.S. Government Printed Office), 1971(b).

————, "Final Population Counts for the 1970 U.S. Census," *U.S. Census Report* PC (V2)–1 (Washington, D.C., U.S. Government Printing Office), 1971(a).

————, "Final Population Counts for the 1970 U.S. Census," *U.S. Census Report* PC (V2)–1 (Washington, D.C., U.S. Government Printing Office), 1971(b).

————, *Statistical Abstract of the United States: 1985* (105th Edition) (Washington, D.C., U.S. Government Printing Office), 1984.

————, *Statistical Abstract of the U.S.: 1986*, 106th edition (Washington D.C., U.S. Government Printing Office), 1985.

————, "Estimates of the Population of the United States, by Age, Sex and Race: 1980 to 1985," *Current Population Reports, Series P-25, No. 985* (Washington, D.C., U.S. Government Printing Office), 1986.

U.S. Bureau of Labor Statistics, *Handbook of Labor Statistics* (Washington, D.C., U.S. Government Printing Office), 1989, pp. 425–426.

U.S. Department of Commerce, *Social Indicators: 1976* (Washington, D.C., U.S. Government Printing Office), 1976.

Zeisel, Hans, *Say It With Figures*, 6th ed. (New York, Harper & Row), 1985.

DIAGRAM 3.1   Creating Distributions

THE PROBLEM

To organize raw data (separate scores) into a frequency distribution or percentage distribution that accurately reflects the data.

1. *Setting Up Classes.* If the data are to be used for *further statistical computation* it is best to either (a) use the raw data rather than create a distribution (computers are helpful in reducing the clerical problems in handling raw scores), or (b) preserve a relatively large number of classes (such as 10 to 15 or more). If the data are for *presentation* only, then fewer classes may be used and the problem is one of setting up classes that accurately portray main differences of importance in the data. Often this takes the form of an interest in grouping classes together so that minor differences are eliminated (the jaggedness of most distributions) and a more smoothed-looking distribution is shown.

   a. *Select the number of categories* you want to use. Usually this should be more than two and probably no more than 15, depending on the purpose of the distribution (see point b, below). Five to ten is a useful range. Call this k.

   b. *Find the interval size needed.* An approximation can be found by:

   $$w = \frac{(\text{highest score}) - (\text{lowest score})}{k}$$

   and take the nearest integer. This is the interval width, w. An odd number for w is preferred because then class midpoints can be arranged so that they are integers, but often this aid to hand computation is not necessary. Widths of multiples of 1, 2, 3, 5, or 10 are common.

   c. *Set up categories.* Start with a lower limit for the lowest class that (a) is equal to or less than the smallest score, and (b) preferably (traditionally) is some multiple of the selected class width w. For example, if the class width was 5, one might begin the lowest class with some value such as 0, 5, 10, or 15. Thus the lower limit of each successive class is a multiple of the class width. It is generally preferred to have classes all of the same width, although sometimes this rule is broken, particularly where class frequencies are very small for large parts of the scale. It is also prudent practice to create categories (especially if later statistical computation is to be carried out) so that there is a specified upper and lower limit for each class including the extreme ones. For presentation purposes sometimes the extreme classes are left "open" (*e.g.,* "65 and over"), but this practice makes it impossible to determine a class midpoint and thus makes the computation of some of the statistics discussed in Chapter 5 impossible.

   d. *Check the categories* to see if they accurately reflect the data they are to contain. That is, does the set of class midpoints fairly accurately reflect the balance of cases in the classes? Ideally, the sum of raw scores would equal the sum of class midpoints times class frequencies for interval variables, where this check is appropriate. If there is gross distortion, reset class limits, change (increase, usually) the number of classes, or change (decrease, usually) the class width.

DIAGRAM 3.1 *(Continued)*

2. *Group Cases Into Classes.* If a computer is not available, the easiest and most accurate way to classify scores in categories is to create a "tally." Be sure to move systematically through the list of scores, tallying each score by making a tally-mark in the appropriate category rather than skipping around through the data to find identical scores. When you have tallied a score put a light line through it to show that it has already been tallied.

3. *Creating The Desired Kind of Distribution.* Most kinds of distributions are created from an initial frequency distribution, so the frequency distribution should be totaled and checked for accuracy.

   a. *Percentage distribution.* Divide each class frequency in a frequency distribution by the total number of scores, and multiply each result by 100. When this kind of distribution is presented, it is good practice to include N, the total number of scores upon which the percentage is based, in brackets or parentheses below the total of the percentages (see Table 3.5).

   b. *Cumulative distributions.* Beginning with the lowest value score category (traditionally), create a cumulative column by entering the class frequency of the lowest class next to that class, the total of the frequencies for the lowest two classes next to the second-lowest class, etc. In the case of a cumulative percentage distribution, the percentage distribution entries are accumulated. The last class (highest score category) should have N or 100% entered next to it (see Table 3.5).

DIAGRAM 3.2   Computing Percentiles

The 10th percentile is the *score* value below which 10% of the scores fall. The *i*th percentile is the *score* value below which *i* percent of the scores fall. There are two typical kinds of percentile problems one might want to solve.

FIND THE PERCENTILE RANK OF A GIVEN SCORE

$$\text{Percentile rank of a given score} = (100)\frac{\text{cum}f \text{ up to and including that score}}{N}$$

The percentile rank of the score "7" in the example below is:

Raw Scores
1
4
4
6
7
9
12
14

$\begin{cases}\text{Percentile rank of the score "7"} = (100)5/8 = 62.5 \text{ since } N = 8 \\ \text{and the score 7 is the 5th one in an array of these scores.}\end{cases}$

FIND THE SCORE AT A GIVEN PERCENTILE RANK

Score at a given percentile is found by:

1. Multiply the percentile by N.
2. Count up an array of scores until that numbered score is found. The score value of that score is the desired value.

The 62nd percentile in the above example is found by multiplying .62 by N, which in this example is 8. The result is 4.96, or, rounded to an integer, 5. We would then begin at the first score in the array above and count up to the 5th place. That fifth score is 7 and that would be the score at the 62nd percentile.

Grouped scores pose a slight problem. Usually in finding the score at a given percentile the class containing this score is identified and then a score value in that class is found by interpolation. The following formula may be used for this purpose and it is discussed later in Chapter 5 and in Diagram 5.1 in connection with the median, which is the 50th percentile.

DIAGRAM 3.2 *(Continued)*

$$\text{Score at a given percentile} = L_P + \left[\frac{(N)(P) - \text{cum}f_P}{f_P}\right]w_P$$

where L_P is the lower real limit of the class containing the Pth percentile score. This is determined by inspection of the cumulative frequency or cumulative percentage distribution;

w_P is the class width of the class containing the Pth percentile;

N is the number of cases (scores);

P is the percentile of interest, expressed as a proportion;

$\text{cum}f_P$ is the cumulative frequency up to but *not* including the frequency in the class containing the Pth percentile;

f_P is the frequency in the Pth percentile class.

Percentile rank of a score at a given percentile also may be found by inspection of a graph of a cumulative percentage distribution, as discussed in the next chapter.

CHAPTER 4

Graphic Presentation
and Analysis

The sex ratio (number of males for every 100 females) declines steadily with increasing age not only in the United States, but in all other societies for which there is information. In the United States, slightly more males are born (104.9 males per 100 females in 1985), but by the early 20s the sex ratio begins to reverse, and for those over the age of 75, in 1985, there were only 54.3 males per 100 females (see Table 3.7).

4.1 RATIOS—A GRAPHIC EXAMPLE

The line graph in Figure 4.1 shows sex ratios by age for each of three years, 1950, 1960, and 1980. In general the line for each of the three years shows a decline in the sex ratio with older age groups, but past age 40 the difference between the three lines shows that there has been a rather marked drop in the sex ratio between 1950 and 1980. Why?

The graphic technique used in Figure 4.1, in this case a line graph, displays changes in the sex ratio very clearly, permitting both overall comparisons and comparisons for specific age segments of the population. Such vivid contrasts help focus further research. Perhaps the greater drop in the mortality rate for females than for males is a factor, and this in turn might be traced to changes in medical techniques and social norms concerning births. The changing sex composition of the post-40 age brackets might also be related to the way sex roles are defined in the United States (*e.g.*, male mortality from wars, or perhaps occupation-related stress and disease). The consequences of such changes in composition would then begin to show up in data as an age cohort. A **cohort** is defined as those people experiencing a similar event in the same period or year—in this case, those *born* in the same year (see Riley, 1973). Such graphic comparisons also help focus research questions about possible *consequences* of an increasing gender imbalance with advancing years for such areas as housing, types of role relationships, and various types of medical and social services that may be used by older age groups in our

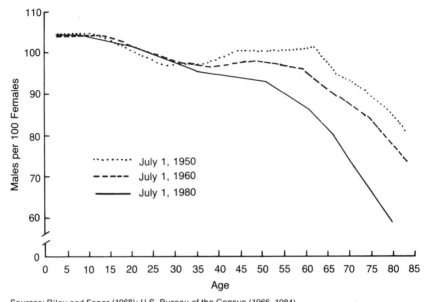

Sources: Riley and Foner (1968); U.S. Bureau of the Census (1966, 1984).
FIGURE 4.1 LINE GRAPH SHOWING THE GENDER RATIO BY AGE FOR THREE YEARS: 1950, 1960, AND 1980 (Total population including Armed Forces overseas)

society. Each of these possibilities suggests further avenues of related research. Graphic presentation in this case probably communicates more relevant information more clearly and accurately than would be immediately evident if the same information were to be presented in a table.

An important aim of statistics is making clear the relevant kinds of information a collection of measures may contain. In a sense, statistical handling of data helps highlight the "information" contained in data so that investigators and their audience may examine it and draw appropriate conclusions in an efficient way. Graphic analysis has much to offer when it comes to clarifying complex relationships, and with the advent of computer-controlled plotting systems, there is an opportunity to exploit this means of analysis more than has been possible in the past. In this chapter we will discuss some of the basic techniques of graphic analysis and illustrate some of the more useful alternatives.*

4.2 BASIC TECHNIQUES FOR GRAPHING DISTRIBUTIONS

Histograms, polygons, ogives, and line graphs are basic graphic procedures used in statistics, and they provide a useful way to introduce some of the techniques and cautions of graphic analysis. The first three of these techniques illustrate how the

* For an excellent reference for graphic presentation, see Schmid (1979, 1983), in the General References.

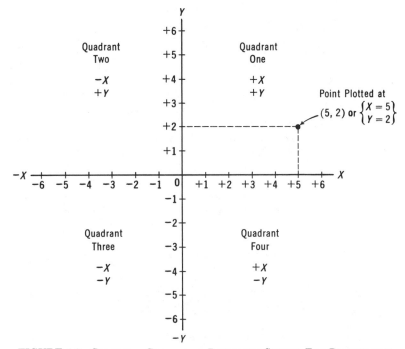

FIGURE 4.2   Cartesian Coordinate Reference System, Two Dimensional

distributions discussed in Chapter 3 may be handled graphically. A later section in this chapter will apply some of these techniques to other styles of graphic presentation and introduce some important and useful alternative approaches to graphic analysis.

Basic to any graphing is the idea of a **reference system** or **coordinate system**. The usual coordinate system for graphing consists of lines or "dimensions" at right angles to each other called a Cartesian coordinate system.* The reference system in Figure 4.2 is two-dimensional, but the idea can be generalized to three or more reference axes or dimensions. Traditionally, the vertical line is called the Y-axis, or **ordinate**, and the horizontal line is called the X-axis, or **abscissa**. This divides a plane into four **quadrants**. The **origin**, or zero point,

---

* Reńe Déscartes (1596–1650), the French philosopher and mathematician, is said to have devised the coordinate system named after him while he was in bed watching a fly. He could locate the fly's travels in terms of three lines perpendicular to the walls and floor. In two dimensions, on a page, points can be exactly located by two axes at right angles to each other and thus by a pair of numbers (X and Y) as shown in the graph in Figure 4.2. Descartes went on to combine graphing and algebra by expressing the location of points in terms of formulas or functions such as $Y = 3 + 2X$, ideas we will use in Chapter 7. Descartes published this important work as an appendix to a 1637 book on the solar system. For an interesting discussion of Descartes' work, see Asimov (1972:83–4). There are other kinds of reference systems, some of which will be introduced later in this chapter.

BOX 4.1 The Theme of Comparison—Making It Graphic

Chapters 1, 2, and 3 introduced some of the language of statistics, the importance of comparisons, and some uses of the ratio as a means by which valid comparisons can be made. This chapter extends this theme to include graphic means for making comparisons, either within one collection of data or between several collections of data. Graphic techniques sometimes make use of ratios discussed in Chapter 3.

Two loose groupings of graphic techniques are discussed:

Group I includes four basic statistical graphs—histogram, polygon, ogive, and line graph. The first three of these are graphic representations of distributions discussed in Chapter 3, and we will use these as a basis for discussing some general techniques of graphing and some cautions.

Group II includes a broader variety of specialized graphic techniques, each of which has important uses in graphic analysis of data in sociology. These techniques include population pyramids, pie charts, statistical maps, scatter plots, and semilogarithmic charts.

---

on both axes is the point where the axes cross, and numerical scales extend outward from this point. Scores upward from the origin on *Y* are positive scores, those downward are negative. Scores to the left on *X* are negative and to the right are positive. Since most social measurements are on scales that extend from zero in a positive direction only, quadrant one is frequently the only one needed, and the other quadrants are then simply omitted from a graph as a matter of esthetics and convenience.

4.2.1 Histograms

A **histogram** is a plot of a frequency or percentage distribution in which the frequency (or percentage) of cases in each category is represented by a bar as wide as the plotted score category on the *X*-axis and to a height that is proportional to the frequency (or percentage) of cases falling in that category as shown on the *Y*-axis. The scale of categories of the variable being plotted is laid out on the *X*-axis *with equal physical distances corresponding to equal differences in scores.* The frequency (or percentage) scale is arranged on the *Y*-axis, again with equal physical distances corresponding to equal frequency (or percentage) amounts. The *Y*-axis always begins with the origin or zero point to avoid distortions in comparing different columns, but the *X*-axis or score axis typically begins with any convenient low score. The objective is to create a figure with a total area corresponding to a total frequency, *N* (or total percentage 100%), that is distributed among the different categories of the variable on the *X*-axis in a way that correctly represents the relative number (percentage) of cases in each category or interval. A step-by-step discussion of the construction of a histogram is presented in Diagram 4.1 (p. 108).

Figure 4.3 illustrates the use of a histogram. In this case it shows the number of U.S. households in 1986 that had different numbers of persons in them. There are

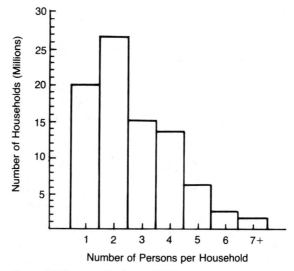

Source: U.S. Bureau of the Census (1985).

FIGURE 4.3   Histogram of Size of Household, United States, 1986

nearly 20 million households of size 1, 26.9 million of size 2, and so forth. The histogram clearly shows that most households were of sizes 1, 2, or 3. The average number of persons per household in the United States in 1986 was 2.71 persons.

Some variations of the basic histogram are used to reflect certain character-istics in the data. For example, if the variable is a *nominal* one, the bars in the histogram may be separated somewhat to visually carry the image that they are separate and distinct categories. This is usually called a **bar chart** or bar graph, and the bars are drawn with equal widths.

For *ordinal* variables, where equal distances are not defined, one may choose to separate bars slightly to emphasize this fact (particularly if the variable is discrete rather than continuous). Alternatively, one might rely on the "stair-step" impression of the histogram to carry this message and still preserve the impression of rank order of categories by placing columns immediately next to each other (incorporating the lines between each column). For ordinal variables, it is customary to adopt a standard category width for plotting purposes, in spite of the fact that distances are not defined. This helps avoid distractions and apparent distortions from arbitrarily varying width. In this respect, one may want to treat ordinal variables by the same rules that apply to plotting interval variables.

For *interval* variables the basic procedures for plotting a histogram in Diagram 4.1 can be followed. A histogram may be particularly appropriate for discrete, interval variables. If the variable being plotted is continuous, one may prefer to turn to another type of basic graphing technique, the polygon.*

A number of variations on the histogram are discussed in Section 4.3.

* However, computer statistical software packages commonly include routines for graphing histograms rather than polygons.

4.2.2   Polygons

The frequency (or percentage) **polygon** is a closed figure connecting points plotted above the *midpoint* of each category at a height corresponding to the frequency (or percentage) of cases in the category. Figure 4.4 shows two percentage polygons of the age distribution of enrolled and nonenrolled people aged 3 to 34 years in the United States in October 1988. Figure 4.4a is the age distribution for those enrolled in school and Figure 4.4b shows the age distribution for those not enrolled. Note that each figure is closed by extending the line down to the *X*-axis at the next midpoint beyond the extreme categories of the distribution. Diagram 4.1 (p. 108) presents step-by-step procedures for constructing a polygon.

A comparison of the two age distributions for those enrolled and those not enrolled, as in Figures 4.4a and 4.4b, shows marked differences in their form. Among those enrolled in school the distribution is concentrated around the 10–13 year ages, with smaller percentages enrolled shown on either side. For those not enrolled, the distribution is concentrated at both ends of the age range and the distribution forms a U-shape.

The polygon gives the visual impression of more gradual shifts in frequency (or percentage) from category to category, whereas the histogram highlights shifts between categories. The frequency (or percentage) polygon also relies on distances between categories, and for this reason it is most appropriate in graphing interval variables.

Figure 4.5 shows an analytic use of the polygon to present the overlap between distributions of income for blacks and for whites in the United States in 1968. The area of overlap of the two distributions is used as one measure of *integration*, which is computed by finding the percentage (or proportion as shown in Figure 4.5) of overlap out of the total area under both curves (*i.e.*, the total

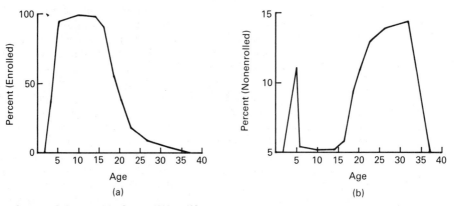

Source: U.S. Bureau of the Census 1990, p. 132.

FIGURE 4.4   (a) Percentage Polygon of the Age Distributions of 3- to 34-Year-Olds Who Were Enrolled in School. United States, 1988; (b) Percentage Polygon of the Age Distributions of 3- to 34-Year-Olds Who Were Not Enrolled in School, United States, 1988

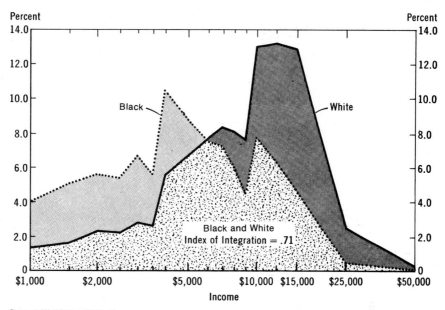

Source: Weitzman (1970:9).

FIGURE 4.5   PERCENT DISTRIBUTION OF BLACK FAMILIES AND WHITE FAMILIES BY TOTAL MONEY INCOME, 1968.

shaded area). In this example, there is a 71% overlap, or the "index of integration" for blacks and whites on family income in 1968 is equal to .71. Complete segregation, where the index of integration is equal to 0.0, would be shown by a graph in which the income distributions of blacks and whites do not overlap at all. Complete integration, where the index of integration would equal 1.0, would be shown by two identical distributions exactly "overlapping" or corresponding to each other. Both polygons are plotted on the same graph for easier comparison, and essentially two bits of information are presented. Ultimately, two measurements had to be made on each family—race and income—for the graph to be prepared. The distribution of income was then plotted separately for the two racial groups.

4.2.3   Ogives

A third basic graphic technique is a plot of a cumulative distribution (either percentage or frequency). This kind of graph shows the frequency (or percentage) of cases falling below the upper boundary of each successive class.* **Ogives** are

---

* Note that we are accumulating from low scores to high scores for consistency throughout this book. Obviously, one could accumulate from high to low, too, in which case statements like this one would have to be reversed or changed to the "lower" boundary. High to low accumulation would be useful in the analysis of waiting times in a queue, for example, where one would be interested in the percentage of people waiting a given number of minutes "or more."

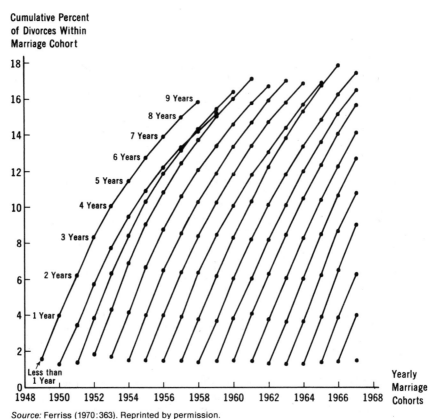

Cumulative Percent of Divorces Within Marriage Cohort

Source: Ferriss (1970:363). Reprinted by permission.

FIGURE 4.6 CUMULATIVE PERCENTAGE OF DIVORCES WITHIN YEARLY MARRIAGE COHORTS, BY DURATION OF MARRIAGE (THROUGH 9 YEARS), 1949–1967.

used with ordinal or interval-level variables to aid in the examination of the overall shape of a distribution, to aid in finding various percentiles, and to aid in the analysis of change. Diagram 4.1 (p. 108) illustrates the steps in drawing an ogive. Notice that the plot points are directly above the *upper boundary* of each class, so that the line connecting them extends from the X-axis up and to the right, ending at the total frequency (or total percentage, 100%, in the case of a percentage ogive).

Figure 4.6 illustrates one use of the ogive in sociology. Abbott Ferriss was interested in the rate of divorce and whether or not the percentage of couples who divorced within the first 9 years of marriage was changing through time (Ferriss, 1970:363). He used a percentage ogive to plot the cumulative percentage of divorces among those married in the same year. That is, Ferriss gathered data on a given marriage cohort, calculating the percentage of that cohort who divorced each year for the first 9 years of their marriage. These percentages were then accumulated and plotted above December 31st of each successive year of marriage. This was done for each of the annual marriage cohorts from 1949 to 1967. For couples married in 1949, for example, about 1.5% divorced in less than a year. By the end of the first year about 4% had divorced, and so on, until by 9 years nearly 16% had divorced. The ogive in this case does not extend to 100% because the base of the percentage was all marriages of a given cohort, a more meaningful figure in this instance, rather than the final number of divorces of a given cohort (which would still be unknown at the time of Ferriss' study).

From the ogives in Figure 4.6, one can see that the percentage of married couples who divorce after a given number of years of marriage is relatively constant from cohort to cohort. There is some suggestion that the percentages might be rising for later-married cohorts (shown by taller ogives in later years), and the 1950 marriage cohort had a lower than usual percentage of divorces after the fifth year of marriage. This later point is shown by the ogive bending to the right rather than extending upward as steeply as other cohort ogives.

BOX 4.2   THE DISTRIBUTIONS BEHIND THE GRAPHS

The basic graphic techniques discussed so far—histograms, polygons, and ogives—are used to present the distributions that were discussed in Chapter 3. Do you understand the following?

Frequency distributions (if not, see Section 3.2.3a)
Percentage distributions (if not, see Section 3.2.3b)
Cumulative distributions (if not, see Section 3.2.3c)
Percentiles (if not, see Section 3.2.3d)

4.2.4   Line Graphs

A fourth type of basic graph is the **line graph**, which shows the value of some dependent variable (scaled along the Y-axis) for each of several categories of another variable, usually an independent variable (scaled along the X-axis). The plotted points are connected by a straight line, and the figure is not closed with the X-axis, since the area under the curve has no particular meaning as it does in the case of the frequency or percentage histogram or the polygon. Diagram 4.1 (p. 108) provides a step-by-step discussion of the way these graphs are constructed.

Figure 4.7 is an example of a line graph, where time is represented on the X-axis. The dependent variable, rate of marital dissolution through death, is plotted on the Y-axis, and the line extends from 1860 to 1964. This is often called a **trend line** because it is a plot of some characteristic through time. The graph shows rather clearly the overall decline in marital dissolutions through death during the century, with minor fluctuations from this trend for most years. The year 1918 stands out as a major departure from the trend. It was during 1918 that there was a major epidemic of influenza, an epidemic that also claimed German sociologist Max Weber in 1920.

Line graphs are particularly appropriate where one of the variables (on the X-axis) is a continuous, interval variable, such as age or time. A bar graph could be

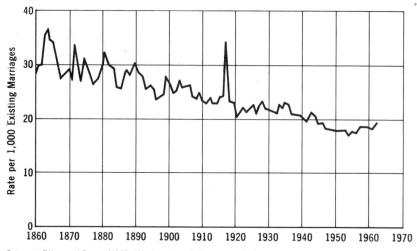

Sources: Riley and Foner (1968:19); Jacobson (1959:118; 1966: Fig. 1). Used by permission.

FIGURE 4.7   Marital Dissolutions by Death, United States 1860–1964 (Includes Armed Forces overseas during 1917–1919 and 1940–1964; excludes civilian populations of Alaska and Hawaii prior to 1960)

used instead of a line graph for situations where the variable on the *X*-axis is nominal or ordinal.

Figure 4.8 provides an example where bars represent the percentage increase in the 65 and older population between 1980 and 1984 for the 10 states with the highest percentage increase. The bars reflect the nominal variable "state" by being

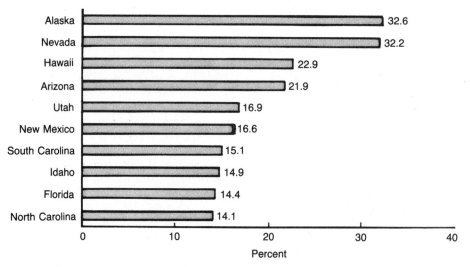

Source: U.S. Bureau of the Census (1985b).

FIGURE 4.8   Ten States with Largest Percentage Increase of Population 65 Years and Over, 1980–1984

separated. Alaska's 32.6% increase is dramatically larger than states lower down on the graph.

4.2.5   Distortions in Graphing

Whichever form of graph is selected, the key criteria should be the accuracy and clarity with which it communicates aspects of the data of interest to an audience or to the investigator. Investigators have a wide variety of techniques from which to choose. To communicate properly requires clear knowledge of what is there to communicate, and of possible misinterpretations and qualifications that should be taken into account, along with a healthy sensitivity for the ways people interpret and understand evidence. Clearly, too, there are vast possibilities for doing mathematically or technically correct manipulations that may intentionally or inadvertently misrepresent the state of affairs being investigated. Some cautions and rules for careful representation are given in this book, but there are always exceptional situations, and the responsibility for clear communication is not solved by mere rules for technical accuracy.

There are several ways of graphing that are numerically accurate but that give a distorted impression of the data. These are illustrated in Figure 4.9. The series of graphs in Figure 4.9a through 4.9c illustrates the distorting effects of extending or shortening one of the axes. Although the data are accurately plotted in each graph, the impression of the steepness of change from category to category is greatly altered. A partial corrective is the three-quarters rule, discussed in Diagram 4.1 (p. 112).

Similar in effect to the distortion involved in changing scale length is the distortion involved in not starting with zero on the frequency (or percentage) scale on the Y-axis in a histogram or polygon. Figure 4.1 illustrates a helpful procedure to use where a graph would be awkwardly tall if all values from zero were included. In Figure 4.1 sex ratios were plotted in the 70 to 110 range, and in order to have the graph large enough so that values could be read off the Y-axis, physical units on the Y-axis were made relatively large. In this case, Riley did not include all score values from zero to 70, because this would have resulted in a graph taller than the printed page (Riley and Foner, 1968). Rather, she skipped some values between zero and 60, but clearly indicated in her presentation that this had been done and the reader should beware. This warning is expressed by "breaking" the graph and the Y-axis to show that part of the table is missing. *In general, it is a poor practice to omit values on the frequency or percentage axis.*

A further potential distortion of graphs is the use of figures that differ in area or volume as well as height. Figure 4.9d illustrates the problem. Although the stick figures are drawn so that the taller is twice as high as the shorter to indicate a doubling in frequency, the figures, if drawn esthetically, also cover an area, and the ratio of these areas is likely to be about 1 to 4 rather than 1 to 2 as intended. It is often not clear to a reader which of these possible comparisons correctly represents the data. If a figure is a square or circle, a doubling of its height or diameter results in an area that is four times larger. If a figure in a graph is illustrated as a cube or sphere, a doubling of its diameter or height results in an

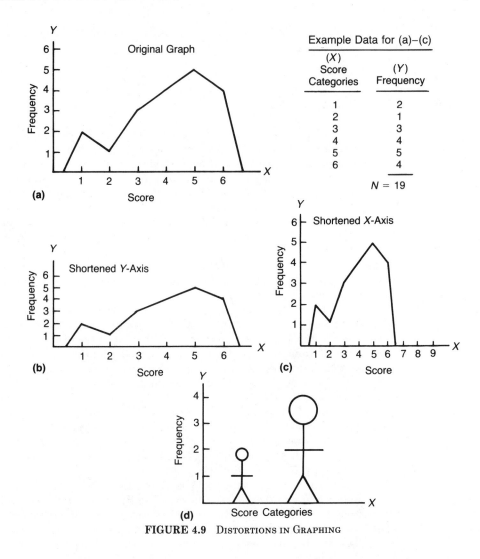

FIGURE 4.9   DISTORTIONS IN GRAPHING

eightfold increase in volume. To avoid this possible distortion, investigators generally prefer to adopt a standard-sized figure to represent a given number or percentage of cases and merely use as many of these figures as is necessary to indicate different category values. Figures 4.10 illustrates this relatively more careful procedure. Called a **pictograph** because of its use of small pictures to represent a standard quantity of the variable being studied, Figure 4.10 shows the relationship between birth and death rates for seven world areas and for the world as a whole. Notice that each figure represents two births or deaths per 1000 population, and that fractional figures are used to represent fractional amounts.

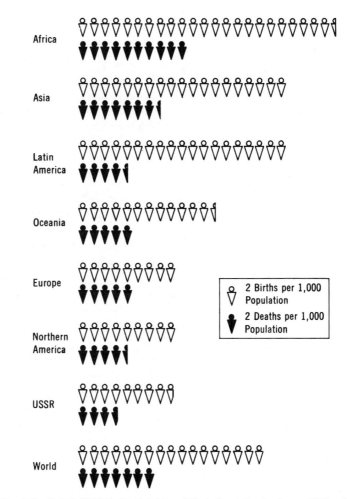

Source: *Population Bulletin* (1971:6). Reprinted from "Man's Population Predicament," *Population Bulletin,* Vol. 27, No. 2 (Population Reference Bureau, April 1971).

FIGURE 4.10   Birth and Death Rates by Area, 1971

4.3   VARIATIONS ON BASIC GRAPHIC TECHNIQUES

A number of other graphic techniques are useful for either analysis of data or presentation of results or both. In this section of the chapter we will illustrate some useful variations of the basic graphing techniques we have been discussing thus far. In the following sections we will discuss techniques involving somewhat different approaches and illustrate new applications of analytic graphing in sociology.

4.3.1   The Bar Chart

The **bar chart** is a variation of the histogram often used for nominal variables or when emphasis is on categories of a variable (see Figure 4.8). Figure 4.11 shows bar charts of "place of physician visits" for four age categories (0–4 years, 5–14, 15–24, and 25 and over). Data from monthly interviews by the U.S. Public Health Service are shown for a year ending June, 1958, in the top graph and for the year 1968 in the bottom graph. In this application each bar represents 100% of the physician visits

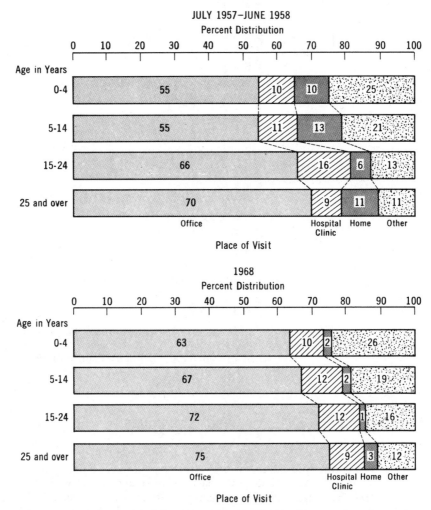

Source: U.S. Department of Health, Education and Welfare (1971:31).

FIGURE 4.11   Percentage Distribution of Physician Visits by Place of Visit, According to Age: July 1957–June 1958 and 1968

made by a given age group in a given year, but the bars are divided differently to show the percentage breakdown of visits by where they occurred—in the physician's office, a hospital, home, or "other" places. For ease in interpretation, percentages are also entered on each of the shaded portions of each bar. Notice that in both graphs, older age groups are more likely to visit physicians in their office, and between 1958 and 1968 there was a shift to greater relative use of the physician's office than alternative sites. Between the two years, home visits nearly disappeared, while hospital visits remained relatively constant both over age groups within a given year and between 1958 and 1968. Generally, this form of graph is adopted to facilitate comparison of the size of separate components within several bars.

Figure 4.12 provides another illustration of a bar chart, where again each bar is the same length, representing 100% of the cases in a category (*e.g.*, number of crimes of a given type known to the police). In this instance, however,

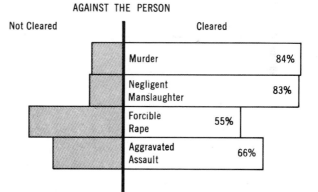

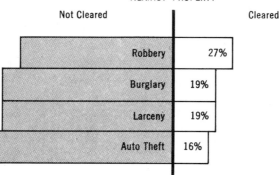

Source: U.S. Department of Justice (1971:32).

FIGURE 4.12   Crimes Cleared by Arrest, United States, 1971

the bars are adjusted so that the part extending to the right of a vertical line represents the percentage of that type of crime that has been "cleared" by an arrest having been made. The top graph presents a group of crimes "against persons" and the bottom graph presents similar data for crimes "against property." It is obvious in these graphs that crimes against persons are much more likely to be cleared by an arrest than are crimes against property—at least among those crimes known to police.

4.3.2   The Population Pyramid

The **population pyramid** is somewhat more complex in design, while again illustrating some simple graphing procedures discussed in connection with the histogram. Figure 4.13 provides an illustration. Each bar in the graph represents the percentage of the population in a specific age-gender category. The bars are organized with those for males on one side and those for females on the other and are stacked from youngest to oldest ages. The population pyramid is used chiefly by demographers to show the overall distribution of a population and to make comparisons of the populations of different countries. Figure 4.14 shows the population pyramid for the United States in 1969 (shaded pyramid) and, on the

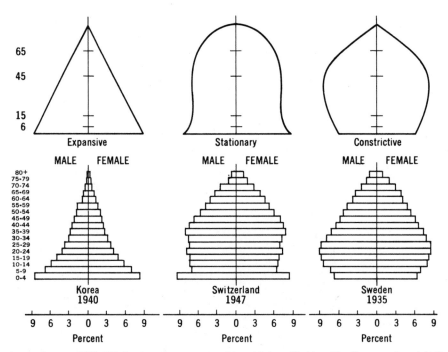

Source: Petersen (1969:628). Reprinted by permission of the publishers. The Macmillan Company. Copyright © William Petersen, 1969.

FIGURE 4.13 THREE ANALYTIC POPULATION PYRAMIDS SHOWING DIFFERENT POPULATION STRUCTURES AND THREE ACTUAL POPULATION PYRAMIDS

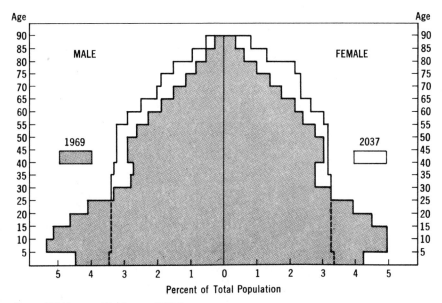

Source: U.S. Bureau of the Census (1970a).

FIGURE 4.14   Population Pyramids for the United States in 1969 and for a Stationary 2037 Projection (By percent of total population)

same axis, a pyramid that is a projection of the population of the U.S. in the year 2037 under the assumption of a stable population just producing enough children to replace itself. Compared with the stable population pyramid, the 1969 pyramid shows a relatively greater percentage of younger people and a relatively smaller percentage of older people. Notice, too, that the imbalance of males and females in the older age brackets is clearly shown, as are the effects of the depression of the 1930s (shown by the "notches" in the 30- and 35-year bars for the 1969 pyramid), and the "baby boom" is shown by the percentage of population in the 20-and-younger age brackets. Sometimes one can see traces of the effects of disease or war on a cohort.

4.3.3   The Pie Chart

A **pie chart** is shown in Figure 4.15, where an entire circle represents the total of some characteristic such as the total world population. In this example, the world population at different dates is shown by larger and larger sized circles (the area representing the changing total size). The "pie" is divided into segments with different kinds of shading, where each segment represents a component of the total. In this figure, the components are four general continental population areas of the world. Each wedge has an angle at the center that represents its relative share of the 360 degrees in a circle. Each 3.6 degrees represents 1% of the total. In 1970, the Americas held about 14% of the world's population and their "wedge" of the pie is represented by 50.4 degrees of the circle ($14\% \times 3.6° = 50.4°$).

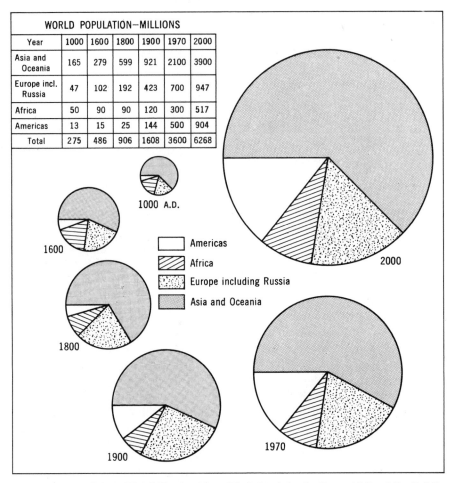

WORLD POPULATION—MILLIONS						
Year	1000	1600	1800	1900	1970	2000
Asia and Oceania	165	279	599	921	2100	3900
Europe incl. Russia	47	102	192	423	700	947
Africa	50	90	90	120	300	517
Americas	13	15	25	144	500	904
Total	275	486	906	1608	3600	6268

Americas
Africa
Europe including Russia
Asia and Oceania

Source: Population Bulletin (1971:9). Reprinted from "Man's Population Predicament," *Population Bulletin*, Vol. 27, No. 2 (Population Reference Bureau, April 1971).

FIGURE 4.15 ONE THOUSAND YEARS OF POPULATION GROWTH

4.3.4   Box and Whisker Diagrams

Box and whisker diagrams (or "boxplots") are useful to display a typical value or average score for a distribution as well as the spread or range of variations of a set of scores.* As shown in Figure 4.16, box and whisker diagrams can be compared to

* The box and whisker diagram idea for displaying distributions was suggested by Tukey (1977). Measures of central tendency and variation, which are usually shown in box and whisker diagrams, are the "arithmetic mean" and the "standard deviation" or the "median" and the "interquartile range" rather than percentiles as discussed here. (These concepts will be developed in the next chapter.) Usually a boxplot has a small line indicating the mean (or

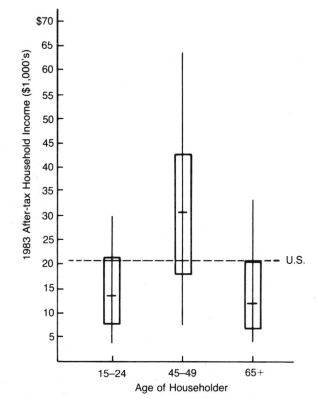

Source: U.S. Bureau of the Census (1985a:13).

FIGURE 4.16 Boxplot of 1983 After-Tax Income for Household by Young, Middle-Aged, and Older Householders, United States, 1985. [Plots extend from 10th to 90th percentiles of households; boxes extend from 25th to 75th percentiles. Median (50th percentile) is shown for each group and for the United States]

see how the same interval/ratio variable is distributed for several different groups of scores. In this illustration we will use the box and whisker plot to show the information from cumulative percentage distributions.

Figure 4.16 shows cumulative percentage distributions of after-tax household income for three different groups of householders in 1983. The household groups are those with young (15–24), middle-age (45–49), and older (65 and over) householders. For each group, the vertical line extends from the group's 10th percentile to its 90th percentile, that is, from the income below which 10% of that

---

(Footnote continued.)
median) with a box extending one standard deviation (or quartile) on each side of the mean (median). Unlike the diagrams above, the usual boxplot has whiskers extending from the highest score down through the box and the mark at the mean (median) on to the lowest score in the distribution. Since box and whisker diagrams are not familiar to most readers, it is helpful to fully label the diagram so that it is clear what is being plotted.

group's incomes fall to the income below which 90% of that group's incomes fall. The box extends from the 25th to the 75th percentile for each group. Finally, the short horizontal mark on the vertical line in the middle of each box is at the 50th percentile income for that group. That is, it is at the middle point where half of the households have less and half have more income. Each whisker (vertical line) shows the range over which the middle 80% of the incomes fall (from 10th to 90th percentiles), and each box shows the range over which the middle 50% fall (from the 25th to 75th percentile). These are both indications of the amount of variability there is in income scores for that particular group (we will later characterize these as the interdecile and interquartile ranges). The middle mark is the 50th percentile, which we will later characterize as the median, one kind of average or typical value that can be computed for a distribution of scores. It is clear that income for households with middle-aged householders extends much higher than the other groups. More than 75% of the elder households and nearly 75% of the younger households have incomes less than the U.S. 50th percentile income. In fact, the 50th percentile for both young and old is a lower income than the 25th percentile for the middle group.

4.4 OTHER APPROACHES TO GRAPHIC ANALYSIS IN SOCIOLOGY

Sometimes graphic analysis will utilize color or combinations of graphs that help highlight contrasts in the data. A number of good examples of color graphs can be found in commercial and, more recently, governmental publications such as *Social Indicators: 1976* (U.S. Department of Commerce, 1977), *General Social and Economic Characteristics, U.S. Summary* (U.S. Bureau of the Census, 1983), or the *Graphic Summary, 1982 Census of Agriculture* (U.S. Bureau of the Census, 1985c). In addition to the methods discussed above for graphing distributions, there are three further graphic techniques that we will introduce here because they offer some special advantages for the examination of data and because they suggest some of the types of contrasts an investigator may be interested in making. These techniques include statistical maps, scatter diagrams, and semilogarithmic graphs.

4.4.1 Statistical Maps

Statistical maps have been widely used in ecological studies in sociology, and three somewhat different approaches will be presented here. A statistical map shows the distribution of a variable over a geographic area. In Figure 4.17, a base map (the outline map of some area) of Chicago was divided into community areas for which data on suicide rates had been gathered for the 1959–1963 period. Often statistical maps report data by states, counties, or census tracts. Each area is shaded or cross-hatched in a way that illustrates the value of the variable in that area. In Figure 4.17, the more dense shading is used for higher suicide rates (12 to 23.1 per 100,000 population) and a *key* explains what each type of shading means.

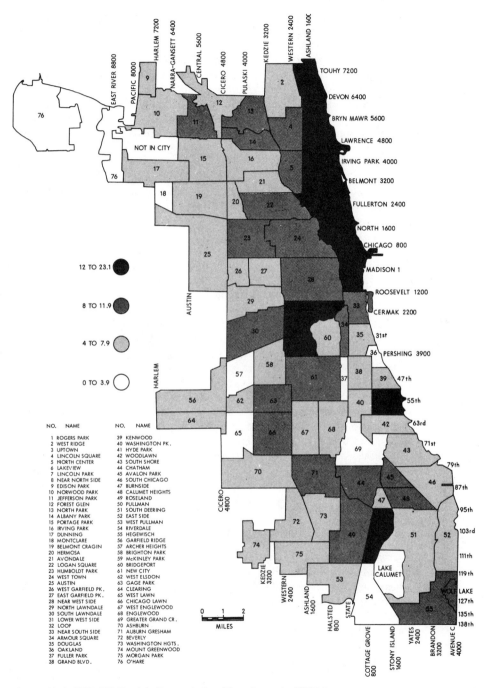

Source: Maris (1969: 139). Reprinted by permission of the author and publishers.

FIGURE 4.17   Suicide Rates by Community Areas, City of Chicago, 1959–1963

Usually, variables with only a few categories are plotted, and shading or coloring is selected to reflect gradations of a variable by brightness or density. The suicide rate map prepared by Maris shows that rates tend to increase toward the center of the downtown area (Maris, 1969). Where other characteristics of community areas are known, clues to the pattern of suicide rates can be found, and changes over time can be studied by comparing similar maps for successive years.

A number of computer programs are available that permit investigators to have their data automatically plotted in some graphic form. This greatly facilitates visual inspection of data and permits investigators to examine even more complex plots that show several variables simultaneously.

The second statistical mapping technique is illustrated in a study by Hoiberg and Cloyd (1971), in which they show the distribution of social class throughout a community by **isolines**. These are lines drawn in such a way that they connect geographic areas with the same values of a variable, creating closed figures that do not cross or meet. Lines are drawn for each of several values of a variable and the values are usually picked at equal intervals along a scale of that variable. Isolines are like lines on a contour map showing the elevation of hills and valleys (or pressure and temperature lines on a weather map) except, in this case, they reflect not elevation but social class (or some other selected social characteristic).

Figure 4.18 was created by laying a grid over a base map of a midwestern community of some 150,000 population. Values of social class were determined for residences in the grid areas and isolines were then drawn. The value of the social class measure through which a given line is drawn is indicated on the line itself and, in this case, larger numbers represent higher social class. Notice that the "peak" of social class appears to be in the central and southern edge of the town. Where isolines are very close together this indicates a rather steep gradient (like a cliff in physical characteristics). Again, as with other statistical maps, this map helps investigators consider a pattern of geographic distribution for some characteristic, and their comparison of this pattern with maps of other characteristics is often useful in producing insight into reasons underlying the pattern. When isolines (a more general term) refer to similar values of *ratios* or *indices* they

---

BOX 4.3 A Pause for Reflection

The expressed purpose of this chapter has been a discussion of different ways to present and analyze data graphically. The hidden purpose, which is also important, has been to provide exercise in the idea of scores and distributions of scores on a variable. You have probably seen examples of most of the graphic techniques presented thus far in the newspaper or in other texts. Most of these techniques are rather direct expressions of ideas contained in some kind of distribution. If you feel somewhat confused at this point, we suggest that you "loop back" for a quick review of distributions in Section 3.2.3. All the techniques described there are useful because they permit some of the kinds of comparisons and contrasts discussed in Section 3.1. The following analytic graphs are also part of this same theme.

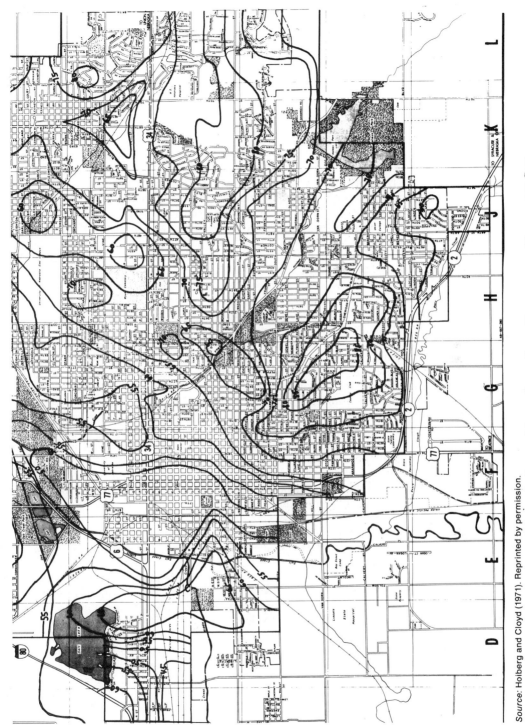

FIGURE 4.18   AN ISOLINE MAP OF SOCIAL CLASS VARIATION IN A MIDWESTERN CITY

are generally called **isopleths**. More precisely, the isoline map shown in Figure 4.18 would be a map where a social class index provided the information for drawing isopleths.

4.4.2 The Scatter Diagram

A particularly useful form of analytic graph is the **scatter diagram**, where individual cases (cities, states, people, organizations, etc.) are plotted as points on Cartesian coordinates. In this graph there are as many dots as there are cases, and the dots are *not* joined by lines. Each dot is placed on the plot in a way that reflects the standing of an observed unit on two variables, one scaled on the Y-axis and the other scaled on the X-axis. We will make extensive use of scatter plots later on, but for now we will illustrate their use in detecting deviant cases and in seeing how cases bunch together in various ways.

Figure 4.19 shows a scatter plot of the 50 states plus the District of Columbia. Each dot represents one of these areas. The dot is placed to show on the X-axis the percentage change in its population overall between 1960 and 1968, and on the Y-axis the percentage change in the state's elderly population aged 65 and older between 1960 and 1968. As shown on the scatter diagram, most states show a total population increase of 10% to 20% and most likewise show a 10% to 20% increase in the elderly population. The dots tend to bunch around the overall U.S. average of an 11% population increase and a 16% increase in the elderly. Yet several states show an unusually large increase in the elderly, or in both elderly and total population. Some of the deviant states are labeled on Figure 4.19. Nevada, Arizona, Florida, Hawaii, and New Mexico show a larger percentage increase in the elderly population than in the total population. Wyoming, North and South Dakota, and West Virginia show a total population decline coupled with a 10% to 15% increase in the elderly population. Only the District of Columbia showed a decline in the elderly and an increase in total population between these two years.

An examination of these deviant cases, which are made evident in an examination of a scatter diagram, may lead to other important insights into migration trends. Likely population processes are relatively obvious for the first two groupings of states noted above. The first group consists of states that are often defined as "retirement areas" with large villages developed mainly for the elderly. The second group of states are those where other population groups are moving out, leaving the elderly behind. The District of Columbia presents an interesting reversal of this pattern of increase in the 65-plus population. Isolation of such deviant cases often serves as the focus for productive future research to understand better and explain the sociological phenomenon of migration, in this instance. Other examples of the use of the scatter diagram will be presented in Chapter 7.

Scatter diagrams are useful to show how individual cases are spread out in a

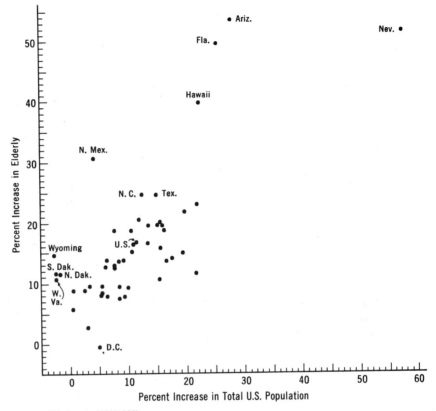

Source: U.S. Senate (1969:326).

FIGURE 4.19   SCATTER PLOT OF PERCENTAGE INCREASE IN TOTAL POPULATION AND IN POPULATION AGED 65 AND OVER BETWEEN 1960 AND 1968 FOR STATES AND THE DISTRICT OF COLUMBIA.

"space" formed by the two dimensions or variables scaled along the *X*- and *Y*-axes. Diagram 4.2 (p. 113) provides a further explanation of the mechanics of creating scatter diagrams.

4.4.3   The Semilogarithmic Chart

The last graphic technique we will discuss here is the **semilogarithmic chart**. This is a graph plotted on rectangular coordinates that are similar to the Cartesian coordinates we have used before. The *X*-axis is a numeric scale often showing time or age. The *Y*-axis, however, is marked off in terms of the logarithms of numbers rather than their arithmetic value. The semilog chart is especially useful in an analysis of *trend lines*, because equal numeric differences between logarithms

TABLE 4.1 AN ILLUSTRATION OF RATE OF CHANGE AS REFLECTED IN DIFFERENCES BETWEEN RAW SCORES AND DIFFERENCES BETWEEN LOGS OF SCORES

*A. Data Illustrating a **Constant Rate** of Increase*

Year	Raw Scores	Amt. of Change	Percent Change	Log of Scores	Amt. of Change in Logs
1930	12.00			1.079	
1935	15.00	+3.00	+25%	1.176	+.097
1940	18.75	+3.75	+25%	1.273	+.097
1945	23.44	+4.69	+25%	1.370	+.097

*B. Data Illustrating an **Increasing Rate** of Decrease*

Year	Raw Scores	Amt. of Change	Percent Change	Log of Scores	Amt. of Change in Logs
1950	8.00			0.903	
1955	6.00	−2.00	−25%	0.778	−.125
1960	4.00	−2.00	−33%	0.602	−.176
1965	2.00	−2.00	−50%	0.301	−.301

*C. Data Illustrating a **Decreasing Rate** of Increase*

Year	Raw Scores	Amt. of Change	Percent Change	Log of Scores	Amt. of Change in Logs
1970	2.00			0.301	
1975	6.00	+4.00	+200%	0.778	+.477
1980	10.00	+4.00	+67%	1.000	+.222
1985	14.00	+4.00	+40%	1.146	+.146

indicate constant *rates of change*.* Table 4.1 and Figure 4.20 illustrate this useful feature of logarithms. Notice that a constant rate of change as shown in Figure 4.20 can be quickly detected on a semilogarithmic chart by a straight line rather than a curved line.

Two studies will serve to illustrate uses of semilogarithmic charts, and Diagram 4.2 (p. 113) provides details about their construction. Figure 4.21 is an example from Schmid contrasting arithmetic and semilogarithmic charts for the number of public and private schools by year from 1929 through 1977 (Schmid, 1986). Given the large differences in number of schools between public and private, in the arithmetic chart (left side of Figure 4.21) most of the trend lines are virtually horizontal and packed together at the bottom of the graph, giving little useful information about trends. Figure 4.21 (right side) shows the same data plotted as a semilogarithmic chart. Here, useful comparative change information

---

* There are charts with logarithmic scales on both axes, and these are called log–log charts. They are used for more complex plots of rates against each other and will not be included in this discussion.

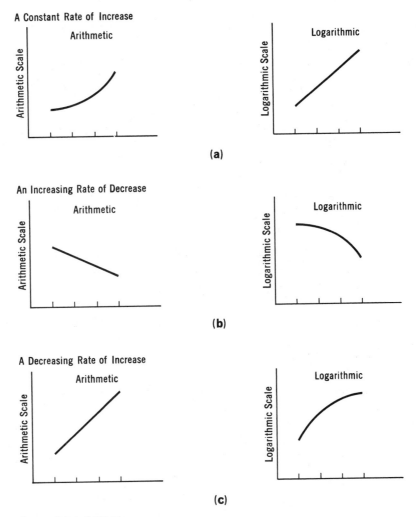

Source: Data in Table 4.1.

FIGURE 4.20   Graphic Illustration of the Difference between an Arithmetic and Logarithmic Chart to Portray Rates of Change

is clearly shown. The number of public elementary schools shows a relatively steady decline from the 1940s through the 1960s, a trend that seems to be starting to level off. In the decade around 1960, private schools and public institutions of higher learning showed sharper rates of increase than other schools. In Figure 4.21 (right side), the fact that this is a semilogarithmic chart can be readily detected by the characteristically unequal spacing on the Y-axis (in fact, here, four-cycle paper had to be used to accommodate the large differences in frequency between the types of schools), and the constant size of unit on the X-axis.

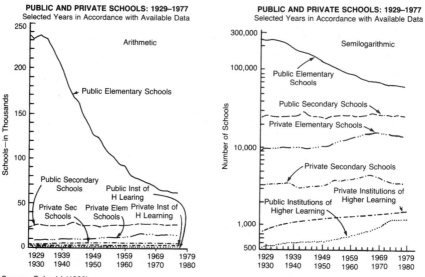

Source: Schmid (1986).

FIGURE 4.21 COMPARISON OF ARITHMETIC (left) AND SEMILOGARITHMIC (right) CHARTS FOR THE NUMBER OF PUBLIC AND PRIVATE SCHOOLS, UNITED STATES, 1929–1977

BOX 4.4 LOGARITHMS AND THEIR USE IN STATISTICS

Common logarithms are a system of exponent numbers for the base number 10, that is, exponential powers to which 10 may be raised to produce any given original number. (Another system that is sometimes used is the Naperian or "natural log" system, which uses a value of approximately 2.718 as its base rather than 10 as used here.) For example, the common logarithm (base 10) of 100 is 2.0 because $10^{2.0}$ is 100; the $\log_{10}$ of 1000 is 3.0, and so on, with the majority of log numbers being a fractional power of 10. Logarithms are convenient for calculation because the process of *adding* these logarithmic exponents amounts to a process of *multiplying* the original numbers produced by the logarithms (and subtracting amounts to dividing). Thus the relatively simple process of adding or subtracting can be used to show more complex multiplying and dividing variations. In statistical graphing, logarithms can thus be used to exaggerate certain effects.

The usual fractional common logarithm has two parts: a whole number to the left of the decimal point called the *characteristic*, and a fractional number to the right of the decimal point called the *mantissa*. The value of the characteristic may be positive or negative and depends on the location of the decimal point in the original number; the value of the mantissa is normally read from a table of common logarithms (usually in reference sections of mathematics texts or in some statistics workbooks). The table below has been developed to show both the characteristic and the mantissa.

BOX 4.4   (*Continued*)

Original Number	Logarithm	Original Number	Logarithm
0.01	− 2.0000	50	1.6990
0.1	− 1.0000	100	2.0000
1	0.0000	200	2.3010
2	0.3010	500	2.6990
5	0.6990	1000	3.0000
10	1.0000	2000	3.3010
20	1.3010	5000	3.6990

Notice that the characteristic for original numbers less than one is negative and is equal to the number of zeros between the decimal point and the first digit, plus 1. For original numbers greater than 1, the characteristic is a positive number that is 1 less than the number of integers in the original number (*i.e.*, 0 for ones, 1 for tens, 3 for thousands, etc.). Thus the complete common logarithm is formulated with the appropriate characteristic for the original number plus its mantissa as shown in a standard reference table. *Antilogarithms* are found by reversing the procedure to find the original number from a logarithm.

Since commercial logarithmic graph paper is available, conversion of statistical scores or frequencies to logarithms by calculation is not normally necessary, and the scale of the original scores can be placed within a labeled sequence in the unequally (*i.e.*, logarithmically) spaced *Y*-axis scale points. Original scores are then plotted directly on the scale.

Logarithms are quite useful in statistics in a number of ways. They simplify the computation of the specialized mean called the geometric mean (see Chapter 5); they greatly aid in the study of trends, as indicated in this chapter; and they are sometimes used to transform scores to simplify a study of relationships between variables (see Chapters 7 and 8). Computer commands can be used to convert scores to logs and back.

accommodate the large differences in frequency between the types of schools), and the constant size of unit on the *X*-axis.

The second example of the use of a semilogarithmic chart is from a study of population growth by Wray (1971) that matched infant death certificates with the corresponding birth certificates for the 1.3 million births in England and Wales in 1949–1950 (an example of the use of recorded data). The graph in Figure 4.22 shows infant mortality ratios on the logarithmic scale on the *Y*-axis. A mortality ratio is the ratio between the mortality rate for infants in a given population subgroup and the overall mortality rate for all infants. The value 100 on the *Y*-axis indicates that the subgroup mortality rate and the overall mortality rate are the same; higher values indicate that the subgroup mortality rate is higher. In this graph, the subgroups that are examined reflect different social classes (I and II are upper classes, III is middle class, and IV and V are lower classes) of the infant's mother. Within social class groupings mortality ratios are computed for

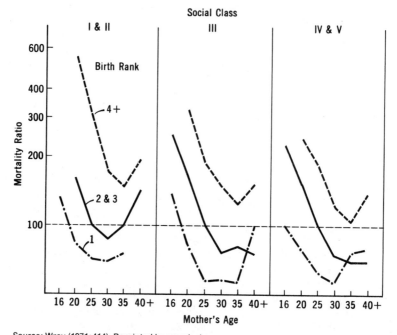

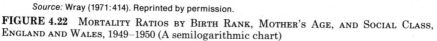

Source: Wray (1971:414). Reprinted by permission.

FIGURE 4.22 MORTALITY RATIOS BY BIRTH RANK, MOTHER'S AGE, AND SOCIAL CLASS, ENGLAND AND WALES, 1949–1950 (A semilogarithmic chart)

infants of different birth order. That is, infants who are firstborn, second- or third-born, and fourth- or later born are handled separately in the graphic presentation. Thus there is a separate line on the graph for each birth order within each of three social classes. Each line shows the mortality ratio of infants in a given birth order and social class category by the age of its mother at the time of its birth. From this semilogarithmic chart a large number of conclusions may be drawn. In virtually all cases the mortality ratio drops with increases in the mother's age until some

---

BOX 4.5 TRENDS AND TREND LINES

The study of *trends*, or change in some characteristic through a period of time, is one of the important areas of sociology. It is an area that poses many problems for research design, for measurement, and for statistical analysis. (For example, see Harris, 1967, for a discussion of several methods for handling change with techniques that are referred to in the chapters of this present volume.)

The study of change via *trend lines* was first discussed in Section 4.2.4. Up to this point we have discussed four methods for describing change: (a) rates (Section 3.2.5a), (b) percentage change (Section 3.2.5b), (c) time-line graphs (Section 4.2.4), and semilogarithmic graphs (Section 4.4.3). In Chapter 5, some methods for "smoothing" trend lines will be discussed.

time around the 30s where the rate of decline slows up and gradually changes direction. Toward the 40s the mortality ratio shows an increasing rate of increase, in most instances, with the increasing age of the mother. The mortality ratio (and, by the way, the infant mortality rate as well) is higher for later-born infants than for earlier-born infants, and this finding holds up within each of the social class categories. Furthermore, looking across the three social class categories for the latest born category of infants, the mortality ratio is higher among infants of upper-class mothers of a given age than it is among infants of lower-class mothers of that age. The investigators conclude that high mortality rates among infants of young mothers with large families occur in all social classes, and the concentration of young mothers with large families in lower social classes cannot totally account for higher mortality rates among infants in lower classes.

4.5   SUMMARY

In this chapter, several basic analytic graphing techniques have been discussed and their principles analyzed. Graphic techniques, properly used, constitute a very useful and powerful means for communicating more involved comparisons in data and for analyzing complex relationships. They are particularly useful in showing the overall form of a distribution (a feature to be discussed at greater length in Chapter 5), in showing the relationships between variables, and in showing trends through time.

It seems to be characteristic of more useful procedures that they also have a potential for distortion as well as a potential for more accurate and helpful displays of interesting features of data. A number of possibly distorting practices have been pointed out in this chapter.

The next chapter returns us again to the basic univariate distribution discussed in Chapter 3, which is graphed by techniques such as the histogram and polygon as shown in this chapter. Chapter 5, however, will approach the description of whole distributions of scores in terms of a few summary index numbers that will communicate key features of a univariate distribution. Later chapters will address similar problems for bivariate and multivariate distributions.

CONCEPTS TO KNOW AND UNDERSTAND

reference system	logarithm
ordinate	cohort
abscissa	population pyramid
origin	pie chart
quadrants	pictograph
histogram	scatter diagram
isoline	statistical map
isopleth	semilogarithmic chart
polygon	trends
ogive	cautions about graphic distortions
line graph	box and whisker diagrams
bar chart	

QUESTIONS AND PROBLEMS

1. Find several examples of graphic presentation of data in popular magazines and newspapers. For each graph describe its strengths and weaknesses in presenting relevant features of the data for the reader. Are there flaws in the graphic presentation? If so, what corrective measures would you suggest, and what possible distortions might your changes promote?

2. If you are assigned a project or piece of research in this course, now is a good time to decide upon the kinds of phenomena you are going to study. This often means selecting an interesting dependent variable. After you have selected your dependent variable and found relevant data, begin your analysis by creating a distribution of the dependent variable. Select an appropriate graphic technique and graph the distribution.

3. Using data in a journal research article, select two scores on each of several cases. Create a scatter diagram with scales for the two variables along the X- and Y-axes. Explain in your own words what it is that the graph shows.

4. Using the data given in problem 5 of Chapter 3 (page 65) and graph paper, construct (a) a histogram, (b) a frequency polygon, and (c) an ogive distribution. What do each of these distributions tell you about the data?

5. The accompanying data give the labor force participation rates of married women in the United States for the years 1948 through 1988. Construct a line

Year	Labor Force Participation Rate	Year	Labor Force Participation Rate
1948	22.0	1969	39.6
1949	22.5	1970	40.8
1950	23.8	1971	40.8
1951	25.2	1972	41.5
1952	25.3	1973	42.2
1953	26.3	1974	43.0
1954	26.6	1975	44.4
1955	27.7	1976	45.1
1956	29.0	1977	46.6
1957	29.6	1978	47.5
1958	30.2	1979	49.3
1959	30.9	1980	50.1
1960	30.5	1981	51.0
1961	32.7	1982	51.2
1962	32.7	1983	51.8
1963	33.7	1984	52.8
1964	34.4	1985	54.2
1965	34.7	1986	54.6
1966	35.4	1987	55.8
1967	36.8	1988	56.5
1968	38.3		

Source: U.S. Bureau of Labor Statistics (1975); U.S. Bureau of Census (1990), p. 384.

graph from these data. What does the line graph tell you about the labor force participation of women over the time period covered by the data? Construct a semilogarithmic graph of these data.

6. Construct bar graphs from the data provided in the accompanying table. Do the graphs reveal any differences in job satisfaction of blacks and whites?

PROPORTION OF MEN HIGHLY SATISFIED WITH JOB, BY TYPE OF OCCUPATION AND SELF-RATING OF HEALTH: EMPLOYED MEN 45–59 YEARS OF AGE, BY COLOR

Type of Occupation and Self-rating of Health	Whites		Blacks	
	Total Number (Thousands)	Percent Who Like Their Job Very Much	Total Number (Thousands)	Percent Who Like Their Job Very Much
White collar	5,065	69	190	68
Excellent	2,094	77	84	66
Good	2,162	64	55	80
Fair or poor	634	57	41	54
Blue collar	5,824	50	738	49
Excellent	1,963	56	254	56
Good	2,557	48	295	46
Fair or poor	1,094	39	164	42
Service	643	51	176	55
Excellent	228	51	68	57
Good	281	60	69	53
Fair or poor	88	34	35	59
Farm	1,097	54	133	36
Excellent	239	68	20	42
Good	468	51	50	37
Fair or poor	352	48	60	29
Total	12,655	58	1,240	51
Excellent	4,543	66	425	58
Good	5,476	55	472	50
Fair or poor	2,168	46	301	43

Source: U.S. Department of Labor (1970).

GENERAL REFERENCES

Cleveland, William S., The Elements of Graphing Data (New York, Macmillan), 1985.
Schmid, Calvin F., Statistical Graphics: Design Principles and Practices (New York, McGraw-Hill), 1983.
Schmid, C. F., and S. E. Schmid, Handbook of Graphic Presentation, 2nd ed (New York, Wiley), 1979.

LITERATURE CITED

Asimov, Isaac, *Asimov's Biographic Encyclopedia of Science and Technology* rev. ed. (Garden City, N. Y., Doubleday), 1972.

Ferriss, Abbott L., "An Indicator of Marriage Dissolution by Marriage Cohort," *Social Forces*, 48 (March 1970), p. 363.

Harris, Chester W. (ed.), *Problems in Measuring Change* (Madison, University of Wisconsin Press), 1967.

Hoiberg, Eric O., and Jerry S. Cloyd, "Definition and Measurement of Continuous Variation in Ecological Analysis," *American Sociological Review*, 36 (February 1971) pp. 65–74.

Jacobson, Paul H., *American Marriage and Divorce* (New York, Rinehart), 1959, p. 118.

————, "The Changing Role of Mortality in American Family Life," *Lex et Scientia*, 3, no. 2 (1966).

Kaufman, Herbert, and David Seidman, "The Morphology of Organizations," *Administrative Science Quarterly*, 15 (December 1970).

Maris, Ronald W., *Social Forces in Urban Suicide* (Homewood, Ill., Dorsey Press), 1969.

Petersen, William, *Population*, 2nd ed. (New York, Macmillan), 1969.

Population Reference Bureau, "Man's Population Predicament," *Population Bulletin*, 27, no. 2 (April 1971).

Riley, Matilda White, "Aging and Cohort Succession: Interpositions and Misinterpretations," *Public Opinion Quarterly*, 37 (Spring 1973), pp. 35–49.

Riley, Matilda White, and Anne Foner, *Aging and Society: An Inventory of Social Research Findings* (New York, Russell Sage Foundation), 1968.

Schmid, Calvin F., "Whatever Has Happened to the Semilogarithmic Chart?", *The American Statistician*, 40, No. 3 (August 1986), pp. 238–244.

Tukey John, *Exploratory Data Analysis* (Reading, Mass., Addison-Wesley), 1977.

U.S. Bureau of the Census, *Current Population Reports*, Series P–25, No. 352 (Washington, D.C., U.S. Government Printing Office), 1966, p. 4.

————, "Projections of the Population of the United States by Age and Sex (interim projections): 1970 to 2020," *Current Population Reports*, Series P–25, No. 448 (Washington, D.C., U.S. Government Printing Office), 1970.(a)

————, "Consumer Income," *Current Population Reports*, Series P–60, No. 72 (Washington, D. C., U.S. Government Printing Office), 1970, Table 6.(b).

————, "Consumer Income," *Current Population Reports*, Series P–60, No. 78 (Washington, D.C., U.S. Government Printing Office), 1971, Table 1.(a).

————, "Metropolitan Area Statistics," *Statistical Abstract of the United States, 1971* (Washington, D.C., U.S. Government Printing Office), 1971. (b)

————, "School Enrollment in the United States, 1971," *Current Population Reports*, Series P–20, No. 234 (Washington, D.C., U.S. Government Printing Office), 1971. (c)

————, "Birth Expectation Data: June 1971," *Current Population Reports*, Series P–20, No. 232 (Washington, D.C., U.S. Government Printing Office), 1972.

————, "General Social and Economic Characteristics, U.S. Summary, Part I, 1980," *1980 Census of Population* (Washington, D.C., U.S. Government Printing Office), December 1983.

————, "Demographic and Socioeconomic Aspects of Aging in the United States," *Current Population Reports*, Series P–23, No. 138 (Washington, D.C., U.S. Government Printing Office), 1984.

————, "After-Tax Money Income Estimates of Households: 1983," *Current Population Reports*, Series P–23, No. 143 (Washington, D.C., U.S. Government Printing Office), 1985, p 13. (a)

————, "State Population Estimates, by Age and Components of Change: 1980 to 1984," *Current Population Reports*, Series P–25, No. 970 (Washington, D.C., U.S. Government Printing Office), 1985. (b)

————, "Part 1: Graphic Summary," *1982 Census of Agriculture* (Washington, D.C., U.S. Government Printing Office), October 1985. (c)

————, *Statistical Abstract of the United States, 1986* (Washington, D.C., U.S. Government Printing Office), 1985, p. 40.

————, *Statistical Abstract of the United States, 1990* (Washington, D.C., U.S. Government Printing Office), 1990, p. 384.

U.S. Bureau of Labor Statistics, *Handbook of Labor Statistics 1975—Reference Edition*, Bulletin 1865 (Washington, D.C., U.S. Government Printing Office), 1975, Table 14, p. 57.

U.S. Department of Commerce, *Social Indicators: 1976* (Washington, D.C., U.S. Government Printing Office), December 1977.

U.S. Department of Health, Education and Welfare: Public Health Service, "Children and Youth: Selected Health Characteristics," *Vital and Health Statistics*, Series 10, No. 62 (Rockville, Md.), 1971, p. 31.

U.S. Department of Justice, *Uniform Crime Reports for the United States* (Washington, D.C., U.S. Government Printing Office), 1971, p. 32.

U.S. Department of Labor, *The Pre-Retirement Years*, Vol. 1, Manpower Research Monograph No. 15 (Washington, D.C., U.S. Government Printing Office), 1970, Table 7.9, p. 216

U.S. Senate, Committee on Aging, "Developments in Aging" (Washington, D.C., U.S. Government Printing Office), 1969, p. 326.

Weitzman, Murray S., "Measures of Overlap of Income Distributions of White and Negro Families in the United States," *Technical Paper* No. 22, U.S. Bureau of the Census (Washington, D.C., U.S. Government Printing Office), 1970, p. 9.

Wray, Joe D., "Population Pressure on Families: Family Size and Child Spacing," Chapter 11, *Rapid Population Growth: Consequences and Policy Implications*, vol. 2 (Baltimore, Md., Johns Hopkins Press), 1971, p. 414. (Prepared under direction of Dr. Roger Revelle.)

DIAGRAM 4.1   How to Construct Basic Graphs

THE PROBLEM

Select an appropriate graphic technique and plot the following data on age of 223 grade-school children. Notice that we will illustrate all of the basic graphic techniques on the same set of data for comparative purposes. In practice, an investigator would choose one appropriate technique and *not* draw all possible graphs.

Age of Grade School Child	Frequency Distribution	Percentage Distribution	Cumulative Frequency Distribution	Cumulative Percentage Distribution
4 years	21	9.4	21	9.4
5	53	23.8	74	33.2
6	69	30.9	143	64.1
7	47	21.1	190	85.2
8	33	14.8	223	100.0
	$N = 223$	100.0%		

Step 1: Using graph paper, lay out the *X*- and *Y*-axes so that the *Y*-axis is three-fourths to equally as long as the *X*-axis. Use standard arithmetic graph paper (10 lines per inch is often a useful type of ruling to use).

Step 2: Mark the score, or *X*-axis. Choose a standard size physical unit to represent a standard size score step. Start near the origin on the *X*-axis with a score that is one score interval lower than the lowest score category to be plotted. This permits you to leave a small space between the *Y*-axis and the left-hand side of the graph you plot. Notice in the examples that 1 year of age is represented by a fixed physical distance in a graph, and that age scores start with age 2 or 3, even though the lowest age category for which there are data is age 4. Label the *X*-axis and create a descriptive heading for the graph (and figure number for reference if there are several graphs).

The remaining steps depend upon the type of graph used.

HISTOGRAM

Step 3: After steps 1 and 2 above, mark the frequency scale on the *Y*-axis. Begin with zero at the origin and mark frequency intervals so that a fixed physical distance stands for a given jump in frequency. Make the height of the *Y*-axis correspond to somewhat more than the maximum frequency in the category with the greatest frequency of cases. Label the *Y*-axis clearly.

Step 4: Using a ruler, construct rectangles each to a height that corresponds to the frequency in a given category as indicated on the frequency axis (see note below). Sides of each rectangle should come down exactly at class

DIAGRAM 4.1 *(Continued)*

boundaries. In this case, there are 21 cases in the age 4 category, so the height of that column is plotted at 21 and the sides of the column come down at 3.5 and 4.5, which are the class boundaries. This is done for each category, although the vertical lines may be omitted or erased, leaving only the outside lines of the enclosed figure.

Note: If a category is twice as wide as most of the others, its frequency and the height of the column should be reduced proportionately (to half) so that the area reflects the number of cases in that interval (*i.e.*, so that the area of each column is proportional to the cases in that interval). For discrete, ordinal, or nominal data, one may want to separate bars somewhat to reinforce this impression of the data.

The *percentage* histogram is constructed in the same way, except that the Y-axis represents percentage points and is as tall as the largest cell percentage.

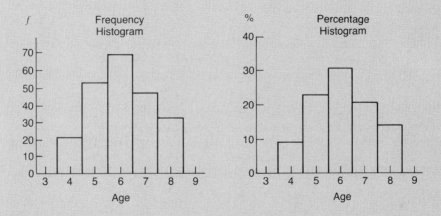

POLYGON

Step 3:   After steps 1 and 2 above, mark the Y-axis in frequency or percentage units in exactly the same way as discussed in step 3 under histograms.

Step 4:   Plot the cell frequency (or percentage) as a dot at a height above the X-axis that corresponds to the frequency (or percentage) in that category as indicated on the Y-axis scale (see note below). The point should be plotted directly above the *midpoint* of the class.

Step 5:   Now plot two added points. Plot a zero at the next lower class midpoint below the lowest plot point on the X-axis. Then plot a zero at the midpoint of what would be the class just higher than the highest plotted point.

Step 6:   Using a ruler, connect all of the dots so that there is a closed polygon— closed with the X-axis.

   Note: If one category is wider than others, reduce the height of that category proportionately. For example, if the category is twice as

DIAGRAM 4.1 *(Continued)*

wide as the others, plot the frequency (or percentage) above the mid-point of the class, but at a height that is equal to one-half of the frequency (or percentage) in that class.

A percentage polygon is constructed in the same way as the frequency polygon, except that the Y-axis is marked in percentages.

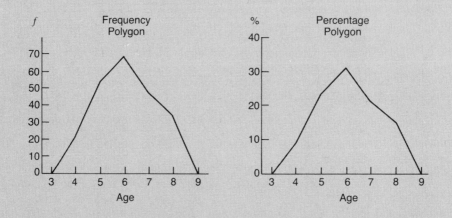

OGIVE

Step 3: After steps 1 and 2 above, mark the frequency or percentage scale on the Y-axis in a way similar to that indicated in step 3 under histogram. The scale should extend up to N, the total number of cases (or 100%, in the case of a cumulative percentage scale).

Step 4: Plot a zero at the *lower* real limit of the first (lowest) score category. Then plot each successive cumulative frequency (or percentage) directly above the upper real limit of the class to which it corresponds (see note below). A cumulative distribution shows the number (percentage) of cases below the upper real limit of each successive class up to N or 100%.

Step 5: Connect all dots with a straight line, ending at the last dot, which is N or 100%. The graph should go up and to the right without any dips. A frequency in the next higher category cannot be any lower than it was at the last category (although it may be the same height if there are no further cases in that next category).

Note: Always plot directly above the upper real limit of a category regardless of the widths of classes. The percentage ogive is the same as the frequency ogive, except that the Y-axis is marked in percentages from zero to 100%. An added use of the percentage ogive is that it permits one readily to find any percentile (*i.e.*, the score below which some percentage of the cases happen to fall).

DIAGRAM 4.1 *(Continued)*

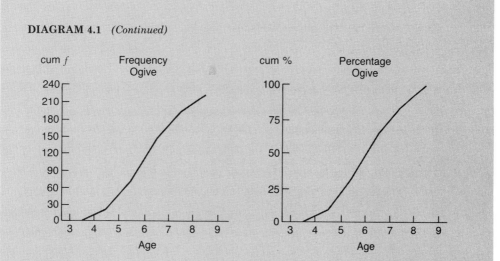

LINE GRAPHS

Step 3: After steps 1 and 2 above, the *Y*-axis is marked in terms of the scale of the dependent variable that is being plotted, and the *X*-axis is marked in terms of the scale of the independent variable. For example, the *X*-axis may represent time or age or regions of the country. The *Y*-axis may be a gender ratio, the number of crimes in a given year, etc. Label each axis. Again, equal physical distances correspond to equal score distances, and the *Y*-axis starts with zero.

Step 4: Plot with a dot the value of the dependent variable above the appropriate category midpoint on the independent variable.

Step 5: Join the dots, using a ruler, without closing the figure with the *X*-axis. Where more than one line is plotted on the same graph, use a different type of line for each (dotted, solid, dot and dash, etc.). Clearly label each line.

An example of a line graph is as follows:

TOTAL BIRTHS EXPECTED PER 1000 WIVES 18–39 YEARS OLD, 1971

Age	*Births Expected per 1000 Wives*
18–24	2,384
25–29	2,617
30–34	2,996
35–39	3,255

Source: U.S. Bureau of the Census (1972).

DIAGRAM 4.1 *(Continued)*

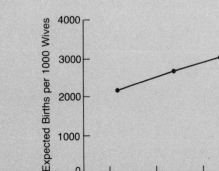

TotaL Births Expected per 1000 Wives 18–39 Years OLD, 1971

CAUTIONS

1. Use the three-fourths rule, creating the *Y*-axis about three-fourths the length of the *X*-axis (or equally long).

2. Always begin the frequency or percentage axis with zero at the origin. The score axis may begin with any convenient score to make an esthetically balanced but accurate diagram.

3. Be sure that equal numeric differences are represented by equal physical distances on all scales.

4. Label the graph adequately to include all scales, the source of data, a heading explaining what is shown, and keys to explain different kinds of lines or symbols, etc.

5. To avoid confusion, do not attempt to show too many different graphs on the same axes.

DIAGRAM 4.2   How to Construct Scatter Plots and Semilogarithmic Graphs

SCATTER DIAGRAMS

Step 1:   *Data* on at least two variables for each case are needed to construct a scatter diagram. For a two-dimensional scatter plot, two scores per case are needed. In the following example, large cities have been measured using a health facility measure (number of hospital beds per 100,000 population in 1970) and an economic measure (that city's per capita personal income as a percentage of the national average per capita income).*

	Selected Cities	No Health Facilities	Economic Measure (%)
a	Youngstown, Ohio	366.8	100
b	Trenton, N.J.	525.1	115
c	Springfield, Mass.	437.5	97
d	St. Louis, Mo.	496.4	108
e	Newark, N.J.	436.9	129
f	New Orleans, La.	507.2	97
g	Huntington, W. Va.	593.5	87
h	Erie, Pa.	397.5	96
i	Cleveland, Ohio	418.3	119
j	Cincinnati, Ohio	370.7	106
k	Chicago, Ill.	436.9	127
l	Charleston, W. Va.	619.1	93
m	Birmingham, Ala.	493.3	87

Source: U.S. Bureau of the Census (1971a, 1971b).

Step 2:   *Set up coordinates* and mark the scale of each variable on one of the axes. Axes should be about the same physical length, and the scales should be marked so that the highest and lowest scores are nearly as far apart as the axis is long. Equal physical units on the scale, of course, should correspond to equal distances on the variable. Label the axes clearly.

   Graph paper is usually used. It is commercially available with different numbers of lines per inch, but for much of the graphing in social science a 10-line-per-inch graph paper is most useful.

Step 3:   *Plot each case* as a dot, located so that it is just opposite the proper point on both axes correctly showing its score on both variables. One dot is plotted per case. Often, in this descriptive usage, the dots will be identified by some sort of coding system such as the small letters used below.

---

* These cities all are Standard Metropolitan Statistical Areas with 200,000 population or more in 1970. All areas chosen have a death rate of 10.0–10.2 (the national death rate is 9.7). The U.S. average on the health facilities measure is 414.6 and for the economic measure it is 100%. Both variables are interval level.

DIAGRAM 4.2 *(Continued)*

Step 4: *Analysis.* The scatter diagram might then, for example, be examined to see to what extent the dots tend to cluster. By looking at other characteristics of cases in a cluster, often additional insight into possible reasons for the clustering can be found. Some investigators used this technique to identify types of cities, for example.

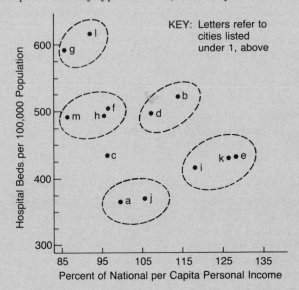

SEMILOGARITHMIC CHARTS

Step 1: *Data* used in semilogarithmic charts are usually measures for a series of different years made on one case, such as a society. There are exceptions, of course, but the usual use of semilogarithmic charts is to show rate of change through time. The following data are median (a kind of average that is discussed in the next chapter) income for families in the United States and these measurements were made for a number of different years. What is the rate of change in median income of families?

Year	Median Income	Year	Median Income
1947	$3,031	1964	6,569
1950	3,319	1965	6,957
1958	5,087	1966	7,500
1959	5,417	1967	7,974
1960	5,620	1968	8,632
1961	5,737	1969	9,433
1962	5,956	1970	9,867
1963	6,249		

Source: U.S. Bureau of the Census (1971a).

114

DIAGRAM 4.2 *(Continued)*

Step 2: *Set up coordinates.* A semilogarithmic chart could be created by marking an arithmetic scale with the logarithms of numbers. Instead, semilog graph paper is available, where the rulings are in terms of original scores rather than their logs, so that scores need not be converted to logs prior to plotting. This is the reason for the unequally spaced appearance of semilog graph paper; it has lines more closely spaced together toward the top of the paper.

Paper is available with several tiers or "cycles" of lines numbered from 1 to 10. One-cycle paper is used to plot scores where the largest is no more than 10 times the smallest. If the largest number is 10 to 100 times the smallest, two-cycle paper is needed, and so on. The X-axis is a standard arithmetic scale (although double-logarithmic paper is available and used for some specialized purposes not discussed in this text).

In the example above, the largest median income figure is less than 10 times the smallest so one-cycle paper is needed. The log scale is labeled in score units, in this case thousands of dollars. Notice that the starting point is some convenient power of ten (.01, .1, 10, 100, 1000, etc.) rather than zero because the rate of increase from zero cannot be determined (recall that percentage change is the difference between a starting and ending figure divided by a starting figure and division by zero is not a defined operation). We chose $1,000 and proceeded upward in $1,000 increments.

The X-axis is labeled in terms of time, in this case years between 1947 and 1970.

Step 3: *Plot the value* of the score above the proper X-axis score. Here, we are plotting each median income figure above the year to which it refers. These dots in the graph on page 116 are connected with straight lines and the figures is *not* closed with the X-axis.

DIAGRAM 4.2 (*Continued*)

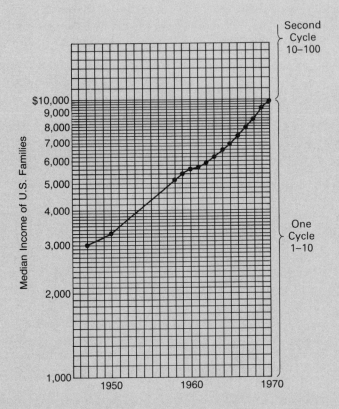

Step 4: *Analysis*. A straight line indicates a constant rate of change. The more steeply the line goes up or down, the more rapid the rate of change, and by inspection of these graphs one can determine whether there is a change at an increasing rate or a decreasing rate of change.

In the example, the rate of increase has been relatively constant, overall, but there have been periods of decreasing rates of increase prior to 1950, just after 1960, and just before 1970. In the mid-1960s there was a slightly increasing rate of increase, as shown by the arching upward of the line into a steeper ascent.

CHAPTER 5

Summary Measures of the Location, Variation, and Form of Univariate Distributions

Because of the differential opportunity structure available to white women as compared to black women, it might be hypothesized that these two categories of women will differ in the ages at which they first marry. Fortunately, the General Social Survey of 1989* questioned a national probability sample of respondents about this very issue. A total of 617 white women reported the age at which they first married, as did 82 black women. Figure 5.1 provides a comparison of percentage polygons of the distributions of ages first wed for both the white and the black women respondents.

 Examination of these two percentage polygons reveals that they differ in a number of respects. First, they differ in their *location* or concentration along the scale of ages. The peak of the distribution for white women is to the right (in a higher age category) of that for blacks. Second, the distribution for white women is more spread out than it is for the black women. Finally, the *forms* of the distributions also differ. The distribution for black women is less lopsided than the one for white women. These three features of distributions are referred to as **central tendency** (or, sometimes, *location*), **variation**, and **form**, respectively.

* See Box 3.5.

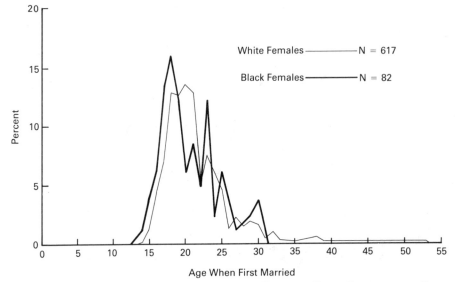

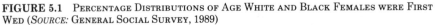

FIGURE 5.1 Percentage Distributions of Age White and Black Females were First Wed (Source: General Social Survey, 1989)

5.1 UNIVARIATE DISTRIBUTIONS—AN EXAMPLE AND OVERVIEW

While the two percentage polygons in Figure 5.1 graphically portray the data for all of the 699 female respondents to the survey and, in fact, are based on all 699 raw scores, they actually provide more detail than we need to draw specific conclusions about differences in the ages at which these white and black women first married.

> **BOX 5.1** The Concept of the Distribution
>
> The idea of a **distribution** of scores—in frequency or percentage form—is basic to the discussion in this chapter. The objective of this chapter is to describe whole distributions in compact ways. The idea of distribution is introduced in Section 3.2.3 and graphic presentations of distributions are shown in Figures 4.3 and 4.4.

The typical interest of investigators is in distributions rather than individual scores. Thus it is very helpful to have ways to describe whole distributions and drop real but nonetheless distracting information about specific individual scores. As we noted in our discussion of distributions of raw scores, one of the helpful procedures in the analysis of data may be selective elimination of distracting detail in order to permit a clearer focus on differences that are important to the research question at hand.

5.1a What Is to Come

Thus it turns out that there are three main ways that distributions differ from each other, and the point of this chapter is to develop accurate ways to measure the features of a univariate distribution for various kinds of data. There are several different approaches to each feature, and the approach one selects in any specific instance will depend on such things as the level of measurement of the variable, the research question in which one is interested, and, sometimes, on the way the variable happens to be distributed. Separate sections of this chapter will deal with approaches to measuring *central tendency* (Section 5.3), *variation* (Section 5.4), and *form* (Sections 5.2 and 5.5) of distributions.

5.2 FORM OF A DISTRIBUTION—I

The feature of a distribution most readily apparent from a histogram or polygon is its form—the overall shape of the distribution. For many purposes a simple verbal description is sufficient, but single summary "indices" can be developed to reflect certain aspects of the form of a distribution. These numeric indices will be presented in Section 5.5 at the end of this chapter after some of the basic notions of location and variation have been developed. For now, we will call attention to aspects of the form of distributions and introduce some of the terminology by which these features can be expressed.

5.2.1 Number of Modes

A first characteristic of the form of a distribution readily picked up from a histogram or polygon is the number of high points or peaks (**modes**) the distribution has. In Figure 5.1, for example, the distribution of black females has two high points with a low between them, and it would be called a **bimodal** distribution. The other distribution in Figure 5.1 is **unimodal** since it has only one main high point. Notice that the determination of the number of peaks depends to some extent upon one's judgment of the importance of differences in the frequency in categories. If several peaks were judged important, comparatively and absolutely, one would refer to a many-moded or **multimodal** distribution. It seems to be true that many distributions actually encountered in research are unimodal, overall.

5.2.2 Symmetry

A second aspect of the form of a distribution is the extent to which it is symmetrical rather than lopsided or skewed. The **skew** in a distribution refers to the trailing off of frequencies toward extreme scores in one direction, away from the bulk of cases. Figure 5.2 shows curves skewed to the left (**negatively**) in Figure 5.2f and to the right (**positively**) in Figure 5.2d.

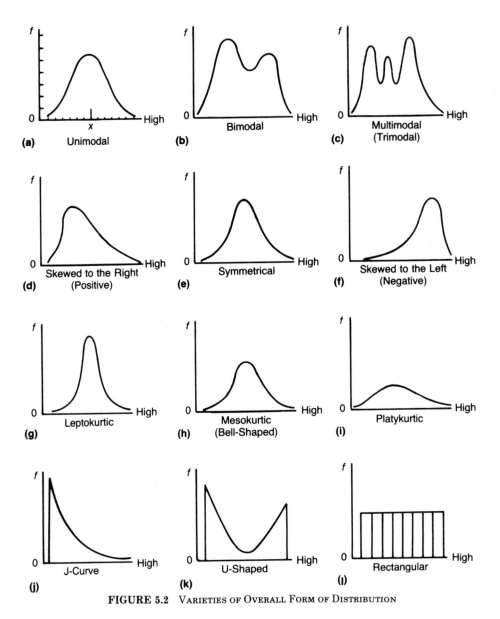

FIGURE 5.2 VARIETIES OF OVERALL FORM OF DISTRIBUTION

A **symmetrical** curve is one in which the two sides of the distribution exactly correspond, if the figure is folded over at its central point. Income is a variable that is usually positively skewed; there are a few extremely high scores (incomes) that are not "balanced" by equally extreme low scores, and most individuals have modest to moderate incomes. Ability or experience test scores very often have a nearly symmetrical shape.*

* The symmetry is generally a function of the way scores are defined.

The distribution for white females in Figure 5.1 is positively skewed.

5.2.3 Kurtosis

Kurtosis refers to the extent to which cases are piled up closely around a point in the distribution or are distributed rather widely among categories. Alternatively, it is interpreted as the relative concentration of scores in the tails of the distribution (relative to the normal distribution). As indicated in Figure 5.2, an unusually concentrated distribution of scores is called a **leptokurtic** distribution, and one that is flatter than "normal" is called a **platykurtic** distribution. A "normally" concentrated distribution of scores is called **mesokurtic**. Here "normal" has a technical and precise meaning that will be discussed later on. For the time being it is sufficient to know the terminology and be able to compare curves in an approximate way. The distribution of black females in Figure 5.1 is relatively leptokurtic.

There are some other terms that refer to the overall shape of a distribution and are handy to know. A *J-shaped* distribution has almost all cases piled up at one end of the scale and then tails off uniformly in one direction from the concentration as indicated in Figure 5.2j. **Rectangular** distributions have equal frequencies in all categories so the distribution has a flat top (Figure 5.2l).* A **bell-shaped** distribution is unimodal and slopes gently off in both directions from the mode to form the general shape of a bell (Figure 5.2h). Finally, a **U-shaped** distribution is bimodal, with modes at both extremes and the low-frequency area in the center of the distribution (Figure 5.2k).

5.3 LOCATION OF A DISTRIBUTION

The location, or *central tendency*, of a distribution refers to the place on the scale of score values where a particular distribution is centered. Consider the following five sets of scores on height of people, measured in feet.

group A	2	3	4	5	6	$N = 5$
group B	2	4	4	4	6	$N = 5$
group C	2	2	2	4	6	$N = 5$
group D	2	4	6	6	6	$N = 5$
group E	5	6	7	8	9	$N = 5$

The five individuals in group C generally have lower heights than those in A or B, and individuals in D and E are generally taller. Notice this is generally true even though some individuals in the group with the tallest heights (E) have heights that are no taller than the tallest individuals in the group with the shortest individuals

---

* It is interesting to note that pseudorandom numbers are a good example of a *rectangular distribution*, where the likelihood of any of the 10 digits from 0 to 9 appearing is equal. Computers sometimes are used to create pseudorandom numbers for random sampling purposes, and one check on such computer programs is to see if they indeed produce a rectangular distribution of numbers. Sampling and the use of random numbers is a topic in the area of inferential statistics (see Chapter 11).

(C). Location is usually measured in terms of some center point or "typical" score in the distribution around which the scores tend to center or cluster in some defined way. An appropriate measure of central tendency would bring out clearly and succinctly the differences in these five distributions. There are three main location measures we will use frequently throughout this book.

5.3.1   Mode (Mo)

The **mode** is the value of the score that occurs most frequently in a collection of scores. It would be represented by the tallest column on a histogram or peak on a frequency polygon. For grouped data, the mode is the midpoint of the *class* that has the largest frequency of cases in it. The mode of group B, above, is the score 4, for example, and the modal category of education for 25 to 29-year olds in Table 5.1 is "4 years of high school." One may choose to designate several modes, as explained in the previous section on the form of a distribution (Section 5.2.1), even though not all "modes" are of exactly equal frequency.

The mode has the virtue of being relatively easily discovered by inspection, and therefore it is often used as a first, quick index to the location of a distribution. In fact, the mode is almost never "computed" by any formula.

The mode can be used with nominal variables (*e.g.*, at birth, the modal gender is "male") since it does not depend upon the ordering of scores. Stated differently, the mode ignores score information about rank order and interval size, even if it is available in the data. Distributions F and G, below, have the same mode, although they are located at seemingly quite different points on a score scale (if the distances between scores are taken to be meaningful):

group F   1   3   3   3   3   4 $N = 6$
group G   3   3   50   75   100   119 $N = 6$

TABLE 5.1   YEARS OF SCHOOL COMPLETED BY SEX AND AGE FOR PERSONS 25 YEARS OLD AND OVER, 1984

| | | | Elementary | | | High School | | College | | Median School |
Sex and Age	All Persons (1,000s)	Total Pct	0–4 yrs	5–7 yrs	8 yrs	1–3 yrs	4 yrs	1–3 yrs	4 years or more	Years Completed
Male	66350	100.0	2.9	5.1	6.5	11.8	34.6	16.1	22.9	12.7
Female	74444	100.0	2.6	4.9	6.6	12.9	41.8	15.6	15.7	12.6
25–29 years	21046	100.0	.8	1.6	1.9	9.8	42.9	21.1	21.9	12.8
30–34 years	19127	100.0	1.2	1.5	1.8	8.3	38.7	21.5	27.0	13.0
35–44 years	30058	100.0	1.4	2.4	2.8	10.2	39.4	18.6	25.3	12.8
45–54 years	22240	100.0	2.2	4.6	5.3	13.9	42.4	14.0	17.6	12.6
55+ years	48324	100.0	5.4	9.6	13.4	15.8	33.8	10.4	11.6	12.2
Total	140794	100.0	2.8	5.0	6.6	12.4	38.4	15.8	19.1	12.6

Source: U.S. Bureau of the Census (1985).

If the variable is ordinal or interval, important locational information will be ignored by using the mode, and this loss of information is usually not desirable. The mode does not require a closed-ended, grouped distribution, provided that the modal category is not the open-ended category (so that a midpoint can be computed). As in G, above, there may be (and generally there are) more scores that are *not* equal to the mode than those that *are* equal to the mode. The mode is merely the particular score that most frequently occurs in the data.

5.3.2   Median (Mdn)

The **median** is the score value below which and above which half the scores in an ordered distribution fall. It is the 50th percentile. In group E, above (in Section 5.3), the median is 7, since this score value exactly divides the odd number of scores ($N = 5$) into high and low halves. Group G in Section 5.3.1 has a median falling between the two center scores, since there is an even number of scores ($N = 6$). In this simple case, the usual procedure is to take the average of the two center scores as the median. Thus, the median would be

$$\text{Mdn} = \frac{50 + 75}{2} = 62.5$$

In a **grouped distribution**, the median is usually computed on the assumption that cases in the category containing the median are evenly distributed throughout that interval (and, in fact, on the assumption that the interval width is a meaningful value). Diagram 5.1 (p. 155) illustrates the steps involved in computing the median from grouped data.

BOX 5.2   PERCENTILES

The 50th percentile is the *score* below which 50% of the scores in a distribution fall. In general, the *n*th percentile is the *score* below which *n* percent of the scores in a distribution fall. You may want to review percentiles by looking back at Section 3.2.3d. Diagram 3.2 illustrates how percentiles may be computed from grouped data. Diagram 5.1 also explains in somewhat more detail the computation of the median itself.

This technique is identical to the one for computing percentiles since, after all, the median is merely the 50th percentile. The explanation of the rationale behind this simple interpolation is given in Diagram 5.1; however, most computations of the median are made by computer directly from raw data. Grouped-data formulas provide a very good approximation if the number of categories is relatively large (15 or more, for example).

From this definition, it is clear that scores must be at least ordinal before the median can be computed, since the concept of a median implies direction (ranking above and below the median). The median is an index of location that does not imply knowledge of distance, however.* This means that it, like the mode, loses some information contained in interval scores. This is sometimes a strength since the median, for example, is little influenced by wild and erratic extreme scores; it is simply the point on the scale of scores dividing all cases into an upper and lower half. Like the mode, the median does not require that extreme classes in a grouped distribution be closed, provided that the median does not fall in one of the open, extreme classes.

To illustrate the median, we have taken data from Reiss concerning the variable, "attitude toward premarital sexual permissiveness"—clearly an ordinal variable (Reiss, 1967:36). Reiss used five attitude questions reflecting permissiveness attitudes, and asked a sample of college students and a national sample of adults to indicate the extent to which they agreed with each of the five questions. An individual's answers were combined, and a "permissiveness score" was assigned using a scaling technique called Guttman scaling. The scores ranged from a conservative or low permissiveness score of 0 to a high permissiveness score of 6. The frequency distribution of students and of adults is given Table 5.2.

TABLE 5.2 FREQUENCY DISTRIBUTION OF ATTITUDES TOWARD PREMARITAL SEXUAL PERMISSIVENESS FOR A COLLEGE STUDENT AND ADULT SAMPLE

College Student Sample

	Permissiveness Scale Types	Frequency	Cumulative Frequency
	0 (low)	127	127
	1	143	270
Median (2.4) →	2	160	430
	3	152	582
	4	135	717
	5 (high)	127	844
		844	

National Sample of Adults

	Permissiveness Scale Types	Frequency	Cumulative Frequency
	0 (low)	517	517
Median (1.6) →	1	154	671
	2	434	1105
	3	70	1175
	4	84	1259
	5 (high)	140	1399
		1399	

Source: Reiss (1967:36). Reprinted by permission of Holt, Rinehart, and Winston, Inc. Copyright ©1967 by Holt, Rinehart, and Winston, Inc.

* The exception is the width of the median interval when the grouped-data formula for the median is used.

From an examination of the cumulative frequency distributions it is clear that slightly over half of the 844 students have permissiveness scores of types 0, 1, or 2. The median, therefore, could be some value between 2 and 3 on the ordinal scale of types of permissiveness, below which half (*e.g.*, 422 students) of the scores fall. *If the grouped formula were used*, the median for these grouped data would turn out to be a value of 2.4.* For the national sample, however, it is clear that subjects are much more concentrated toward the low-permissiveness scale types, and in fact nearly half (48%) have scale-type scores of 0 to 1. Again, we could compute an interpolated value between 1 and 2 as the point on the scale below which (about) half of the cases fell. If this were computed by the grouped formula, the value would be 1.6 for the median (see Diagram 5.1). These two values can be compared; the national sample of adults is "located" at a lower point on the permissiveness scale than are the college students. The median is a useful measure of location for a distribution of cases on an ordinal variable.

Table 5.3 shows another use of the median, in this instance, used on an interval variable—income—which usually has a skewed distribution with extreme or erratic scores. Table 5.3 is the income distribution for the United States as a whole and for household heads who have attained differing amounts of formal education. Notice that the *median* income for all households in 1983 was $20,885 and that the median increases markedly as the amount of formal education increases. In fact, for households whose head had 4 or more years of college, the median income ($34,709) was about four times the median income for families whose head had completed no more than 7 years of elementary school. The median greatly aids in making these kinds of contrasts between the location of different income distributions. In each case, half the individuals have scores below the median income for that group, and half have scores above.

5.3.3   Arithmetic Mean ($\bar{X}$)

The common average or **arithmetic mean** is simply the sum of all scores divided by their number, as in Formula 5.1[†] You have used this average for some time in computing your grade point average, batting averages, etc.

* It should be noted that the Guttman scaling procedure can result only in integers as score values. Thus the score 2.4 is not defined within the operations of that procedure. Nevertheless, the variable "permissiveness attitudes" is defined as a continuous variable, so that fractional values are conceptually meaningful. Thus the score of 2 would be assumed to have a lower limit of 1.5 and an upper limit of 2.5. Use of the grouped data formula also implies that the width of the median interval is a meaningful concept. This is not defined in an ordinal variable, so the median computed by this formula should be considered an approximation. Investigators generally are willing to make this approximation, and they use fractional values for the median of ordinal variables or of interval variables that are defined as discrete. Other than the assumptions about the category containing the median, the median uses only the ranked aspect of scores.

[†] Two comments are relevant. First, the symbol for the arithmetic mean is traditionally $\bar{X}$ or μ (the Greek letter *mu*). The X-bar symbol is used for the statistic "the arithmetic mean computed on sample data" and the symbol μ is used for the parameter "the arithmetic mean computed for the population." Since sample data are generally used in sociology, we will use symbols appropriate for sample statistics. In the section of the book on inferential statistics, where the distinction becomes critical, the two different symbol usages will be made plain and
(Footnote continues.)

TABLE 5.3 TOTAL INCOME OF HOUSEHOLDS BY EDUCATIONAL ATTAINMENT OF HOUSEHOLDER, 1983

Householder Education Level	Income for All Hsehlds (1,000s of $)	Total (%)	Percentages of Households by Income (In Dollars)											Mdn ($)	X̄ ($)
			Under 5,000	5,000 to 9,999	10,000 to 14,999	15,000 to 19,999	20,000 to 24,999	25,000 to 29,999	30,000 to 34,999	35,000 to 39,999	40,000 to 49,999	50,000 to 74,999	75,000 and over		
Elementary															
0–7 years	6,895	100.	23.9	30.0	16.5	11.0	6.5	4.1	2.5	1.8	2.2	1.3	.1	9,221	12,772
8 years	5,964	100.	17.9	24.1	18.9	11.8	8.2	6.1	4.4	2.9	3.1	2.2	.4	11,811	14,483
Total	12,859	100.	21.2	27.3	17.6	11.4	7.3	5.0	3.4	2.3	2.6	1.7	.2	10,370	14,076
High school															
1–3 years	10,975	100.	15.7	21.3	16.9	12.4	9.8	7.5	5.5	3.5	4.2	2.4	.6	13,705	17,456
4 years	30,416	100.	7.1	12.5	14.3	13.7	12.8	10.7	8.3	6.2	7.6	5.4	1.1	20,800	23,688
Total	41,391	100.	9.4	14.9	15.0	13.3	12.0	9.9	7.5	5.5	6.7	4.6	1.0	18,883	22,035
College															
1–3 years	13,766	100.	5.9	9.5	11.5	11.9	12.0	10.9	9.8	7.5	10.2	8.1	2.6	24,606	27,836
4 or more	17,390	100.	2.5	4.2	6.1	9.0	9.4	9.5	10.0	8.4	13.9	17.8	9.3	34,709	39,861
Total	31,156	100.	4.0	6.5	8.4	10.3	10.6	10.1	9.9	8.0	12.2	13.5	6.3	29,982	34,548
All Households	85,406	100.	9.2	13.7	13.0	12.0	10.8	9.2	7.8	5.9	8.1	7.4	2.8	20,885	25,401

Source: U.S. Bureau of the Census (1984b).

(5.1)
$$\bar{X} = \frac{\Sigma X_i}{N}$$

It has some other interesting and useful characteristics beyond its value as the most commonly known and widely used measure of central tendency.

To start with, it is another example of the statistical use of the ratio as an aid to valid comparison. The aggregate or total of scores is "standardized," so to speak, in terms of the number of scores that are included in the sum. This permits one to compare "averages" for groups of different sizes, whereas a direct comparison of totaled-up scores would be misleading. Sometimes, however, the number of scores contributing to a sum is not the only source of distracting differences one may want to take into account in making comparisons of central tendency. In arriving at a sum, each score contributes a different amount depending upon its numerical value. Big scores count more than small ones, and this means that extremely large (or small) scores will tend to have a stronger influence on the mean than more modest, single scores. The arithmetic mean is "pulled" toward unbalanced, extreme scores in a distribution. To state it differently, the mean is pulled toward the tail of skewed distributions; it is pulled higher in positively skewed distributions and lower in negatively skewed distributions. The median, on the other hand, uses the numerical value of the score merely to establish the rank order of a point so that the "upper" and "lower" half of the scores can be identified. Extreme scores would have a minimal effect on the median (each score has the same weight in determining ranking) but a larger effect on the arithmetic mean (where each score contributes different amounts according to its magnitude).

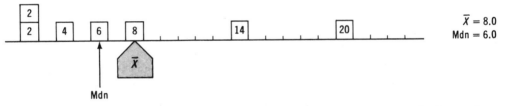

The arithmetic mean is like a balancing point or fulcrum of a lever where uniform blocks are placed on the lever at distances representing their score value. Extreme scores are farther out in both directions on the lever. If the top score in the diagram above were larger, say 30 instead of 20, the balancing point would move to a higher score point on the scale, as shown on page 128.

(*Footnote continued.*)
observed carefully. Second, we should note that the summation operator, sigma, correctly includes limits of summation (see Box 3.2) written below and above the symbol as follows:

$$\bar{X} = \frac{\sum_{i=1}^{N} X_i}{N}$$

However, in statistics, summations are nearly always over all N cases, so we will simplify our notation by omitting the limits where this is true. Limits will be introduced when they are needed for clarity.

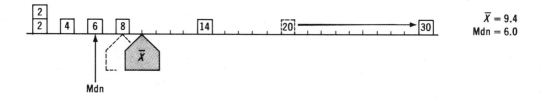

$\bar{X} = 9.4$
Mdn = 6.0

If the extreme score were 200 instead of 30, the fulcrum would have to move to the right even farther as illustrated below, a value for the mean that is quite different from the bulk of scores and perhaps gives too much weight to the single extreme score, 200. The median, on the other hand, would be the same (Mdn = 6), in all three of these situations, since the center of the distribution in the 50%–50% sense has not changed.

$\bar{X} = 33.7$
Mdn = 6.0

200

Mdn

If one were to compute the difference between the arithmetic mean and each of the scores in the group upon which it was based, as in Table 5.4, and sum these differences algebraically, another interesting and useful property of this measure of location would be evident. The algebraic sum of deviations of scores from their arithmetic mean is 0, and this is always the case, since the mean is defined as the value around which scores balance or cancel each other's effect. Notice that the sum of deviations from the median bears out this rule. It is not 0. On the other hand, the median has an interesting property of being the point on the scale for which the sum of absolute deviations, ignoring minus signs, is a minimum value. In Table 5.4, the sum, 34, is smaller than the sum of absolute deviations from the mean, which is 36 in this case.

It is also a property of the arithmetic mean that it is the value for which the sum of the squared deviations of the scores from the mean is a minimum value. The sum of the squared deviations of the scores from any other point will be larger. The significance of this property for measures of variability will be discussed in Chapter 12.

For grouped distributions, the arithmetic mean can be computed only if the distribution is "closed" so that midpoints can be computed for each and every class (so that the total of all scores can be computed). This is because the midpoint is used to represent the value of each of the scores in a class, since information on the exact value of scores is lost when a grouped frequency distribution is created. Diagram 5.1 (p. 157) explains the computational procedures involved for the mean for grouped data.

The arithmetic mean for household income is shown in Table 5.3 in addition to the median income. Notice that in each case, the mean is higher than the median, even though it, too, increases as one moves toward higher educational attainment

TABLE 5.4 DEVIATION OF SCORES FROM THE ARITHMETIC MEAN AND FROM THE MEDIAN OF A DISTRIBUTION

	Raw Scores X_i	Difference $(X_i - \bar{X})$	Absolute Difference $|X_i - \bar{X}|$	Difference $(X_i - Mdn)$	Absolute Difference $|X_i - Mdn|$
	2	$2 - 8 = -6$	6	$2 - 6 = -4$	4
	2	$2 - 8 = -6$	6	$2 - 6 = -4$	4
$N = 7$	4	$4 - 8 = -4$	4	$4 - 6 = -2$	2
	6	$6 - 8 = -2$	2	$6 - 6 = 0$	0
	8	$8 - 8 = 0$	0	$8 - 6 = +2$	2
	14	$14 - 8 = +6$	6	$14 - 6 = +8$	8
	20	$20 - 8 = +12$	12	$20 - 6 = +14$	14
	$\Sigma X_i = 56$	0	36	$+14$	34

$$\bar{X} = \frac{56}{7} = 8.0$$

$$Mdn = 6$$

categories. This difference between the mean and median indicates a positive skew in the distribution of income; some extremely high scores are evident.

5.3.4 Special Kinds of Means

There are other measures of central tendency designed for special situations, although most of these measures do not find frequent use in the social sciences. One might, for example, weight scores unequally, before combining them into a mean, by multiplying each score times some previously arranged weighting factor. If, for example, a score were simply too wild or unusual, it might be considered to be due to errors in recording information or making measurements and, accordingly, weighted 0 (*i.e.*, dropped). If this were done, the base of the ratio, the number of scores, would usually be reduced accordingly. Other weighting schemes take account of over- or underrepresented segments of samples, weighting these segments accordingly to produce a **weighted mean**.

Two somewhat different kinds of means are the **geometric mean** and the **harmonic mean**, neither of which is widely used as a measure of central tendency, but both of which have specific uses. The geometric mean is the nth root of the product of all scores (or, using logarithms, the antilog of the average logarithm of the scores). It is used in averaging rates or scores where one expects a constant rate of change.

(5.2)
$$\text{Geometric mean} = \sqrt[N]{(X_1)(X_2)\cdots(X_N)}$$

We will use this average in Chapter 7 (Section 7.3.2d) to combine two summary statistics. The geometric mean is also used in Chapter 16 in the basic formula (16.4) for a hierarchical model.

The harmonic mean is the reciprocal of the average reciprocal of scores and it is used to average ratios in which the numerators are constant but the denominators vary.

(5.3)
$$\text{Harmonic mean} = \frac{N}{\Sigma(1/X_i)}$$

Both the geometric and the harmonic mean turn out to be somewhat smaller than the arithmetic mean.*

5.3.5   Selecting the Most Appropriate Measure of Central Tendency

Generally, in analyzing the distribution of a variable, only one of the possible measures of central tendency would be used. Its selection is largely a matter of judgment based upon the kind of data, the aspect of the data to be examined, and the research question. Some of the points that might be considered are the following.

Central tendency for interval data is generally represented by the arithmetic mean, which takes into account the available information about distances between scores. For ranked data, the median is generally most appropriate, and for nominal data, the mode or modes.

If there are several modes, then the mode may be useful alone, or in addition to one of the other two measures, even for ordinal or interval data. If the distribu-

---

BOX 5.3   FEATURES OF A DISTRIBUTION

At this point you have made it through essentially two-thirds of the different kinds of features of a univariate distribution one might describe: form and central tendency. The third feature, variation, will be discussed next, followed by a bit more on form. Check yourself at this point to see if you are ready to go on. The ideas of *variables* and *level of measurement* are discussed in Sections 2.1 and 2.1.1. The different kinds of distributions are defined and illustrated in Section 3.2.3 and the following few pages. The graph in Figure 4.5 illustrates differences in form and location, and Figure 5.1 illustrates differences in form, location, and the degree of concentration of scores. After a comment on the use of the arithmetic mean to "smooth" curves, our next main topic will be measures of the variability of scores.

---

tion is badly skewed, one may prefer the median to the mean, because the median would not be affected as much by unusual extreme scores. For this reason, for example, the median income of people is usually reported rather than the arithmetic mean.

---

* Wallis and Roberts (1956) discuss these means and provide several suggestive examples of their application. A somewhat more detailed, illustrated discussion of the geometric and harmonic means is presented in Newman (1956).

If one is interested in prediction, the mode is the best value to predict if an *exact* score in a group has to be picked. More cases occur at the value of the mode than at any other *single* score value one could pick. The median has the property of exceeding and being exceeded by half of the scores, so the result of guessing the median would be an overprediction as frequently as an underprediction. Overall, the median produces the smallest absolute error (*i.e.*, the sum of absolute deviations is less from the median than from any other point). The arithmetic mean is the score around which the aggregate of deviations (algebraically considered) is 0, so the result of guessing or predicting the value of the mean would always be an algebraic deviation that balances out. In each case, a certain kind of predictive error is minimized, and one would pick a measure of central tendency to use in prediction depending upon the kinds of predictive error one wants to minimize.

Finally, it should be noted that it is possible to have median or arithmetic mean measures that do not correspond to any specific score in a distribution. There may be a gap at that point where the median or mean happens to fall, but the score at that gap would still be the result of the computation. This is, of course, quite appropriate, since we are trying to develop an index that characterizes the *distribution* of scores—not a single number that will be close to each individual case. The mode, of course, is an actually observed score, and the most frequent one, at that.

The arithmetic mean is probably the most important measure of central tendency for interval data, since it appears repeatedly as a part of the logic of other statistical techniques you will encounter in this book, techniques such as the variance and standard deviation, standard scores, correlation, and regression. Furthermore, the sample mean is important as an estimator of the population mean, and the population mean is a key parameter in many common population distributions (especially the normal distribution).

5.3.6   Smoothing Trend Lines: The Moving Average

In Chapter 4 we introduced the idea of a line graph that might be used to plot the value of some variable through time. This would result in a trend line, but often the trend line shows a confusing picture. In addition to the general trend, most plots of real data show a saw-toothed pattern of cyclic or minor variations that tend to obscure the overall trend. This is shown in Figure 5.3 by the dotted lines, showing components of population change by month for the 1980 to 1984 period in the United States. Investigators who want to show the general trends may want to "smooth out" some of the minor ups and downs. The arithmetic mean is sometimes used to smooth trend lines, and in this application it is called a **moving average**.

A *moving average* is an average of a fixed number of scores over successive periods of time. In Figure 5.3, the heavy line represents the result of using successive 12-month periods. In our example in Table 5.5, annual net growth rate figures are given, one for each year between 1935 and 1984. We would create a moving average by deciding upon a fixed time period over which to average, say, 5 years. We would then add together and average the rates for the first 5 years

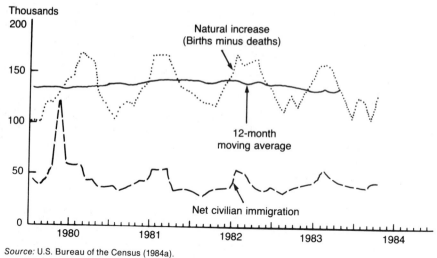

Source: U.S. Bureau of the Census (1984a).

FIGURE 5.3 COMPONENTS OF POPULATION CHANGE BY MONTH, JANUARY 1, 1980, TO MAY 1, 1984

(*i.e.*, for 1935 through 1939), and place this value, 8.2, at the midpoint of that time span, the year 1937. This is the first of our 5-year averages.

The second 5-year average is created by dropping the earliest year's rate and adding in the following year's rate. Thus the rate for 1935 would be dropped, and the rate for 1940 would be added, and again these five figures would be averaged; then the resulting average would be placed by the year that is midway in this 5-year period. Successive sets of 5 years are averaged in this way until the last period, 1984, is reached, as shown in Table 5.5. Notice that the two years at the extreme ends of the distribution do not have 5-year average figures, because they are not the middle year for an average. The moving averages are then plotted, and the result is a smoothed trend line.

The longer the period over which the average is taken, the smoother the trend line. By careful choice of the time period over which averages are taken, an investigator can remove or smooth out cyclic or seasonal variations in data. This is very clearly indicated in Figure 5.3, where the 12-month moving average removes the seasonal fluctuation in the rate of natural increase over a year. The result shows the trend as it changes year by year, with seasonal variation removed.

Moving averages tend to "anticipate" marked changes in the direction of the unsmoothed graph; that is, the moving average begins to drop or rise somewhat before the point where a marked shift occurs. You will recognize this as one of the properties of the arithmetic mean; it tends to be influenced in the direction of extreme scores. This method of smoothing trend lines is useful in analytic graphing.

TABLE 5.5 Illustration of a 5-year Moving Average for Annual Rates of Net Growth, United States, 1935–1984

Year	Net Growth Rate	5-Year Moving Average
1935	8.0	—
1936	7.1	—
1937	7.9	8.2
1938	9.1	8.4
1939	8.9	9.1
1940	9.2	10.0
1941	10.3	10.8
1942	12.7	11.4
1943	13.1	
1944	11.5	
1980	11.1	
1981	10.0	
1982	9.7	9.8
1983	9.1	—
1984	9.2	—

Source: U.S. Bureau of the Census (1971, 1985).

5.4 VARIATION

Adams (1953), in a study of the occupational origins of physicians, sent mailed questionnaires and did interviews with physicians in four northeastern cities. Their "occupational origin" was measured by using the North-Hatt occupational prestige scale on their reported father's occupation. Needless to say, physicians tend to come from families where the father also has a relatively highly prestigious occupation (lawyer, banker, physician, etc.). Separating his data by the date of birth (age) of sampled physicians, Adams noted that the prestige of father's occupation tended to decrease. For physicians born between 1875 and 1895 the mean of father's prestige was 79.8, and it dropped to 74.6 for the later period, 1895 to 1920. He also found an interesting second feature in the data, namely that there seemed to be more variation in the background of physicians born more recently. That is, their mean prestige rating not only dropped somewhat through time, but there was more variation in father's occupational prestige scores; they were not as similar to each other at the end of the time period as they were before. How can this feature be measured?

The third of the three characteristics of a univariate distribution, for which we are developing simple measures, is the characteristic of spread, or dispersion of scores. Some groups of scores are wide-ranging; others are more bunched together, as the following illustrations show, quite apart from their location or the distinctive features of the distribution's form.

group A	2	3	4	5	6	$N = 5$
group B	5	6	7	8	9	$N = 5$
group C	51	52	53	54	55	$N = 5$
group D	52	53	53	53	54	$N = 5$
group E	47	50	53	56	59	$N = 5$

In the first three groups (A–C), each has the same amount of variability between scores. Notice that the scores are all consecutive integers. Group D, however, has less variability since the scores are more closely grouped, many of them identical in value. Group E is even more dispersed than group C, yet C, D, and E all have the same mean and median, 53. (As far as the other measure of central tendency is concerned, only group D could be usefully described as having a mode that is 53.)

There are several different approaches to the measurement of variability in a group of scores, and the basic distinction between measures appropriate for nominal data and interval data is needed here as well. Variation in ordinal data is often treated by techniques used for interval data even though the distance idea is not defined. Ordinal data could also be treated by measures of variation appropriate to nominal data.* Diagram 5.2 (p. 158) provides the step-by-step procedures for computing each measure. We will begin with some of the measures for interval data.

For interval scores where the idea of distance on a score scale has meaning, there are two general approaches to the measurement of variation in a distribution. First, one could think in terms of the range of the scale over which the scores in a distribution actually fall. Second, one could think of measures of variation describing the extent to which scores in a distribution differ from some single score, such as a central tendency score for the distribution.

---

* To the authors' knowledge, there is no measure of variation designed specifically for ordinal data. If each score in a set of ordinal data is unique—that is, there are no tied scores—then the spread of the set is dependent upon the number of scores there are. Thus, as many ranks are represented as there are cases or observations. If there are ties and the ordinal data are grouped into a frequency distribution, then the use of the usual measures of variability leads to the implicit assumption that the ordinal categories are separated by equal distances. Of course, ordinal measurement does not justify such an assumption.

5.4.1   Range

The **range** is simply a number representing the difference in value of the highest and the lowest scores in a distribution. More accurately, it is the difference between the lower real limit of the lowest score category and the upper real limit of the highest score category in a distribution.

(5.4)
$$\text{Range} = U - L$$

In Formula 5.4, U refers to the upper real limit of the highest class in the distribution, and L refers to the lower real limit of the lowest class. Usually, the range, like the mode, is quickly computed by inspection of a distribution, and this is one of its chief claims to fame.

Unfortunately, the range only depends upon the two extreme scores in a distribution, and ignores completely the way the data between these extremes are distributed. In practical research and everyday life we often expect that extreme scores will be undependable. They may be due to some gross measurement error or, perhaps, a very rare and unusual score.

For example, if we were interested in studying multiple births, we might examine birth records. In most modest-sized samples of these records, we would undoubtedly find scores from 1 to 3 (single birth to triplets) or a range of 3. With much larger samples, or by unusual luck in small samples, an investigator might observe a range of 5 (from single birth to a quintuplet birth). Only seven quintuplet births (where all survived infancy) are known to have occurred in the world in the decade of the 1960s.

If sample size is increased, the crude range can only stay the same or increase, regardless of the concentration or dispersion of the rest of the scores. For these reasons, a more refined measure of variation is generally preferred.*

5.4.2   Interquartile Range (Q)

The **interquartile range** is defined as the range that includes the center 50% of cases in a distribution or the distance between the first and third quartiles.

(5.5)
$$Q = Q_3 - Q_1$$

Although, again, this range measure is based on the difference between two points in a distribution, these points are determined in a way that is sensitive to the concentration in the data themselves.

The first quartile is the point on the score scale below which 25% of the cases fall, and the third quartile is the point on the score scale below which 75% of the cases fall. These are determined in ways quite similar to the way the median (point below which 50% of the cases fall) is computed [see Diagram 5.2 (p. 158)] and the discussion of percentiles in Chapter 3].

---

* The range is often a useful measure of variability for small samples because it is easy to compute and, in the case of small samples, is an efficient estimate of the standard deviation (discussed in Section 5.4.4 below).

The interquartile range avoids the exclusive use of the two extreme scores and it is thus less subject to the erratic variation in extreme scores. Distance between other points on a distribution may be used instead of quartiles. For example, the interdecile range would be the difference between the 9th and the 1st decile (the difference between the point below which 10% of the cases fall and the point below which 90% fall). The interdecile range was used in the box and whisker diagrams in Figure 4.16. The interquartile range is shown there as well. Sometimes the semiinterquartile range is used, which is simply half the interquartile range. Its virtues and properties are the same as those of the interquartile range itself.

5.4.3 Average Absolute Deviation (AD)

Another kind of measure of variation measures the deviation of scores from a central point in a distribution, usually the arithmetic mean. Although seldom used, the **average absolute deviation** illustrates nicely the principles involved in this approach.

A portion of Table 5.4, reproduced below, provides a good illustration of the logic of this dispersion measure.

Raw Scores X_i	Absolute Difference $|X_i - \bar{X}|$	
2	6	$N = 7$
2	6	
4	4	$\bar{X} = \dfrac{56}{7} = 8.0$
6	2	
8	0	
14	6	
20	12	
56	36	

You will recall from our discussion of the arithmetic mean that one of its properties is that the sum of algebraic deviations of scores from it is always 0. Thus, as a measure of dispersion, the average deviation shifts to the use of the absolute value of the deviation of each score from the mean, as shown above.* The average absolute deviation is simply the arithmetic mean of absolute deviations, another ratio.

(5.6)
$$AD = \frac{\Sigma|X_i - \bar{X}|}{N}$$

$$= \frac{36}{7} = 5.1$$

* Absolute deviations ignore the direction and thus the sign of the deviations of scores from the mean.

On the average, these seven scores deviate 5.1 points from the arithmetic mean of this distribution. The larger the average deviation, the more variability there is between scores in the distribution. If all the scores are identical to the mean, the numerator becomes zero and the average absolute deviation is, appropriately, zero. The average deviation can become quite large; in fact, its upper limit depends upon the unit of measurement and the magnitude of variation in the data themselves.

While the average deviation is easily interpreted and relatively easily computed, another measure is generally preferred, simply because of its mathematical uses in other areas of statistics. This measure is the variance, or its square root, the standard deviation.

5.4.4   Variance and Standard Deviation (s^2 and s, respectively)*

The **variance** and **standard deviation** are similar to the average deviation in that differences between the mean and each score are used, but instead of taking the absolute value of these deviations, the square of the deviation is used. This has a similar effect in side-stepping the zero-sum-of-deviations property of the arithmetic mean, and results in a measure of dispersion for interval data that has wide applicability and some interesting connections with other topics in statistics.

The variance is simply an average of squared deviations of scores from the arithmetic mean, and the standard deviation is the square root of the variance.[†]

(5.7)
$$s^2 = \frac{\Sigma(X_i - \bar{X})^2}{N} \quad \text{(variance)}$$

(5.8)
$$s = \sqrt{s^2} \quad \text{(standard deviation)}$$

Using the illustrative data below from a portion of Table 5.4, the variance and standard deviation can be computed. Diagram 5.2 (p. 159) presents the procedures for computation from grouped data.

* Again, we should note that there are, conventionally, two sets of symbols for the variance and for the standard deviation, one for the statistic (s^2 and s) and one for the parameter (the lowercase Greek letter *sigma*, σ^2 and σ). In this part of the book we will use the symbols for sample statistics (s^2 and s), whereas in the parts on inferential statistics, both sets are introduced when contrasts between them are important.
† Although the variance is defined as the simple "average" of squared deviations from the mean (and thus it has N in the denominator as in Formula 5.7), this is a biased estimate of the population standard deviation if Formula 5.8 is used on sample data from small samples. Since we are dealing in the area of description rather than inference at this point, we will use the formula with N in the denominator. It simplifies some later computations in this book, and the "unbiasing" can be better explained and treated in inferential statistics (see Chapter 12). The square root of the variance is called the "standard deviation," a term coined by Karl Pearson in 1894. As an aside, the sample variance as computed in Formula 5.7 yields a slight *under*estimate of the population variance, especially for small, probability samples. This underestimate bias is usually corrected by using $N - 1$ in the denominator for the sample variance rather than N. In most sociological studies the sample size is sufficiently large so that the effect of this correction is nil.

	Raw Scores X_i	Differences $(X_i - \bar{X})$	Squared Differences $(X_i - \bar{X})^2$
	2	$(2 - 8) = -6$	$(-6)^2 = 36$
	2	$(2 - 8) = -6$	$(-6)^2 = 36$
$N = 7$	4	$(4 - 8) = -4$	$(-4)^2 = 16$
	6	$(6 - 8) = -2$	$(-2)^2 = 4$
$\bar{X} = 8.0$	8	$(8 - 8) = 0$	$(0)^2 = 0$
	14	$(14 - 8) = +6$	$(6)^2 = 36$
	20	$(20 - 8) = +12$	$(12)^2 = \underline{144}$
	56	0	272

$$s^2 = \frac{272}{7} = 38.9$$

$$s = \sqrt{38.9} = 6.2$$

In the study of the occupational origins of physicians discussed at the beginning of this section, Adams actually chose the standard deviation as a measure of variation to show what has been happening from 1875 to 1920 in terms of variation. Table 5.6 shows his data. The standard deviation shows the variation in physician's father's occupational prestige in units of the North-Hatt occupational prestige scale. In the earliest period the standard deviation was 7.5 North-Hatt units, whereas in the last period, 1915–1920, the standard deviation was nearly double, 13.6 units. The average standard deviation prior to 1895 was 7.4 and the average standard deviation from 1895 on was 11.3.* There is clearly an

TABLE 5.6   Trend in the North-Hatt Occupational Prestige Scores for the Occupation of Fathers of Physicians by Date of Birth of Physicians

Date of Birth	Number	Mean Prestige Rating	Standard Deviation of Prestige Rating
1875–1879	9	78.7	7.5
1880–1884	9	77.9	6.7
1885–1889	7	81.6	8.8
1890–1894	15	80.9	7.0
1895–1899	19	74.8	10.4
1900–1904	22	74.6	8.7
1905–1909	24	74.2	14.2
1910–1914	23	74.0	10.6
1915–1920	9	76.4	13.6
Overall	137	76.1	10.7

Source: Adams (1953: 406). Reprinted by permission.

* These averages cannot simply be computed by adding up standard deviations and dividing by their number, because each standard deviation is based on a different number of cases. The averages quoted in the text are weighted averages.

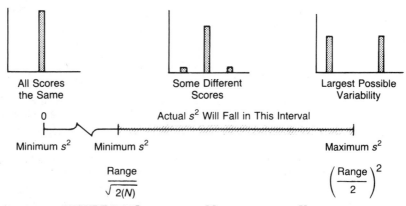

FIGURE 5.4   Limits on the Magnitude of the Variance

increased variation in the prestige backgrounds of physicians in addition to a slightly declining average level of father's occupational prestige. This finding is backed up by other data Adams collected. Physicians have a 93 on the 100-point North-Hatt rating scale, and if many physicians had fathers who were also physicians, the average prestige score would be higher (and the standard deviation smaller). There has been, however, a decline in the percentage of physicians who come from families where their father is also a physician. For physicians born in the 1870–1879 period, 22% came from physician families; the figure drops to 9.4% from physician families among those born in the 1910–1920 period.

Apart from its interpretation in terms of sheer difference in magnitude, the standard deviation has a number of interesting properties and may be interpreted in at least two different ways. To start with, the standard deviation and variance will both be 0 where all scores have the same value (*i.e.*, the value of the mean), and they reach a maximum magnitude for a given set of data when scores are divided between the extreme ends of the scale. These conditions are illustrated in Figure 5.4. Given some difference between scores, the value of the **variance** can be interpreted in terms of a scale extending from a minimum possible value that equals the range divided by the square root of twice the sample size, to a maximum that is the square of half the range. The **standard deviation** could also be interpreted in terms of a scale extending between values that are simply the square root of these values. For most curves, the range is approximately equal to six times the standard deviation as a rough rule of thumb.

The variance and standard deviation can be interpreted in a somewhat different way, too: in terms of the area under a curve within z standard deviations of the mean of a distribution.* This central area, between z standard deviations below the mean and z standard deviations above the mean, is always, and for any

* The number of standard deviations a score is from the mean of its distribution is called a "standard score" or z-score. If all scores in a distribution were expressed in terms of the number of standard deviations they are from their mean, the distribution of z-scores that would result would have a mean of 0 and a standard deviation of 1. This follows simply from the definition of the standard score or z-score. We will discuss standard scores somewhat further later in this chapter (Section 5.4.7).

shape of distribution, at least as follows:

$$\begin{array}{l} \text{Minimum percentage of cases} \\ \text{within } z \text{ standard deviations} \\ \text{above and below the mean} \\ \text{of a distribution of any shape} \end{array} = 100\left[1 - \left(\frac{1}{z^2}\right)\right]$$

Thus at least 75% of the area under a curve will fall within 2 standard deviation units of the mean: $100[1 - (1/2^2)]$. This is called Tchebycheff's* theorem (or Tchebysheff's inequality), and it has applicability especially in inferential statistics. In descriptive statistics it provides a useful interpretation of the standard deviation. For many curves, particularly those more concentrated around the mean, the percentage of cases within a given number of standard deviation units is much larger than the minimum given by Tchebycheff's theorem. For curves that are unimodal and symmetrical, for example, the minimum percentage within $(\bar{X} + z)$ and $(\bar{X} - z)$ is at least as follows:

$$\begin{array}{l} \text{Minimum percentage of cases} \\ \text{within } z \text{ standard deviations} \\ \text{above and below the mean} \\ \text{of a unimodal, symmetrical} \\ \text{distribution} \end{array} = 100\left[1 - \left(\frac{4}{9}\right)\left(\frac{1}{z^2}\right)\right]$$

The minimum percentage of cases within 2 standard deviations above and below a unimodal, symmetrical distribution would be 89% rather than 75%, which is the general figure for any shaped distribution.

In fact, the special bell-shaped curve called the normal curve has considerably more area within z standard deviations of the mean, as is shown in Table 5.7. The normal curve has 95% of its area within 2 standard deviations above

TABLE 5.7 Percentage of Area Under a Curve between the Mean Plus z and the Mean Minus z Standard Deviations

Within (z) Standard Deviations from $\bar{X}$	Minimum for any Distribution (Chebyshev's Theorem)	Minimum for any Unimodal and Symmetrical Distribution	Value for a Normal Distribution
.67	0%	1%	50%
1.00	0	56	68
2.00	75	89	95
3.00	89	95	99.7
4.00	94	97	99.99 ..
10.00	99	99.6	99.999 ...

* Also spelled Chebyshev.

and below the mean. The normal curve will be discussed further in Section 5.5 on the form of a distribution, since it is an important shape that is used in other topics in statistics.

5.4.5   Index of Dispersion (D)

The measures of variation discussed above all imply a knowledge of distance and thus are most appropriate for interval measures. The idea of variation in scores is not limited, however, to interval data. The index of dispersion is suggested as a measure of variation for nominal (or ordinal) variables by Hammond and Householder (1962).*

The **index of dispersion** is a ratio whose numerator and denominator are counts of the number of pairs of scores. The denominator is the maximum number of unique pairs that could be created out of the scores such that each member of the pair is from a different category. It turns out that this condition is met where the N scores are evenly distributed among the k categories into which the scores are grouped. This corresponds exactly with our idea of maximum variability in a nominal (and ordinal) variable.

To illustrate the logic behind this measure, suppose we consider the following 9 scores on marital status, classified into three categories ($k = 3$):

	f
Married	5
Single	3
Other	1
Total	9

If variability among the 9 scores were a maximum, there would be three scores in each of the three categories. The number of pairs of cases where the members of a pair came from *different* categories would be 27. That is, three pairs could be formed by pairing each of the three married people with the single people. Since there are three married people this means $3 \times 3 = 9$ pairs. The same pairing could occur between married and "other" and between single and "other" categories, yielding a total of $9 + 9 + 9 = 27$. In this example, however, cases are not evenly distributed over categories; they are more concentrated than that. The 5 married people could be paired with the 3 single (15 pairs) and with the 1 "other" person (5 pairs), and the single people could be paired with the "others" (yielding 3 pairs), or a total of 23

---

* Senders (1958, pp. 78–84) takes a somewhat different approach in a measure she calls "relative uncertainty," based on the amount of information needed to determine category placement of a case. It is computed as

$$H_{\text{rel}} = \frac{-\Sigma p_i \log_2 p_i}{\log_2 k}$$

where p_i is the proportion of cases in the ith category of a univariate distribution, and k is the number of categories in the variable. It varies from 0, if all cases fall into one category, to 1.0, where cases are evenly spread over all k categories. The logic of H_{rel} is similar to that of a measure of association for bivariate distributions, the uncertainty coefficient, discussed in Section 7.2.2. Another measure of variation for nominal variables is called the index of qualitative variation (IQV) in Mueller et al. (1977).

pairs. The D statistic is simply a ratio of the number of different pairs that could be made out of the data at hand, compared with the maximum number of unique pairings that could be created if cases were evenly spread over all available categories. Here $D = .85$.

If all scores were in a single category of a variable that has several possible categories, then there is maximum concentration or minimum variability, and D would equal 0. On the other hand, if cases were evenly distributed among the possible categories, there would be maximum variability, and the numerator and denominator of the D ratio would be the same, and D would equal 1. D varies, then, from 0 to 1 and is a useful measure of variation for nominal or ordinal variables.

The computational formula for D is given below and the step-by-step procedures for computing D are discussed on page 160 in Diagram 5.2.* Box 5.4 provides a small computer program to compute D.

(5.9)
$$D = \frac{k(N^2 - \Sigma f_i^2)}{N^2(k - 1)}$$

where N is the number of scores, k is the number of categories of the variable into which data might be classified (apart from whether all categories are used or not), f_i is the frequency of cases in the ith category; and the summation is over the squared category frequencies of all categories.

Table 5.8 illustrates an application of the index of dispersion for the distribution of alcohol consumption by each of six different age groupings for the United States in 1983. The National Center for Health Statistics conducted a survey of health practices in which questions on alcohol usage were asked. The sample included noninstitutionalized adults, aged 20 and over throughout the United States. Alcohol consumption categories were developed from self-reports of quantity and frequency of drinking in the most recent 2-week period in which alcohol was consumed. Using frequency information on drinking for each age category (*note:* this can be gotten by multiplying percentages by the frequency upon which 100% is based at the bottom of each age column), an index of dispersion can be computed using Formula 5.9.

The distribution of alcohol consumption could be called bimodal for all groups, with one mode at "abstainer" and the other at "light drinker." The most prominent mode differs by age group. It falls at light drinking in the first two age categories and abstainer for the later age categories. The index of dispersion (listed along the bottom of Table 5.8) also differs by age category; it is highest for the 45–54 group, indicating that individuals are spread relatively uniformly across

---

* Maurice G. Kendall defined a measure of variation that examined all pairs of scores in a set of data. The measure is equal to one-half of the average squared difference between all possible pairs of scores. This one can be computed as

$$\text{Variation} = (N^2 - f_1^2 - f_2^2 \cdots - f_k^2)/2N^2$$

where f_1^2 is the frequency squared in category one of the variable and the square of the frequency in each of the k categories is subtracted in the numerator. This is a slightly different expression of Formula 5.9, and it has several useful implications for sociological analysis, as shown in Hawkes (1971).

BOX 5.4   COMPUTING THE INDEX OF DISPERSION

The *D* statistic is not yet included in many statistical program packages, although it is a very useful statistic for variation in nominal and ordinal variables. The following program in BASIC will compute *D* for you.

```
10  PRINT "Computes D, the Index of Dispersion"
15  INPUT "Number of categories = "; K
20  LET N = 0
25  LET F1 = 0
30  PRINT; "Enter category frequency (or %)"
35  PRINT; "(one per line, touch return)"
40  FOR L = 1 TO K
45  LET F = 0
50  INPUT F
55  LET N = N + F
60  LET F1 = F1 + F ↑ 2
65  NEXT L
70  LET D = (K*((N ↑ 2) − F1))/((N ↑ 2)*(K − 1))
75  PRINT "D ="; D
80  PRINT
85  PRINT "next problem"
90  GOTO 15
100 END
```

---

all available categories, and drops markedly for the oldest two groups, which are more concentrated in the abstainer end of the distribution.

The *D* statistic (with slightly different symbols used to express it) was proposed by Rushing and Davies (1970) as a measure of the concept "division of

---

TABLE 5.8   PERCENTAGE DISTRIBUTION OF ALCOHOL CONSUMPTION BY AGE, UNITED STATES, 1983

Alcohol Consumption	Age (Years)					
	20–34	35–44	45–54	55–64	65–74	75+
Lifetime abstainer	26.3	28.2	32.8	36.3	45.3	61.1
Former drinker	3.5	6.3	8.1	9.4	12.2	8.9
Lighter drinker	33.7	31.9	28.7	27.3	22.8	14.5
Moderate drinker	26.1	23.1	19.1	17.3	11.7	10.9
Heavy drinker	10.4	10.4	11.3	9.8	8.0	4.7
Totals	100.0	100.0	100.0	100.0	100.0	100.0
N (in 1,000's)	(59.9)	(28.9)	(22.2)	(22.0)	(16.4)	(9.5)
Index of dispersion (*D*)	92.1	93.8	94.3	93.2	88.5	73.1

Source: National Center for Health Statistics (1986).

labor." The concept of division of labor refers to the difference or variability among individuals in their sustenance activity. Generally, this can be thought of in terms of the different kinds of occupations there are in a society, and the extent to which individuals are spread out among them rather than all concentrated in, say, laboring or farming or housewifing. Division of labor, then, could be measured by D, using information about the number of individuals in different occupational categories. The more evenly spread they are among the different possible occupations, the greater the division of labor. The amount of division of labor could be expressed for societies, cities, states, and within organizations. It could also be expressed for different groups of people (e.g., men and women in different age cohorts). Besides noting the relative concentration of individuals within occupational categories, it would be useful to note the number of different categories (k) that exist for given organizations or societies. The Rushing and Davies suggestion indicates a rather important usage of statistics: to measure interesting characteristics of social organization.

5.4.6   Selecting the Most Appropriate Measure of Variability

Several criteria might enter into the selection of a measure of variability. One, of course, is the meaning of the scores one has. For interval scores, the standard deviation (or variance) would ordinarily be chosen. Since range-based measures tend to be sensitive to only two scores or points in the distribution, they would probably not be selected if the standard deviation could be computed. If a distribution is severely skewed, so that the mean is thought not to give an appropriate indication of central tendency, the range-based measures (*e.g.*, the interquartile range) may then be preferred.

Ordinal data present something of a problem. The index of dispersion is not sensitive to the ordering of categories implied in ordinal variables; thus it loses some information. On the other hand, measures that rely on distances, such as the interquartile range or the standard deviation, imply information in data that are not defined into the scores. The usually recommended halfway measure is to use the median for central tendency and the interquartile range for variation of ordinal data, interpreting the interquartile range as the range of ranked categories that includes the middle 50% of cases.

For nominal, and perhaps ordinal variables, the index of dispersion provides a nice solution to the problem of measuring the variation in scores. D has another feature that recommends its use: it can be interpreted more readily because it varies on a scale from 0 (for no variation) to 1.0 (which corresponds to the maximum amount of variation possible in the data at hand). Although the other measures of variation indicate greater amounts of variation by larger magnitudes, the maximum possible magnitude of the variation measures, other than D, varies depending upon the size of the score units used (*e.g.*, years or decades) and the spread of the scores. Zero may indicate no variation (for the range and for the variance and standard deviation), but beyond that the magnitude of the number has little meaning in an absolute sense.

BOX 5.5   REVIEWING FEATURES OF A DISTRIBUTION

Form, central tendency, and variation, the three features of a univariate distribution, may each be measured in several different ways. Thus far only verbal descriptions of the **form**—kurtosis (Section 5.2.3), symmetry (Section 5.2.2), and number of modes (Section 5.2.1)—of a distribution have been discussed, but the next section (Section 5.5) will introduce some indices for these features as well. For the **central tendency** of a distribution we have discussed the mode (Section 5.3.1), the median (Section 5.3.2), and the arithmetic mean (Section 5.3.3), plus a couple of special types of means that are not too frequently seen. Finally, **variation** measures included the range (Section 5.4.1), the interquartile range (Section 5.4.2), the average absolute deviation (Section 5.4.3), the variance and standard deviation (Section 5.4.4) and the index of dispersion (Section 5.4.5). Box and whisker graphs graphically portray measures of central tendency and form.

An important point you should be clear about is the three different ways that the standard deviation (and variance) can be interpreted: (a) in terms of sheer magnitude (Section 5.4), (b) in terms of the minimum and maximum values it could take for a given set of data (Section 5.4.4), and (c) in terms of the proportion of the area under curves between points defined in terms of the number of standard deviations on either side of the mean (Section 5.4.4).

5.4.7 Standard Scores

Early in Chapter 3 we discussed the various types of comparisons that one might well want to make, statistically . Many of the procedures in Chapters 3, 4, and 5 were designed to facilitate the group–group or group–standard types of comparison. One could also make use of what we have described up to this point to indicate the relative standing of an individual in a group. One of these ways is to compute an individual's *percentile rank* (*i.e.*, the percentage of all scores equal to or less than that score). Another way to make this individual–group comparison is to create **standard scores**, which are also called **z-scores**. A standard score is merely the number of standard deviation units an individual falls above (or below) the mean of the group.

(5.10)
$$z = \frac{X_i - \bar{X}}{s}$$

The standard score takes out the effect of the mean (by subtraction) and expresses the difference in standard deviation units (by dividing by the standard deviation). It is a simple rescaling of scores so that the mean of z-scores is 0 and the standard deviation is 1—a change from dollars and years, for example, to "standard" units.*

---

*An interesting property of z-scores, which will be used later on in Chapter 7 when we discuss a correlation coefficient, is that the sum of squared z-scores always equals the number of cases, N (i.e., $\Sigma z^2 = N$).

Because of this, one could compare z-scores of an individual on different distributions (*i.e.*, the person may be 2 standard deviations above the mean on one score distribution and only 1 standard deviation above on another). More will be said about standard scores later in this chapter.

5.5 FORM OF A DISTRIBUTION—II

Thus far, we have described the form of a distribution only in terms of some general verbal labels that pointed to symmetry, kurtosis, and number of peaks in a general way. In this section we will introduce some approaches to summary measures of form that might be expressed as single numbers, which are similar in purpose to the summary index numbers used to describe central tendency and variation in a distribution. In the process, we will discuss the idea of standard scores (z-scores) and special kinds of curves such as the standard normal curve.

As a matter of fact, we already have a number of tools at hand that would be useful in creating an index of skewness. If the mean and median are different, for example, we would conclude that the distribution is skewed in the direction of the mean and, for the same range of scores, bigger discrepancies would indicate more skewness. We could also examine differences between quartiles in a distribution. If the distance between the first and second quartiles is greater than the distance between the second and third, we would conclude that the distribution is negatively skewed. In this section, however, we will develop more useful and interpretable measures of skewness and kurtosis for interval variables.

5.5.1 The Central Moment System

When one is dealing with interval-level data, it is often useful to describe the data in terms of their balance around some central point. The arithmetic mean, for example, is the point around which the algebraic "balance" of scores is perfect in the sense that the algebraic sum of deviations of scores is 0. Deviation of scores from the mean of a distribution is often expressed by the small letter x.

$$x = (X_i - \bar{X})$$

The **first moment** about the arithmetic mean, or *first central moment*, is simply the average of the first power of deviations from the mean, or

$$m_1 = \frac{\Sigma x}{N}$$

Since the sum of deviations around the mean is always 0, the first moment is always 0, a defining characteristic of the mean. If higher powers of deviations about the mean are examined, however, additional information about a distribution is revealed. The **second central moment**, for example, is the variance

$$m_2 = \frac{\Sigma x^2}{N}$$

Statisticians generally consider only the first four moments. The **third** and **fourth central moments** are simply the averages of the third and fourth powers of deviations from the mean.

$$m_3 = \frac{\Sigma x^3}{N}$$

$$m_4 = \frac{\Sigma x^4}{N}$$

The advantage of each of the moments stems from two factors: (a) the fact that even powers have the effect of eliminating negative signs but odd powers preserve negative signs in the numerator of the moments (above), and (b) the fact that higher powers tend to emphasize larger deviations from the mean, as illustrated below:

(x) Deviation	Power to Which x Is Raised			
	1st	2nd	3rd	4th
-9	-9	$+81$	-729	$+6561$
$+1$	$+1$	$+1$	$+1$	$+1$
$+1$	$+1$	$+1$	$+1$	$+1$
$+2$	$+2$	$+4$	$+8$	$+16$
$+5$	$+5$	$+25$	$+125$	$+625$
Totals	0	$+90$	-594	$+7204$

The third moment provides an index of skewness because it is an odd moment; thus if the high and low scores do not balance out around the mean (as they do not in the example above), it will not be equal to 0. Also, it is a higher moment, and thus it emphasizes the extreme deviations there may be from the arithmetic mean. The fourth moment is an even moment, thus it does not distinguish between deviations above or below the mean. It is a higher moment, so that it emphasizes the deviation of scores that fall in the tails or extremes of a distribution. Thus the fourth moment is useful in measuring the degree of kurtosis in a distribution.

None of the moments are relative measures. That is, they range in magnitude over a scale that starts at 0 and extends upward (and downward in the case of m_3) to a value that depends upon the units of measurement and the variability of scores. For this reason, skewness and kurtosis measures are sometimes created in such a way that the range of index values is over a defined scale. This would permit one to compare skewness and kurtosis scores regardless of units of measurement. Two such measures, Sk and Ku, are sometimes created out of the moment system:

(5.11)
$$Sk = \frac{m_3}{m_2^{3/2}} \quad \text{(skewness)}$$

(5.12)
$$Ku = \frac{m_4}{m_2^2} \quad \text{(kurtosis)}$$

5.5.1a Skewness (Sk)

The *skewness* measure, *Sk*, is the ratio of the third central moment divided by the cube of the square root of the second central moment. Since the second central moment is the variance, its square root is the standard deviation (*s*); therefore, the denominator of Formula 5.11 can be expressed more simply as s^3. Thus, the formula becomes

$$Sk = \frac{m_3}{s^3}$$

If the distribution is symmetrical, the third moment will be 0 and thus the *Sk* measure will be 0. If the distribution is skewed to the right, *Sk* will be a positive value, and if it is skewed to the left, *Sk* will be negative. The magnitude of *Sk* expresses the relative amount of skewness and can be compared between distributions of different units of measurement. An *Sk* value of $+1$ represents rather extreme positive skewness and a value of -1 rather extreme negative skewness.

5.5.1b Kurtosis (Ku)

The *kurtosis* measure, *Ku*, is the ratio of the fourth moment to the square of the second moment, and it too is a relative measure. Small values of *Ku* indicate a platykurtic (flatter than normal) distribution, and high values indicate a leptokurtic distribution. The normal or mesokurtic distribution has a *Ku* value of 3 (although some programs subtract 3 to set *Ku* to 0 for a mesokurtic distribution).

5.5.2 The Standard Normal Distribution

The "normal curve" is only one possible shape of a distribution, but it is frequently used in statistics because it usefully describes a large number of chance distributions that are explored at great length in the field of inferential statistics (see Chapter 12 and the chapters following). Here it is useful to mention for two reasons, first, because it is the mesokurtic normal curve and, second, because it is helpful in explaining the usefulness of the standard deviation as a measure of variation. Platykurtic and leptokurtic distributions are, quite aptly, less peaked and more peaked than this normal one. As noted above, the normal curve has a *Ku* value of 3.0 and, because it is a symmetrical curve, it has a *Sk* value of 0. It is a unimodal, bell-shaped distribution that can be precisely described in terms of the following formula:

(5.13)
$$Y = \frac{1}{s\sqrt{2\pi}}\, e^{-\frac{1}{2}\left(\frac{X-\bar{X}}{s}\right)^2}$$

where *Y* is the value on the ordinate of a graph; *X* is the score value on the abscissa, *e* is the base of the natural logarithms (2.71828), raised to the power indicated by the somewhat complicated exponent attached to it; and π is pi (3.14159).

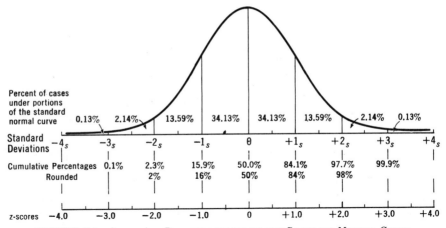

FIGURE 5.5   AREAS AND PERCENTAGES UNDER THE STANDARD NORMAL CURVE

Notice that the formula includes s, the standard deviation, as well as the mean. It is not at all necessary to remember this formula; it is only introduced to indicate that the normal curve is a specific form of curve that is a function of the mean and standard deviation.

5.5.2a   The Standard Normal Curve

There is a family of normal distributions each having different means and standard deviations and with scores expressed in orginal units (*e.g.*, tons, pounds). In statistics, however, it is often useful to think of a normal curve whose scores are expressed as standard scores (z-scores) rather than in terms of original units. Such a normal curve is called a *standardized normal curve*. It has a mean of 0, a standard deviation of 1.0, and an area that is set as equal to 1.0. Figure 5.5 shows a standard normal curve.

Notice that for any normal distribution the percentage of area under the curve falling between a point that is 1 standard deviation below the mean ($z = -1$) and 1 standard deviation above the mean ($z = +1$) is about 68% of the total area under the curve. Since 50% of the area under the curve is on either side of the mean (it is a symmetrical curve), there is only about 16% of the area under a curve falling above a z-score of $+1.0$.

5.5.2b   Area under a Normal Curve

Since the normal curve is used rather frequently in inferential statistics, tables of areas under the normal curve have been prepared. Table B in the Appendix of this book is such a table. The first column in Table B is the z-score value. Notice that it is necessary to list only positive z-scores, since we know that the normal curve is symmetrical and thus identical values would be given for negative z-scores. The second column in Table B shows the proportion of the area under the entire normal curve that falls between the mean and a point that is a given number of standard deviation units away in a given direction. Column 3 in Table B gives the proportion of the area under the curve falling beyond a given z-score in one tail of the distribution. With this table, it is possible to compute the proportion of the area under a normal curve between any two z-scores.* The following table lists some of the more frequently used z-score values from Table B and corresponding percentages of the area under a normal curve.

	Percentage of the Total Area In:	
z-score	One-Tail	Both Tails Combined
1.64	5%	10%
1.96	2.5%	5%
2.33	1%	2%
2.58	0.5%	1%

To illustrate the use of the table of the normal curve, let us consider the case of an achievement test with normally distributed scores, a mean of 50, and a standard deviation of 15. What proportion of the scores in the distribution will fall above the score of 66? To solve this we compute the z-score for 66:

$$z = \frac{66 - 50}{15} = \frac{16}{15} = +1.07$$

Looking up this value in the z-score column of Table B in the Appendix, we take the value in column C, which is the proportion beyond a given z-score. For the z-score of $+1.07$, this table value is .1423. Slightly more than 14% of the area under the curve is above this z-score or, in terms of the problem, slightly more than 14% of the scores in this test distribution are above 66. Using similar procedures, we could compute the percentage of the area under a curve "below" a given z-score or between z-scores of different values, or we could compute the scores at the first and third quartiles.†

Notice in Figure 5.5 that the tails of the distribution do not touch the baseline. No matter how far the tails of the distribution are extended to the right or the left, although they get closer and closer to the baseline, they never touch. This property of the standard normal distribution is called *asymptosis*. It is obvious from this property that the standard normal distribution is a theoretical distribution rather than an empirical one; nevertheless, it is a very useful one, as you will see. Because of this property of asymptosis, the entries in column B of Table B in the Appendix (areas between the mean and z) never quite reach .5 and the entries in column C (areas beyond z) never quite reach 0.

* Tables are also available that give the height of the ordinate on a normal curve at various points represented by z-scores. [See, for example, Hagood and Price, (1952:558), listed in General References].

† Note that the relationship between standard scores and areas in Table B holds only for normal distributions. These areas will not be accurate for nonnormal distributions, although the amount of error involved will depend upon how far a distribution departs from the normal distribution. See Table 5.7 for areas within z-score values for other shapes of distributions.

5.5.3   Other Mathematical Descriptions of Form

Although the standard normal curve is useful in statistics and one that can be described mathematically, it is but one form of curve. It was once thought that many characteristics were distributed in this way, but it is clear that this is not particularly true. The main use of the normal curve will be as a description of the probability of a certain kind of sampling outcome. For most other distributions, a description in terms of central tendency, form, and variation adequately expresses what the distribution is like. Where a mathematical description of a curve is possible, a clearly powerful tool for description is available. Examples in the social sciences include a curve describing the likelihood of the end of wars and strikes, the Lorenz curve in demography, and the J-curve concerning adherence to social norms. Some examples of these may be found in various sociology texts.

5.6   SUMMARY

In this chapter we have discussed ways of describing distributions of univariate data. Three overall features of distributions were described—*form, central tendency*, and *variation*—and several indices of each feature were presented. The form of a distribution varies in terms of *kurtosis, skewness*, and number of *modes* or peaks, and it is often simply described in terms that are broad and descriptive, such as a bell-shaped distribution. Measures of kurtosis and skewness were developed and mathematical formulas that can be used to describe precisely the shapes of distributions. Measures of central tendency and variation were also discussed, and some of the reasons one might prefer one measure over another were presented. Typically, investigators would describe their data in terms of one index of variation, one index of central tendency, and measures of form (kurtosis, skewness, and modes). The specific measures used depend on the type of data and research interests.

Comparisons of individuals and between individuals and their group are facilitated by using standard scores, and, if the curve happens to be a normal curve, tables of areas under the normal curve permit one to convert a z-score readily into an expression about the proportion of cases in a group falling below it. In any case, measures presented in this chapter are highly useful in making valid comparisons between individuals, groups, or outside standards as dictated by one's problem.

Thus, looking back at the comparison of ages at which white and black women first wed (Figure 5.1), we can describe the distribution for black women as bimodal (with a primary and a secondary mode), as compared to a unimodal distribution for white women. However, the central tendency (as represented by the mode) was two years higher, at 20, for white women than it was for black women (whose primary mode was 18). Furthermore, the distribution for white women was more variable and more positively skewed than the one for black women, while the distribution for black women was more leptokurtic. Numeric indices for all these features would have greatly aided our overall comparisons and would have helped considerably in drawing clearer conclusions about the differences between these distributions.

The next section of this book also deals with description, but instead of describing and contrasting univariate distributions, it will confront the problem of describing sets of distributions where each individual is simultaneously characterized by scores on two different variables. This is called a bivariate distribution. Although descriptive tools vary from those we have covered, the essential principles remain: the selective reduction of detail with more pointed and powerful summary measures; the utility of the idea of a ratio; the description of location and variation in a dependent variable across subgroups defined in terms of the second variable; and, finally, the importance of identifying clearly the contrasts and comparisons one wants to make.

CONCEPTS TO KNOW AND UNDERSTAND

three features of a distribution
  central tendency (location)
  variation (dispersion)
  form
kurtosis
  platykurtic
  mesokurtic
  leptokurtic
unimodal, bimodal, multimodal
median
other specialized means
range
interquartile range
average absolute deviation
variance, standard deviation

index of dispersion
moving average
skewness
  negative
  positive
  symmetry
arithmetic mean
mode
standard scores (z-scores)
standard normal distribution
area under a normal curve
  (Appendix Table B)
central moment system
asymptosis

QUESTIONS AND PROBLEMS

1. Select data on an interesting variable from U.S. Census sources available to you or from a computer data set. Compute appropriate measures of central tendency, variation, and form. Justify your selections and explain in your own words what each of the measures means.

2. Sometimes it is helpful at this point to set up a chart that organizes the main differences between the measures included in this chapter. Computing formulas and interpretations could be entered on such a chart for each measure. See if you can set up a chart that helps you select and interpret measures.

3. Sociology often deals with the idea of "normal" or "average" and with the idea of similarity and difference. Find one research article in a sociology journal that uses statistical measures of central tendency and variation to discuss these

concepts. Explain how the authors use and interpret these measures in their work.

4. Using the scores from Problem 4 or 5 of Chapter 3, compute the arithmetic mean, the median, the mode, and the standard deviation. What do the three measures of central tendency tell you about the form of the distribution? Does the standard deviation adequately represent the variability of the distribution? If so, why? If not, why not? Compute the interquartile range. How does it compare with the standard deviation as a measure of variability for this distribution. Create a box and whisker graph of these data.

5. The accompanying table shows annual international suicide rates (per 100,000 population) for 33 countries. Compute the appropriate measures of central tendency and variability for these data. What do these measures tell you about the central tendency, dispersion, and form of the distribution?

ANNUAL INTERNATIONAL SUICIDE RATES (per 100,000 population)

Nation	Rate	Nation	Rate
Hungary	36.9	Australia	11.6
E. Germany	30.5	Bulgaria	11.6
Czechoslovakia	24.7	Singapore	10.9
Finland	24.0	Hong Kong	10.6
Denmark	23.8	New Zealand	9.0
Austria	21.9	Norway	9.0
Sweden	20.3	Portugal	8.6
W. Germany	19.9	Scotland	8.4
Switzerland	19.2	Netherlands	8.2
France	15.4	England & Wales	7.7
Belgium	15.4	Israel (Jews only)	6.8
Yugoslavia	13.8	Italy	5.8
Luxembourg	13.4	N. Ireland	4.5
Iceland	13.2	Spain	4.4
Canada	12.2	Greece	3.0
Poland	11.7	Ireland	3.0
United States	11.7		

Source: World Health Organization 1975 (data for 1970, 1971, 1972, or 1973).

6. Using the formula for a standard score and Table B in the Appendix, find how many of the suicide rates in Problem 5 would be expected to be as high as or higher than 19.2 if the distribution of suicide rates were normal. How many in the actual distribution of scores are as high as or higher than 19.2? How closely do the two numbers correspond? Now find out how many suicide rates would be expected to be as low as or lower than 6.8 if the distribution were normal. How many in the actual distribution are as low as or lower than 6.8? How closely do the two numbers correspond? What can you conclude about the distribution from the analysis you have just performed?

GENERAL REFERENCES

Hagood, Margaret Jarman, and Daniel O. Price, *Statistics for Sociologists*, rev. ed. (New York, Henry Holt and Company), 1952.
See especially Chapters 9 and 14.

LITERATURE CITED

Adams, Stuart, "Trends in Occupational Origins of Physicians," *American Sociological Review*, 18 (1953), pp. 404–409.

Hammond, Kenneth R., and James E. Householder, *Introduction to the Statistical Method* (New York, Knopf), 1962, pp. 136–142.

Hawkes, Roland K., "Multivariate Analysis of Ordinal Measures," *American Journal of Sociology*, 76 (March 1971), pp. 908–926.

Mueller, John H., Karl F. Schuessler, and Herbert L. Costner, *Statistical Reasoning in Sociology* (Boston, Houghton Mifflin), 1977.

National Center for Health Statistics, C. A. Schoenborn, and B. H. Cohen, "Trends in Smoking, Alcohol Consumption, and Other Health Practices among U.S. Adults, 1977 and 1983," advance data from *Vital and Health Statistics*, No. 118, DHHS Publ. No. (PHS) 86–1250 (Hyattsville, Md., Public Health Service), June 30, 1986.

Newman, James R., *The World of Mathematics*, Vol. 3 (New York, Simon and Schuster), 1956, pp. 1489–1493.

Reiss, Ira L., *The Social Context of Premarital Sexual Permissiveness* (New York, Holt, Rinehart and Winston), 1967, p. 36.

Rushing, William A., and Vernon Davies, "Note on the Mathematical Formalization of a Measure of Division of Labor," *Social Forces*, 48 (March 1970), pp. 394–396.

Senders, Virginia L., *Measurement and Statistics* (New York, Oxford University Press), 1958.

U.S. Bureau of the Census, "Estimates of the Population of the United States and Components of Change: 1940 to 1971," *Current Population Reports*, Series P–25, No. 465 (Washington, D.C., U.S. Government Printing Office), 1971.

————, "Estimates of the Population of the United States to May 1, 1984," *Current Population Reports*, Series P–25, No. 954 (Washington, D.C., U.S. Government Printing Office), 1984.(a)

————, "Money Income and Poverty Status of Families and Persons in the United States: 1983," *Current Population Reports*, Series P–60, No. 145 (Washington, D.C., U.S. Government Printing Office), 1984. (b)

————, *Statistical Abstract of the United States: 1986* (106th ed.), (Washington, D.C., U.S. Government Printing Office), 1985.

Wallis, W. Allen, and Harry V. Roberts, *Statistics: A New Approach* (New York, The Free Press of Glencoe), 1956, pp. 226–230.

World Health Organization, *World Health Statistics Annual* (GENEVA, 1975).

DIAGRAM 5.1 Measures of Central Tendency for Grouped Scores

MODE (Mo)

The mode in the case of grouped distributions is the midpoint of the class that has the highest frequency of cases in it.

MEDIAN (MDN)

Step 1: Create a cumulative frequency distribution, starting with the lowest score category as shown in the illustration below.

Scores	f	cumf	Real Limits	Class Width
22–26	18	88	21.5–26.5	5
17–21	21	70	16.5–21.5	5
12–16	26	49	11.5–16.5	5
7–11	15	23	6.5–11.5	5
2–6	8	8	1.5–6.5	5
	$N = 88$			

Step 2: Find $N/2$, the *number of cases* that fall below the median score. Here this is 44 cases (88/2).

Step 3: Using the cumulative frequency distribution of step 1 and the results of step 2, find the category within which the median falls. This is the category with a cumulative frequency equal to or greater than (but closest to) $N/2$.

In this example, the class 12–16 is the median category because the 44th case falls in that class. The "cum$f$" of 49 is just bigger than $N/2 = 44$.

Step 4: If the cumulative frequency in a class exactly equals $N/2$, then the upper class boundary is the value of the median.

If the median falls someplace within a category, as is usually the case, it is traditional to "interpolate" to find a median value within the median class on the assumption that cases are distributed in an even (rectangular) fashion within the median interval.

The logic is that we want to locate a median score within the median class, which is a certain distance into that class. The distance depends upon the proportion of the frequency in the median class that needs to be added to the cumulative frequency below the median class in order to equal $N/2$, or the number of cases that should fall below the median score. This proportion is found by asking how many additional cases are needed:

$$\frac{N}{2} - \text{cum}f_{mdn}$$

(where cumf_{mdn} is the cumulative frequency up to but *not* including the

DIAGRAM 5.1 *(Continued)*

frequency in the median category), or

$$\frac{88}{2} - 23 = 21$$

This is .81 or 81% of the cases in the median category (21 divided by 26, which is the number of cases in the median category itself, f_{mdn}). Since the median class is 5 score units wide, 81% of the way through that width is

$$.81 \times 5 = 4.0$$

Adding these 4.0 points to the lower real limit of the median class yields a median of 15.5,

$$11.5 + 4.0 = 15.5$$

a score below which 50%, or 44, of the 88 cases should fall. These relationships are shown in the "offset" bar-chart diagram.

In practice, the formula for computing the median of grouped data is used directly. It is a specific case of percentiles (discussed in Section 3.2.3d).

$$\text{Mdn} = L_{mdn} + \left(\frac{\frac{1}{2}N - \text{cum}f_{mdn}}{f_{mdn}}\right)w$$

where L_{mdn} is the lower limit of the median class; N is the number of cases; $\text{cum}f_{mdn}$ is the cumulative frequency up to *but not including* the frequency in the median class; f_{mdn} is the frequency in the median class; and w is the width of the median class interval.

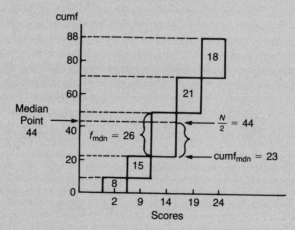

Graphically, the median could be found from a percentage ogive (plot of a cumulative percentage distribution) by extending a line from the 50%

DIAGRAM 5.1 *(Continued)*

point to the ogive and dropping down to the score axis to read off the 50th percentile or median score:

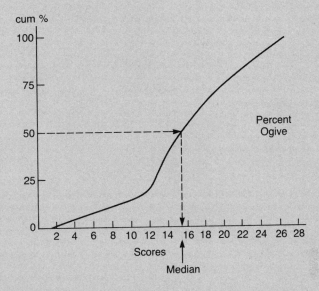

ARITHMETIC MEAN ($\bar{X}$)

Step 1: Create a column of class midpoints. Scores in a class will be treated as if they have the value of the midpoint of a class since no other information on their specific value is known.

Scores	X_i	f_i	$f_i(X_i)$
22–26	24	18	432
17–21	19	21	399
12–16	14	26	364
7–11	9	15	135
2–6	4	8	32
		$N = 88$	1362

Step 2: Multiply, for each class, the class midpoint times the frequency in that class, $f_i(X_i)$.

Step 3: Sum the products in step 2, $\Sigma f_i X_i$, and compute the mean by dividing by N.

$$\frac{\Sigma f_i X_i}{N}$$

The mean is thus 15.5, since $1362/88 = 15.5$.

DIAGRAM 5.2   Computing Measures of Variation and Form for Grouped Scores

As is true of central tendency, measures of variation and form may be computed on grouped frequency distributions. These procedures are illustrated below using the set of scores on number of items correct on a 20-item exam in arithmetic. The 32 scores are shown in a grouped distribution.

RANGE (FORMULA 5.4)

$$Range = U - L$$

where U is the upper real limit of the highest class, and L is the lower real limit of the lowest class.

$$Range = 17.5 - 1.5 = 16$$

INTERQUARTILE RANGE (FORMULA 5.5)

$$Q = Q_3 - Q_1$$

where Q_3 is the *score value* below which 75% of the cases fall, and Q_1 is the score value below which 25% fall.

Quartiles are found in the same way that any percentile is found. This is discussed with respect to the median (the second quartile) in Diagram 5.1 and Diagram 3.2.

(D1)
$$Q_3 = L_{Q_3} + \left(\frac{\frac{3}{4}N - \mathrm{cum}f_{Q_3}}{f_{Q_3}}\right)w$$

(D2)
$$Q_1 = L_{Q_1} + \left(\frac{\frac{1}{4}N - \mathrm{cum}f_{Q_1}}{f_{Q_1}}\right)w$$

where L_{Q_3} and L_{Q_1} are lower limits of the classes containing the third and first quartiles; N is the number of cases; w is the width of the class containing the quartile of interest; $\mathrm{cum}f_{Q_3}$ and $\mathrm{cum}f_{Q_1}$ are cumulative frequencies of cases *up to but not including* cases in the class containing the quartile being computed; f_{Q_3} and f_{Q_1} are frequency of cases in the quartile class being computed. In these data:

Grouped Distribution

	Class	X_i Midpoint	f_i	$\mathrm{cum}f_i$	f_iX_i	$f_iX_i^2$
	2–3	2.5	2	2	5.0	12.50
	4–5	4.5	3	5	13.5	60.75
Class containing Q_1 →	6–7	6.5	5	10	32.5	211.25
	8–9	8.5	9	19	76.5	650.25
Class containing Q_3 →	10–11	10.5	7	26	73.5	771.75
	12–13	12.5	4	30	50.0	625.00
	14–15	14.5	1	31	14.5	210.25
	16–17	16.5	1	32	16.5	272.25
	Total		$\Sigma f_i = 32$		$\Sigma f_iX_i = 282.0$	$\Sigma f_iX_i^2 = 2{,}814.00$

DIAGRAM 5.2 *(Continued)*

Step 1: Find the class containing the quartile of interest by using a cumulative frequency distribution and finding the class below which the $3N/4$th case would fall for the third quartile, or the $N/4$th case for the first quartile. For the first quartile that class is the 6–7 score class (since one-fourth of the 32 scores is 8, and the 8th score falls in the 6–7 class). The third quartile is the 10–11 score class because it contains the 24th score out of 32.

Step 2: Substitute values in the formulas for Q_3 and Q_1 and compute the estimated third and first quartiles.

$$Q_3 = 9.5 + \left(\frac{\frac{3}{4}(32) - 19}{7} \right)(2) = 10.9$$

$$Q_1 = 5.5 + \left(\frac{\frac{1}{4}(32) - 5}{5} \right)(2) = 6.7$$

Step 3: Compute the interquartile range.

$$Q = 10.9 - 6.7 = 4.2$$

A similar procedure could be used to compute other percentiles (or other "fractiles"), and other ranges, such as the interdecile range, could be used.

Variance and Standard Deviation (Formulas 5.7 and 5.8)

The definitional formulas for variance and standard deviation are as follows:

$$s^2 = \frac{\Sigma(X_i - \bar{X})^2}{N} \quad \text{variance* (Formula 5.7)}$$

$$s = \sqrt{s^2} \quad \text{standard deviation (Formula 5.8)}$$

These formulas are rarely used in computational work because they require the extra step of computing the arithmetic mean and finding the deviation of each score from the mean—steps that may result in extensive rounding error. The computational formulas given below make use of category midpoints and frequencies.

---

* See the first two footnotes in Section 5.4.4 in this chapter. The variance given here is a "biased" estimate of a population variance. Where inferences to population variances on the basis of sample variances is of interest, the unbiased sample variance is used. It simply substitutes $N - 1$ in place of N in the denominator of the variance. Where N is large, this makes little numeric difference, but the resulting variance is referred to as "unbiased." This topic is explored in the field of inferential statistics (see Chapter 12). Here we will use the somewhat simpler formula given above.

DIAGRAM 5.2 *(Continued)*

The formula for the variance is:

(D3)

$$s^2 = \frac{\Sigma f_i X_i^2 - (\Sigma f_i X_i)^2 / N}{N}$$

where $\Sigma f_i X_i^2$ is the square of the category midpoint times the category frequency summed up over all categories, and $(\Sigma f_i X_i)^2$ is the sum of category midpoint and category frequency products, squared.

In this example, these two sums can be found by creating two columns of products as shown above. The result is

$$s^2 = \frac{2814 - (282)^2 / 32}{32} = \frac{328.88}{32} = 10.3$$

$$s = \sqrt{10.3} = 3.2$$

INDEX OF DISPERSION (FORMULA 5.9)

$$D = \frac{k(N^2 - \Sigma f_i^2)}{N^2(k - 1)}$$

where k is the number of categories of the variable; Σf_i^2 is the sum of squared frequencies; and N is the number of cases.

In the example below, the scores for this ordinal variable would be grouped into a frequency (or percentage) distribution prior to computing D, and a new column of f_i^2 values would be created and summed as follows:

Satisfaction with Work	f_i	f_i^2
1 very satisfied	2	4
2 .	7	49
3 .	3	9
4 .	9	81
5 .	5	25
6 very dissatisfied	4	16
	30	184

D is then computed as

$$D = \frac{6(30^2 - 184)}{30^2(6 - 1)} = \frac{4296}{4500} = .95$$

Part III
DESCRIPTIVE STATISTICS:
TWO VARIABLES

CHAPTER 6

Cross-Classification
of Variables

Sociologists concerned with urban life styles have theorized that people living in larger cities will have greater tolerance of racial and ethnic differences than people living in smaller cities or rural areas. This is thought to be due to a number of features of the urban environment such as the greater impersonality, greater variety of contacts between people, greater concern for utility of others, and the greater use of universalistic standards for making judgments and decisions. Does the evidence support this prediction? To examine this question, Fischer (1971) used data from five Gallup polls taken between 1958 and 1965 in the United States.

6.1 BIVARIATE DISTRIBUTIONS: AN EXAMPLE

In these polls, the question was asked, "If your party nominated a generally well-qualified man for president and he happened to be a (Negro or Jew or Catholic), would you vote for him?" Tolerance was measured by counting the number of these three minority groups for which the respondent would agree to vote (*i.e.*, none, one, two, or all three). Table 6.1 presents the overall frequency and percentage distribution of tolerance scores for the 7,714 cases in the combined polls.

It is interesting to note that about 41% of the respondents would vote for all three minority groups—that is, they show high tolerance for ethnic and racial minorities as measured by this question.* This overall distribution does not permit one to say whether the theory is supported, however, simply because it does

---

* It is important to distinguish between tolerance views expressed in polls and behavior indicative of tolerance, which may be expressed in the privacy of an election booth or in other situations.

TABLE 6.1 TOLERANCE SCALE FOR 7,714 RESPONDENTS OF FIVE GALLUP POLLS TAKEN BETWEEN 1958 AND 1965 IN NATIONAL SAMPLES OF THE UNITED STATES

Tolerance Level	f	%
3 (High)	3,147	40.8
2	2,317	30.0
1	1,229	15.9
0 (Low)	1,021	13.2
Totals	7,714	99.9%

Source: Fischer (1971:849). Reprinted from Fischer, Claude S., "A Research Note on Urbanism and Tolerance," *American Journal of Sociology,* 76 (March, 1971) p 847–856, by permission of the University of Chicago Press. Copyright 1971 by the University of Chicago and Published by the University of Chicago Press.

not provide necessary information to make a comparison between cities of different sizes. We need to know whether tolerance scores are distributed differently for people residing in places of different sizes, and if they are, we also need to see whether the percentage of respondents in the high-tolerance categories of the tolerance scale is higher in larger cities compared to smaller ones. We need to compare this univariate distribution of tolerance scores for small, medium, and large cities to see if the differences are in the predicted direction.

Table 6.2 provides the detail we need. Here four separate univariate distributions (both frequency and percentage) are shown: one for those respondents who live in rural areas, one for small places, one for medium places, and one for large cities. In each case, however, we are looking at the distribution of the dependent variable, tolerance scores.* Comparing the "high-tolerance" end of the scale of tolerance, for example, we see that the predictions seem to be borne out: 26.2% have "high tolerance" among rural and country people; this percentage increases to 36.0% for people living in places of under 25,000 population; it increases again to 43.4% for people living in the next larger category of cities; and it reaches 55.5% among people living in the largest cities. A comparison of the "low-tolerance" category reflects this trend also. Smaller places have a larger percentage of people with "low tolerance," and the percentages decrease as city size increases. Thus far, then, it seems to be the case that the theory holds for these data.† Notice that we arrived at the conclusion by comparing percentages rather than absolute frequencies, since it is clear that the number of respondents differs between tables we want to compare. Chapter 3 discussed this use of

* Recall from Chapter 2 that a dependent variable is either the primary variable in whose variation one is interested or the effect or outcome of variation in an independent or causal variable. The roles variables play in a study depends upon the investigator's theoretical reasoning about what influences what.
† Actually, Fischer presents a more exciting analysis, tracing out differences in the distribution of tolerance by such other variables as occupation, race, region, and religion. These appear to be important conditions that govern the distribution of tolerance, and this leads to important questions about the explanatory importance of city size alone.

TABLE 6.2 TOLERANCE SCALES FOR RESPONDENTS IN TABLE 6.1

A. *Tolerance Scale for Farm or Country Residents*

Tolerance Level	f	%
High	597	26.2
2	700	30.7
1	497	21.8
Low	485	21.3
Totals	2,279	100.0%

B. *Tolerance Scale for Subjects Living in Towns Under 25,000*

Tolerance Level	f	%
High	483	36.0
2	396	29.5
1	272	20.3
Low	191	14.2
Total	1,342	100.0%

C. *Tolerance Scale for Subjects Living in Places 25,000 to 500,000*

Tolerance Level	f	%
High	739	43.4
2	536	31.5
1	241	14.2
Low	185	10.9
Total	1,701	100.0%

D. *Tolerance Scale for Subjects Living in Cities over 500,000*

Tolerance Level	f	%
High	1,328	55.5
2	685	28.6
1	219	9.2
Low	160	6.7
Totals	2,392	100.0%

Source: Fischer (1971:849, Table 2). Reprinted from Fischer, Claude S., "A Research Note on Urbanism and Tolerance," *American Journal of Sociology*, 76 (March, 1971) p 847–856, by permission of the University of Chicago Press. Copyright 1971 by the University of Chicago and Published by the University of Chicago Press.

percentages as one means by which valid comparisons could be made in just such circumstances.*

Up to this point we have discussed the problem of comparing tolerance scores for each category of city size as a comparison of several univariate distributions. The dependent variable is common to each of the distributions, and separate tables are distinguished in terms of categories of the independent variable, size of place. It happens to have four categories, and thus there are four separate distributions of the dependent variable to examine. A more efficient way to reach conclusions under these conditions would be to combine all of the separate tables (*i.e.*, Tables 6.1, 6.2A, 6.2B, 6.2C, and 6.2D) into one overall table, as shown in Table 6.3. This table permits us to compare the separate groups more readily, and it leads to a type of summary of the whole table in terms of a relationship between the two variables in general. Table 6.3 is known as a **bivariate percentage distribution** because it permits one to examine the percentage distribution of one variable (the dependent variable) within the different categories

* We might also have made the comparisons from table to table in terms of some set of measures of central tendency, form, and variation, as discussed in Chapter 5. Certainly, a comparison of medians, for example, would have reduced some of the detail involved in comparing all of the percentages shown in Table 6.2.

TABLE 6.3 PERCENTAGE DISTRIBUTION OF TOLERANCE SCORES FOR RESPONDENTS IN A COMBINED SAMPLE CONSISTING OF FIVE GALLUP POLLS OF U.S. NATIONAL SAMPLES CONDUCTED BETWEEN 1958 AND 1965, BY SIZE OF PLACE OF RESIDENCE

| Tolerance Level | Size of Place of Residence | | | | Total |
	Farm & Country	Under 25,000	25,000 to 500,000	Over 500,000	
High	26.2%	36.0%	43.4%	55.5%	40.8%
2	30.7	29.5	31.5	28.6	30.0
1	21.8	20.3	14.2	9.2	15.9
Low	21.3	14.2	10.9	6.7	13.2
Totals	100.0%	100.0%	100.0%	100.0%	99.9%
	(2279)	(1342)	(1701)	(2392)	(7714)

Source: Fischer (1971). Reprinted from Fischer, Claude S., "A Research Note on Urbanism and Tolerance," *American Journal of Sociology*, 76 (March, 1971), p 847–856, by permission of the University of Chicago Press. Copyright 1971 by the University of Chicago and Published by the University of Chicago Press.

of the other variable. The comparison ideas behind such cross-classifications are the basis of analysis in attempts to develop theoretical statements about the relationship of variables and the conditions under which these occur.

6.2 CONDITIONAL DISTRIBUTIONS

A bivariate distribution, such as that shown in Table 6.3, permits one to examine not only the overall distribution of some dependent variable, but also some of the conditions that may be thought to influence how that variable is distributed. The theory suggested that under certain conditions tolerance would be higher than under other conditions. In Table 6.3 these conditions corresponded to different sizes of place of residence, but one could think of other conditions that may have a bearing upon the level of tolerance. A table merely puts together a related set of **conditional distributions** and an overall total distribution of some dependent variable.

Our objective in this chapter is to explore some of the characteristics of bivariate distributions or cross-classifications of two variables. The following chapter will discuss some of the more useful index numbers that can be used to summarize variables that are related to each other. This is a road we have already traveled once before in previous chapters. In univariate descriptive statistics we started with a set of raw scores or a distribution and then asked if there were ways it could be summarized in terms of a few overall index numbers. We developed several measures for each of the three features of a distribution: central tendency, form, and variation. Our objective in bivariate description is quite similar, and, in many respects, simpler, more interesting, and more useful in sociological inquiry.

6.3   HOW TO SET UP AND EXAMINE TABLES

A cross-classification of two variables requires a table with rows and columns. The categories of one variable are labels for the rows, and the categories of the other variable are labels for the columns. Usually, where there is a dependent variable, it is used as the row variable, and the independent variable is used as the column variable, but this tradition is sometimes broken.

6.3.1   Creating a Bivariate Frequency Distribution

To illustrate how a table is constructed, Table 6.4 shows a tolerance score and a gender score for each of 13 fictitious individuals. The bivariate distribution is set up as shown, with the categories of tolerance (here two categories, high and low) down the **stub** or side of the table and the categories of gender across the **heading** at the top.

TABLE 6.4

A. Frequency Table

Score Key	Case I.D.	Tolerance Score	Gender Score
M = male	A	L	M
F = female	B	L	M
H = high	C	H	M
L = low	D	H	M
	E	H	F
	F	L	F
	G	H	F
	H	H	M
	I	L	F
	J	L	M
	K	H	F
	L	H	F
	M	H	F

B. Table of Tolerance Level by Gender

Tolerance Level	Gender		Row Totals
	Male	Female	
High	||| 3	|||| 5	8
Low	||| 3	|| 2	5
Column Totals	—	—	—
	6	7	13

Source: Fictitious data.

This table is a **2 by 2 table**, or **fourfold table**, because it has two rows and two columns (or four cells in the **body** of the table where the rows and columns intersect). Tables can, of course, have any number of rows and columns (in general we refer to an r by c table, where r refers to the number of rows and c to the number of columns); r and c depend upon the number of categories that are distinguished for row and column variables.

The problem now is to count the number of cases that have various possible combinations of values on the two variables and to enter these totals into the table to form a bivariate frequency distribution. Notice that 3 of the 13 cases in this sample are males who are "high" on tolerance, 3 are males who are "low" on tolerance, 5 are females who are "high" on tolerance, and 2 are females who are "low" on tolerance. These numbers are written in the boxes or cells in the body of the table corresponding to the appropriate row and column labels shown in Table 6.4. Sometimes it is helpful to create a tally within each cell as a workmanlike way to assure accuracy.

Each of the boxes in the table is called a **cell**, and the frequency in a cell is called a **cell frequency**. Cell frequencies are sometimes symbolized by the small letter n_{ij}, where the first subscript (i) indicates the number of the row and the second subscript (j) indicates the number of the column, as follows:

	Column 1	*Column 2*	*Row Totals*
Row 1	n_{11}	n_{12}	$\sum_j n_{1j}$
Row 2	n_{21}	n_{22}	$\sum_j n_{2j}$
Column totals	$\sum_i n_{i1}$	$\sum_i n_{i2}$	N

They indicate the number of cases in the total sample that fall in a certain category of the row and column variables as indicated by the row and column labels.* The cell frequencies indicate the number of cases with two characteristics simultaneously. Row and column totals each add up to 13, the total number of cases there were. The cell frequencies constitute the conditional distributions, and the row and column totals reflect the marginals or univariate distribution of each variable.

6.3.2 Traditions of Table Layout

As mentioned above (Section 6.3), tables usually are set up so that the dependent variable is the one with categories listed down the *stub*, or left side of the table, and the independent variable is listed across the top in the *heading*. This convention, of

---

* Some authors use a different set of symbols for row and column totals, where a dot is used in place of a subscript for totals. Thus $n_{.1}$ would be the sum of column 1 over all of the rows in the table (the row subscript is replaced by a dot) instead of $\sum_i n_{i1}$, and $n_{1.}$ would symbolize the total of row one, instead of $\sum_j n_{1j}$. One could use n rather than N or $\Sigma\Sigma\, n_{ij}$ to indicate the grand total number of cases.

course, is not always kept, but it does tend to aid the examination of conditional distributions in each column to have it set up this way. Table 6.3 illustrates proper labeling of a table. Notice that low categories of the independent variable, where there are low categories on that variable, are listed at the left and the high categories at the right. For the dependent variable the high categories are at the top of the table and the low categories are at the bottom. This is similar to the labeling of other graphs, although in the case of tables the convention is not as rigidly adhered to, and the investigator would do well to double-check the table layout before proceeding to make any interpretation.

A table usually has a title that lists the dependent variable, whether the table contains frequencies or percentages (or some other measure), the independent variable(s), and the kind of case upon which the measurements were taken. Table 6.3 contains data on 7,714 individuals. If the table is a percentaged table, it is important to indicate the base upon which the percentage was computed in brackets, at the bottom by the column total percentages*; when this is done, cell frequencies may be omitted from a percentaged table. The source of data is indicated, typically, in a footnote to the table, and both the stub and heading are clearly marked with the variable and the name of each of the categories of each variable.

Table 6.4 is a frequency table. It has categories of the tolerance variable down the side (the stub), and categories of the variable, gender, across the top. In this case, tolerance played the role of dependent variable.

Notice that cell frequencies in columns are summed, and the sums are put at the bottom of the table. Rows are also summed, and the totals put at the right side. These row and column totals are called **marginals**, or simply row totals and column totals, and they are merely the univariate distribution of each variable separately.

If the table shows percentages it is called a bivariate percentage distribution, and if frequencies are shown, it is called a bivariate frequency distribution.

6.3.3   Percentaged Tables

Probably the most often used type of table is the percentaged table. Its value lies in the way it helps one to make comparisons across the conditional distributions one wants to compare. The basic rule for computing percentages in a table is as follows:

Compute percentages in the direction of the independent variable.

This means that percentages should sum up to 100% for each category of the independent variable. For tables set up such as Table 6.3, the percentaging rule leads to computation with column totals as the base of the percentage: thus column percentages add up to 100% for each column. If the independent variable and the dependent variable were switched around, the percentages would have to be run in the other direction. There are three ways that a table can be percentaged,

---

* This is true if column totals are the bases of percentages. If rows sum to 100%, then row total frequencies are given.

TABLE 6.5 Illustration of Different Ways Percentages Can Be Computed on Tables

Original Frequency Distribution from Table 6.4

	Gender		
Tolerance Level	Male	Female	Total
High	3	5	8
Low	3	2	5
Total	6	7	13

A. *Percentaging to Column Totals as the Base*

	Gender		
Tolerance Level	Male	Female	Total
High	50%	71%	62%
Low	50	29	38
Total	100%	100%	100%

B. *Percentaging to Row Totals as the Base*

	Gender		
Tolerance Level	Male	Female	Total
High	38%	62	100%
Low	60%	40	100%
Total	46%	54	100%

C. *Percentaging to Overall Grand Total as the Base*

	Gender		
Tolerance Level	Male	Female	Total
High	23%	39%	62%
Low	23%	15%	38
Total	46%	54	100%

as shown in Table 6.5, using the hypothetical data from Table 6.4. Tables could be percentaged with *column totals* as the base of percentages, with *row totals* as the base of percentages, and with the *grand total* as the base of percentages. Since the dependent variable is down the stub of Table 6.4, the proper table to examine to see what differences there may be between categories of the gender variable would be that with column totals (the number of males or the number of females) as the base of the percentages. One wants to contrast the distribution of the dependent variable between men and women, and the only way to do this is to take out the effect of different numbers of men and women by percentaging down (in the direction of the independent variable). This type of operation permits one to make comparisons in the *other direction*. **Comparisons are made in the opposite direction from the way percentages are run.**

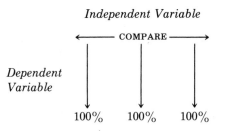

Comparisons are made in a percentaged table by examining differences between percentages. In Table 6.5A, for example, the difference between percentage "high" on tolerance among men and women is 21% (71% − 50% = 21%). This value is called **epsilon**, the percentage difference in a table, symbolized by the Greek letter ϵ. For tables larger than a 2 by 2 table, there are a number of percentage contrasts or epsilons that may be computed and used in interpretation. Epsilon will be discussed further later on in this chapter.

Sometimes an investigator will compute percentages, as in Table 6.5C, with the total number of cases (*N*) as *the base for all cell percentages*. Where this is done, we no longer can compare conditional distributions, but we can express the percentage of cases that have each of the different combinations of characteristics labeled by the rows and columns.

If it is not clear which variable is dependent or independent, or if we could think of the data in both ways, we might compute percentages to *both row and column totals* (as in Tables 6.5A and 6.5B) and *examine each table*. Table 6.5A would permit us to say that females are more likely to be high on tolerance than are males. Table 6.5B would permit us to say that high-tolerance people are more likely to be female than are low-tolerance people—a subtle shift with worlds of import, as we shall soon see.

As shown in Table 6.5, percentaging *down* permits an examination of any influence gender may have on the distribution of tolerance; percentaging *across* shows the possible recruitment pattern into tolerance levels from each gender, and percentaging to the *grand total* permits us to examine the joint percentage distribution of tolerance levels and gender.

6.3.3a   Effect of Sampling Decision on Percentaging Tables

Sometimes investigators will decide to draw a sample in such a way that there are, say, equal numbers of cases in each study category or condition they want to examine. This means, of course, that the marginals for that variable do not necessarily reflect or represent the way that variable is distributed in the population from which the sample was drawn. This happens, for example, in situations where the phenomenon being studied is quite rare. Only a small percentage of the population of the United States is classified as alcoholic, and it would be easy to draw a sample and completely miss this kind of person. If one were interested in comparing alcoholics and nonalcoholics on some characteristic, the whole study would collapse if alcoholics were not sampled in sufficient numbers to analyze. To protect themselves, investigators may sample in a way that is called disporportionate *stratified sampling* (sampling techniques are introduced and discussed in Chapter 11). That is, the investigator may divide a population list up into alcoholic and nonalcoholic and deliberately draw an equal (or specified) number of cases from each condition for the sample.

This procedure results in a distribution of alcoholism that may not reflect the way that variable is distributed in the population as a whole (*i.e.*, alcoholics would be overrepresented in the sample). This procedure would mean that alcoholism would be an "unrepresentative" variable. Because of this essentially arbitrary changing of the distribution of a variable in a sample, we must follow procedures that take this into account; that is, the table must be percentaged so that the category totals of the unrepresentative variable are used as the base of percentages. This is called **percentaging in the direction of the nonrepresentative variable**. If the nonrepresentative factor happens also to be an independent variable, then the percentages we have to examine are indeed the percentages we need to use in drawing conclusions. If, however, we stratify or make nonrepresentative in some way a dependent variable, then the percentages have to be run the "wrong" way for this analysis, and we are unable to make the contrasts we need to make.

6.3.3b Illustration of Percentaged Tables

Table 6.6 provides data on the bivariate distribution of income and ethnicity for families in the United States in 1983. The table clearly shows differences in the distribution of income for the different ethnic groups. Whites are less likely to be in the lowest income categories as compared to those of Spanish origin or blacks. Epsilon, the difference between percentages, is 11.5% between whites and those of Spanish origin and 19.0% between whites and blacks. This table would be quite useful to answer the question: What is the consequence of ethnicity for the distribution of income? Had the percentages been run the other way, we could

TABLE 6.6 PERCENTAGE DISTRIBUTION OF TOTAL MONEY INCOME FOR FAMILIES BY ETHNIC BACKGROUND OF HOUSEHOLDER, 1983

Income	White	Spanish	Black	Total
$35,000 and over	31.5	15.1	13.2	29.6
$25,000 to 34,999	20.3	14.3	14.0	19.5
$15,000 to 24,999	23.7	27.0	21.5	23.4
$7,500 to 14,999	16.0	23.6	23.7	16.8
Under $7,500	8.5	20.0	27.6	10.7
Total	100.0	100.0	100.0	100.0
N (millions)	(53.9)	(3.6)	(6.7)	(62.0)

Source: U.S. Bureau of the Census (1985).

TABLE 6.7 Percentage Indicating Support for a Woman President in Gallup Polls in the United States between 1937 and 1969, by Sex

"If your party nominated a woman for President, would you vote for her if she were qualified for the job?"

	Men				Women			
Year	Would	Would Not	No Opinion	Total*	Would	Would Not	No Opinion	Total
1937	27%	69%	4%	100%	40%	57%	3%	100%
1945	29	58	13	100	37	51	12	100
1949	45	50	5	100	51	46	3	100
1955	47	48	5	100	57	40	3	100
1963	58	37	5	100	51	45	4	100
1967	61	34	5	100	53	44	3	100
1969	58	35	7	100	49	44	7	100

*Epsilons between Men and Women on Percentage Saying They **Would** Vote for a Woman for President*

Year	ε[†]
1937	+13%
1945	+ 8
1949	+ 6
1955	+10
1963	− 7
1967	− 8
1969	− 9

Source: Data from Erskine (1971:278).

* Total number of cases was not given in the source.

† Epsilons here are computed by subtracting the percentage who would vote for a woman president among men from that among women in any given year. Minus signs indicate that the male percentage is higher and plus signs indicate that the female percentage is higher.

answer the question: What is the consequence of different income levels for the ethnic distribution of the United States? The first question is probably much more sensible and useful to answer than the second, given the likely influence of the two variables on each other; thus percentaging down, as has been done in Table 6.6, would be the appropriate way to percentage this table. Notice that the last column has the overall percentage distribution of income for all families.

Table 6.7 provides a somewhat different example of percentaged tables that also highlights one of the problems involved in comparison of percentages. These data are from a summary of questions asked in a number of public opinion polls concerning the role of women. One of the questions asked over several years by the Gallup Poll is: "If your party nominated a woman for President, would you vote for her if she were qualified for the job?" The response categories are, basically, "yes" and "no," and the percentage giving these two responses out of the total number of men and women sampled is shown in the table.

Notice that for men and, to some extent, for women the percentage saying that they would vote for a woman increases over the years between 1937 and 1969. The percentage saying they would not vote for a woman also tends to decline. Using these data, we could answer the question: "What is the difference in percentage agreeing to vote for a woman from year to year?"

Notice, too, that the comparison is contaminated by the troublesome "no opinion" or "no response" category. Where percentages include these cases, the percentage comparisons of voting intention between one year and another are distorted because they also reflect differences in the level of "no opinion." This is particularly important to watch for in comparisons involving the year 1945, where the percentage of men and women giving no opinion increased a very great deal. Among men, there is little difference in the percentage who would vote for a woman between 1937 and 1945, but there is a big drop between these years in percentage saying they would not vote for a woman. Apparently, overall, there was a shift to "no opinion" from the "would not vote for a woman" category. Between 1945 and 1949, however, the percentage with "no opinion" drops back to a modest level, and there is a corresponding increase in the percentage who would vote for a woman. In this case, then, epsilons computed between years on one response category of the dependent variable (*e.g.*, willingness to vote for a woman as president) would be misleading without taking account at the same time of the shift in the "would not" and "no opinion" categories.* We need some descriptive device that will permit an overall examination of a table rather than only a contrast of individual pairs of percentages (although these, too, are interesting for some purposes).

BOX 6.1 TAKING STOCK

Thus far in Chapter 6 we have discussed the way in which bivariate distributions may be created (Sections 6.1, and 6.2, and 6.3.1) and the use of percentages to examine the distribution of a variable within categories of some other variable (Section 6.3.3). Percentaging rules that are useful in this process are: (a) percentage in the direction of the independent variable and compare in the other direction (Section 6.3.3); (b) percentage in the direction of the nonrepresentative variable, if any (Section 6.3.3a); and (c) percentage to the grand total to examine the joint percentage distribution (Section 6.3.3). These uses of percentages should be understood before you go on into some further applications of these rules under somewhat more detailed conditions, and before we discuss the measures of association that follow.

6.3.4 More Complex Conditional Distributions

It is often the case that investigators look at percentage tables that are a good deal more complex than those we have just been examining. That is, they may be

---

* Other procedures for handling the "no response" or "no opinion" categories in research of this sort are discussed by Davis and Jacobs (1968) (see General References).

interested in conditional distributions that are distinguished by more than one independent variable. We will have more systematic things to say about such tables later on in this book, but for now we should indicate how we might begin to approach the interpretation of such a table.

Often a table is set up so that the focus is on either the *dependent variable alone* under various kinds of conditions, or on some basic set of conditional distributions—*a basic table*—that is examined in another set of conditional situations.

Table 6.8 illustrates a more complex table of the first kind. Four variables, education of mother, education of father, I.Q. of the child, and gender of the child, define the conditions within which a student's plans regarding college can be examined. The dependent variable is, as the title suggests, the percentage *planning on going to college*. Actually, there could be two tables presented here instead of one. One table would show the percentage *planning* on college and the other would show the percentage *not planning* on college. Since these two percentages would total to 100%, it is sufficient to look at only one of the two possible tables. The other table would be a mirror image of the first. There is a relatively orderly way to go about reading such a table, and we will discuss these procedures here.

The **first step** in examining a table as complex as this one is to **identify clearly the dependent variable** and the kind of *unit* (*i.e.*, family, person, parent) upon which it has been measured. In this table it is "expressed plans to go to college," which is the dependent variable asked of individual people who are the school-aged sons and daughters of parents who have differing amounts of formal education.

The **second step** is to **identify the other variables** involved in the table and the categories of each. In the education variables, three categories are used in this table, and they are labeled "low," "medium," and "high." One education variable has to do with father's education and one has to do with mother's education. There is also an intelligence variable that has three values: "high," "middle," and "low." Finally, female students and male students are distinguished.

Third, the structure of the table should be clarified. In the case of Table 6.8, categories of gender are reflected in the left two strips of percentages down the table. The right strip down the table shows percentages for all cases, regardless of gender, in the total sample. Father's education appears in rows of the table; three rows for father's education within the "high intelligence" category of students, three in the "middle intelligence" category, and three within the "low intelligence" category. The bottom general row of numbers combines together students of all intelligence categories, but distinguishes (in rows, again) the father's education. Mother's education is indicated across the top of the two gender and total strips down the table, and there are "total" columns that summarize over father's education or over mother's education, etc. In Table 6.8, rows of figures around the bottom and right sides of the table indicate totals, and more detailed breakdowns are concentrated in the rows and columns in the middle and upper-left areas of the table. Having identified the structure of the table, we can examine percentages.

The *percentage* 89.3% in the upper-left corner of the table means that 89.3% of students plan on going to college among those students who are (a) male, (b) of high intelligence, (c) have a father with "high" education, and (d) have a

TABLE 6.8 PERCENTAGE WHO PLANNED ON COLLEGE BY SEX, INTELLIGENCE, FATHER'S EDUCATION, AND MOTHER'S EDUCATION*

Father's Education	Males Mother's Education				Females Mother's Education				Total Mother's Education			
	High	Middle	Low	Total	High	Middle	Low	Total	High	Middle	Low	Total
a. High Intelligence												
High	89.3(186)	87.8(115)	55.6(36)	85.2(337)	86.0(186)	74.5(102)	68.6(35)	80.5(323)	87.6(372)	81.6(217)	62.0(71)	82.9(660)
Middle	85.7(63)	66.0(241)	53.5(144)	64.7(448)	62.7(75)	50.0(238)	40.5(131)	49.3(444)	73.2(138)	58.0(479)	47.3(275)	57.1(892)
Low	55.1(78)	54.8(210)	45.3(415)	49.2(703)	62.0(100)	37.3(220)	28.1(438)	35.2(758)	59.0(178)	45.8(430)	36.5(853)	42.0(1461)
Total	80.4(327)	66.3(566)	47.9(595)	62.0(1488)	74.5(361)	49.5(560)	33.1(604)	48.9(1525)	77.3(688)	57.9(1126)	40.5(1199)	55.4(3013)
b. Middle Intelligence												
High	71.0(69)	50.0(84)	36.1(36)	55.0(189)	79.5(78)	56.9(58)	38.2(34)	63.5(170)	75.5(147)	52.8(142)	37.1(70)	59.1(359)
Middle	51.1(47)	39.5(218)	30.9(139)	37.9(404)	49.0(49)	31.7(205)	17.1(152)	28.3(406)	50.0(96)	35.7(423)	23.7(291)	33.1(810)
Low	36.0(50)	32.7(211)	26.8(598)	28.8(859)	37.2(78)	24.1(232)	17.1(679)	20.3(989)	36.7(128)	28.2(443)	21.6(1277)	24.2(1848)
Total	54.8(166)	38.4(513)	27.9(773)	34.7(1452)	56.1(205)	31.1(495)	17.9(865)	27.1(1565)	55.5(371)	34.8(1008)	22.7(1638)	30.8(3017)
c. Low Intelligence												
High	48.6(35)	32.1(53)	21.4(28)	34.5(116)	50.0(34)	26.2(42)	30.8(26)	35.3(102)	49.3(69)	29.5(75)	25.9(54)	34.9(218)
Middle	43.3(30)	16.6(181)	15.8(101)	18.9(312)	43.3(30)	20.6(136)	13.4(142)	19.5(308)	43.3(60)	18.3(317)	14.4(243)	19.2(620)
Low	27.3(44)	14.2(211)	0.5(765)	11.3(1020)	30.8(39)	8.3(193)	7.6(887)	8.5(1119)	28.9(83)	11.4(404)	8.5(1652)	9.8(2139)
Total	38.5(109)	17.3(445)	10.6(894)	14.8(1448)	40.8(103)	14.8(371)	8.9(1055)	12.5(1529)	39.6(212)	16.2(816)	9.7(1949)	13.6(2977)
d. Total												
High	80.0(290)	63.5(252)	39.0(100)	67.1(642)	80.2(298)	59.4(202)	47.4(95)	67.9(595)	80.1(588)	61.7(454)	43.1(195)	67.5(1237)
Middle	65.0(140)	43.0(640)	35.4(384)	43.1(1164)	54.6(154)	36.6(579)	23.1(425)	34.0(1158)	59.5(294)	40.0(1219)	28.9(809)	38.6(2322)
Low	42.4(172)	33.9(632)	23.7(1778)	27.4(2582)	47.5(217)	23.9(645)	15.3(2004)	19.6(2866)	45.2(389)	28.8(1277)	19.2(3782)	23.3(5448)
Total	65.8(602)	42.6(1524)	26.4(2262)	37.4(4388)	63.7(669)	34.1(1426)	17.8(2524)	29.5(4619)	64.7(1271)	38.5(2950)	21.8(4786)	33.3(9007)

Source: Data from Sewell and Shah (1968:198, Table 2).
* The base of percentages is shown in parentheses by each percentage. Data are based on a large, randomly selected cohort of Wisconsin high school seniors who were followed for a 7-year period.

mother with "high" education. Pick another percentage and express its meaning in your own words.

At this point, it is a good idea to **find the overall percentage** planning on college regardless of education of parents, I.Q., or gender; that is, find the *overall* distribution of the dependent variable without any conditional distribution being considered separately. That figure can be found in Table 6.8 in the lower-right corner, at the bottom of the total column that summarizes all of the conditional distributions. This figure is 33.3%. Then we can look at the distribution of this variable by each of the other variables one at a time, making contrasts between the percentages, perhaps in terms of epsilon. For example, 67.5% of those with high father's education plan on college, but only 23.3% of those with low father's education have similar plans. For mother's education the figures are 64.7% and 21.8%, respectively. For males, overall, the percentage is 37.4% and for all females it is 29.5%. Finally, among those with high measured intelligence, 55.4% plan on college, but only 13.6% of those with low measured intelligence plan to attend college.

The next step would be to **look at the various combinations of variables** two at a time, three at a time, and four at a time to see what the effect of various conditional distributions is. The highest percentage planning on going to college is the 89.3% in the upper-left corner of the table, and this corresponds to high-I.Q. males with parents each of whom has high education.

At the other extreme, 7.6% of the female students with low intelligence and with parents who have low educational backgrounds plan on going to college. Notice that percentage planning on college increases uniformly as one moves from the less advantaged conditions to the more advantaged. Each improvement in advantage seems to be related to an increase in the percentage planning on college.

We could draw several conclusions from these data. First, it seems to be the case that the higher the educational background of the mother or of the father, the higher the percentage of students who plan on college becomes. This will later be called a *positive* relationship, since increases in one variable tend to go along with increases in the other. Males are more likely to plan on college, and the higher the measured intelligence, the higher the percentage planning on college. These findings seem to hold up even within various categories of other variables shown in the table. Furthermore, all of these conditions seem to be cumulative in the sense that individuals with two sources of "advantage" do better than those with only one, and so forth.

At this point we should recall our initial purpose. We wanted to examine the distribution of a *dependent* variable under conditions that are defined in terms of other variables singly or in combinations.

Although many investigations call for the examination of conditions based on one variable at a time (*i.e.*, by gender or separately by education, etc.). many of our core interests are posed in terms of more complex conditions defined in terms of more than one variable, as was the case in Table 6.8. In many instances the examination of percentaged tables provides the kind of summary and the kinds of contrasts we need for the problem at hand. They are easily understood and direct.

On the other hand, it is clear that even simple percentaged tables may contain more detail than we want, and some of these may interfere with the, analysis. The solution, as in the case of univariate summary statistics, is to try to create overall indices that summarize the aspect of the distributions we are most interested in examining.

6.4 FOUR CHARACTERISTICS OF AN ASSOCIATION

Going back to a bivariate distribution such as that shown in Table 6.3, we can think of that distribution as a relationship between two variables. Suppose we want to know how the distribution of the dependent variable varies as we move from category to category of the other variable. The way two variables relate to each other is called an **association** between the variables. In Table 6.3, as city size increased, the percentage of individuals showing higher tolerance increased. The two variables were associated in that particular fashion.

We can speak of the association of any two variables and describe that association in terms of a percentaged table, as we have shown. There are other ways to summarize the association, however, and, in fact, there are four characteristics of an association that we will single out for summary, just as there are three characteristics of an univariate distribution that we summarized in terms of different index numbers (*i.e.*, central tendency, variation, and form). The four aspects of a bivariate association are:

1. Whether or not an association *exists*.
2. The *strength* of that association.
3. The *direction* of the association.
4. The *nature* of the association.

Each of these characteristics will be discussed in turn, and in the next chapter we will develop several alternative measures of them. In fact, we will create a single number that will be used to describe the first three features of an association listed above and in some cases a simple formula can be used as an efficient description of the last.

6.4.1 Existence of an Association

An association is said to exist between two variables if the distribution of one variable differs in some respect between at least some of the categories of the other variable. This rather general statement can be pinned down in a number of ways, the first of which we have already discussed. If, after computing percentages in the appropriate direction in a table, there is *any* difference between percentage distributions, we would say that an association exists in these data. In the table, below, the distribution of education is slightly different for men compared with women. We know this by percentaging in the direction of the independent variable and comparing across.

Education	Men	Women	Total
High	40%	38%	38%
Low	60	62	62
Total	100%	100%	100%
	(43)	(56)	(99)

In the next table, however, there is *no* association between "toenail length" and "education," and this is shown by the fact that there is no difference in the percentage distribution of education (the dependent variable), regardless of the category of the independent variable within which we examine the dependent variable.

	Toenail Length		
Education	Short	Long	Total
High	33%	33%	33%
Low	67	67	67
Total	100%	100%	100%
	(521)	(1756)	(2277)

In the following table it is clear that there *is* an association between social class and the number of arrests, because the percentage distributions, comparing across the way percentages were run, are different.

Number of Arrests	Social Class		
	Low	Medium	High
None	16%	28%	45%
Few	18	18	35
Many	66	54	20
Total	100%	100%	100%
	(129)	(129)	(73)

Recall that there is a name for these comparisons: **epsilon** (ε), which is the percentage difference computed across the way percentages were run in a table. In a table where *all* the epsilons are 0, there is *no* association. If any epsilon is non-0, there is an association in the data even though we may not choose to consider the very small differences important enough to talk about.

The second way to tell whether or not there is an association in a table is to compare the **actual observed table frequencies** with the frequencies we would expect if there were no association, or **expected frequencies**. If the match between actual data and our model of no association is perfect, then there is no association in the actual data between the two variables that were cross-tabulated in the table.

6.4.1a No-Association Models

A **model of no association** can be set up for a specific table as follows. Usually, in setting up a model of the way frequencies in a table should look if there were no association, we assume that the marginal distribution of each variable is the way it is in the observed data table and that the total number of cases is the same. The problem is to specify the pattern of cell frequencies in the body of the table in a way that shows no association. As an example, suppose the marginals for variables X and Y are as follows:

		(X)	
(Y)	Low	High	Total
High	a	b	57
Low	c	d	50
Total	34	73	107

The problem is to find a pattern of frequencies for cells a, b, c, and d such that they exhibit no association between X and Y. The reasoning goes like this. If there is no association in the table, then the ratio of "high" cell frequencies for variable Y as related to the corresponding column totals should be the same throughout the table, as it is in the overall distribution of Y itself, namely 57 to 107. In the table above, we would expect 57/107ths of the 34 cases in the "low" category of X to be in the "high" category of Y. Furthermore, we would expect the same ratio, 57/107ths of the 73 cases in the high column of X, to be in the top row. This would mean that, relatively speaking, there is no difference between the proportion of cases in the top row for any column of the table.

$$\frac{57}{107}(34) = .533(34) = 18.1 \text{ cases } expected \text{ in cell } a$$

$$\frac{57}{107}(73) = .533(73) = 38.9 \text{ cases } expected \text{ in cell } b$$

Given that one of the above cell frequencies in a 2 by 2 table is computed, the other expected cell frequencies could be determined by subtraction. The resulting table of expected cell frequencies (expected if there were no association between the two variables X and Y for these 107 cases) is shown below.

"EXPECTED" CELL FREQUENCIES

		(X)	
(Y)	Low	High	Total
High	18.1	38.9	57.0
Low	15.9	34.1	50.0
Total	34.0	73.0	107.0

This is a hypothetical tabulation showing no association, and thus fractional frequencies are acceptable.

Expected cell frequencies (f_e) can be computed for a given cell by multiplying the row total for that cell by the column total for that cell and dividing by N, which is the operation explained above.

(6.1)
$$f_{e_{ij}} = \frac{(n_i.)(n_{.j})}{N}$$

where $f_{e_{ij}}$ refers here to the expected cell frequency for the cell in the ith row and jth column of the table; $n_i.$ is the total for the ith row and $n_{.j}$ is the total for the jth column; and N is the total number of cases. An expected cell frequency is computed (or found by subtraction) for each cell in the table.

Now the difference between the table of observed data and the model we could construct of how this table would look if there were no association can be compared. This comparison is made by subtracting an expected cell frequency, f_e, from the corresponding observed cell frequency, f_o. The difference is called **delta**, and in this text we will symbolize delta with the uppercase Greek letter delta (Δ). For a given cell,

(6.2)
$$\Delta = f_o - f_e$$

A delta value can be computed for each cell in a table, regardless of the size of the table. If any of the deltas are *not* 0, then there is at least some association shown in the table. Whenever all deltas are 0, all epsilons will also be 0. Later we will discuss summary measures of association based on these ideas.

In summary, whether or not an association exists in a table of observed frequencies can be exactly determined in two ways that yield the same conclusions. One way is to compute percentages in one direction and compare across in the other direction, using epsilon. The other way is to create a table of expected cell frequencies and compare the observed and expected cell frequencies, cell by corresponding cell, using delta. If all of the epsilons that can be computed in a table, or if all of the delta values for a table, amount to 0, then there is no association between the two variables cross-tabulated in the bivariate distribution. This is called **statistical independence**. If, on the other hand, there is any epsilon or any delta that is not 0, then there is an association in the observed frequency table, however slight or large that association might be.

6.4.2   Degree (Strength) of Association

Where the differences between percentages (epsilons) are large or where the deltas are large, we speak of a strong **degree of association** between the two variables; that is, the dependent variable is distributed quite differently within the different conditional distributions defined by the independent variable. This can be contrasted with a weak association, where there is very little difference or where the epsilons and deltas are very small, approaching or equaling 0.

Often investigators use epsilon (or delta) as a crude first indicator of the strength of association. The problem with both delta and epsilon is that it is difficult to determine what a given-sized delta or epsilon means, other than that there is some association in the table. The reason for this is that both delta and epsilon values for any cell(s) can vary from 0 or near-0 up to a magnitude that is not, in general, fixed. They are not "normed" or standardized. Later, in this chapter and in the next, the problem of creating good standardized measures of the strength of association will be discussed and several alternative measures will be described. Suffice it to say here that some tables show a strong relationship between independent and dependent variables, and some show a weak association or no association at all.

6.4.3 Direction of Association

Where the dependent and independent variables in a table are at least ordinal variables, it makes sense to speak about the **direction** of an association that may exist. If the tendency in the table as shown by the percentage distribution is for the higher values of one variable to be associated with the higher values of the other variable (and the lower values of each variable also tend to go together), then the association is called a *positive* association. Height and weight tend to have a positive association, since taller persons tend to be heavier, in general, across the people in a general population.

On the other hand, if the higher values of one variable are associated with lower values of the other (and the lower values of the first with higher values of the second), the association is said to be *negative*. Sociologists generally expect that the higher the educational level of people, the lower their degree of normlessness will be—a negative association.

The association between city size and tolerance scores (see Table 6.3) is positive because the larger the city, in general, the higher the tolerance level becomes (*i.e.*, the higher the percentage of people who have high tolerance scores). The older a person's age, in general, the fewer the years left until retirement, a negative association.

6.4.4 Nature of Association

Finally, the **nature** of an association is a feature of a bivariate distribution referring to the general *pattern* of the data in the table. This is often discovered by examining the pattern of percentages in a properly percentaged table. Often the pattern is irregular, and an investigator would cite many epsilons in describing where the various concentrations of cases are in the different categories of the independent variable. Sometimes there is a rather uniform progression in concentration of cases on the dependent variable as we move toward higher values of the independent variable. If, with an increase of one step in one variable, cases tend to move up (or down), a certain number of steps on the other variable we might call the nature of the association "linear." That is, the concentrations of cases on the dependent variable (the mode, for example) tend to fall along a straight line that could be drawn through the table.

The nature of association will be discussed at length in the next chapter. Simple linear associations have an intrinsic interest to investigators as one of the simplest natures of association, but some associations are curvilinear in nature or of some more complex patterning. In most cases the nature of association will be determined from a percentage table or a scatter plot, but in some cases nature can be described in terms of an equation.

At this point we should pause to examine several tables and describe them in terms of these four features of an association. Table 6.9 presents a series of examples together with brief summary statements.

TABLE 6.9A PERCENTAGE DISTRIBUTION OF TOTAL MONEY INCOME FOR FAMILIES BY ETHNIC BACKGROUND OF HOUSEHOLDER, 1983

Income	White	Spanish	Black	Total
$35,000 and over	31.5	15.1	13.2	29.6
$25,000 to 34,999	20.3	14.3	14.0	19.5
$15,000 to 24,999	23.7	27.0	21.5	23.4
$ 7,500 to 14,999	16.0	23.6	23.7	16.8
Under $7,500	8.5	20.0	27.6	10.7
Total	100.0	100.0	100.0	100.0
N (millions)	(53.9)	(3.6)	(6.7)	(62.0)

Source: U.S. Bureau of the Census, (1985).

1. *Existence of association.* Table 6.9A is percentaged down in the direction of the independent variable, ethnicity, so that comparisons may be made across. The percentages across are different so that there is an association evident in the table. Compare any row, say the row for incomes of $35,000 or more; percentages range from a high of 31.5% down to 13.2%, all different from 29.6%, the total for that income category.

2. *Strength of association.* Overall, if ethnicity makes any difference in the distribution of family income, we should find fairly substantial epsilon's. Here, the white–black epsilon for $35,000 and over incomes is 18.3. That for Spanish–black for the same income category is 1.9 and for white–Spanish is 16.4. These are all less than 100% but substantially larger than 0.

3. *Direction of association.* Here, ethnicity is a nominal variable so that it is impossible to talk about direction of association.

4. *Nature of association.* To see the pattern of association most clearly, it is helpful to underline the highest percentages in each row-wise comparison. Here, for the highest income category, 31.5% is clearly the largest percentage and is underlined. In the next row, 20.3% is largest and is underlined; 27.0% is underlined in the third row. We will underline both the 23.6% and 23.7% in the fourth row because they are essentially equivalent in magnitude. Finally, 27.6% is underlined in the bottom row. Notice that we are underlining the percentages from the body of the table, not the marginal distribution. The nature of association is the pattern of white's having higher percentages in the highest two income groups, compared

with the other two ethnic groups; Spanish have higher percentages in the next group and are tied in percentages in the $7,500 to 14,999 income category. Finally, the black group has highest percentages in the lowest income category. The pattern of high and low percentages from comparisons across the way the percentages were run is the nature of the association of ethnicity and income for families in 1983.

TABLE 6.9B PERCENTAGE DISTRIBUTION OF FATHER'S OCCUPATION BY OCCUPATION OF 30–59-YEAR-OLD CHILD, SWEDEN, 1977

Occupation of Father	Occupation of Child				
	Farmer	Worker	Entrepreneur	Middle Class	Not Known
Middle Class	2	10	13	29	14
Entrepreneur	6	10	23	15	10
Worker	14	52	38	39	39
Farmer	77	24	21	14	21
Not Known	1	4	5	3	16
Total	100.	100.	100.	100.	100.
N	(241)	(2964)	(525)	(2557)	(166)

Source: Adapted from Sundstrom (1986:369).

1. *Existence of association.* Table 6.9B presents the results of a Swedish survey of occupations of adult (30–59-year-old) children and their fathers. The table is percentaged in the direction of the child's occupation to show the distribution of fathers' occupation. Comparing across, the percentages are different; thus there is an association in the table. Note that this way of percentaging the table permits one to make statements about background occupational experience of children (*e.g.*, their father's occupation). Percentaging the table the other way would permit one to say something about the distribution of children's occupations for fathers in different occupations. Percentaging to the total would permit statements about the percentage of people who, for example, stayed in the same occupation that their father had. The way percentages are run permits quite different kinds of comparisons.

2. *Strength of association.* In this table, the highest epsilon ought to be seen in comparing extreme categories of child's occupation for the highest (or lowest) category of father's occupation. Taking the middle-class fathers, the farmer to middle-class epsilon is 27%. For fathers who are farmers, the same epsilon comparison is 63%. These are both very substantial epsilons. The association is quite strong.

3. *Direction of association.* As an aid in finding the direction of an association between variables each of which is at least ordinal, a useful procedure is to make comparisons across the way percentages are run, underlining the highest percentage for each comparison. In this table we could make four comparisons (aside from the not known category that the author provides here). For the middle-class row the highest percentage is "middle class." For the

entrepreneur row, it is "entrepreneur"; it is "worker" for the worker row, and "farmer" for the farmer row. Notice that the highest percentage is for father having the same occupation as the child, or no social mobility between generations. One could draw a diagonal line through the underlined percentages in the table. In this case, the line would extend from the "farmer"–"farmer" or lower-status occupational category to the highest category combination, "middle class"–"middle class." This indicates a positive association: the higher the occupational status of the father, the higher the child's occupational status tends to be. In this "occupational mobility" table, people in the upper-left area above the diagonal are downwardly mobile (fathers had higher status occupations than the child), and people in the lower-right area under the diagonal are upwardly mobile (child has a higher status occupation than father).

    4. *Nature of association.* In this table, the nature of association (*i.e.*, the pattern of concentration in the table) tends to be almost linear. There is a relatively uniform shift toward higher-status child's occupation with shifts upward in the category of father's occupation. There are no "reversals" in this general trend of concentration. Because the variables are ordinal, it would be more appropriate to speak of this nature of association as "monotonic" rather than linear. If distances were defined, then one could determine whether in fact there is a constant amount of shift in values of one variable, given a fixed amount of difference in the other. In ordinal variables, one can only say that the value of one variable remained the same or shifted in a fixed direction with increases in the other variable—a monotonic nature of association. Contrasted with this type of nature are those such as the one shown in Table 6.9C.

TABLE 6.9C   Percentage Distribution of Body Weight by Age for Persons 20 Years and Older

Percentage Above or Below Desired Weight	Age					
	20–34	*35–44*	*45–54*	*55–64*	*65–74*	*75+*
30% or more above	10.7	16.7	21.8	21.3	21.5	11.6
20–29.9% above	7.8	11.0	13.3	13.9	13.3	11.2
10–19.9% above	16.5	21.5	23.1	23.3	22.0	20.1
5–9.9% above	12.6	13.1	13.1	12.9	12.3	15.2
Plus or minus 4.9%	27.6	22.5	18.1	18.9	18.1	21.5
5–9.9% below	11.5	8.0	5.7	5.0	6.2	8.4
10% or more below	13.3	7.2	4.9	4.7	6.6	12.0
Total	100.0	100.0	100.0	100.0	100.0	100.0
N (1000's)	(59.9)	(28.9)	(22.2)	(22.0)	(16.4)	(9.5)

Source: National Center for Health Statistics (1986).

    1. *Existence of association.* In Table 6.9C, age is treated as the independent variable and the amount by which one's weight is above or below the desired weight is the dependent variable. Comparing across, there is a percentage difference, thus there is an association shown in the table.

2. *Strength of association*. Among all the possible percentage comparisons, the strongest percentage difference should be evident in comparing the extreme categories of age for the extreme categories of percentage above or below desired weight. Taking the top row, the overall epsilon is only .9 and in the bottom row it is only 1.3. There are indeed small percentage differences. Yet, there are larger percentage differences in the table, for example, the difference between the 20–34 and 45–54 age categories for the top row of the table. As we will see, the pattern of association in this table is irregular, making the assessment of strength of association more complex. Even at its best, however, the percentage differences are rather small, suggesting a weak association between the two variables.

3. *Direction of association*. As before, we will underline the highest percentages in each comparison, underlining more than one where percentages are very close in magnitude. Table 6.9C shows a pattern that moves generally from upper right to lower left, between overweight associated with older ages and underweight with younger ages, a generally "positive" association.

4. *Nature of association*. Although the overall pattern of association indicated by drawing a diagonal line through the underlined percentages (or the middle of several underlined percentages in a given comparison) is linear, there are other patterns that need to be examined. Notice that percentages are essentially tied from 45–74 years old for the top two rows, but there is a clearer concentration of high percentage in one column for the bottom three rows. There is a broader area of high percentages for overweight rows than for underweight rows. Notice, too, that the percentages along the bottom three or four rows drop down as one moves across from low to high age categories and then the percentages begin to rise again. If the second "high" is italicized for each of the bottom three rows, one can see a more complex pattern emerging from the table. The nature of association begins to appear "curvilinear." Underweight is concentrated in the lowest age bracket and to some extent in the oldest age bracket, while overweight is more likely found in middle-age categories. It was this curved nature of association that made an assessment of strength of association difficult to determine if we made epsilon comparisons as if we expected a monotonic relationship. We will have more to say about this later, but suffice it to say that one needs to be aware of the nature of association in selecting measures of strength of association.

6.5 CREATING MEASURES OF ASSOCIATION

It is possible to create single summary index numbers that will indicate the existence, degree, and direction of association all in the same number. Typically, these measures are set up so that they will vary along a scale from a minus value, indicating a negative relationship, to 0, indicating no association, to a positive value, indicating a positive association. The larger the magnitude of the index number, the stronger the association.

Ideally, a measure of association would also have the characteristic of varying between two *fixed limits* such as −1.00 and +1.00, where 1 indicates a **perfect association** or a maximally strong one in some sense (either positive or negative), and 0 indicates statistical independence.

SCALE OF STRENGTH OF ASSOCIATION WITH FIXED LIMITS OF −1.00 AND +1.00

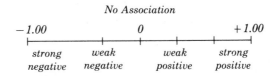

A measure of association that varies over a scale with fixed limits, such as that shown above, is called a **normed** or **standardized measure of association**. Its value lies in the fact that normed measures can be validly compared. If a normed measure of association computed on one table was −.56, and the same kind of measure computed on another table was −.13, both could be interpreted as falling between the highest possible negative value of −1.00 and 0 (no association). They could be compared. The first table has a stronger negative association (about four times stronger) than the second, and the difference between the two measures is substantial. If the limits on a scale were not fixed or known, then an index number such as −.56 would be very difficult to interpret, and it would be essentially impossible to use the index in making comparisons of strength of association between two different tables. Since comparison and interpretation are important to sociologists, normed measures of association are important.

Table 6.10 illustrates the problem of norming with epsilon, an unnormed measure of association. Notice that the marginals of a table may be such that an epsilon of 100% could not be achieved regardless of the way cell frequencies are rearranged.

To create a measure of association that is normed and interpretable, we need to define 0 and 1, that is, we need to define what no association and perfect association are supposed to mean. We have already discussed two ways to identify the situation of no association, or statistical independence, in a table, using epsilon or delta. The definition of the opposite extreme is less clear-cut.

6.5.1   In Pursuit of a Normed Measure of Association

Measures of association turn out to be rather simple ratios designed to be sensitive to changes in the strength of association and, in some cases, to the direction, nature, and role played by independent and dependent variables. Often the ratios suffer from the usual ailments of ratios: the numerator or denominator may fluctuate because of some irrelevancy, such as the number of cases in the table or the number of rows and columns—things that are generally of no interest when one wants to compare strength of association, and things that prevent the valid comparisons needed in a study. As in the other approaches to refining ratios, better measures of association become better by excluding irrelevant influences on the numerator and denominator.

One set of measures of association is based upon the difference between observed and expected cell frequencies in a table, which we have called delta. Most of the delta-based measures (phi-square, Pearson's C, Tschruprow's T, and Cramer's V) are not widely used because of the problems they *do not* adequately take into account. On the other hand, there are not very many measures of

TABLE 6.10 Epsilon, an Illustration of the Norming Problem in Developing Measures of Association

Two illustrative tables are given below. The first shows a 2 by 2 table with balanced marginals (both row and column totals are the same) and it is possible in this particular table to fix cell frequencies so that, computing percentages down, an epsilon could equal 0 or 100 for this table. Thus, the achieved epsilon of 17% can be interpreted as falling on the scale from 0% to 100%. In the second illustration, however, it is *not* possible to adjust cell frequencies (also given fixed marginals) so that 0 or 100 could be achieved for epsilon. The epsilon in this second table, 19%, thus has to be interpreted in terms of a more limited scale extending in this case from 3% to 83%. It would not be appropriate to compare the two epsilons because they not only reflect the pattern of frequencies in a table, but are also constrained by the way the marginals are distributed, and this confounds the intended comparison. About all one is able to say is that epsilon indicates that there *is* an association in both original tables.

Illustration 1

	Weakest Possible Association				Original Table				Strongest Possible Association		
	X				X				X		
	Low	*High*			*Low*	*High*			*Low*	*High*	
Y High	9	6	15	Y High	10	5	15	Y High	15	0	15
Low	6	4	10	Low	5	5	10	Low	0	10	10
	15	10	25		15	10	25		15	10	25
	$\varepsilon = 0\%$				$\varepsilon = 17\%$				$\varepsilon = 100\%$		

Illustration 2

	Weakest Possible Association				Original Table				Strongest Possible Association		
	X				X				X		
	Low	*High*			*Low*	*High*			*Low*	*High*	
Y High	8	7	15	Y High	9	6	15	Y High	13	2	15
Low	5	5	10	Low	4	6	10	Low	0	10	10
	13	12	25		13	12	25		13	12	25
	$\varepsilon = 3\%$				$\varepsilon = 19\%$				$\varepsilon = 83\%$		

association for nominal variables, so certain of these measures (e.g., *C*, *T*, and *V*) are familiar to and are used by investigators in certain situations. Here we will omit them, but a study of their differences nicely illustrates attempts to develop measures that are normed at 0 for no association and 1.00 for perfect association of nominal variables.*

* The sum of all deltas in a table [*i.e.*, the sum of all $(f_o - f_e)$ for all cells in a table] does not take account of the magnitude of expected frequencies and, because marginals are set, it always sums to 0. A chi-square first squares delta and then divides delta squared by f_e,

(Footnote continues.)

But norming is only part of the problem of developing a summary index number to stand for the association in a table. There are at least two other desirable characteristics for such measures. *First*, the number should be interpretable in some intuitively useful sense; *second*, the meaning of 1.0, the "perfect association" norm, should be definable. In the first instance, most of the delta-based measures of association can only be thought of as a magnitude; the bigger the number the stronger the association. They *cannot* be interpreted, for example, as the percentage of variation in one variable explained by another, nor can they be interpreted as the proportion of predictive errors that may be reduced by prior knowledge of one of the variables—interpretations that are especially useful for substantive investigators. More will be said about this criterion, and the next chapter will develop a group of measures called *proportionate reduction in error* (PRE) measures of association, which have as one of their strengths this kind of ready operational interpretation.

The other criterion for a good measure of degree of association is that the limits be defined in a meaningful way. We have already established the idea of

---

(Footnote continued.)
summing over this figure for all cells:

$$\chi^2 = \sum \frac{(f_o - f_e)^2}{f_e}$$

The maximum chi-square is $N(k-1)$, where k is the number of rows or columns, whichever is smaller. For a table with two rows or columns, the maximum chi-square is N. Chi square is incorporated into formulas for delta-based measures of association. A phi-square divides chi-square by N, but the maximum of 1.00 is achieved only for tables with two rows (or columns) and can exceed 1.00 for larger tables:

$$\phi^2 = \frac{\chi^2}{N}$$

Pearson's C is the square root of chi-square divided by chi-square plus N. Although it can never exceed 1.00, it also can never reach 1.00:

$$C = \sqrt{\frac{\chi^2}{\chi^2 + N}}$$

Tschruprow's T replaces Pearson's chi-square plus N in the denominator by a *degrees of freedom* (*df*) figure: $df = (r-1)(c-1)$, where r is the number of rows and c is the number of columns in the table:

$$T = \sqrt{\frac{\chi^2}{N(df)}}$$

T has an upper limit of 1.00 for any table that is square (where $c = r$).

Finally, Cramer's V replaces Tschruprow's denominator with N times t (where t is the smaller of $r-1$ or $c-1$):

$$V = \sqrt{\frac{\chi^2}{Nt}}$$

V is properly normed between 0 and 1.0, and these extremes can be achieved for any table; thus V can be compared between tables. V, however, poses other problems of interpretation discussed in the text. An extended discussion of these measures appeared in earlier editions of this book (editions 1 and 2, 1974, 1976, and 1980).

statistical independence, or no association. In fact, all the measures of association discussed here are always zero when there is no association in a table.* What about the upper extreme? What does a *perfect association* mean? What is the pattern in a table when a normed measure of association equals 1.0?

6.5.2   Meanings of Perfect Association

In this chapter we will limit our attention to the 2 by 2 table in considering alternative meanings of **perfect association**. In the next chapter, this idea will be extended to larger tables. For now, however, we can focus on two somewhat different notions of perfect association: a more restrictive or stringent definition, and a less restrictive one.

6.5.2a   Restrictive Meaning of Perfect Association

The first model, the *more restrictive* meaning of perfect association, is one where all of the cases in a table are in one diagonal of the table. Stated differently, each value of one variable is associated with only one single value of another variable, so that for any category of the independent variable, only one cell of the dependent variable is non-0. In a 2 by 2 table, this means that two diagonal cells have frequencies, and the other two cells do not. This condition for a positive and a negative association is indicated by X's (for some frequency) and 0's (for no frequency) in the *upper* set of tables in Figure 6.1. The upper-left table in Figure 6.1 indicates a perfect negative association, and the upper-right table indicates a perfect positive association. The middle table shows symbols for cell frequencies that are traditionally used in expressing formulas for measures of association for a 2 by 2 table.

If we were testing a theory that stated that the larger the city size the higher the degree of tolerance, this model of perfect association would seem to reflect a situation where this hypothesis held exactly. All individuals living in large cities would have "high" tolerance scores, and all individuals living in smaller cities would have "low" tolerance scores. Any deviation from this pattern would be less than a perfect association. The *more restrictive* definition of a perfect association would require that all the data be concentrated in one diagonal of the table, as in Figure 6.1.

* Although it is the case that measures we have discussed will be 0 when a table shows no association (when one variable is statistically independent of another), it is not always the case that they will be 0 only when two variables are statistically independent of each other. As we shall see later, measures of association are defined to be sensitive to certain features of an association, often depending upon the level of measurement, for example. Variables may be statistically independent with respect to categorical or ranking features, but not independent with respect to the distance features defined into interval variables. For this reason, selection of an appropriate measure of association again depends critically upon meanings defined into variables and variation in data that an investigator is interested in examining. This will be discussed later, in this chapter and the next, as measures of association are presented and contrasted.

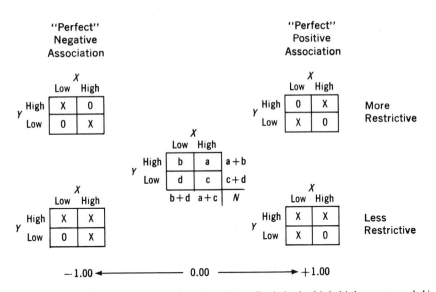

In the notation shown in the middle table, cell *a* is in the high-high corner and *d* is in the low-low corner; thus cells *a* and *d* are on the positive diagonal. This is important simply because it assures the correct sign in computations which follow.

FIGURE 6.1 TWO MODELS OF PERFECT ASSOCIATION FOR 2 × 2 TABLES

Phi, one of the delta-based coefficients mentioned above, is a normed measure of association for a 2 by 2 table in this more restrictive sense. The *phi coefficient* of a 2 by 2 table can be computed as follows:

(6.3)
$$\phi = \frac{ad - bc}{\sqrt{(a + b)(c + d)(b + d)(a + c)}}$$

where *a*, *b*, *c*, and *d* refer to cell frequencies in a 2 by 2 table as indicated in Figure 6.1, and the denominator of ϕ is the square root of the product of the marginals.

Phi can be interpreted as the degree of diagonal concentration, and it will have values between -1.0 and $+1.0$ for ordinal variables. The values of 1.0 will be reached only if all cases are concentrated in the main diagonal of the table, and the magnitude of this coefficient indicates the extent to which this condition is exhibited in the table. For nominal variables, of course, one would use only the magnitude of the coefficient and drop the sign.*

---

* As we will show in the next chapter, the 2 by 2 table is an interesting situation because many measures of association that are different for larger tables turn out to have the same numeric value for a 2 by 2 table. Thus $\phi = T = V = r$. Pearson's *r* is a measure of association to be developed in the next chapter.

6.5.2b   Second Model of Perfect Association

For a 2 by 2 table, the second model of perfect association, the *less restrictive* one, is illustrated by the *lower* set of tables in Figure 6.1. Here the requirement is that only one of the four cells need have a 0 frequency and the other three, including the two on the main diagonal in the table, may have any frequency. Where might this definition be useful? Consider two illustrations.

Social science investigators often measure attitudes in terms of a series of individual items that are carefully graded to indicate different degrees of that attitude. For example, in a study of permissiveness attitudes regarding sex, an individual might be asked to indicate agreement with statements such as these:

a) I believe that petting is acceptable for the female before marriage when she is engaged to be married.
( ) Agree
( ) Disagree
b) I believe that petting is acceptable for the female before marriage even if she does not feel particularly affectionate toward her partner.
( ) Agree
( ) Disagree

Now, if these questions do indeed reflect differences in degree of permissiveness attitudes, and if people realize this and respond accordingly, we would not expect anyone to disagree with the first question and agree with the second. If a person's permissiveness attitudes are sufficient to accept or agree with the second statement, then that person should also agree with the first. This expected state of perfect association between answers to these two attitude items would be shown in a table, as follows:

"More Permissive" Attitude Item (b)	"Less Permissive" Attitude Item (a)	
	Disagree	Agree
Agree	0	X
Disagree	X	X

where X indicates some cell frequency

The scale items may not be graded as we thought, or people responding to the items may not understand them in this light, so any given set of responses may not match this idea of a perfect logical relationship between these "strong" and "weak" attitude items. A measure of association that used this less restrictive definition of perfect association as its definition of 1.0 would be the one we would need to measure the distance between the data we observe and this idea of perfect association.

Another example is more of a philosophical distinction between "cause" and "effect" conditions. If one is interested in factors that influence the state of other variables, one may think of two kinds of effect. One situation is where the effect does not occur unless and until a given factor is present in a certain amount (called a *necessary* condition). The other situation is where the outcome may occur for other reasons, but it will never be the case that the outcome will be absent if a given factor is present to a certain extent (called a *sufficient* condition).

These states can be illustrated in terms of the two tables below. The first would be the model of a necessary condition and the second a model of a sufficient condition. The third table below is a model of a condition that is *both* necessary and sufficient. It is only this third table that technically meets the requirements of a cause and effect relationship because the effect *always* occurs when the cause is present and *never* occurs when the cause is absent.

MODEL OF A "NECESSARY" CONDITION

	Cause	
Effect	*Absent*	*Present*
Present	0	X
Absent	X	X

where X indicates some cell-frequency

MODEL OF A "SUFFICIENT" CONDITION

	Cause	
Effect	*Absent*	*Present*
Present	X	X
Absent	X	0

where X indicates some cell frequency

MODEL OF A CONDITION THAT IS BOTH NECESSARY AND SUFFICIENT

	Cause	
Effect	*Absent*	*Present*
Present	0	X
Absent	X	0

where X indicates some cell frequency

To what extent does the real world behave the way we reason that it should? A measure of association using the less restrictive model of perfect association would permit one to contrast actual data with the models for either a necessary or a sufficient condition to help provide the answer.

6.5.2c Yule's Q

Yule's Q is one measure of association for a 2 by 2 table that uses the less restrictive definition of perfect association.* In a 2 by 2 table, Yule's Q is computed by finding the cross-products, the product of the cell frequencies on one diagonal, and the product of the cell frequencies on the other diagonal in the table.

BOX 6.2 THE THEME

Thus far in this chapter we have (a) discussed the idea of association (end of Section 6.1; Section 6.2; Section 6.4) and how percentages may be computed to examine an association (Section 6.3.3); (b) introduced four features of an association—*existence, degree, direction,* and *nature* (Section 6.4); and (c) considered how to define no association (6.4.1a) and the more and less restrictive models of perfect association (Section 6.5.2). Measures of association that, in one index number, can summarize the existence/degree/direction of association are simple ratios. The problem is to select numerator and denominator in such a way that (only) those features of an association that are of interest are revealed. Finally, and most importantly, we ended up with two properly normed measures for 2 by 2 tables, each "normed" at a different but useful definition of perfect association. Other normed measures will be presented in Chapter 7, but two other topics—one caution and one distinction—will be discussed next.

(6.4)
$$Q = \frac{ad - bc}{ad + bc}$$

where a, b, c, and d are cell frequencies as shown in Figure 6.1.

If one of the cells on one of the diagonals is 0, then the Q value will be either $+1.0$ or -1.0, depending upon the direction of association. In any event, the value 1.0 means that the data meet the less restrictive criteria for patterns of frequencies required for a perfect association. Notice that the Q and ϕ coefficients have the same numerator, but the denominators differ and produce measures that norm at different definitions of perfect association. If ϕ is 1.0, Q will also be 1.0, since the data will at least show the less restrictive pattern of perfect association. On the other hand, if $Q = 1.0$, it may be the case that $\phi \neq 1.0$, because the criterion for perfect association for ϕ is more stringent, calling for a different pattern of concentration of cases in the 2 by 2 table. As in the case of ϕ, Q can be used for nominal variables, but the sign of Q would not be

---

* Actually, for a 2 by 2 table, Yule's Q is identical to gamma, a coefficient to be introduced in the next chapter. Gamma is suitable for any size of table, including a 2 by 2 table. The different name introduced here is again tradition and a usage frequently found in the literature. The formula for Q with its particular notation is not useful for larger tables, but here it nicely indicates one important difference between measures of degree, direction, and existence of association.

used, only its absolute magnitude.* Q can reach 1.0 for any 2 by 2 table that is identical in pattern with its model of perfect association.

6.6   GROUPING ERROR

Before we proceed, we should update our attention to grouping error. You will recall that grouping error was one possibly distorting factor involved when we wished to group data into a smaller number of categories that might be handy for presentation purposes (see Section 3.2.3a). In univariate statistics, if the midpoint of a grouped class did not reflect the character of scores in that class, some grouping error would result. The same type of problem may happen where a bivariate distribution is grouped into fewer categories. Table 6.11 illustrates this possibility. Notice that the original 2 by 3 table is grouped in two ways to create a 2 by 2 table. We might want to group data if we preferred to use 2 by 2 measures of association (usually, we would be well advised to pick a measure that works on the table we have), or we might have relatively few cases in some rows or columns) of a larger table that might suggest that less precision is called for. In any event, as Table 6.11 shows, different groupings are not necessarily equivalent.

In the first 2 by 2 table (6.11B) the degree of association is rather strong and positive; in Table 6.11C, the association is 0. Which accurately represents the relationship between income and education in the original table? Neither one does, and the distortion is called *grouping error*. Generally, it is a useful practice to retain as much precision as possible and to be reluctant to group more precise data.†

6.7   SYMMETRIC AND ASYMMETRIC MEASURES OF ASSOCIATION

One of the distinctions between the different measures of association that we have discussed was whether or not they are properly normed. Another distinction was between measures that define perfect association in a more rather than less restrictive fashion. There is a third distinction between measures of association that should be mentioned here. That is the distinction based upon whether or not the measure distinguishes between independent and dependent variables.

---

* It is true, of course, that the sign of Q or ϕ resulting from their computation on the association of nominal variables would indicate which categories of one variable go with which categories of the other variable. Direction of association would not have meaning, however, and one would have to examine the table layout carefully to determine and express what the sign means. For this reason, Q and ϕ computed on nominal variables are expressed on a scale extending in magnitude from 0 to 1.0.

† Even where it is decided to group data into a smaller table, it is advisable to preserve at least three categories in each variable. The reason for this is that one is able to identify curvilinear natures of association if at least three categories are preserved. With only two categories, if there is an association, it will always appear to be linear.

TABLE 6.11 AN ILLUSTRATION OF GROUPING ERROR IN BIVARIATE DESCRIPTION

A. Original 2 × 3 Table of Frequencies

	Education			
Income	Low	Medium	High	Total
High	5	2	8	15
Low	2	3	1	6
Total	7	5	9	21

B. The Same Data With "High" and "Medium" Categories of Education Grouped Together to Create a 2 × 2 Table

	Education		
Income	Low/Med	High	Total
High	7	8	15
Low	5	1	6
Total	12	9	21

$$Q = +.70$$
$$\phi = +.33$$

C. The Same Data With "Medium" and "Low" Categories of Education Grouped Together to Create a 2 × 2 Table

	Education		
Income	Low	Med/High	Total
High	5	10	15
Low	2	4	6
Total	7	14	21

$$Q = 0$$
$$\phi = 0$$

Symmetric measures of association do not distinguish between independent and dependent variables, but merely address themselves to the strength (and direction) of relationship between pairs of variables. The role of variables in a bivariate distribution does not matter as far as the computation of these measures is concerned. Measures such as Yule's Q and the phi coefficient (as well as C, T, V, and some others) are all symmetric measures of association.

Asymmetric measures of association, by contrast, do require a distinction between independent and dependent variables for their computation. Usually, they are oriented toward measuring the usefulness of the independent variable in predicting values of the dependent variable. In general, two different asymmetric coefficients can be computed on a single bivariate table. One of these coefficients measures the value of predicting Y from a knowledge of variable X and the other measures the value of predicting X from a knowledge of Y. Most of the measures of association of this type will be discussed in the next chapter (*e.g.*, lambda, Somers' *d*, eta, etc.) but we have used a measure in this chapter that illustrates the asymmetric idea. That measure, crude (not normed) though it may be, is called **epsilon** (ε), and it is simply the difference between percentages. If percentages are computed down, ε is simply the difference between percentages compared across in the other direction. Traditionally, ε is computed as the difference between the extreme corners of a table, as shown below, but it is clear that many ε's can be computed to compare cells of particular interest.

Here ε would equal $+25\%$, the difference between 40% in the upper-right corner (the high–high corner of this table), and 15% in the upper-left corner (the high–low corner of this table).* The implication is that social class makes a difference in electoral activity. The reason for taking this particular difference is that if one expects cases to pile up on the main diagonal as the association becomes stronger, the smallest cell should be in one corner and the biggest in the other, so that this difference is expected to be the biggest in the table.

ELECTORAL ACTIVITY OF RESPONDENTS IN A CROSS-SECTIONAL SAMPLE OF THE UNITED STATES, BY SOCIAL CLASS[†]

Electoral Activity	Social Class			Total
	Low	Medium	High	
High	15%	24%	40%	26%
Medium	25	31	31	29
Low	60	45	29	45
Total	100%	100%	100%	100%
	(857)	(871)	(871)	(2599)

Overall $\varepsilon = 40\% - 15\% = 25\%$

If percentages were computed in the opposite direction (by rows rather than columns as shown here), the percentage comparison would be made the other way, and it would be a different value, as shown below. The implication here is that electoral activity makes a difference in the distribution of social class.

* We could have taken the overall difference between 60% and 29% in the bottom row of the table, following the same logic. The ε is likely to be different, of course, and this is one reason why other measures of the overall association in a table are generally preferred.

[†] Table reprinted from Jae-On Kim, "Predictive Measures of Ordinal Association," *American Journal of Sociology*, 76 (March 1971), p. 900, by permission of the University of Chicago Press. Copyright 1971 by the University of Chicago and Published by the University of Chicago Press.

SOCIAL CLASS OF RESPONDENTS IN A CROSS-SECTIONAL SAMPLE OF THE UNITED STATES, BY
ELECTORAL ACTIVITY

Electoral Activity	Social Class			Total
	Low	Medium	High	
High	18%	31	51	100% (687)
Medium	28%	36	36	100% (756)
Low	45%	34	21	100% (1156)
Total	32%	34	34	100% (2599)

Overall epsilon = 45% − 18% = 27%

Epsilon, then, is an asymmetric measure because it is a comparison made
after we percentage in the usual manner—in the direction of the independent
variable. As we noted earlier, ε suffers from norming problems, but it will serve
here as a rough and ready measure of overall differences and as an example of an
asymmetric measure. Normed asymmetric measures will be presented in the next
chapter.

6.8   SUMMARY

A number of key steps in statistical description have been taken in this chapter,
and these will be elaborated in the next chapter and later in this book.

One of the first ideas was that of *association* between two variables. This
amounted to a systematic way of examining the differences in the distribution of
one variable that are associated with differences in a second variable. As values of
one variable increase, there is a tendency for an increase (or decrease) in the scores
cases have on a second variable. Stated more generally, a table is said to show some
association if a variable is distributed differently within the various categories
of some other variable.

An association can be described in terms of four features—*existence, degree,
direction*, and *nature* of an association. These may be examined by percentaging a
table properly (and three rules for percentaging were given). Alternatively, the
first three features—existence, degree, and direction—may be measured in terms
of a variety of different ratios called *measures of association*. A single measure of
association may be computed that (a) will indicate whether or not there is an
association in the table (measures are generally 0 if there is no association);
(b) will indicate the degree or strength of association (by a magnitude that is usually between 0 and 1.0); and (c) will show direction (by a plus or minus sign), although this is used only for bivariate distributions of ordinal or interval variables.

To permit valid comparison of measures of association computed on different tables and to aid in the interpretation of these measures, investigators
generally prefer *normed* measures of association. This means that the numerator and denominator of the measure of association are selected so that the

limiting values of the measure could be -1.0 or 0 or $+1.0$ for any table, regardless of the number of cases it contains, the number of rows or columns, or the way the marginals happen to be distributed.

Distinctions among measures of association can be made in terms of (a) whether or not they are normed, (b) the model of perfect association and no association the measure assumes, (c) whether the measure is asymmetric or symmetric, (d) whether it is most appropriate for cross-classifications of nominal, ordinal, or interval variables, and (e) the size of table for which it may be computed. Chapter 7 will discuss measures of association for nominal, ordinal, and interval cross-classifications.

The *nature* of association in a table refers to the patterning of concentration. It is often examined using a properly percentaged table, but Chapter 7 will discuss the use of a regression equation to describe the nature of association between interval variables. Sometimes this patterning is monotonic or linear or curvilinear, but often the pattern is more complex. In any event, the description of the existence, degree, direction, and nature of association between pairs of variables provides a powerful means for describing how variables relate to each other and this, after all, is the central activity of any scientific inquiry.

CONCEPTS TO KNOW AND UNDERSTAND

bivariate distribution; conditional
    distribution
  heading
  stub
  body
  cell
  cell frequency
  2 by 2 or fourfold tables
  $r \times c$ tables
  marginals
  percentaging rules
epsilon (percentage difference)
how to examine complex conditional
    distributions
association
  existence

degree
direction
nature
expected cell frequencies
delta
statistical independence
norming measures of association
definition of perfect association
  more restrictive
  less restrictive
phi coefficient
Yule's Q coefficient
grouping error in tables
asymmetric and symmetric measures
    of association

QUESTIONS AND PROBLEMS

1. Using data from an article or from a piece of research you have been asked to conduct, select a series of cases that have been measured on two variables of interest to you. Construct a bivariate frequency distribution and compute percentages in the appropriate direction. Set up the table with labeling in a form suitable for presentation. Write a paragraph describing the relationship between the pair of variables shown in this table.

2. Select an interesting percentaged table from a journal article. Compute percentages in each of the three possible ways (to row totals, column totals, and the grand total). Then, in your own words, express what it is that percentages in each of these three tables tell about the existence, degree, direction, and nature of association in the table.

3. Look through one of the professional journals for an example of a properly percentaged table and for a table that was not properly percentaged for the kinds of conclusions the author wanted to make. Discuss the use of percentaged tables in both articles.

4. Find one example of the use of Yule's Q and phi in journals or texts. Examine each usage and discuss the merits of selecting Q or ϕ in each case.

5. The data in Tables A and B are from a study of the relationship between economic strain and the consumption of alcohol to relieve distress. Compute percentages in the appropriate direction for each table and use epsilons to interpret the data. What conclusions do the data warrant about the relationship between economic strain and the disposition to use alcohol to relieve distress?

TABLE A ECONOMIC STRAIN AND LEVEL OF ANXIETY (Frequencies)

| | Anxiety | | |
Economic Strain	Intense	Moderate	Low
Severe	41	27	32
Moderate	51	46	48
Little	56	80	98
None	190	310	691

Reprinted from Pearlin, Leonard I., and Clarice W. Radabaugh, "Economic Strains and the Coping Functions of Alcohol," *American Journal of Sociology*, 82 (November, 1976), by permission of the University of Chicago Press. Copyright 1976 by the University of Chicago and Published by the University of Chicago Press.

TABLE B LEVEL OF ANXIETY AND USE OF ALCOHOL FOR CONTROL OF DISTRESS (Frequencies)

| Disposition to Use Alcohol for Distress Control | Anxiety | | |
	Intense	Moderate	Low
Strong	77	66	104
Weak	61	75	138
Minimal	198	329	623

Reprinted from Pearlin, Leonard I., and Clarice W. Radabaugh, "Economic Strains and the Coping Functions of Alcohol," *American Journal of Sociology*, 82 (November, 1976), by permission of the University of Chicago Press. Copyright 1976 by the University of Chicago and Published by the University of Chicago Press.

6. Compute the percentages in Tables A and B in the opposite direction from that in which you computed them in Problem 5. Again use epsilons to interpret the data. Now what conclusions do the data warrant?

7. The following data are from a study of political leadership in Chile. Compute ϕ and Yule's Q for these data. How do they compare? Which is more appropriate for the analysis of these data? Why?

TABLE C FREQUENCIES OF LARGE CORPORATION EXECUTIVES WHO HELD NATIONAL POLITICAL OFFICE, BY POLITICAL OFFICEHOLDING IN IMMEDIATE FAMILY

Political Officeholders in Immediate Family	Number of Political Offices Held	
	None	One or More
Yes	39	29
No	126	35

Source: Zeitlin et al. (1976:1019).

GENERAL REFERENCES

Davis, James A., and Ann M. Jacobs, "Tabular Presentation," in David L. Sills (ed.), *International Encyclopedia of the Social Sciences*, Vol. 15, (New York, Macmillan and The Free Press), 1968, pp. 497–509.
  This article discusses appropriate ways to set up tables, how to handle "don't knows," and how to use percentages in examining tables.

Riley, Matilda White, *Sociological Research I: A Case Approach* (New York, Harcourt Brace Jovanovich), 1963.
  Unit 8 presents brief summaries of several important sociological studies, and commentary discusses their examination of the relationship among variables.

Rosenberg, Morris, *The Logic of Survey Analysis* (New York, Basic Books), 1968.
  Chapter 1 discusses the meaning of relationships.

Weiss, Robert S., *Statistics in Social Research: An Introduction* (New York, Wiley) 1968.
  See especially Chapter 4 on tables and Chapter 9 on association.

Zeisel, Hans, *Say It with Figures*, sixth ed. (New York, Harper & Row), 1985.
  Chapter 1 discusses the use of percentages, Chapter 2 presents examples of percentaging rules, and Chapter 3 handles the problem of "don't knows" or "no responses." More complex tables are discussed in Chapter 4. This is a very thoughtful and readable treatment of the use of percentages in tables.

LITERATURE CITED

Erskine, Hazel, "The Polls: Women's Role," *Public Opinion Quarterly*, 35 (Summer 1971), p. 278.

Fischer, Claude S., "A Research Note on Urbanism and Tolerance," *American Journal of Sociology*, 76 (March 1971), pp. 847–856.

Kim, Jae-On, "Predictive Measures of Ordinal Association," *American Journal of Sociology*, 76 (March 1971), p. 900.

National Center for Health Statistics, C. A. Schoenborn, and B. H. Cohen, "Trends in Smoking, Alcohol Consumption and Other Health Practices among U.S. Adults, 1977 and 1983," advance data from *Vital and Health Statistics*, No. 118, DHHS Publ. No. (PHS) 86–1250, (Hyattsville, Md., Public Health Service), June 30, 1986.

Pearlin, Leonard I., and Clarice W. Radabaugh, "Economic Strains and the Coping Functions of Alcohol," *American Journal of Sociology*, 82 (November 1976).

Sewell, William H., and Vimal P. Shah, "Parents' Education and Children's Educational Aspirations and Achievements," *American Sociological Review*, 33 (April 1968).

Sundstrom, Gerdt, "Intergenerational Mobility and the Relationship between Adults and Their Aging Parents in Sweden," *The Gerontologist*, 26 (No. 4, 1986) p. 369.

U.S. Bureau of the Census, "Money Income of Households, Families, and Persons in the United States: 1983," *Current Population Reports*, Series P-60, No. 146 (Washington, D.C., U.S. Government Printing Office), 1985.

Zeitlin, Maurice, W. Lawrence Neuman, and Richard Earl Ratcliff, "Class Segments: Agrarian Property and Political Leadership in the Capitalist Class of Chile," *American Sociological Review*, 41 (December 1976).

CHAPTER 7

Measures of Association for Nominal, Ordinal, and Interval Variables

The ecology of Chicago has long been a source of attention since Park, Burgess, and other early investigators at the University of Chicago began to study the patterning of urban phenomena (*e.g.*, social class, ethnic organization). Continuing this research, Albert Hunter (1971) examined ecological trends in Chicago, using census data from the 1930s to the 1960s based on 75 community social areas into which the city had been divided by sociologists in the 1920s. Among the interests Hunter had in this longitudinal study was whether there has been increasing or decreasing segregation in the city over these years. One would expect greater assimilation to occur the longer a group has been in an area, so that the area could no longer be distinguished in terms of the kinds of jobs or income level or family status of people living there compared to others in the city in general.

Ecological research in the United States has found three main differences between urban areas: social rank, family status, and segregation or ethnic status of the people living in an area. Hunter used percentage black and percentage foreign-born in each of the 75 Chicago areas to measure segregation or ethnic status. Percentage of females employed was used as a measure of family status, and median dollar value of homes was used as a social rank measure. Four other variables were measured but are not reported here. The basic statistical measure used to describe the relationship between ethnic status and each of the other variables was a correlation coefficient or measure of the existence, strength, and direction of association between pairs of variables.

Hunter's reasoning was that if there is low segregation (*i.e.*, high assimilation) then there should be little or no correlation between the percentage black in an area, for example, and the percentage of females who are employed— one of the family status measures he used. Ethnic status should not make a difference in the distribution of that particular variable, nor should it be correlated with the median value of homes if there is no segregation. If

TABLE 7.1 MEASURES OF THE BIVARIATE ASSOCIATION BETWEEN EACH OF TWO SEGREGA-TION VARIABLES (PERCENTAGE BLACK AND PERCENTAGE FOREIGN-BORN) AND MEASURES OF STATUS IN 75 COMMUNITY AREAS OF CHICAGO FOR THE YEARS 1930–1960

A. Correlations between Percentage Black and the Percentage Females Employed*

1930	.46
1940	.42
1950	.13
1960	−.03

B. Correlations between Percentage Black and the Median Value of Homes*

1930	−.09
1940	−.17
1950	−.22
1960	−.18

C. Correlations between Percentage Foreign-born and Percentage Females Employed*

1930	−.28
1940	−.13
1950	.06
1960	.12

D. Correlations between Percentages Foreign-born and Median Value of Homes*

1930	−.31
1940	−.41
1950	−.15
1960	−.01

E. Correlations between Percentage Black and Percentage Foreign-born*

1930	−.52
1940	−.63
1950	−.71
1960	−.77

Source: Data from Hunter (1971:437, Table 5).
* Correlation coefficients are Pearson product-moment correlations (*r*), described at the end of this chapter. These coefficients vary between −1.00 and +1.00 and, in general, they can be interpreted in terms of their magnitude—the larger the absolute value of the coefficient, the stronger the association or the more closely related are the two variables being compared.

correlations become smaller in magnitude over time, there is evidence of decreasing segregation or increasing assimilation. If, on the other hand, there has been no drop or if there is a shift from a lower to a higher correlation through time, then there is evidence of either no change in segregation or increasing segregation in urban Chicago. The data in Table 7.1 indicate some interesting increases and decreases that may be due in part to migration from urban to suburban areas that occurred across the nation during this time.

Notice, in Table 7.1, that the correlation measure for the bivariate relationship between percentage black and percentage females employed in these 75

Chicago areas was +.46 in 1930, and that the correlation drops steadily in magnitude over time to +.42 in 1940, +.13 in 1950, and −.03 (smaller in magnitude but a shift to a negative rather than positive direction of association) in 1960.

BOX 7.1   ASSOCIATION

Double-check yourself. If you are not really clear about the following ideas, go back for a review.

Association of two variables (Section 6.4)
Direction of association (Section 6.4.3)
Degree of association (Section 6.4.2)
Nature of association (Section 6.4.4)
Normed measure of association (Sections 6.5 and 6.5.1)
Statistical independence (end Section 6.4.1a)
Perfect association (Section 6.5.2)

The purpose of the current chapter is to develop several measures of association that are normed, are interpretable in an intuitively meaningful way, and are appropriate for variables defined at various levels of measurement and related in certain characteristic ways.

Comparison of these correlation coefficients indicates that percentage black in an area is becoming less associated with percentage of women employed, and this can be interpreted to mean that assimilation has occurred in this aspect of family status for this ethnic group. Notice that just the opposite pattern characterizes the association of percentage black and median value of homes. The small negative association in 1930 becomes larger through time, although there is a reduction to −.18 in 1960. A decline in correlations from 1930 to 1960 is more characteristic of foreign-born than of black, indicating a more general pattern of assimilation for the foreign-born group. Notice, too, that the association between the two segregation measures, percentage black and percentage foreign-born, for these 75 areas is getter stronger through time and that it is a negative association. Where one group is concentrated in an area, the other group, increasingly, is absent. This is what one would expect if the foreign-born were moving to suburban areas, leaving higher percentages of black families behind in the central city neighborhoods.

Now for some comments on the coefficients we have been examining. No comparison such as those Hunter made would be possible unless the measures of association he used were comparable, that is, properly normed so that correlations of a given magnitude have the same meaning. This is an important point, because we want to compare across time and across different combinations of variables to reach a conclusion. Second, notice that the numbers may be positive or negative, showing direction of association (*e.g.*, the *higher* the percentage black in an area, the *higher* the percentage of employed women—a positive association—or, the *higher* the percentage of foreign-born in an area the *lower* the percentage of black—a negative association). Magnitude of the number varied from a small value, indicating a weak association between the cross-classified variables (a value of 0 would mean no association), to a large value, approaching 1.0, indicating a strong association.

This chapter will introduce three families of measures of association that Hunter might have used for the purpose of validly comparing strength and direction of association between variables. There are several other types of measures we could find and use.* The purpose of this chapter is to present some of those measures used frequently by sociologists and to illustrate the straightforward logic underlying a normed measure of association. We will present measures suitable for the association of nominal, ordinal, and interval variables.

7.1 PROPORTIONATE REDUCTION IN ERROR MEASURES (PRE)

All the measures we will discuss in this chapter are of a type called "proportionate reduction in error" measures, or **PRE** measures. They are all relatively simple ratios of the amount of error made in predicting under two situations: *first*, the situation where there is no more information than simply the distribution of the dependent variable itself, and, *second*, a situation where there is additional knowledge about an independent variable and the way the dependent variable is distributed within the categories of that independent variable. PRE measures simply state the proportion by which one can reduce errors made in the first situation by using information from the second situation, above.

$$\text{PRE} = \frac{\text{Reduction in errors with more information}}{\text{Original amount of error}}$$

The problem of prediction is a common one to the sciences, so it makes some sense to focus a measure of association on the idea of making accurate predictions of the values of some dependent variable. If theoretical knowledge leads us to say that people with a higher social class standing in a society will feel less alienated than those with lower class standing, we are saying, in effect, that knowledge of social class score differences will permit us to make more accurate predictions of differences in alienation scores. If *all* the errors of prediction can be eliminated by basing predictions on social class, then there is a perfect association between these two variables, and our theoretical basis for expecting this outcome is supported. On the other hand, if the association between social class and alienation is poor or nonexistent, then that fact would be indicated in a measure of association that expresses the proportion of the original predictive errors that can be avoided by virtue of the additional knowledge about social class—in this case little or none.

There are three things we might be interested in predicting, depending, essentially, upon the definition of the variables involved in a problem. For *nominal* variables, we are usually interested in predicting the category or exact score of the

---

* Among the variety of measures for special situations, a good sampling is presented in an elementary statistics book by Freeman (1965) (see General References).

dependent variable. Often it is sufficient to focus prediction on the most typical or *modal* value of the dependent variable. Alternatively, we could be interested in predicting the distribution of the variable across its categories.* If the dependent variable is *ordinal*, then we are probably interested in predicting *rank order* of pairs of scores on the dependent variable (although we could think of other possibilities, such as predicting its distribution or the median). Finally, if the dependent variable is an *interval* variable, we would probably be interested in predicting the arithmetic *mean* of that dependent variable. These alternatives are shown in Table 7.2.

Generally, we make errors in predicting modes or rank orders or means, but with the proper theory and proper information we can cut down the amount of predictive error we would otherwise make. Table 7.2 indicates the way in which we might make predictions both with and without information. The PRE measures of association we will discuss below are simply a contrast between the errors made in using rule 1 and those made in using rule 2 to predict the mode or rank order or mean of a variable that interests us. Usually, the contrast will be formed as follows:

$$\text{(7.1)} \qquad \text{PRE} = \frac{(\text{Errors made using rule 1}) - (\text{Errors made using rule 2})}{(\text{Errors made using rule 1})}$$

7.2   MEASURES OF ASSOCIATION FOR NOMINAL VARIABLES

PRE measures for each of the situations illustrated in Table 7.2 will be discussed in turn in the following sections.

7.2.1   Lambda (λ_{yx})

The **lambda** measure of association (also called Guttman's coefficient of predictability), λ_{yx}, is an asymmetric measure of association especially suited to bivariate distributions where both variables are interpreted to be nominal variables. It is a measure that very nicely illustrates the logic of PRE measures.

Suppose that we are interested in marital status as a dependent variable, and we are interested in making predictions about the marital status of individuals

* In the case of nominal variables, we will also discuss a measure of association that focuses upon predicting the category into which a case may fall, whether or not this category is the modal category. This measure focuses, in a sense, upon the *distribution* across categories of the dependent variable rather than upon an *optimal* single category. The various coefficients are introduced here, not only because they find important uses in sociology, but because they highlight the notion that there are many things about a dependent variable that one might be interested in predicting. The choice among available measures of association involves not only technical or computational matters, but also a basic understanding of what one wants to predict, and this in turn is tied directly to the substantive logic of the research problem itself.

TABLE 7.2 DEVELOPMENT OF PRE MEASURES OF ASSOCIATION*

Feature of the Dependent Variable to be Predicted	Rules Used in Making Predictions	
	Rule 1: Minimum Guessing Rule	Rule 2: Improved Guessing Rule
A. Modes (alternatively, the distribution) (Nominal variables)	For each case to be predicted, predict the modal category of the variable, overall. (Alternatively, predict the category placement at random.)	For each case, first determine the category of the independent variable into which the case falls; then predict the mode of the dependent variable for that category. (Alternatively, predict category placement within categories of the independent variable.)
B. Rank Order (Ordinal variables)	For each pair of cases for which rank order on a variable is to be predicted, determine which is ranked higher by a flip of a coin (random selection.)	For each pair of cases for which rank order on a dependent variable is to be predicted, first determine the ordering on an independent variable, and then, if the overall association is positive, predict "same ordering" on the dependent variable. If the overall association is negative, predict "reverse ordering" on the dependent variable.
C. Mean (Interval variables)	For each case to be predicted, predict the overall mean of the dependent variable.	For each case to be predicted, either (a) determine the category of the independent variable it falls in and predict the mean of that category, or (b) develop a regression equation that permits you to compute a predicted value for the dependent variable, given the score on the independent variable.

* There are a number of good references for further discussions of PRE measures of association; see, for example, Costner (1965) and Kim (1971).

who are household heads in the United States.* Given the information in Table 7.3 about how marital status is distributed in the United States, we would do best to predict that household heads are married. That is, if we know that the modal marital status is "married," then the most rational single score we could predict for

* The U.S. Census defines a head of household as one person, usually the person regarded as the head of the family members, but women are not classified as heads if their husbands are resident members of the family.

TABLE 7.3 Frequency Distribution of Marital Status of Household Heads in the United States in 1970 by Type of Household (in thousands)

| | Type of Households | | | | |
| | Male Headed | | Female Headed | | |
Marital Status of Head	Related Children under 18	No Related Children under 18	Related Children under 18	No Related Children under 18	Total
Married	25,776	19,214	313	198	45,501
Separated	79	502	998	425	2,004
Divorced	74	946	1,135	1,105	3,260
Widowed	181	1,199	942	6,457	8,779
Single	64	2,302	349	2,113	4,828
Total	26,174	24,163	3,737	10,298	64,372

Source: U.S. Bureau of the Census (1971a:17, Table 6).

a head of household would be that the person is married. We will be correct more frequently than if we picked any other single category of marital status for our prediction. Information on the overall distribution of marital status for household heads in the United States in 1970 is shown in the total column at the right of Table 7.3.

If we were to guess that the head of the household is married before knocking on the door for an interview with each of the 64,372 (thousand)* households in the U.S., we would be right 45,501 times and wrong 18,871 times (*i.e.*, 64,372 − 45,501 = 18,871). This is the total number of errors of prediction we would make if we merely predicted the overall mode of the marital status of the head of household variable. How many of these errors could be eliminated if we had more information to start with? (See rule 2 information listed in Table 7.2).

The body of Table 7.3 shows the distribution of marital status of the head of household for separate categories of the "type of household" variable. A quick glance at this table suggests that, indeed, marital status is distributed differently depending upon whether the household is headed by a male and includes related children 18 years of age, is male-headed with no related children under 18, is female-headed with related children under 18, or is female-headed with no children under 18. How much is this added information worth? If we knew the "type of household" before making our prediction of modal marital status, would we be able to refine our prediction and make fewer errors in predicting scores on the dependent variable?

In this example, the answer is "yes." If we knew that the head of household is a male and that the household contained related children under 18 years of age, we would certainly predict the modal marital status for the head of that kind of

* In Table 7.3, the number of households is expressed in terms of thousands of households—the unit of measurement. It is important, of course, to be consistent in using the same unit throughout our computations and to interpret the result with this unit in mind.

household—that he would be "married." We would be right 25,776 times out of 26,174 households of this type. We would also predict "married" for the male-headed-no-young-children households—again the mode for that category—and we would be right 19,214 times out of 24,163. If we knew, however, that the household head was female and that there were under-18, related children in the household, we would predict that the household head is "divorced," the mode of marital status for that type of household. We would be right 1,135 times out of 3,737 households of this type. Finally, if the household head were female and if there were no under-18, related children, we would do best to predict "widowed" as the household head's marital status. Here we would be right 6,457 times out of 10,298.

Have we improved our predictive power by virtue of this added information? We can determine the answer by adding up the correct predictions resulting from within-category prediction (*i.e.*, using rule 2 to predict) and contrasting that with the overall frequency in the modal category of the marital status variable.

Within-Category Modal Frequency m_y	Category of the Independent Variable(s)
25,776	Households with male heads and related children under 18 years of age.
19,214	Households with male heads and no related children under 18 years of age.
1,135	Households with female heads and related children under 18 years of age.
6,457	Households with female heads and no related children under 18 years of age.
$\Sigma m_y = 52,582$	Total of within-category modal frequencies

The more refined prediction would lead to 52,582 correct predictions, which is some 7,081 fewer errors than would be made if we had merely predicted the overall mode of marital status (in which case we would be correct only 45,501 times out of 64,372). That amounts to a reduction of 37.5% in the errors made in predicting marital status of households. This value is called **lambda** (λ_{yx}), and it is a simple substitution of total errors and reduction in errors in the earlier generalized PRE formula. More specifically, λ_{yx} is computed as follows:*

(7.2)
$$\lambda_{yx} = \frac{\Sigma m_y - M_y}{N - M_y}$$

* Lambda could be expressed in terms of the format given earlier for PRE measures as

$$\lambda_{yx} = \frac{(N - M_y) - (N - \Sigma m_y)}{N - M_y}$$

where the first term in the numerator is the number of errors made using rule 1, and the second term is the number of errors made using rule 2. Simplifying the numerator, as shown in the text above, yields a numerator that is the number of *nonerrors* under rule 2 (Σm_y) minus the number of *nonerrors* under rule 1 (M_y). The denominator is the number of errors under rule 1.

where N is the total sample size; M_y is the *overall* modal frequency of the dependent variable, Y (45,501 in this case); and Σm_y is the sum of modal frequencies on the dependent variable, Y, *within* separate categories of the independent variable, X (52,582 in this example). Applied to the current example, the computation is as follows:

$$\lambda_{yx} = \frac{52,582 - 45,501}{64,372 - 45,501} = \frac{7,081}{18,871} = .375$$

The numerator expresses the reduction in error with improved information and the denominator expresses the error with minimum information. Notice that the symbol for lambda is a lowercase Greek letter and that it has two subscripts: λ_{yx}. The first of these subscripts indicates which variable is the *dependent* variable (variable Y, traditionally), and the second subscript indicates which variable is the *independent* variable (variable X, traditionally).

Lambda$_{yx}$ is the asymmetric measure of degree of association that expresses the proportionate reduction in errors made in predicting modal values of variable Y when prior information about variable X is available to use in refining predictions of modal scores on the dependent variable. Lambda$_{xy}$ simply reverses the role of the two variables, predicting X from information about Y. Asymmetric measures of association, unlike symmetric measures, always must be labeled in this way because the magnitude of the number depends upon which variable is predicted from which.

Suppose, to turn the predictive problem in Table 7.3 around, we were interested in predicting scores on "type of household." How much improvement in prediction would result from using marital status as the predictor variable? The formula for lambda is the same as before, with the exception that the x subscript is substituted for the y subscript (and vice versa) in the previous example. We are interested in predicting modes on type of household, overall, and then within categories of the marital status variable. The computations are as follows:

	m_x
$M_x = 26,174$	25,776
	998
	1,135
	6,457
	2,302
$\Sigma m_x =$	36,668

$$\lambda_{xy} = \frac{36,668 - 26,174}{64,372 - 26,174} = \frac{10,494}{38,198} = .275$$

Marital status permits one to reduce errors in predicting type of household by 27.5%. The difference between λ_{yx} and λ_{xy} in this example is 10%. In general, the two different coefficients need *not* be the same for any given table, and they are interpreted quite differently. Using lambda, one could ask which variable(s) permit the greatest reduction in errors in predicting modes of some specific dependent variable, and lambda is a measure of association that helps one assess the utility of certain additional information.* Note that some of the difference between the two lambdas in a table may result from the precision of measurement of the predictor variable. If one wants to predict a dependent variable that has five categories by using a predictor variable that has only four categories, then one could predict only four different modes, not five. Where the predictor variable is more precise than the dependent variable, then more exacting predictions can be made. This is one reason why investigators attempt to preserve more rather than fewer categories of variables that are to be used in statistical analysis. Note that, since lambda is suitable for nominal variables, rows or columns may then be reordered without affecting the magnitude of lambda.

Lambda$_{yx}$ varies in magnitude from 0.0 to $+1.0$, and it can take on these values for any table regardless of size or marginals. Assuming that, overall, there is some range of scores on the dependent variable, a perfect association is defined as a condition in which all of the cases in each category of the independent variable fall into only one category of the dependent variable (the modal category).† Lambda$_{yx}$ is 0 where the *same* modal prediction is made within all categories of the independent variable as would be made if the overall mode were predicted. Here, quite literally, the extra information about the independent variable adds nothing in refining predictions of the mode of the dependent variable. Table 7.4 illustrates a situation in which λ_{yx} is 0, but from an examination of percentages one can see that an association exists in a sense other than the meaning involved in predicting modes.

The problem with the data in Table 7.4 is that the distribution of the dependent variable (type of impairment) is decidedly skewed. Six out of every ten impairments are of the orthopedic type. Consequently, the modes within each category of the independent variable are likely, also, to be in the orthopedic category of the dependent variable. In general, lambda is a poor choice as a measure of association for data in which there is pronounced skewness in the distribution of the dependent variable. It is only sensitive to modes.

It should be noted that lambda tells only part of the story of the association between two nominal variables. In addition to knowledge of the *strength* of association, an investigator would want to examine the *nature* of the association. In this case nature would probably best be discussed in terms of the pattern of percentages in a properly percentaged table such as Table 7.4.

---

* Some investigators compute a "symmetrical" lambda coefficient, which is a kind of average of the two asymmetric lambda coefficients shown here. This is believed to be of limited value and its computation will not be discussed. The formula for the symmetrical lambda, however, is

$$\lambda = \frac{\Sigma\, m_y + \Sigma\, m_x - M_x - M_y}{2N - M_x - M_y}$$

† If all scores in the sample are identical, then a perfect prediction could be made with only the overall mode, and λ_{yx} would equal 0, because additional information about an independent variable cannot reduce predictive errors at all. There would be no errors to reduce, in this instance, and thus the independent variable would not be worth anything in prediction of the mode of the dependent variable.

TABLE 7.4 PERCENTAGE DISTRIBUTION OF TYPE OF IMPAIRMENT BY AGE, UNITED STATES, JULY 1966–JUNE 1967

(Y) Type of Impairment	Age(X)			
	Under 15	15–24	25 and over	Total
Visual	12.6	10.4	15.2	14.5
Hearing	18.0	11.7	24.4	22.7
Speech	21.0	3.4	1.6	3.1
Orthopedic	48.4	74.5	58.8	59.7
Total	100.0	100.0	100.0	100.0
N(1000's)	(2,601)	(3,952)	(32,560)	(39,113)
		$\lambda_{yx} = .00*$		

Source: U.S. Dept. H.E.W., 1971:15, Table 3. Data are from household interviews with a national sample of children and youth aged 25 or under from civilian, noninstitutionalized population by the U.S. Census. Impairments are defined as chronic or permanent defects which cause a decrease or loss of ability to perform various functions.
* Note that although λ_{yx} is 0, there *is* a difference in percentages compared across rows of this table, and thus an association not detected by a measure concerned with predicting modes. Here again it is important to select measures that are sensitive to desired features of data.

7.2.2   The Uncertainty Coefficient, U

When a dependent variable has a skewed distribution as in Table 7.4, it is desirable to have a measure of association for nominal variables that uses information about the whole distribution of the dependent variable, not just the modal category as in the case of lambda. Two measures are available to measure the reduction in uncertainty achieved by predicting the **distribution** of the dependent variable. We will discuss one of these measures, the **uncertainty coefficient**, in some detail here. The other coefficient, tau-y, uses a similar measurement philosophy, but it takes a different approach to calculation.*

The uncertainty coefficient is a normed, asymmetric, PRE measure of association whose formula is similar in form to those of other PRE measures of association. The formula is as follows:

(7.3)
$$U_{yx} = \frac{h_y - h_{y \cdot x}}{h_y}$$

* Goodman and Kruskal's tau-y is based on discriminating between pairs of cases, while the uncertainty coefficient is based on predicting categorical placement of cases. Tau-y varies from 0.00, for no reduction in uncertainty to 1.00, for complete reduction of uncertainty in predicting a dependent variable from knowledge of an independent variable. In most (not all) cases, tau-y is slightly higher in magnitude than U. Tau-y extends the univariate index of dispersion to a bivariate situation. See the discussion of tau-y in the third edition of this text. Note that in the third edition a distinction between tau-y and the uncertainty coefficient as computed above was not made. Both approaches are useful in measuring relationships in variability of nominal variables. We have changed our emphasis to the uncertainty coefficient because it is more consistent with expressions of relationship contained in hierarchical modeling discussed later in the book (Section 16.2) and because it is available for use as part of such popular computer statistical packages as SPSS and SAS.

where U_{yx} is the coefficient in which y is the dependent variable and x is independent. The numerator is a measure of the reduction in uncertainty of prediction of y based on knowledge of x (i.e., the improvement in prediction), and the denominator is the original amount of uncertainty in the dependent variable considered alone. It is the calculation of uncertainty (h) that we will now explain.

7.2.2a   Concept of Uncertainty

To explain the concept of uncertainty, we will start with a univariate example in which we have a group of 20 individuals measured on their employment status. Suppose we are only interested in whether they are employed or not employed. That is, we would like to locate them in one of these two categories of the variable, *employment status*. To start with, suppose we know that all 20 individuals have the same status, but we don't know whether they are employed or not.

Our objective is to identify correctly the employment status of any one of the 20 cases. We can make as many guesses as we wish and, after each guess, we will be told whether or not we are correct. **How many guesses will we need to identify correctly the employment status of any one of the 20 individuals?** Here, with two categories of employment status, the answer is that we need only one guess. If we guess "employed" and are told we are correct, then we know the person is employed. On the other hand, if we guess "employed" and are *in*correct, then we know the person must be unemployed. It takes only one guess. Furthermore, since all 20 individuals are in the same employment category, once we have established the status of the first individual we also know the status of the other 19.

Employment Status	Frequency
Employed	0
Not employed	20
Total	20

Now, suppose that we had defined employment status as a variable with four categories: employed full time; employed part time; unemployed but looking for work; and not in the labor force. Again, assume that we know that all 20 cases fall in the same category. How many guesses will be needed to locate any given case correctly? First, we could guess the person is "employed." If we are right, then we need one more question: "Is the person employed full time?" If we are right again, we know the person is a full-time worker. If wrong on the second question,

Employment Status	Frequency
Employed full time	0
Employed part time	0
Unemployed	0
Not in the labor force	20
Total	20

we know that the person must be a part-time worker. On the other hand, if we were wrong on the first employment question, then we know the person is not employed, but we now need to ask, "Is the person in the labor force?" Overall, we need two yes–no questions answered to identify the employment status of a person in this distribution of four categories. Here, too, the two questions necessary to locate one case are sufficient to determine the location of the other 19 since they all share the same status.

In the first example, when there were two categories of response, it took the answer to one question for us to resolve our **uncertainty** about employment status; thus, in the technical terminology used, we had **1 bit** of uncertainty. In the second example, in which there were four categories, it took answers to two questions to discover the employment status of the person. Therefore, there were **2 bits** of uncertainty involved.

The relationship between the number of categories of response and the number of answers needed to resolve the uncertainty can be symbolized as follows:

(7.4)
$$k = 2^h$$

where k is the number of response categories for a variable and h is the power to which 2 needs to be raised to equal k. The base 2 is used because we are using yes–no, or *binary* questions. The quantity h is the building block for our measure of uncertainty. Senders (1958) notes that h, in information theory, the source of this measure, is referred to as a *bit*, which is short for "binary digit." That explains the terminology, "1 bit" or "2 bits" of uncertainty above.

7.2.2b   The Use of Logarithms

In the two examples just considered, the numbers of bits of uncertainty were integers, 1 bit or 2 bits, because the number of categories of response for the variable used was first two then four. When the number of categories is an exponentially increasing function of the number 2, h will be an integer. Otherwise, h will be a fractional power of 2. Consequently, it is convenient to express the formula for h in logarithmic form with the logarithm to the base 2 to facilitate solving for h. Since, by Formula 7.4, $k = 2^h$, h can be expressed logarithmically as follows:

(7.5)
$$h = \log_2 k$$

Suppose that the 20 cases are distributed equally over the four categories of employment, rather than all being located in the same category. One-fourth of the 20 cases, or 5 cases, would be in each employment category, as shown below:

Employment Status	Freq.	Prop.	No. of Guesses	Prop × Guesses
Employed full time	5	.25	2	.50
Employed part time	5	.25	2	.50
Unemployed	5	.25	2	.50
Not in labor force	5	.25	2	.50
Totals	20	1.00		2.00

BOX 7.2   THE BASE OF LOGARITHMS

Logarithms are useful at several points in statistics and you may want to refresh your memory on their use. Box 4.4 and section 4.4.3 describe logarithms that use base 10, expressing the exponent to which 10 must be raised to equal some number. This was useful in examining rates of change. Here base 2 is used because we have two-state, yes–no or binary questions to ask, and it is the exponent that expresses the number of "bits" of uncertainty in establishing the location of a case in a distribution. The uncertainty coefficient, as you will see, uses logarithms in the numerator *and* denominator, so the particular base of logarithms cancels out. Thus, for the final computation of U, the base of logarithms used does not make a difference. Most computer programs readily compute both base 10 or "common" logarithms and natural logarithms, which use the constant e (2.71828...) as a base.

Common logs (base 10) may easily be converted to logs to the base 2 by multiplying them by the reciprocal of $\log_{10} 2$ (3.322), and natural logs (base e) may be converted to logs to the base 2 by multiplying them by the reciprocal of $\log_e 2$ (1.443).

Here, again, we need two guesses to locate a particular individual's employment status, but since only 25% of the cases are in any given category, the two particular questions and answers needed to eliminate the uncertainty about each individual's location would differ according to the individual's employment status. For example, to establish that an individual was employed full time, the following two questions and answers would be necessary:

Question 1: "Is the person employed?" Answer: "Yes."

Question 2: "Does the person work full time?" Answer: "Yes."

On the other hand, to establish that the individual was employed part time, the answer to question 1 would have to be "yes," but the answer to question two would have to be "no."

There are, as a matter of fact, four combinations of questions and answers needed to locate all 20 cases in their correct categories. For each employment category there are 2 bits of uncertainty (or 2 bits of information necessary) to establish the location of a case. However, the 2 particular bits necessary to place each individual depend on the category. Since there are 25% of the cases in each category, it is necessary to compute an average number of bits over the four categories. This is done by multiplying the number of bits for a category by the proportion of cases in that category and summing these products. This, in effect, produces a weighted average of bits of uncertainty for the table. Notice that, for the table above, the weighted average of proportions times guesses in the last column still sums to 2 bits, the same result we got when all cases were in a single response category. This is so because the cases were evenly distributed over the four response categories: it will not generally be true.

The general formula for computing the weighted average of guesses necessary to locate the cases in a distribution is as follows:

(7.6)
$$h = -\sum p_i \log_2 p_i$$

where h is the weighted average of the number of guesses and p_i is the proportion of cases in the ith category. This formula applies no matter how the cases are distributed over the categories of the variable of interest and no matter the number of categories into which the cases have been divided.

To illustrate the use of this formula, suppose that we have the following distribution of 20 cases:

Employment Status	Freq.	Prop.	$-p_i \log_2 p_i$
Employed full time	10	.50	.5000
Employed part time	5	.25	.5000
Unemployed	4	.20	.4644
Not in labor force	1	.05	.2161
Totals	20	1.00	1.6805

The entries in the last column of this table come from the Table of Selected Values for $-p_i \log_2 p_i$ in Box 7.3. The total in the last column is equal to h; therefore, we need, on average, 1.6805 bits of information to locate a case accurately. This is less than the 2 bits we found where the cases were evenly distributed over the four categories. The smaller value of h reflects the fact that we can reduce our uncertainty in guessing by guessing "employed full time" first, because half of all of the cases are in that category.*

7.2.2c The Bivariate Situation

To apply the concept of uncertainty to the bivariate situation, it is necessary to introduce a predictor (independent) variable. Dividing the 20 cases in the previous example by gender, we now wish to determine whether knowing a person's gender will improve our prediction of the dependent variable, employment status. In this example, the 20 cases are distributed such that 75% are men and 25% are women, but the overall distribution of employment status is the same as it was before. This bivariate distribution in both frequency and proportion form

* Note that Senders' univariate measure of variation H_{rel}, mentioned in Chapter 5 (p. 141), uses the same approach as the uncertainty coefficient. It is calculated as the ratio of the uncertainty in a given distribution (in bits, or h) divided by the maximum potential uncertainty (in bits, or h), where cases are evenly distributed over all available categories. As an example, the H_{rel} for the data distributed over the 4 categories in the table above would be 1.6805/2.000 = .84. The H_{rel} measure varies between 0.0, where there is no variation (all cases are in one category), to 1.00, where cases are evenly distributed over all categories (maximum variation).

BOX 7.3 TABLE OF SELECTED VALUES FOR $-p_i \log_2 p_i$

Tables are available to look up this calculation (see Senders, Table A, 1958), but more frequently one would use a computer program such as SPSS or SAS for the calculation of U. Illustrative values, useful for the examples covered here, are:

Proportion	$\log_2$	$-p_1 \log_2 p_i$
.00	−0.0000	.0000
.05	−4.3219	.2161
.07	−3.8365	.2686
.10	−3.3219	.3322
.15	−2.7370	.4105
.20	−2.3219	.4644
.25	−2.0000	.5000
.27	−1.8890	.5100
.30	−1.7370	.5211
.33	−1.5995	.5278
.35	−1.5146	.5301
.40	−1.3219	.5288
.45	−1.1520	.5184
.50	−1.0000	.5000
.55	−0.8625	.4744
.60	−0.7370	.4422
.65	−0.6215	.4040
.67	−0.5778	.3871
.70	−0.5146	.3602
.75	−0.4150	.3113
.80	−0.3219	.2575
.85	−0.2345	.1993
.90	−0.1520	.1368
.95	−0.0740	.0703
.999	−1.4434	.0014

is as follows:

Employment Status	Frequency			Proportion		
	Men	Women	Total	Men	Women	Total
Employed full time	10	0	10	.67	.00	.50
Employed part time	1	4	5	.07	.80	.25
Unemployed	4	0	4	.27	.00	.20
Not in labor force	0	1	1	.00	.20	.05
Totals	15	5	20	1.01	1.00	1.00
Gender proportions	(.75)	(.25)	(1.00)			

Here we would improve our prediction somewhat by knowing the gender of the person before making our guesses. For example, if the person were a woman, we would be better off guessing that she is employed part time. If the person were a man, we would be correct more often if we first guessed the man to be employed full time. Other guesses would follow our knowledge of how men and women are distributed on this variable.

The overall uncertainty in the dependent variable (here symbolized as h_y) has already been computed to be 1.6805 using Formula 7.6. This is the total amount of uncertainty that we seek to reduce by introducing a predictor variable.

To calculate uncertainty for the rest of the table, we would first convert the frequencies to proportions, as shown. Then we would compute the uncertainty measure, as we did before (using Formula 7.6 and the partial table of h values in Box 7.3) separately for each category of the independent variable (gender).

	For Men		For Women	
Employment Status	*Prop.*	$-p_i \log_2 p_i$	*Prop.*	$-p_i \log_2 p_i$
Employed full time	.67	.3871	.00	.0000
Employed part time	.07	.2686	.80	.2575
Unemployed	.27	.5100	.00	.0000
Not in labor force	.00	.0000	.20	.4644
Totals	1.00	1.1657	1.00	.7219

Since 75% of the cases were men and 25% were women, we will use these proportions to weight the h's for men and women to create a weighted sum, $h_{y \cdot x}$. Thus,

$$\text{For men:} \quad 1.1657 \times .75 = .8743$$

$$\text{For women:} \quad .7219 \times .25 = \underline{.1805}$$

$$h_{y \cdot x} = 1.0548$$

$$h_y = 1.6805$$

We can compute the uncertainty coefficient using Formula 7.3, which is as follows:

$$U_{yx} = \frac{(h_y - h_{y \cdot x})}{h_y}$$

where $h_{y \cdot x}$ is the weighted sum of h's computed within each category of the independent variable and h_y is the overall uncertainty computed from the marginal totals of the dependent variable. The numerator is the improvement in prediction (or reduction of uncertainty) due to the independent variable. The denominator is the original uncertainty available to be reduced.

Substituting the h's computed above into Formula 7.3 produces the following result:

$$U_{yx} = \frac{1.6805 - 1.0548}{1.6805}$$

$$= \frac{.6257}{1.6805} = .37$$

This means that there is a 37% reduction in uncertainty in predicting the dependent variable, employment status, using prior knowledge of the independent variable, gender, as compared to the overall level of uncertainty of predicting employment status.

The uncertainty coefficient, U_{yx}, is an asymmetric measure of association that is useful for nominal variables. As in the case of lambda, U_{yx} is generally different than U_{xy}. Which variable one wishes to predict determines which of these coefficients one computes. There is also a symmetric version of the coefficient that takes a value intermediate to the values of the other two.

U_{yx} is zero when the independent variable provides no help in reducing uncertainty and 1.00 when uncertainty is completely eliminated by knowledge of the independent variable. The relative magnitude of the uncertainty coefficient can be compared between tables because it is normed. Unlike lambda, the uncertainty coefficient takes account of the whole distribution of the dependent variable, not just its mode. The extent to which uncertainty can be reduced depends, of course, on the number of categories in the predictor variable compared to the predicted variable. One generally cannot accurately predict for four categories of a dependent variable, for example, when the independent variable only has two categories. Furthermore, predictor variables that are not distributed in a way that parallels the distribution of the dependent variable will not be able to completely reduce errors in predicting category placement of cases on the dependent variable. These, of course, are situations that one would want a measure of association to reflect.

7.2.2d Comparison of U with Lambda

To compare the uncertainty coefficient with lambda, we can subject the data in Tables 7.3 and 7.4 to analysis with the uncertainty coefficient since we have already computed lambda for both. In actual practice, we would use the computer to determine the values of these uncertainty coefficients, but for the sake of explanation we will compute the coefficients in the same manner as we did in the example above.

Converting the Table 7.3 entries from frequencies to proportions and using Formula 7.6 to compute h_y and $h_{y \cdot x}$, we arrive at the following results:

$$h_y = 1.3998$$

$$h_{y \cdot x} = .8417$$

Substituting these values into Formula 7.3, we find that the uncertainty coefficient is the following:

$$U_{yx} = \frac{1.3998 - .8417}{1.3998}$$

$$= \frac{.5581}{1.3998}$$

$$= .399$$

Therefore, the uncertainty coefficient reduced prediction errors in Table 7.3 by 39.9%, compared to lambda, which reduced prediction errors by 37.5%.

When we computed lambda for Table 7.4, it equaled 0 because the modes fell into the same category of the dependent variable for each of the three categories of the independent variable. In other words, the independent variable contributed nothing toward reduction of errors in predicting the mode of the dependent variable.

Using Formulas 7.6 and 7.3 on the data in Table 7.4, we will now compute the uncertainty coefficient for purposes of comparison. Use of Formula 7.6 results in the following values for h_y and $h_{y \cdot x}$:

$$h_y = 1.4887$$

$$h_{y \cdot x} = 1.4503$$

Substituting these values into Formula 7.3 produces the following uncertainty coefficient:

$$U_{yx} = \frac{1.4887 - 1.4503}{1.4887}$$

$$= \frac{.0384}{1.4887}$$

$$= .026$$

Therefore, the uncertainty coefficient reduces prediction errors by 2.6%. While the coefficient is not 0, as is lambda, the reduction in prediction error is still minimal. Nevertheless, this example illustrates the fact that the use of the uncertainty coefficient is not ruled out, as is lambda, by skewed distributions of dependent variables.

7.3   MEASURES OF ASSOCIATION FOR ORDINAL VARIABLES

Prediction of scores on ordinal variables is somewhat different from prediction of scores on nominal variables. Since we are interested in the *ranking* of scores on ordinal variables, it is useful to think of *pairs* of observations. It takes at least two scores before the idea of "rank" is meaningful. For a measure of association we are interested in the rank of pairs of cases on *two* ordinal variables, since we are interested in whether or not knowledge of the rank ordering of pairs of cases on one variable is useful in predicting their rank order on the other variable. If the knowledge of ranking of pairs on one variable is of no use in predicting rank order on the other variable, then we would like an ordinal measure of association to equal 0. This situation is equivalent to predicting the rank order of cases **randomly** by tossing a coin; if "heads" guess that the highest ranked case on one variable is highest on the second variable, and if "tails" guess that the highest ranked case on the first variable is lowest on the second variable.

Two other prediction rules may prove to be useful, however, in reducing errors in predicting rank order of pairs of observations on a dependent variable.

These rules correspond to situations where there is (a) a positive association, or (b) a negative association between two variables. In the first instance, we would predict **same rank order** on the *second* variable as the pair had on the *first* variable. The other rule is predicting the **opposite rank order** of cases on the second variable as opposed to the first. Thus, if Johnson is higher on social class than Jones, we could probably use the *same rank order* rule to predict their rank in terms of prestige of occupation, the second variable, since these two ordinal variables are positively related. Likewise, we could reduce predictive errors in predicting the rank order of their scores on "anomie" by using the *opposite rank order* rule, since social class and anomie are negatively related. To express how good these rank-order prediction rules may be, we need to consider the distribution of different kinds of pairings of scores on two variables.

7.3.1 Types of Pairs

The total number of possible, unique pairs of cases that can be formed from N cases can be computed as follows:

$$T = \frac{N(N-1)}{2}$$

where N is equal to the total sample size.* With five cases, ten unique pairings are possible; with $N = 10$, 45 unique pairings of cases can be distinguished. Furthermore, if these T unique pairs are measured on two ordinal variables, there are only five possible patterns of ranking pairs of cases on these two variables:

1. **Concordant pairs** (N_s) are pairs that are ranked in the same order on both variables.
2. **Discordant pairs** (N_d) are pairs that are ranked one way on one variable and the opposite way on the other variable.
3. Pairs *tied on the independent variable* (X), but not tied on the dependent variable (Y). These are symbolized T_x.
4. Pairs *tied on the dependent variable* (Y), but not tied on the independent variable (X). These are symbolized T_y.
5. Pairs *tied on both variables*, symbolized, T_{xy}.

These five types of pairs exhaust the possibilities, and the sum of pairs of these types equals T, the total number of possible, unique pairs of cases.

To illustrate these different kinds of pairs, consider the following six individuals, cases A through F, who are measured on the two ordinal variables, "social class" (X) and "alienation" (Y):

---

* The symbol T is the total number of unique pairs in a set of N cases. It has nothing to do with a different measure of association, Tschruprow's T, mentioned in the last chapter. Unfortunately, traditional symbolism is not altogether consistent, although in context the meanings are generally quite clear.

Individuals	(X) Social Class	(Y) Alienation
A	Upper	Moderate
B	Middle	Moderate
C	Upper	High
D	Lower	Moderate
E	Middle	Low
F	Lower	Moderate

These six cases can be combined to make up $N(N-1)/2$ or 15 unique pairs as listed below:

Pairs	2nd member is ___ on X	2nd member is ___ on Y	Type of Pair
A, B	Lower	Tied	T_y (tied on Y)
A, C	Tied	Higher	T_x (tied on X)
A, D	Lower	Tied	T_y
A, E	Lower	Lower	N_s (concordant pair)
A, F	Lower	Tied	T_y
B, C	Higher	Higher	N_s
B, D	Lower	Tied	T_y
B, E	Tied	Lower	T_x
B, F	Lower	Tied	T_y
C, D	Lower	Lower	N_s
C, E	Lower	Lower	N_s
C, F	Lower	Lower	N_s
D, E	Higher	Lower	N_d (discordant pair)
D, F	Tied	Tied	T_{xy} (tied on both X and Y)
E, F	Lower	Higher	N_d

Note that out of 15 different pairings in these data, there are 5 concordant pairs (N_s) and only 2 discordant pairs (N_d); thus, the "same rank" guessing rule would be the one to use since this indicates a positive association. If one were using variable X (social class) to predict variable Y (alienation), predictions of rank on Y could be made for any pair that was ranked differently on X. That would include all the N_s, N_d, and T_y pairs above [i.e., all except the A–C, B–E, and D–F pairs, which are tied on the predictor variable (X), meaning that no rank on alienation (Y) could be predicted from this information].

These data for the six individuals usually would be seen in table form, as indicated below:

Alienation	Social Class			Totals
	Lower	Middle	Upper	
High	0	0	1	1
Medium	2	1	1	4
Low	0	1	0	1
Totals	2	2	2	6

The procedures used to calculate the number of pairs of each type first puts data in a table format, and the calculations are carried out by computer. Diagram 7.1 (p. 267) illustrates the computational procedures involved in finding the number of these different types of pairs for a given set of data.

The difference between the frequency of concordant (N_s) and discordant (N_d) pairs is a measure of which rule is the most accurate predictor of rank order. If there are more N_s pairs than N_d pairs, the *same rank order* rule would be more accurate. If there is a preponderance of N_d pairs in a given set of data, then the *opposite rank order* rule would serve best. In fact, the greater the preponderance of concordant (or discordant) pairs, the better the appropriate rank-prediction rule. If there is no difference between the number of concordant and discordant pairs, then neither rule is better and we would not improve our rank-order predictions over simply using a random guess or flip of a coin.

7.3.2 Measures of Association

For ordinal variables, **measures of association** can be created as simple ratios of the various types of pairs distinguished here. In every case, however, the numerator of the ratio turns out to be the same, $N_s - N_d$, the preponderance (if any) of like-ranked or opposite-ranked pairs in a set of data. Each of these measures is a PRE measure, indicating the proportionate reduction in error that could be achieved by using one of the rank-order prediction rules ("same rank" or "opposite rank"), as compared to errors involved in making a random guess about the rank ordering of pairs. The different ordinal measures of association we will discuss have different denominators. That is, they differ in terms of the kinds of pairs "at risk" or for which a prediction is attempted. Let us start with the most intuitively meaningful measure of the group, Kendall's tau-*a*.*

7.3.2a Tau-*a*

Tau-*a* (t_a) is defined as the preponderance of concordant (or discordant) pairs out of all possible unique pairs in the data.

(7.7)
$$t_a = \frac{N_s - N_d}{T}$$

Looking back at the example in Diagram 7.1, the tau-*a* (t_a) coefficient would be

$$t_a = \frac{31057 - 14705}{101025} = \frac{16352}{101025} = +.16$$

Diagram 7.1 illustrates how the components of t_a would be computed. This coefficient can vary from -1.0 to 0 to $+1.0$, depending on whether the association is negative or positive. An association of 0 indicates an even split between

---

* An interesting discussion of the relationship between some of the ordinal measures of association is contained in Somers (1962), as well as original work by Kendall, Kruskal and Goodman, and Kruskal, which are cited in Somers' article.

concordant and discordant pairs (and thus that neither predictive rule will help in reducing predictive errors over errors one would expect to make just by chance). A t_a of 1.0 would indicate that all of the possible pairs are of one kind (either concordant or discordant depending upon the sign of t_a). This is a symmetric coefficient since no distinction is made between independent and dependent variables in the computation of N_s, N_d, or T, and it is appropriate for any size table or any number of ranks on either of the two ordinal variables. Unfortunately, if there are ties, as there usually are, t_a cannot reach the magnitude of 1.0, because the denominator, which includes ties, will always be greater than either N_s or N_d. Thus, tau-a is not usually recommended for use.

7.3.2b   Gamma (G)

One solution to the problem of actually obtaining a coefficient equal to 1.0 when there are ties is to eliminate the ties from the denominator as well. **Gamma** (G) is a frequently used, symmetric measure for association of two ordinal variables that does just that.* The numerator is the same as that used for t_a and the denominator is simply the sum of pairs that are ranked differently on both variables.

(7.8)

$$G = \frac{N_s - N_d}{N_s + N_d}$$

Gamma (G), like t_a, is a symmetric measure. Unlike t_a, it can always achieve the limiting values of -1.0 or $+1.0$ regardless of the number of ties. In fact, it would be possible to have a G of $+1.0$ (or -1.0) based on only one pair, where all the rest of the pairs are tied on one or both variables. This may be an undesirable property of G, considering it as an overall characterization of the proportionate reduction in errors in predicting rank order of one variable based on a knowledge of ranking on the other variable. It is interesting to note that G in the 2 by 2 case is the same as Yule's Q, which was introduced in the last chapter (Section 6.5.2c), except for the traditionally used symbolism. G is a generalized version of Yule's Q for $r \times c$ tables.

Using the illustration from Diagram 7.1, G would be computed as follows:

$$G = \frac{31{,}057 - 14{,}705}{31{,}057 + 14{,}705} = \frac{16{,}352}{45{,}762} = .36$$

G can be interpreted as the proportionate reduction in errors in predicting ranking that would result from using the "same" (or "opposite") ranking rule rather than randomly predicting rankings among pairs that are ranked differently on both of the two variables in the table.

7.3.2c   Sommers' d_{yx}

If one distinguishes between independent and dependent variables a somewhat different ordinal measure would be appropriate. If one were predicting

---

* Sometimes gamma is symbolized by the lower-case Greek letter (γ). G is used here.

the ranking of cases on a dependent variable (variable Y), using variable X as the independent or predictor variable, one would make a prediction about ranking not only for the pairs that are ranked differently on each variable (*i.e.*, the N_s and N_d pairs), but also for the T_y cases, which are different on the predictor variable but tied on the dependent variable. The difference on the independent variable permits a prediction even in the T_y cases, although the prediction may not work out, since there may not be any difference on the dependent variable. The denominator of an association measure, then, should include *all* the pairs for which a prediction would be risked. This is essentially the definition of this next ordinal measure of association, Somers' d_{yx}.

(7.9)
$$d_{yx} = \frac{N_s - N_d}{N_s + N_d + T_y}$$

Notice that the numerator is again the difference between concordant and discordant pairs. Somers' d_{yx} is an asymmetric measure, since it does take account of which variable is the predictor and which is the predicted. Ties on the predict*ed* variable are *in*cluded, and ties on the predict*or* variable are *ex*cluded. Like G, it can vary from -1.0, indicating a perfect negative association (*i.e.*, a situation where all pairs are discordant), to $+1.0$. It can be interpreted as the proportionate reduction in errors in predicting ranking on a dependent variable resulting from using the ("same" or "opposite") predicting rule, rather than chance prediction among pairs that are ranked differently on the independent variable. Here again, as was true of other asymmetric measures such as lambda (λ_{yx} and λ_{xy}), there are two Somers' d's that can be computed on any table, depending upon which variable is treated as the independent variable. In general, too, these values may be (and generally will be) different for any given table.

7.3.2d Tau-*b*

An investigator might be interested in a measure of degree of association that is symmetric but, unlike G, takes account of ties on one or the other variable (but not ties on both, T_{xy}). Ties on both variables are considered to be trivial, since they correspond to the number of pairs that could be created out of cases that are identical on both variables and that thus fall within the same cell of a table. In most tables, cell frequencies are greater than 1, and it is not the absolute size of any cell frequency, but the *pattern* of frequencies in different cells that is what one means by the association in a table.

Tau-*b* (t_b) is a kind of average of the two Somers' d's that could be computed on a given set of data. In fact, it can be expressed as the square root of the product of these two d's (an example of the use of the geometric mean from Section 5.3.4).

$$t_b = \sqrt{d_{yx} d_{xy}}$$

It is usually computed directly from computations including the number of each type of pairs, expressed as follows:

(7.10)
$$t_b = \frac{N_s - N_d}{\sqrt{(N_s + N_d + T_y)(N_s + N_d + T_x)}}$$

It can take on values from -1.0 to $+1.0$, depending on the direction of association, and its magnitude indicates strength of association. Tau-b cannot achieve a magnitude of 1.0 if a table is not square (*i.e.*, if $r \neq c$) since in this case there would have to be more pairs tied on one variable (the one with the fewer categories) than on the other.* Unfortunately, the more complicated the denominator the more difficult it becomes to express a clear operational definition in a PRE sense, and this is true of t_b. It is, however, one of the more useful of the ordinal, symmetric measures, and is superior to tau-a because it does take account of nontrivial ties in expressing the relationship between two variables.

For the illustrative data in Diagram 7.1, t_b would be computed as follows:

$$t_b = \frac{31{,}057 - 14{,}705}{\sqrt{(31{,}057 + 14{,}705 + 21{,}614)(31{,}057 + 14{,}705 + 21{,}677)}}$$

$$= \frac{16{,}352}{67{,}407.5} = .24$$

7.3.2e   Comparison of Gamma, Somers' d_{yx}, and Tau-b

Like all properly normed measures, these ordinal measures are affected by features of the data that affect the things they take into account, which in the case of ordinal variables are types of rankings (or ties). For example, the greater the precision of a variable (*i.e.*, the more categories it has), the larger the potential number of differently ranked pairs one can make out of a given number of cases (and the smaller the potential number of tied ranks could be, simply because the cases could be spread over more possible categories). Differences in precision of two variables usually is reflected in a marked difference between Somers' d_{yx} and d_{xy}, with the more precise variable showing up as the better predictor. Similar differences occur because of the distribution of cases across categories of a variable, when data are concentrated in certain categories of a variable rather than spread out more evenly. More even distribution of cases on a predictor variable (*i.e.*, greater variability) generally improves the possibilities for accurately predicting a dependent variable. These are all "legitimate" data effects that are reflected in coefficients that take account of ranking of pairs (similar effects occur for nominal and interval/ratio association measures, too, but here we will illustrate the effects for ordinal data). It is helpful to understand patterns in the data to which the association measures are sensitive. Some of these are shown in Table 7.5.

All the preceding measures of association for ordinal variables were simply ratios created out of the number of pairs of the several types we

---

* Another measure of association that *can* achieve 1.0 for tables where the number of rows and columns is not equal is called tau-c. In this formula, m is the smaller of the number of rows or the number of columns.

$$t_c = \frac{2m(N_s - N_d)}{N^2(m - 1)}$$

As is true of tau-b, this measure, although normed, is hard to interpret operationally, and tau-c is less frequently used than the other tau measures.

TABLE 7.5 COMPARISONS OF GAMMA, SOMERS' d_{yx}, AND TAU-b (HYPOTHETICAL DATA)

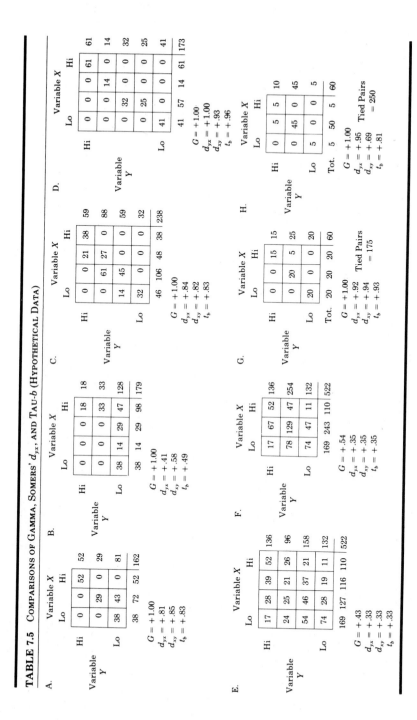

distinguished. In each case the balance of concordant and discordant pairs was evaluated in terms of the number of pairs "at risk," so to speak—that is, the number of pairs that constituted a potential number of errors or correct predictions that might have been made. The essential difference between the measures derives from what pool of differences one would be interested in predicting.

Tau-a takes as its pool all possible pairs; gamma, only untied pairs; Somers' d_{yx}, untied pairs and pairs tied on the dependent variable; and tau-b, untied pairs and ties on variable Y or variable X, but not on both.*

Costner (1965) takes the position that any measure that includes ties of any kind in its pool is not properly a PRE measure because a tie cannot be clearly counted as either a correct or an erroneous prediction and only those two categories are permissible for PRE measures. On the other hand, if one takes the position that the pool of potential errors should include all those for which a prediction is likely to be made, then it is reasonable to include ties. For example, if a pair is untied on the independent variable, a prediction is possible. Although concordance or discordance would be predicted, the possible outcomes would be concordance, discordance, or a tie on the dependent variable. These are, in fact, the possibilities that are included in the pool for Somers' d_{yx}. Since tau-b is a symmetric measure, it makes provision for a tie to occur on either variable Y or variable X.

What consequences do these differences in denominators (pools) have for the resulting measures of association? It is worth spending some time to examine these consequences because they are important both in helping us decide which of the measures of association to use for a particular situation and in helping us interpret what the resulting coefficient means for our data analysis.

Tau-b requires the most restrictive meaning of perfect association. That is, it reaches a value of $+1.00$ or -1.00 when all the frequencies fall on the diagonal of a square table. If the table is not square, perfect association is not possible. Somers' d_{yx} will reach $+1.00$ or -1.00 either when all of the frequencies in a square table are on the diagonal or in a table that is not square when their progression on the scale of the dependent variable is *monotonic*—that is, where the frequencies for the dependent variable progress in a stepwise fashion such that a single value of the dependent variable (Y) cannot take more than one value on the independent variable (X), but two different values of the dependent variable may take the same value on the independent variable. Table 7.5D is an example of a table of frequencies that is characterized by a monotonic progression of the dependent variable (Y). Note that if X is taken as the dependent variable d_{xy} is not equal to 1.00. That is so because two different values of Y would predict the same value of X. Gamma takes the least restrictive meaning of perfect association of the three measures we are considering. This is so because all sorts of ties are ignored in its computation. Tables 7.5A through 7.5D illustrate four different patterns of frequencies, all of which result in gammas of $+1.00$. Notice that Somers' d_{yx} and tau-b are high for those patterns in which the frequencies all fall close to the

---

* Because of the problem of interpreting tau-a when ties occur, we will not consider it further here.

diagonal and are low for Table 7.5B, where the frequencies are off the diagonal and, incidentally, form an L-shaped pattern. Also, take note of the fact that gamma can reach 1.00 even though a table is not square (Tables 7.5B and 7.5D).

In all the tables of Table 7.5 the value of tau-b is intermediate to the values of d_{yx} and d_{xy} because, mathematically, the magnitude of tau-b is equal to the geometric mean of d_{yx} and d_{xy}.

7.3.2f   More on Grouping Error

Collapsing a table (i.e., decreasing the number of rows and/or columns by combining categories) is generally not a good idea because it loses information about rank differences and, if this is done differently on one of the variables than the other, grouping error may be a result. For example, Table 7.5F has the same data as 7.5E except the middle two categories of each variable have been combined; note the differences in each of the measures of association, most striking for gamma, which focuses only on the differently ranked pairs, N_s and N_d. Differences in marginal distributions (i.e., Tables 7.5G and 7.5H) may also be created by grouping as well. Differences in concentration have different effects on the two Somers' d's that can be computed, reflecting the ties on the dependent variable (which are in the denominator of Somers' d). In both Tables 7.5G and 7.5H, gamma is +1.00, reflecting the fact that it ignores ties of any kind (ties are primarily what *differences in marginal distributions* are dealing with in these tables). In Table 7.5G, gamma ignores 175 tied pairs in its computation, and in Table 7.5H, 250 tied pairs are ignored. It is possible, in extreme cases, for gamma to be based on a minority of available data. In using gamma, it is advisable to take account of the proportion of ties; where this is large, gamma probably should not be used (or should be used with caution).

BOX 7.4   ORGANIZING

Often it helps to organize statistical measures in terms of some of their main differences—those differences that might lead to the selection of one measure in preference to another for some particular problem. One organizing scheme is suggested here, and the six PRE measures we have discussed thus far in this chapter are entered for illustration. Your own scheme might have more detailed classifications. Selection of an appropriate measure is discussed later in this chapter.

Level of Measurement	Symmetric	Asymmetric
Nominal		Uncertainty coefficient (U_{yx}) Lambda (λ_{yx})
Ordinal	Kendall's t_a Kendall's t_b Gamma (G)	Somers's d_{yx}

TABLE 7.6   Rank of Selected Sociology Departments on Productivity of Doctorates and Representation on the Editorial Board of a Major Sociology Journal

Sociology Department	(1) Rank on Number of PhDs Produced 1964–1968	(2) Rank on No. of Editorial Positions on the ASR for 1948–1968	Difference between Ranks	
			D	D^2
Chicago	1	1	0	0
Columbia	2	3	−1	1
Wisconsin	3	4	−1	1
Minnesota	4	5	−1	1
UCLA	5	8	−3	9
Berkeley	6	6.5	−0.5	0.25
Michigan	7	9.5	−2.5	6.25
Ohio State	8.5	6.5	+2	4
Washington (Seattle)	8.5	9.5	−1	1
Harvard	10	2	+8	64
			$\Sigma D = 0$	$\Sigma D^2 = 87.5$

Source: Data from Rossi (1970) (column 1), and from Yoels (1971) (column 2). Tied ranks are averaged.

7.3.2g   Spearman's rho (r_s)*

The last ordinal measure we will deal with is interesting, because it takes a different approach to the problem of measuring the direction and strength of association. It is primarily used where rankings of individual cases on two variables are available so that rankings range from 1 to N for each variable. Table 7.6 provides an example of two rankings of the same set of sociology departments. One ranking is in terms of degree productivity and the other is in terms of positions held on the editorial board of the *American Sociological Review*, official journal of professional sociologists in the United States.

The argument is that the representation of schools on the editorial board of a major journal merely reflects the productivity of these departments. If that is so, then there should be a perfect association—exactly the same ranking—of the departments on both variables, representation and production.

Spearman's rho (r_s) is a measure of association for ordinal variables based on the difference between ranks. If there is no difference, then D will equal 0. Since the sum of the differences between ranks is always 0, as illustrated by Table 7.6, differences between ranks are squared before summing. In the case of comparisons in ranking of these 10 schools, the sum of squared differences between ranks (called ΣD^2) is 87.5. Since this is different from 0, we know that the two variables are not identically ranked. But we do not know how to interpret this figure, because we would expect it to vary with the number of individuals ranked in the first place. We could, however, create a ratio of the ΣD^2 obtained and the

* Sometimes rho is symbolized by the lowercase Greek letter (ρ). r is used here.

maximum possible ΣD^2 that could be achieved for a given number of ranked individuals. This maximum ΣD^2 is this: $N(N^2 - 1)/3$, where N is the number of cases ranked. Then, in order to make it possible for a minus sign to indicate opposite ranking and for the magnitude of 1.0 to be a maximum degree of association, the formula for r_s is written as follows:*

(7.11)
$$\text{Spearman's } r_s = 1 - \frac{6\Sigma D^2}{N(N^2 - 1)}$$

Rho (r_s) will have a value of $+1.0$ for a perfect match of ranks, and a value of -1.0 if the ranks are exactly opposite. A r_s of 0 indicates no systematic ordering or, rather, no rank pattern between the two variables. In the case of Table 7.6, r_s is as follows:

$$r_s = 1 - \frac{6(87.5)}{10(10^2 - 1)} = 1 - \frac{525}{990} = 1 - .53 = +.47$$

Intermediate values of r_s can be interpreted in terms of their relative magnitude, but r_s does not have a PRE interpretation. However, r_s^2 has a PRE interpretation (for ranks) equivalent to that of r^2 (discussed in Section 7.4.2), although there is no prediction equation equivalent to the regression equation (discussed in Section 7.4.1) for r_s.

It should be noted that r_s, like gamma, loses its effectiveness as a measure of association as the number of tied ranks increases.

7.4 MEASURES OF ASSOCIATION FOR INTERVAL VARIABLES

As you will recall from the discussion of univariate statistics, the arithmetic mean of an interval variable is a useful prediction because the mean has the property that the algebraic sum of deviations of actual scores from it is 0. A measure of how badly wrong this prediction is can be derived from these deviations. The variance (or its square root, the standard deviation) is one such measure that expresses the amount of scatter of scores around this mean.

Thus, as a minimum, one could predict the mean of a dependent variable and measure "errors" made in that prediction in terms of the familiar variance (s^2) and this, in fact, constitutes the minimum guessing rule for interval variables shown in Table 7.2. How much better can we do? Is there any way that scores on an independent variable could be used to improve the prediction of a dependent variable? As you might suspect by now, the answer is "yes," although the solution to the problem is new to our line of discussion thus far.

---

* The formula for Spearman's rho (r_s) is simply a Pearsonian r computed on ranks. Since the ranks for both variables extend from 1 to N, we know that the sum of each variable is $N(N + 1)/2$ and the mean of each variable is $(N + 1)/2$. The sum of squares becomes $N(N + 1) \times (2N + 1)/6$, and the variance for each variable is $(N^2 - 1)/12$. Substitution into the formula for Pearson's r yields the Spearman's r_s formula for the relationship between X ranks on two variables for N cases. The derivation is nicely illustrated in Hammond and Householder (1962). Note that tied ranks are averaged.

BOX 7.5   INTERVAL VARIABLES AGAIN

The following section deals with *measures of association* and a description of the *nature of association* for interval variables. It is helpful, at this point, to recall some of the concepts from univariate description that were used to summarize interval variables. In particular, you should feel quite comfortable with the following:

Arithmetic mean (Section 5.3.3)
Variance and standard deviation (Section 5.4.4)
Scatter diagram (Section 4.4.2)

Suppose that we were able to derive a formula that would describe the way the mean of variable Y varied as one moved up the scale of variable X. This would, in effect, be a mathematical description of the *nature* of the relationship between two variables, and it would also permit us to "compute" for each case an estimate of its score on the dependent variable from information about its score on the independent variable. With a predicted score (called Y', or Y-prime) and an actually observed score (Y), we could then ask how accurate the prediction equation is. This might take the form of a measure of association (usually called a *correlation coefficient* where the variables are interval level) that would express the amount by which predictive errors could be reduced, given the prediction equation rather than the overall mean of the dependent variable to use in predicting. This is precisely what we will do to create the next measure of association, called **Pearson's product–moment correlation coefficient**, r. Its square (*i.e.*, r^2) will indicate the proportionate reduction in errors resulting from a use of the predictive equation. To develop this idea, however, we need to step back and develop a formula for describing the nature of the relationship between two interval variables so as to predict the dependent variable.

7.4.1   Regression Equations

Suppose we start with a small collection of data where two scores are measured on each of six cases.

Case	(X) Years of Education	(Y) Income in ($1000)
A	1	2
B	2	4
C	3	6
D	4	8
E	5	10
F	6	12

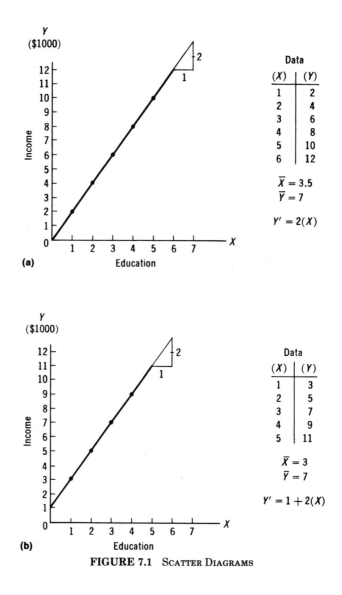

FIGURE 7.1 SCATTER DIAGRAMS

These scores could be plotted on a graph, as in Figure 7.1a, where values of the independent variable appear along the X-axis and values of the dependent variable appear along the Y-axis. This is one use of a **scatter diagram** discussed earlier in this book (see Section 4.4.2). Each case is represented by a point on the graph placed above the X-axis and across from the Y-axis at the point that represents that individual's score on each variable. These points can be located by a pair of scores (*e.g.*, 1,2 where the first number is the X-score and the second is the Y-score for that point).

7.4.1a   Slope

In this example, it is clear that the two variables are related in a rather simple and obvious way. In fact, one could predict the Y-score exactly from a knowledge of the X-score merely by doubling the X-score for these cases. This relationship is expressed in the following equation, where Y' (called Y-prime) is an estimated or predicted value for that individual case on the Y-variable.

$$Y' = 2(X)$$

The predictions fall along a straight line, which is drawn in for Figure 7.1a. The 2 in this equation is called the **slope** of the regression line, and it means that for a 1-unit increase along the X-axis there will be a predicted increase of 2 units on the Y-scale. This slope is illustrated in the small triangle shown on the graph (Figure 7.1a). Notice that predictions derived from this equation describe a **straight line**, one of the simplest (*i.e.*, easiest to describe) ways in which two variables can be related. One could say that these variables are related in a "linear" fashion; a straight line very nicely describes their relationship. This line is called a linear **regression** line, a regression of the dependent variable, Y, based on the independent variable, X.

7.4.1b   Y-Intercept

Let us examine another set of data.

Case	(X) Years of Education	(Y) Income in ($1000)
A	1	3
B	2	5
C	3	7
D	4	9
E	5	11

Here again we have a very small set of scores for illustrative purposes and, again, we can create a scatter diagram of these five cases, as shown in Figure 7.1b. The Y-scores for these five cases can also be accurately predicted by a simple formula:

$$Y' = 1 + 2(X)$$

Given an X-score, we can predict a Y-score for that case by merely doubling the X-score and adding a constant, 1. Here, too, the equation describes a simple straight line, and one would again say that the relationship between these two variables is of a linear nature. The formula above describes the specific nature of this relationship for these data. Here, again, the 2 expresses the number of units change in Y given a single unit change in X. The constant, 1, indicates the point at which the straight line crosses the Y-axis in this scattergram in Figure 7.1b and is called the **Y-intercept**.

7.4.1c   Formula for a Straight Line

The general formula for any straight line is expressed as follows:

(7.12)
$$Y' = a_{yx} + b_{yx}(X)$$

where Y' is the predicted or computed value of the dependent variable; a_{yx} is the constant or Y-intercept for this equation; and b_{yx} is the slope coefficient (also referred to as the **b-coefficient**). The first subscript of the a_{yx} and b_{yx} values is the dependent variable, and the second subscript refers to the independent variable. It is immediately clear that these coefficients distinguish between the independent and dependent variables and we might, in fact, expect a different equation if values of X were to be predicted from values of Y. As in the case of asymmetric measures of association, there are two regression equations that could be computed to describe the nature of the relationship between the pair of variables in a scattergram. More will be said on this later.

Notice that the a_{yx} and b_{yx} coefficients in the last two examples are positive values. This need not be the case for all sets of data. In fact, where the *b-coefficient* is negative, that indicates that the variables are negatively related, since an increase of 1 unit in the X-value would signal a negative change in the predicted Y-score.

Not all associations between variables are described very well by a straight line, although many come relatively close to this nature of association. A straight line represents in many respects the simplest relationship we could express, and it is usually the one expressed in theoretical statements about how sociological variables are related. As a matter of fact, some relationships between variables are better described by curved rather than straight lines, but these raise the problem of discovering, by some curve-fitting method, what the proper formula for the curve might be, as well as what the particular form of curve might mean, theoretically. Although the problems are not particularly different, we will limit our attention to the simple straight-line relationship and the development of the linear regression equation.

Figure 7.2 illustrates the more typical scatter diagram. In this instance, individual cases are represented by the placement of the small numbers on the scatter diagram. The number itself indicates how many cases fall at that same point. The relationship is between two different measures of social class. One measure was developed by Hollingshead using occupation and education; the other social class score was developed by Duncan from income and education for different occupation groups (see Haug and Sussman, 1971). One would expect that the two measures would be relatively closely related, since they are supposed to measure essentially the same phenomenon. In actuality, however, there is a considerable amount of scatter even though, overall, the data cluster fairly well around a linear regression line. The problem is to place the linear regression line in such a manner that it fits the data as well as possible.

7.4.1d   Best-fitting Line

The criterion of "fit" of a regression line is still how well the dependent variable can be predicted by the equation the line represents. Let us take another

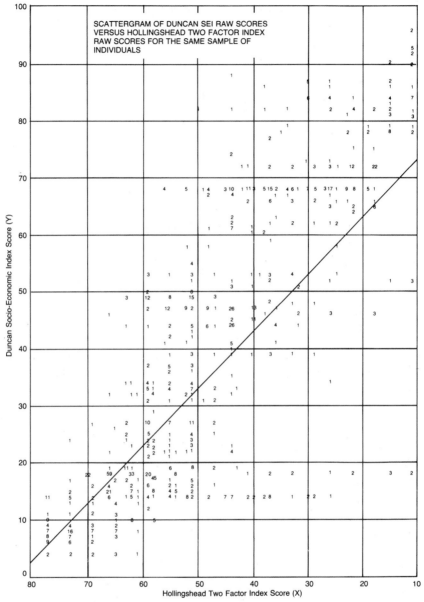

Source: Haug and Sussman (1971). Used by permission.

FIGURE 7.2 Scatter Diagram Showing the Relationship between Hollinghead's Two-factor Index and Duncan's Socioeconomic Index as Measures of Social Class Standing

set of data, where a small number of data points are scattered somewhat, and try to discover how one might derive a best-fitting straight line.

The set of five scores shown in Figure 7.3 could be predicted in terms of several different equations. Let us take three to show the kind of error that the best-fitting line would help eliminate. First, we could simply predict the mean of Y for every case. This is the regression line shown in Figure 7.3a. Second, we could use the prediction $Y' = 2.5 + .83X$, which is an arbitrarily picked equation (for illustrative purposes) that is *not* the best fitting, but it is better than the overall mean of Y as a prediction. Any number of other inadequate equations could be chosen rather than this one, of course. Third, we could use the best-fitting regression equation for these data, which happens to be $Y' = 1.1 + 1.3X$.

7.4.1e   Standard Error of the Estimate

Notice in Figure 7.3a that the same prediction will be made for each case, and the inaccuracy of this prediction will be described by subtracting the *predicted* Y-score from the *actual* Y-score, squaring the difference, and dividing this sum by N, the number of cases. You will recall that this is simply the formula for the variance of a set of scores. In Figure 7.3a this variance can be symbolized by s_y^2 to indicate clearly that it is the variance of the Y-variable.

$$s_y^2 = \frac{26.0}{5} = 5.2$$

Using the second prediction equation shown in Figure 7.3b, a different predicted value of Y is shown for each different value of X. Again, we can describe the accuracy of prediction by subtracting the predicted from the actual Y-score, squaring the difference, summing, and dividing by N.* In this case, since the predicted value of Y depended on the value of X, we will symbolize the variance by $s_{y \cdot x}^2$, which is called the *error variance of the estimate*; the square root of this value is called the **standard error of the estimate**.

(7.13)
$$s_{y \cdot x}^2 = \frac{\Sigma(Y - Y')^2}{N}$$

Notice that the subscript indicates not only the dependent variable, Y, but the dot followed by the X indicates that different values of X are reflected in predictions upon which this error figure is based. The dotted lines in Figure 7.3a, b, and c represent the error of predicted value (on the regression line) and the actual score. These lines get shorter as the prediction gets better. This improvement in prediction is also illustrated by the decrease in the variance figures we have created.

---

* At this point your attention should again be called to the fact that a better estimate (unbiased estimate) of the population variance includes $N - 1$ in the denominator of the variance computed from sample data. The same principle applies to $s_{y \cdot x}^2$, which would have $N - 2$ in the denominator.

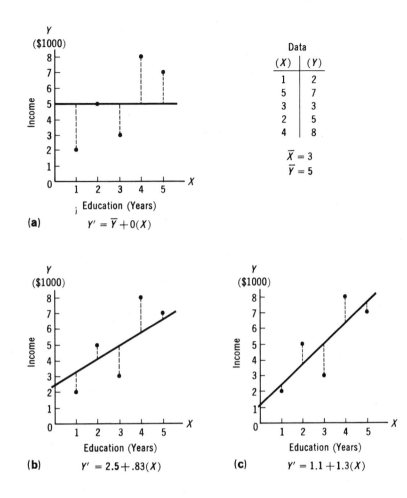

(a) $Y' = \bar{Y} + 0(X)$

(b) $Y' = 2.5 + .83(X)$

(c) $Y' = 1.1 + 1.3(X)$

Data

(X)	(Y)
1	2
5	7
3	3
2	5
4	8

$\bar{X} = 3$
$\bar{Y} = 5$

Measurement of Prediction Errors

	Actual Data		Prediction #1 $Y' = \bar{Y} + 0(X)$			Prediction #2 $Y' = 2.5 + .83(X)$			Prediction #3 $Y' = 1.1 + 1.3(X)$		
	(X)	(Y)	Y'	(Y−Y')	(Y−Y')²	Y'	(Y−Y')	(Y−Y')²	Y'	(Y−Y')	(Y−Y')²
	1	2	5	−3	9	3.33	−1.33	1.77	2.4	− .4	.16
	5	7	5	+2	4	6.66	+ .34	.12	7.6	− .6	.36
	3	3	5	−2	4	5.00	−2.00	4.00	5.0	−2.0	4.00
	2	5	5	0	0	4.17	+ .83	.69	3.7	+1.3	1.69
	4	8	5	+3	9	5.83	+2.17	4.71	6.3	−1.7	2.89
Sum	15	25	25	0	26.0	25.00	.0	11.29	25.0	.0	9.10
Avg	3	5	5	0	5.2	5.0	0	2.26	5.0	0	1.82

FIGURE 7.3   FINDING THE BEST-FITTING STRAIGHT LINE TO REPRESENT DATA

7.4.1f Least-squares Criterion

Finally, using the third prediction equation shown in Figure 7.3c, we notice that the dotted lines representing inaccuracy in prediction are generally shorter than those for Figures 7.3a and b. This is reflected in a smaller error variance of the estimate, which is again simply the sum of squared differences between actual and predicted scores averaged over the number of cases. This third prediction equation ($Y' = 1.1 + 1.3X$) produces the smallest variance. It happens to be the best prediction, and the regression line in Figure 7.3c is called the "best-fitting regression line of Y on X." It is *best* in the sense that the sum of squared deviations of scores around this line is the smallest it could be for any straight line and, thus, this is called the **least-squares** regression line. It is important to notice that the deviations are all figured in terms of Y-scores, since that is the variable we are interested in predicting. Figure 7.4 includes the least-squares regression line for the situation where we are interested in predicting X-scores instead. Notice that a different equation is needed to minimize errors in predicting X from a knowledge of Y. It is generally true that two different regression equations can be computed for a given bivariate distribution, depending on which variable is to be predicted.

Computation of the coefficients for the least-squares regression line is illustrated in Diagram 7.2 (p. 272). The numerator for b_{yx} is a value that expresses how well the two variables go together, and it is formed from the sum of the product of deviations of each score from its mean. The denominator is the sum of squared deviations for the independent variable.

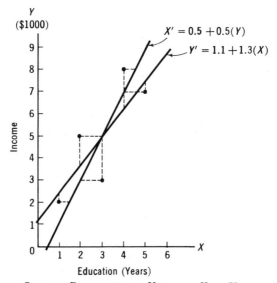

FIGURE 7.4 Least-Squares Regression of Variable X on Variable Y Taken from Data in Figure 7.3, Superimposed on the Regression Line of Y on X

(7.14)
$$b_{yx} = \frac{\Sigma(X - \bar{X})(Y - \bar{Y})}{\Sigma(X - \bar{X})^2}$$

A more convenient formula for hand computation, using only the raw scores themselves, is given in Diagram 7.2.

7.4.1g   The Y-Intercept Formula

Since the best-fitting regression line always goes through the point where the means of both variables intersect (note that this is true of Figures 7.3c and 7.4), the a_{yx} coefficient can be computed by substituting these mean values in the regression equation and solving for a_{yx}, as follows:

(7.15)
$$a_{yx} = \bar{Y} - b_{yx}(\bar{X})$$

We have now established a simple formula that describes the **nature** of association between two variables and permits us to use information about the independent variable to reach a better prediction of the dependent variable. The next problem is to devise a measure of degree of association that will express the proportionate reduction in predictive errors that this formula permits.

7.4.2   Product–Moment Correlation Coefficient (r)

The least-squares regression line computed in Figure 7.3c permitted us to predict scores on the dependent variable, Y, with somewhat greater accuracy than we could achieve merely by predicting the overall mean of Y. We could say that the regression line helps us "explain" some of the variation in the dependent variable. This left, typically, some of the variation in Y unexplained, and the total of **unexplained** and **explained variation** equals the **total variation** of Y around its mean. This can be shown for Figure 7.3c as follows:

Actual Scores X Y	Predicted Y-Scores (Y')	Unexplained Variation $(Y - Y')^2$	Explained Variation $(Y' - \bar{Y})^2$	Total Variation $(Y - \bar{Y})^2$
1 2	2.4	.16	6.76	9
5 7	7.6	.36	6.76	4
3 3	5.0	4.00	.0	4
2 5	3.7	1.69	1.69	0
4 8	6.3	2.89	1.69	9
15 25	25.0	9.10	16.90	26

$$\bar{Y} = 5$$

$$
\begin{array}{ccccc}
26 & = & 9.10 & + & 16.90 \\
\Sigma(Y - \bar{Y})^2 & = & \Sigma(Y - Y')^2 & + & \Sigma(Y' - \bar{Y})^2 \\
\dfrac{\text{Total}}{\text{Variation}} & = & \dfrac{\text{Unexplained}}{\text{Variation}} & + & \dfrac{\text{Explained}}{\text{Variation}}
\end{array}
$$

The object, of course, is to explain as much of the variation as possible. The Pearson product–moment correlation coefficient expresses how well the linear regression line explains the variation in the dependent variable as follows:

$$r = \sqrt{\frac{\text{Explained variation}}{\text{Total variation}}}$$

$$= \sqrt{\frac{\Sigma(Y' - \bar{Y})^2}{\Sigma(Y - \bar{Y})^2}}, \quad \text{which also equals} \quad \sqrt{\frac{\Sigma(X' - \bar{X})^2}{\Sigma(X - \bar{X})^2}}$$

The square of the Pearsonian correlation coefficient (called the *coefficient of determination*) could also be expressed in terms of variances:

(7.16)

$$r^2 = \frac{s_y^2 - s_{y \cdot x}^2}{s_y^2}$$

$$= 1 - \frac{s_{y \cdot x}^2}{s_y^2}$$

The numerator in the first version of the equation represents the explained variation using the informed guessing rule. The denominator represents the overall variation.*

The Pearsonian correlation, r, is a PRE measure. If squared (r^2), it expresses the proportionate reduction in error in predicting scores for the dependent variable, given the best-fitting linear regression equation rather than the overall mean to use in prediction. Since the regression of Y on X and the regression of X on Y both have the same amount of scatter around their respective regression lines, the same correlation coefficient will result from either predictive equation. Thus r is a *symmetric* measure of degree of correlation. Expressed differently, r^2 indicates the proportion of the variation in one variable that is explained by its linear association with the other variable.

Usually, the correlation coefficient is expressed as r, the square root of r^2, although its interpretation is most useful in its r^2 form as indicated above. The actual computational formula of r is given in Diagram 7.2 (p. 273).

Pearson's r is a symmetric measure of correlation between two interval variables. Its values vary from -1.0 to 0 to $+1.0$, indicating direction and strength of association. Figure 7.5 shows some of the values of r and their related scatter diagrams. Notice in Figure 7.5 that a Pearson r may be quite small, not only where there is a random scattering of cases around the mean of the dependent variable, but also where there is an obvious association, yet the best-fitting *straight line* is no better than simply the mean of the dependent variable. Where the data are of a nature that is not too close to linear, an alternative, either eta (symbolized E_{yx}, to be discussed in Section 7.4.5), or a curvilinear regression equation would be more

---

* The ratio shown in the second formula above $\left(\dfrac{s_{y \cdot x}^2}{s_y^2} \right)$ is called the *coefficient of nondeter-mination*, or *K*-squared, since it expresses the proportion of total variation that is unexplained by the regression equation.

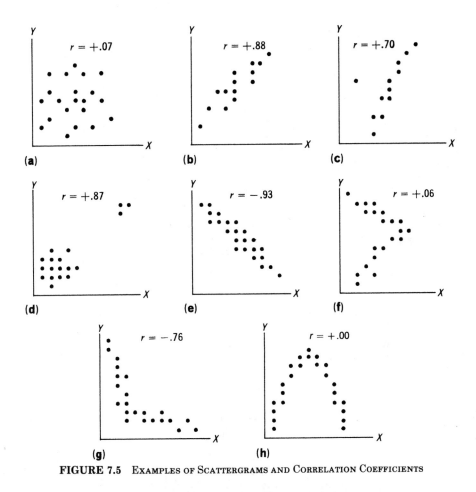

FIGURE 7.5   EXAMPLES OF SCATTERGRAMS AND CORRELATION COEFFICIENTS

appropriate.* Figure 7.6 shows the relationship between r and r^2, that is, the proportion of variation explained for different values of r.

7.4.3   Another Way to Express Correlation and Regression

To get a better grasp of the meaning of correlation and regression, it might be helpful to look at them from a somewhat different perspective. Suppose we are interested in determining whether there is a relationship between the number of

---

* In some cases it is possible to find an equation that expresses in mathematical terms the nature of curvilinear associations. If such an equation can be found, its "fit" to the data can be described by the correlation coefficient, r, by simply substituting the predicted value from the curvilinear equation for the predicted value from the simple linear equation in formulas for Pearson's r, above.

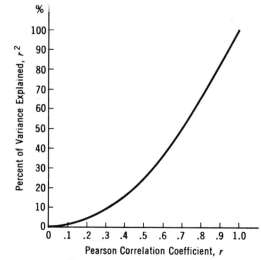

FIGURE 7.6   PERCENTAGE OF VARIATION EXPLAINED BY CORRELATIONS OF DIFFERENT SIZES

Years of Education (X)	Monthly Rental Value of Residence (dollars) (Y)	$X - \bar{X}$ (x)	$Y - \bar{Y}$ (y)	xy
12	400	−1.5	27.5	−41.25
10	325	−3.5	−47.5	166.25
16	450	2.5	77.5	193.75
8	350	−5.5	−22.5	123.75
14	200	0.5	−172.5	−86.25
20	525	6.5	152.5	991.25
11	375	−2.5	2.5	−6.25
15	250	1.5	−122.5	−183.75
16	675	2.5	302.5	756.25
13	175	−0.5	−197.5	98.75
$\bar{X} = 13.5$	$\bar{Y} = 372.5$			+2012.5
$s_x = 3.29$	$s_y = 144.24$			
$N = 10$				

If there is a correlation between these two variables, then it should be reflected in the deviations of the scores for each from their respective means. If the correlation is positive, then scores on variable X (years of education) that deviate in a positive direction from their mean should be paired with scores on variable Y (monthly rental value) that also deviate in a positive direction from their mean; furthermore, X-scores that deviate in a negative direction should be paired with Y-scores that deviate in a negative direction. Thus, if families with better educated heads tend to live in residences with higher monthly rental values, then for each pair of scores the head of household's years of education should be higher than the mean for that variable and the monthly rental value of the family's residence should be above the mean on that variable. And, if the head's years of education is lower than the mean, then the monthly rental value should also be below the mean. In the third column of the table are entered deviations of head of household's years of education from the mean years of education [$x = (X - \bar{X})$], and in the fourth column, the deviations of the monthly rental value of the family's residence from the mean of that variable [$y = (Y - \bar{Y})$].

Column five of the table contains the products of the multiplication of each X deviation by its corresponding Y deviation (called *cross-products*). These cross-products are summed algebraically at the bottom of the table. If the sign of the sum of these cross-products is positive, then there is a positive correlation between the variables; if the sign is negative, then there is a negative correlation. Furthermore, the magnitude of the sum should tell us something about the strength of the relationship. The greater the magnitude, the stronger the correction is (whether positive or negative).

If the sum of the cross-products is divided by N (the number of cases in the sample), the resulting figure is the **covariance**. The covariance is literally the average magnitude of the cross-products of the X and Y deviations. The covariance for the data we are considering is $+201.25$. Since the sign is positive, we know that there is a positive correlation between the two variables. But is 201.25 large in magnitude or not? It is quite a bit larger than the mean value for variable X, but it is smaller than the mean value for variable Y. The problem is that the possible upper limit of the covariance depends on the units of measurement of the variables being correlated. In this case we are correlating years of education with dollars—two very different kinds of units. By comparing the means of the two distributions we can see that the magnitudes of the dollar values are considerably greater than those of years of education. Furthermore, from the standard deviations we can see that monthly rental values of residences are much more variable than are years of education of heads of households. We find ourselves in a situation of trying to relate two variables that are very different from each other both in central tendency and in variability. It is like trying to compare apples and oranges.

There is a way out of this dilemma, however. We can convert the values on both variables to **standard scores** (z-scores) and thus convert them into comparable units. When the values on both variables are expressed as standard scores they become comparable because, as you may recall from Section 5.4.7, the mean of a set of standard scores is always 0 and the standard deviation is always 1. The values for the X variable can be converted into standard score form by using the formula

$$z_x = \frac{X - \bar{X}}{s_x} = \frac{X - 13.5}{3.29}$$

For the values of the Y variable, a similar formula is used:

$$z_y = \frac{Y - \bar{Y}}{s_y} = \frac{Y - 372.5}{144.24}$$

The table below presents the same 10 pairs of scores as were presented in the previous table, but, in this case, the deviations of scores from their means are expressed as standard scores rather than simply deviation scores.

Years of Education (X)	Monthly Rental Value of Residence (Dollars) (Y)	z_x	z_y	$(z_x)(z_y)$
12	400	−0.46	0.19	−0.09
10	325	−1.06	−0.33	0.35
16	450	0.76	0.54	0.41
8	350	−1.67	−0.16	0.27
14	200	0.15	−1.20	−0.18
20	525	1.98	1.06	2.10
11	375	−0.76	0.02	−0.02
15	250	0.46	−0.85	−0.39
16	675	0.76	2.10	1.60
13	175	−0.15	−1.37	0.21
				+4.26

Data from table, p. 242.

BOX 7.6 STANDARD SCORES

If you are a bit rusty on z-scores, refer back to Section 5.4.7. Briefly, a standard score is the number of standard deviation units above or below the mean that a score falls. The mean of a distribution of z-scores is therefore 0, and the standard deviation of a distribution of z-scores is 1.

Column three consists of standard scores for the 10 values on the X variable, and column four consists of those for the Y variable. Column five consists of cross-products of the standard scores for the X and Y variables. If the sum of the cross-products of these standard scores is divided by N, the result is a sort of standardized covariance. This standardized covariance is, in fact, Pearson's product–moment coefficient of correlation. That is,

(7.17)

$$r = \frac{\Sigma z_x z_y}{N}$$

In the case of the data under consideration, the r that results from using this formula is

$$r = +\frac{4.26}{10} = +.43$$

where the $+4.26$ is the sum of the cross-products of the standard scores of the X and Y variables.

Since z_x can be expressed as x/s_x and z_y as y/s_y (x and y are deviations), we can substitute those expressions into Formula 7.13 as follows:*

$$r = \frac{\Sigma \left(\dfrac{x}{s_x}\right)\left(\dfrac{y}{s_y}\right)}{N}$$

Then the two standard deviations s_x and s_y may be put in the denominator with N without changing the value of the formula, as follows:

(7.18)
$$r = \frac{\Sigma\,xy}{Ns_x s_y}$$

The resulting formula is one that is the basis for the computing formula for r that is given in Diagram 7.2 (p. 273).

Formula 7.17 makes it clear that the computation of r involves the transformation of the pair of scores from their original measurement units (such as years and dollars) to standard score units. You will recall from univariate statistics that the sum of squares of z-scores equals the total number of cases, N. If a case has a score that is in the same relative position on both the X and Y variables (and thus the same-sized z-score on each variable), the correlation coefficient will equal 1.0, since the numerator will then sum to N. To the extent that the standing of individuals is different on the two variables (as reflected in different z-scores), the numerator of the equation above will not reach N and the correlation will be less than 1.0. If one set of the z-scores tends to be negative while the other tends to be positive, then the correlation coefficient is negative. The point is that here again a simple ratio expresses one feature of the relationship between two variables, a kind of ratio that was used (with somewhat different components, to be sure) for the

---

* From Formula 5.7, we can rewrite the denominator of Formula 7.18 as follows:

$$N \times \sqrt{\frac{(\Sigma\,x^2)}{N}} \times \sqrt{\frac{(\Sigma\,y^2)}{N}} = \sqrt{(\Sigma\,x^2)(\Sigma\,y^2)}$$

Substituting this version of the denominator into Formula 7.18 results in the following formula:

$$r = \frac{\Sigma\,xy}{\sqrt{(\Sigma\,x^2)(\Sigma\,y^2)}}$$

which is the immediate basis for the computational Formula (D.3) (p. 273) and its simplified version on page 274.

arithmetic mean, the variance, ordinal measures of association, and now an interval measure of association.

The standard score form of the simple regression equation is also interesting to note.

$$z'_y = r(z_x)$$

Here the estimated value of the z-score on variable Y can be computed from a knowledge of the z-score on variable X, if an adjustment is made. The adjustment corresponds to multiplication of the z_x score by a constant, which (for one independent variable only) is the Pearson correlation coefficient, r. This is called a **standardized regression equation**, which we will return to again in Chapter 9.

One final expresssion for the correlation coefficient will help show similarities between this measure and some of the other symmetric measures. The Pearson correlation coefficient can also be expressed as a kind of average* of two asymmetric regression coefficients, b_{yx} and b_{xy}, as follows:

(7.19)
$$r = \sqrt{(b_{yx})(b_{xy})}$$

You will recall that this way of forming a symmetric correlation coefficient out of two asymmetric measures was also possible for tau-b, which was expressed in terms of Somers' d as follows:

$$t_b = \sqrt{(d_{yx})(d_{xy})}$$

7.4.4 Assumptions and Interpretations

The regression equation and correlation coefficient have provided an opportunity to show the interconnectedness of several ideas we have discussed in bivariate and univariate statistics. We have shown some of this by providing alternative formulas and their underlying reasoning for each of the main ideas discussed: the regression coefficients, the standard error (i.e., standard deviation around the regression line), and the correlation coefficient.

The essential interpretation and underlying assumptions implied in using the linear regression equation and correlation coefficient are straightforward. If both variables can be interpreted as interval variables, the linear regression equation can be computed as one possible description (linear) of their relationship. It is an asymmetric description in the sense that it matters which variable is dependent and which is the predictor variable. The least-squares regression equation, in general, minimizes the squared deviation of predicted from actual scores on the dependent variable. In its raw score form, the Y-intercept coefficient, a_{yx} is the value of Y when X is 0. The b-coefficient, where scores are in the raw score form, expresses the *impact* X has on Y. It is the predicted number of (positive or negative) units of change in Y (in terms of Y's unit of measurement) for each 1-unit increase in X (in terms of X's unit of measurement). When scores are in z-score form, then the Y-intercept is always 0 (and thus is dropped from the equa-

---

* This average is an example of the geometric mean discussed in Chapter 5.

tion). The slope coefficient is called a standardized b-coefficient (or b^* or "beta") and it expresses the amount of (plus or minus) change in the z-score of Y given a z-score increase in the value of the independent variable, X. It is clear that a straight line may not be a very good description of the nature of data (a curvilinear or irregular pattern may be evident from a scatterplot).

The amount of variation around the regression line can be expressed by the *standard error of the estimate*, which is a single number expressing in general how much variation there is around the predicted values. It is a standard deviation, except that at each point on the X-axis it uses the predicted value of Y instead of the overall mean of Y as its centering point. If, for each value of X, the spread of Y-scores is not equal (a condition referred to statistically as a lack of **homoscedasticity**), then the standard error of the estimate can only be interpreted as an overall measure of variation around the regression line rather than a figure that applies to the variation at any given point.

Information on variation around the regression line, such as that summarized by the standard error of the estimate, is used in the *correlation coefficient*, a normed (-1.0 to $+1.0$), symmetric measure of association. When r is squared, it can be given a PRE interpretation as the "proportionate reduction in error in predicting values of one variable by using the regression equation (and, thus information about the predictor variable) rather than the overall mean of the predicted variable." The correlation coefficient is appropriate for interval variables and it indicates how completely the regression line describes the bivariate relationship. One reason r might be low is that a straight line is not a very good description of the nature of the relationship: there may be no relationship or it may be of a different form than the regression equation expresses.* Sometimes one or a few deviant scores (called *outliers*) may have an overpowering effect on the computation of the regression equation and correlation coefficient, since, like the arithmetic mean, extreme scores have an impact reflecting their extremeness.

---

* Sometimes in statistics books an assumption underlying Pearson's r is said to be that "the two variables are related in a bivariate normal distribution," referring to the specific mesokurtic, unimodal normal distribution discussed earlier, applied to the relationship of two variables. Where X and Y are related in a bivariate normal distribution, the relationship will be linear. However, if two variables are related in a linear fashion that does not imply that they must be related in a bivariate normal way. Also, if the distribution of Y-scores for each value of X is normal, then properties of the normal curve can be used, such as approximately 68% of the predictions of an actual value of Y will be within ± 1 standard error of estimate of that value. Normality assumptions are important for making inferences from regression and correlation coefficients computed on a sample to population descriptions, discussed later in inferential statistics. If the data being correlated constitute a random sample of some population and our intention is to generalize to the population (see Chapter 16), then it is assumed that variables X and Y are both normally distributed random variables. If this assumption is tenable, then inferences made from the sample data allow us to make statements about the independence or dependence of variables X and Y in the population. A 0 correlation between X and Y assures their independence from each other only when X and Y jointly form a bivariate normal distribution.

When inferences are made to population data using the regression equation, it is assumed that the dependent variable is a random variable, but values of the independent variable are assumed to be constants. Thus specific values of the independent variable may be entered into the regression equation to predict a value of the random dependent variable with a range of prediction error measured by the standard error of the estimate.

The correlation coefficient is a measure of the usefulness of the regression line (*i.e.*, are scores closely scattered around the regression line or not?). The b-coefficient, on the other hand, expresses the specific direction and level of impact that differences in the independent variable have on the dependent variable if the data were to be described in a linear fashion.*

7.4.5 The Correlation Ratio, Eta (E_{yx}), a Measure of Association for Nonlinear Relationships

If a straight line is not a very useful description of the nature of the association between two interval variables, and if there is no other formula for a curved line available, one might fall back on a procedure of making separate predictions of the mean of the dependent variable within categories of the independent variable. This situation is illustrated in Table 7.7.

Here the dependent variable is the number of wage earners in a family and the independent variable is family income, divided up into a series of categories (52 families are in the sample). The overall mean number of wage earners for these data is 1.7, and this would constitute the minimum prediction rule we could use. On the other hand, we could improve our prediction by guessing the category mean for those cases falling in a given income category.

The prediction errors are calculated as the squared difference between prediction and actual score on the *Y*-variable. Since there are, again, two predictions, the *overall mean* and the *mean within categories*, we have two error figures to contrast. Eta squared (E_{yx}^2) is simply a ratio of these quantities, as defined below:[†]

$$E_{yx}^2 = 1 - \frac{\text{Variance of } Y \text{ within categories of } X}{\text{Overall variance of } Y}$$

(7.20)
$$E_{yx}^2 = 1 - \frac{\Sigma(Y - \bar{Y}_i)^2}{\Sigma(Y - \bar{Y})^2}$$

---

* For the situation where there is one independent and one dependent variable and scores are expressed as z-scores, the numeric value of r is the same as b* in the standardized regression equation. The student is encouraged to keep the two ideas separate, however, since the similarity of values only occurs in this two-variable situation. Chapter 9 will return to the notion of a predictive equation and a correlation measure that picks up on the ideas developed in this chapter, but for situations where several predictor variables are used.
† The following formula for eta (E_{yx}) is a more useful computational version.

$$E_{yx}^2 = 1 - \frac{\Sigma Y^2 - \Sigma n_k \bar{Y}_k^2}{\Sigma Y^2 - N\bar{Y}^2}$$

where $n_k \bar{Y}_k^2$ is the product of the number of cases times the squared mean of the *k*th subcategory, and these products are then summed over all *k* subcategories. ΣY^2 is simply the overall sum of squared scores and $\bar{Y}$ is the overall mean of *Y*.

TABLE 7.7   NUMBER OF EARNERS IN 52 FAMILIES IN THE UNITED STATES IN 1970 BY FAMILY INCOME

			(X) Family Income			
Under $1,500	$1,500– 2,999	$3,000– 4,999	$5,000– 7,999	$8,000– 11,999	$12,000– 49,999	$50,000 and up
1	0	0	0	0	1	1
1	0	1	1	1	1	
	2	1	1	1	1	
		1	1	1	1	
		2	1	1	2	
			2	1	2	
			2	1	2	
			2	2	2	
			3	2	2	
				2	2	
				2	2	
				3	2	
				3	2	
				4	2	
					2	
					3	
					3	
					7	
$N_i = 2$	3	5	9	14	18	1
$\bar{Y}_i = 1$	0.7	1	1.4	1.7	2.2	1

$N = 52$
$\bar{Y} = 1.7$
N_i and $\bar{Y}_i$ refer to category size and mean.
N and $\bar{Y}$ refer to the overall size and grand mean.

Source: Data are fictitious in this example, but they follow the distribution of number of earners and family income for families in the United States in 1970. Each figure is approximately 1 million families (see U.S. Bureau of the Census, 1971b). The actual average number of earners per family for the above income categories from lowest to highest are: .76, .69, 1.05, 1.44, 1.73, 2.19, and 1.97. The overall average number of earners per family in the United States in 1970 was 1.68.

Table 7.8 illustrates the logic of the calculations. Number of wage earners in each of the 52 families is listed, grouped by income categories. Table 7.8 lists the two possible predictions: the overall or "grand" mean number of wage earners for all families (*i.e.*, 1.7), which corresponds to the minimum guessing rule in this case, and the mean number of wage earners within each category separately, which corresponds to the more informed guessing rule. From this we can compute the deviation of each score from the grand mean (and the squared deviation), as well as the deviation of each score from its category mean Y_j (and its squared deviation). The sum of the latter two columns gives the overall variability of Y (67.88 for these data) and the variability of Y within categories of X (56.29 for these data).

TABLE 7.8 Actual and Predicted Number of Family Earners by Income Category

Income Category	(Y) Number of Earners Per Family	Predictions Overall Mean	Category Mean	$(Y - \bar{Y})$	$(Y - \bar{Y})^2$	$(Y - \bar{Y}_i)$	$(Y - \bar{Y}_i)^2$
Under $1,500	1	1.7	1	− .7	.49	0	0
	1	1.7	1	− .7	.49	0	0
$1,500–2,999	0	1.7	.7	−1.7	2.89	− .7	.49
	0	1.7	.7	−1.7	2.89	− .7	.49
	2	1.7	.7	+ .3	.09	+1.3	1.69
$3,000–4,999	0	1.7	1	−1.7	2.89	−1.0	1.00
	1	1.7	1	− .7	.49	0	0
	1	1.7	1	− .7	.49	0	0
	1	1.7	1	− .7	.49	0	0
	2	1.7	1	+ .3	.09	+1.0	1.00
$5,000–7,999	0	1.7	1.4	−1.7	2.89	−1.4	1.96
	1	1.7	1.4	− .7	.49	− .4	.16
	1	1.7	1.4	− .7	.49	− .4	.16
	1	1.7	1.4	− .7	.49	− .4	.16
	1	1.7	1.4	− .7	.49	− .4	.16
	2	1.7	1.4	+ .3	.09	+ .6	.36
	2	1.7	1.4	+ .3	.09	+ .6	.36
	2	1.7	1.4	+ .3	.09	+ .6	.36
	3	1.7	1.4	+1.3	1.69	+1.6	2.56
$8,000–11,999	0	1.7	1.7	−1.7	2.89	−1.7	2.89
	1	1.7	1.7	− .7	.49	− .7	.49
	1	1.7	1.7	− .7	.49	− .7	.49
	1	1.7	1.7	− .7	.49	− .7	.49
	1	1.7	1.7	− .7	.49	− .7	.49
	1	1.7	1.7	− .7	.49	− .7	.49
	1	1.7	1.7	− .7	.49	− .7	.49
	2	1.7	1.7	+ .3	.09	+ .3	.09
	2	1.7	1.7	+ .3	.09	+ .3	.09
	2	1.7	1.7	+ .3	.09	+ .3	.09
	2	1.7	1.7	+ .3	.09	+ .3	.09
	3	1.7	1.7	+1.3	1.69	+1.3	1.69
	3	1.7	1.7	+1.3	1.69	+1.3	1.69
	4	1.7	1.7	+2.3	5.29	+2.3	5.29
$12,000–49,999	1	1.7	2.2	− .7	.49	−1.2	1.44
	1	1.7	2.2	− .7	.49	−1.2	1.44
	1	1.7	2.2	− .7	.49	−1.2	1.44
	1	1.7	2.2	− .7	.49	−1.2	1.44
	2	1.7	2.2	+ .3	.09	− .2	.04
	2	1.7	2.2	+ .3	.09	− .2	.04
	2	1.7	2.2	+ .3	.09	− .2	.04
	2	1.7	2.2	+ .3	.09	− .2	.04
	2	1.7	2.2	+ .3	.09	− .2	.04
	2	1.7	2.2	+ .3	.09	− .2	.04
	2	1.7	2.2	+ .3	.09	− .2	.04
	2	1.7	2.2	+ .3	.09	− .2	.04
	2	1.7	2.2	+ .3	.09	− .2	.04
	3	1.7	2.2	+1.3	1.69	+ .8	.64
	3	1.7	2.2	+1.3	1.69	+ .8	.64
	7	1.7	2.2	+5.3	28.09	+4.8	23.04
$50,000 and up	1	1.7	1.0	− .7	.49	0	0
					$\Sigma(Y - \bar{Y})^2 = 67.88$		$\Sigma(Y - \bar{Y}_i)^2 = 56.29$

Source: Data in Table 7.7.

Substituting these figures into Equation 7.20 yields eta squared and its square root, eta:

$$E_{yx}^2 = 1 - \frac{56.29}{67.88}$$

$$= 1 - .83$$

$$= +.17$$

$$E_{yx} = \sqrt{.17} = +.41$$

This is essentially the same type of measure as r, except that the mean within categories of the independent variable, rather than a regression equation, is used to form the predicted Y-score. Unlike r, however, there are two etas depending upon which variable is predicted from which. E_{yx} is asymmetric. E_{yx}^2 indicates the proportionate reduction in error in predicting the Y-scores if category means rather than the grand mean are predicted. E_{yx} varies in magnitude from 0 (in which case the grand mean is as useful in prediction as category means) up to $+1.0$, which indicates that category means permit exact prediction. The independent variable may be nominal, ordinal, or interval just as long as the dependent variable is interval. Thus, the sign of the association does not have meaning, or, in other words, E_{yx} will always be a positive value. Since r and E_{yx} have essentially the same form, and differ only in the source of the refined prediction, they can be compared directly. If, for a set of data where both measures are appropriate, E_{yx} is *larger* than Pearson's r, then one can infer that category means do not fall along a simple straight line and thus, to some degree, the *nature of association is curved* or different from a straight line. If they are identical, then it is clear that subcategory means fall exactly along a least-squares regression line, such as we have developed here. (E_{yx} will never be smaller in magnitude than r.) This provides a rather handy basis for making inferences about the nature of an association.

7.5   THE CORRELATION MATRIX

Because of the interest in comparison, it is not unusual for investigators to compute a number of similar association measures, showing the relationship between all possible pairs of a set of items. An example of such a correlation matrix is shown in Table 7.9, where Pearsonian correlation coefficients are used. A matrix consisting of gammas (G's) or any other appropriate measure of association could be used, depending upon the variables being correlated.

In Table 7.9, Marsh and Stafford (1967) present a correlation matrix showing the correlation between answers to each pair of eleven questions given male professionals with an M.A. or Ph.D. degree. This was a part of a larger 1962 survey of the United States. Some respondents were from academic settings and some were from private industry. Agreement with certain of the 11 items was thought to reflect a "professional" orientation and agreement with some of the other items was thought to reflect an "acquisitive" attitude. Examination of the correlation matrix helped Marsh and Stafford see to what extent these two groups of

TABLE 7.9 MATRIX OF CORRELATION COEFFICIENTS BETWEEN ELEVEN MEASURES OF ATTITUDE TOWARD WORK

(Items) Attitudes	1	3	10	12	13	11	9	4	5	7	8
1 Opportunity to be original and creative		.36	.29	.42	.49	.14	.25	.31	.24	.18	.27
3 Relative independence in doing my work			.31	.36	.36	.11	.26	.23	.25	.17	.21
10 Freedom from pressures to conform in my personal life				.35	.31	.26	.37	.17	.34	.23	.33
12 Freedom to select areas of research					.47	.12	.19	.18	.24	.20	.21
13 Opportunity to work with ideas						.29	.31	.29	.24	.24	.26
11 Opportunity to work with people							.43	.39	.35	.44	.27
9 Pleasant people to work with								.29	.45	.36	.39
4 A chance to exercise leadership				correlation					.26	.34	.31
5 A nice community or area in which to live				of items 4						.45	.41
7 Social standing and prestige in my community				and 11							.36
8 A chance to earn enough money to live comfortably											

Source: Marsh and Stafford (1967). Used by permission. The study is based on an April, 1962, self-administered questionnaire given to some 51,505 professional, technical, and related workers in the United States. This investigation selected 13 professional fields and limited the analysis to males who had M.A. or Ph.D. degrees. The correlations above are Pearsonian r's between responses to the different items on attitude toward work.

items—"professional" and "acquisitive"—appeared in their data. Their basic hypothesis in this part of their study was that professionals in academic settings would be more likely to agree with "professional" items and disagree with "acquisitive" items. They thus expected to show that professional and intellectual values of work would serve as "compensation" for the lower incomes academics generally receive as contrasted to professionals in industry. An examination of the correlation matrix in Table 7.9 helped them identify items that could be used to measure these two attitudes.

Table 7.9 is organized so that each row refers to a different item (the numbers at the left are item numbers in the original questionnaire). Columns also refer to the same items, with each item (numbered as they are at the left of the table) being in a different column. The number at the intersection of a given row and column is a correlation coefficient showing the correlation between the items indicated by row and column headings. The items in Table 7.9 appear to fall into two clusters, one being a "professional" cluster including items about independence, creativity, and working with ideas, and the other being a cluster dealing with "acquisitive" items, such as pleasant work, community settings, and prestige.

For convenience, Marsh and Stafford organized the attitude items into these two clusters and drew lines around the coefficients that refer to the inter-correlation of items within each cluster. The expectation is that items within a cluster would be more highly related than they would be with items not in the cluster. This appears to be the pattern in Table 7.9. Patterns in matrices such as this will be discussed later in this text. For now, the important point is to call your attention to the *comparative* use of measures of association. Notice that the Pearsonian correlation coefficient, *r*, which was used in this instance, is a symmetric measure. Thus, only half of the matrix in Table 7.9 needs to be given, because the other half would be identical (*i.e.*, the correlation between item 3 and item 1 is the same as the correlation between item 1 and item 3).

7.6   SELECTING A USEFUL MEASURE OF ASSOCIATION

There have been a number of criteria discussed that distinguish between the different measures of association and, in general, these lead to the selection of one measure rather than another for a particular problem. In a given problem, certain features of these distinctions may take on more importance for the comparisons that are to be made and, thus, these criteria would override other kinds of differences.

7.6.1   Symmetry and Asymmetry

One of the important kinds of differences between coefficients is the way independent and dependent variables are handled. In problems where explanation and prediction of a dependent variable are of particular interest, such as that shown in the following arrow diagram, an asymmetric measure, if one is available, would be most useful.

If we are interested simply in the way variables covary or relate to each other,

---

BOX 7.7   THE THEME

In Chapters 6 and 7 we have attempted (a) to show how the various measures of association are relatively straightforward solutions to the problem of describing relationships between variables, and (b) to highlight some of the differences between them that may correspond closely to the substantive problem being investigated.

At this point, you should avoid looking for the easy or oversimplified ways to choose an appropriate measure. It might seem easy to recommend only one or two measures out of the variety we have discussed, but the decision on appropriate measures is not that mechanical. The selection depends on the substantive problem being investigated, as well as some "technical" characteristics of each measure. Part of the skill in creating and using statistical information lies in an understanding of the relationship between problem and tools.

---

which might be the case when we examine the relationship between indicators of the same concept, then a symmetric measure provides this information.

Interrelation of social class measures.

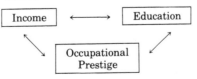

7.6.2 Level of Measurement

The meaning defined into variables also provides an important basis for selecting among the various measures. If only the categorical aspect of a variable is meaningful, then measures suitable for nominal variables would be most appropriate. Diagram 7.3 (p. 275) organizes the measures we have discussed, in this and the last chapter, in terms of level of measurement. As in the case of univariate descriptive statistics, one might find occasion to use lower-level measures on data defined at a higher level of measurement, but this would result in the loss of information contained in the data—a result that is usually undesirable.

In some cases, level of measurement does not make much difference. In a 2 by 2 table, for example, Pearson's *r*, tau-*b*, and the phi coefficient are numerically identical. Spearman's rho is a Pearsonian correlation (a measure designed for interval variables) computed on ranks that may be assigned in the process of measuring an ordinal variable. Generally, however, level of measurement *is* an important consideration in selecting and certainly in interpreting association measures.

7.6.3 Nature versus Strength of Association

Increasingly, sociology is addressing itself to the description of the amount of change a unit change in a given independent variable will likely produce in a dependent variable. "Powerful" variables in this sense are those that produce great effect, and this, of course, has both theoretical and practical importance. If this is the interest, then **regression coefficients** become more interesting than correlation coefficients, or **percentaged tables** more interesting than the various measures of association. Association measures express how well the variables go together, in the sense of scatter around a regression line, or errors in predicting a mode or rank order. A "clean" prediction might be possible in this sense, when a predictive variable is (relatively) so "powerless" that changes in it may not reflect or cause much change in the dependent variable. These two interests go together, of course, but in any given problem one may be more important than the other, and this may influence the selection of measures. Regression coefficients are particularly useful if the main focus is on the potency of a variable.* On the other hand, if the nature of association is not relatively linear, then one either would have to find a curvilinear regression equation that fits the nature of the data or shift to some measure that is not tied to a specific nature of association.

7.6.4 Interpretation

Finally, measures of association differ in terms of the features of the association toward which they are most sensitive. Some measures (λ_{yx}, r, E_{yx}, for example)

* For other models expressing the impact of changes in one variable upon another, see Coleman (1964), Chapters 6 and 7.

are oriented toward predicting an optimal or central value of a dependent variable. Others (*e.g.*, U_{yx}) are oriented toward predicting the distribution of a categorical dependent variable. Some measures (*e.g.*, r, G) contrast observed data with a specific model of perfect association or independence. Some of these differences are organized in Table 7.10.

Many of the measures discussed have PRE interpretations, which makes them more convenient to use in research. Selection among these measures depends upon what one wants to treat "at risk." For asymmetric measures, such as Somers' d_{yx}, the at-risk pairs are pairs that are distinguishable on the independent variable. For gamma (G), they are pairs that are ranked differently on both X and Y variables. Other measures of association do not have PRE interpretations (*e.g.*, C, t_c) but can be interpreted in terms of magnitude between limiting values (often 0 and 1.00).

Finally, some of the measures of association we have discussed are most suited to tables that have specific types of layout. Yule's Q and the phi (ϕ) coefficient from Chapter 6 are oriented toward 2 by 2 tables, unlike most of the other measures. Tau-b (t_b) cannot achieve 1.0 in a table that is not square, and measures such as tau-b and Spearman's rho, for example, are especially sensitive to ties in data.

Another aspect of interpretability of a coefficient is its familiarity to an audience. At this point in sociology, coefficients such as Pearson's r, G, t_b, lambda (λ_{yx}), percentage difference (epsilon, ϵ), and Somers' d_{yx} are relatively frequently seen in print. If an appropriate measure for one's data happens to be among these, the more widely known and used measure would probably be chosen.

As you might suspect, there is no flat rule that predetermines which coefficient *must* be used. Usually, the choice is clear-cut—we would prefer a PRE measure, appropriate for the level of measurement of variables in the study, asymmetric if the independent/dependent role of variables is important, and sufficiently widely known so that it can be used to communicate information about the data. But often the choice is a matter of balance and judgment between coefficients that offer somewhat different strengths. The reasoning process in selecting a coefficient is closely tied to the logic and purpose of a research problem. Probably, the only flat rule is that it is *not* appropriate to compute all possible measures and select one measure for the reason that it is, numerically, the largest!

7.7   SOME CAUTIONS ABOUT THE INTERPRETATION OF ASSOCIATION MEASURES

An important purpose of correlation coefficients is, of course, to aid in comparison. Generally, we want to know the proportionate reduction in error that can be achieved in some defined respect by using one variable or another. Usually (or it is hoped), the contrasts are suggested by some prior reasoning. We expect social class rather than toenail length to have something to do with the distribution of anomie, and a comparison of correlation coefficients bears this expectation out.

TABLE 7.10 CHARACTERISTICS OF SELECTED MEASURES OF ASSOCIATION

Measure	Level of Measrmt.	Table Size	Range	Definition of a Perfect Assn.	Formula	Interpretation
ϵ (epsilon)	Nominal	$r \times c$	0 up to some x, $x \leq 100\%$	More restrictive	Usually, difference of two extreme corner percentages in properly percentaged table	A rough degree of association measure for a table. With care, one can compare across tables of the same size and similar marginals. In a 2×2 table, an epsilon of 0 means no association. Asymmetric.
Yule's Q	Ordinal (nominal if sign dropped)	2×2	-1 to $+1$ (0 to $+1$)	Less restrictive	$Q = \dfrac{ad - bc}{ad + bc}$ with table (columns Lo, Hi; rows Hi: b, a; Lo: d, c)	Symmetric measure identical to gamma. Probability of like (opposite) ranking on two variables among cases ranked differently on both variables.
ϕ (phi)	Ordinal (nominal if sign dropped)	2×2	-1 to $+1$ (0 to $+1$)	More restrictive	$\phi = \dfrac{ad - bc}{\sqrt{(a+b)(c+d)(a+c)(b+d)}}$	Degree of diagonal concentration. A symmetric measure.
λ_{yx} (lambda)	Nominal	$r \times c$	0 to $+1$	Y all one category for any given X	$\lambda_{yx} = \dfrac{\Sigma m_y - M_y}{N - M_y}$	An asymmetric measure indicating the percentage improvement in predictability of the dependent variable with information about the independent variable classification.

Measure	Level	Table type	Range	Assumption	Formula	Description
U (uncertainty)	Nominal	$r \times c$	0 to +1	Y all one category for any given X	$U_{yx} = \dfrac{h_y - h_{y.x}}{h_y}$	PRE in predicting category placement. Asymmetric.
G (gamma)	Ordinal	$r \times c$	−1 to +1	Less restrictive	$G = \dfrac{N_s - N_d}{N_s + N_d}$	A symmetric measure indicating the relative preponderance of like (unlike) ranked pairs among pairs ranked differently on both variables.
t_b (tau-b)	Ordinal	$r \times c$	−1 to +1	More restrictive	$t_b = \dfrac{N_s - N_d}{\sqrt{(N_s + N_d + T_y)(N_s + N_d + T_x)}}$	Symmetric. For square tables.
d_{yx} (Somers')	Ordinal	$r \times c$	−1 to +1	More restrictive	$d_{yx} = \dfrac{N_s - N_d}{N_s + N_d + T_y}$	Asymmetric. PRE among predicted pairs.
E_{yx} (eta)	Interval-dependent variable (other may be nominal)	$r \times c$	0 to +1	Line of category means	$E_{yx}^2 = 1 - \dfrac{\Sigma Y^2 - \Sigma n_k \bar{Y}_k^2}{\Sigma Y^2 - N\bar{Y}^2}$	PRE. Asymmetric.
r_{yx}	Both interval	$r \times c$	−1 to +1	linear	$r = \sqrt{1 - \dfrac{S_{y.x}^2}{S_y^2}} = \dfrac{\Sigma z_x z_y}{N}$	PRE. r^2 is the proportion of variation in one variable explained by linear association with the other. Symmetric.
r_s (Spearman's rho)	Ordinal	Two sets of N ranks $2 \times c$	−1.0 to +1	Identical ranks on two variables	$r_s = 1 - \dfrac{6\Sigma D^2}{N(N^2 - 1)}$	Symmetric. Sensitive to large differences in ranks. D is difference between the two ranks for a given subject.

7.7.1   Causation and Association

There is sometimes a temptation to attribute more meaning to a measure of association than it really contains. This is particularly likely if the reasoning behind the investigation comes from an interest in explaining the causes of some phenomenon. If knowledge of social class reduces the errors of predicting alienation by quite a large amount, the temptation is to say that social class influences or causes alienation, simply on the basis of the correlation coefficient. Upon reflection, this statement is not appropriate at all. The correlation may well be the same between doorknob size and education of head of household, and we would hardly want to conclude on this basis that knob size has anything particularly to do with education. Even asymmetric measures only state how measured variables relate in the data at hand.

The investigators' theory, however, may lead them to predict that there will be a certain kind of association in the data because of some hypothesized causal link between variables. The data may yield correlations that turn out to be close to predictions, and because of this investigators may feel that their faith in the theory is well founded, that one of the variables does cause the other to change as the theory stated, and that this is what the static correlation coefficients mean. The cause–effect argument is clearly *in the investigators' theory*, however, and *not* something that is included in the correlation coefficient itself. *Correlation is not causation.* More evidence than merely a correlation coefficient is needed to begin to have confidence in a conclusion that one variable influences another under certain conditions.*

7.7.2   Ecological Fallacy

There are some interpretative problems that arise essentially because investigators are not clear about the kind of subject or case they are examining. This may be an important source of difficulty in sociology, since the field deals with many different levels of units, from dyads to individual people, to individual roles or self-concepts, to small groups or societies or groups within groups. Correlations between variables measured on *groups* are *not the same* as correlations measured on *individuals*, in spite of the fact that those individuals may form the groups. The error of inferring how two variables are related among *individuals* by examining the correlation of similar variables measured on *groups* is called the **ecological fallacy.**†

A classic example of this problem is illustrated in an article by Robinson (1950) in which he shows the relationship between "foreign-born" and "illiter-

---

* Although one would expect a correlation between a causal variable and its effect, there are several other types of information that need to be examined before a casual interpretation of data begins to become useful. One needs to know time order (that the cause occurred before the effect) and that the association could not be "explained away" by other factors. The next chapter deals with some of these questions.

† Some of the problems in inference between different levels of phenomena are nicely discussed in a research methods book by Matilda White Riley (1963; see especially unit 12).

acy." The correlation between percentage foreign-born and percentage illiterate
was computed to be $-.62$, with these two variables measured on each of several regions of the United States. The correlation used geographic regions as
the case. This means that the higher the percentage of foreign-born in a region, the
lower the percentage of "illiterate" we would expect. Does this mean that foreign-
born *people* are less likely to be illiterate? No! The correlation between foreign-
born or not, and illiterate or not, for individuals in the United States was $+.12$, a
low, positive association. In fact, there is little reason why there should be the same
correlation between these variables at the group and individual levels. To start
with, the variables are different. In one case they are percentage concentration of
foreign-born or illiterates, and in the other case they refer to whether or not an
individual is or is not foreign-born or illiterate. Second, it is quite possible that all
of the foreign-born people are illiterate, or none of the foreign-born people is
illiterate, in any given area with a given percentage of foreign-born and illiterate.
The moral is the same one raised early in this volume: we must be clear about the
meaning of statistical observations before statistical results can be meaningfully
interpreted.

7.7.3   Built-in Correlations*

One final set of alternative interpretations of measures of association should be
mentioned before we close this chapter. The possibilities to be mentioned here flow
fairly directly from the meaning of specific measures of association we have
discussed, but they might be pulled together as a set of possible interpretations of
association. A measure of association may be low, not because two variables are
unrelated, but because they are not related in the way to which a given measure of
association is sensitive. This was pointed out for Pearson's r, where strong
curvilinear associations may show up as low correlations because r is measuring fit
to a simple straight line. It is usually a good idea to examine the nature of an
association via a scattergram or table.

An association could be "built-in," if the original *observations are not
independent.* For example, if a study of income and education of people were
designed so that husbands and wives were both measured, then it is likely that
some association between education and income would be observed, simply
because these variables are probably related for husbands and wives. We would be
able to handle the problem in this instance by separating out husbands and wives
or by treating them as a pair. Built-in correlations could occur in other ways, too—
for example, in correlations between a score that is derived from answers to all
items and a score that is derived from answers to a subset of the same items.

It is possible to have a correlation affected by a few very extreme cases (as
shown for Pearson's r in Figure 7.5d), where, except for these extremes, there may
be little or no association (or a very strong association) in the data. Correlations
based on small numbers of cases are especially vulnerable to these kinds of
aberrations. It is possible, too, that an association may appear very weak (or

---

* Chapter 6 of Edwards (1984) is a particularly good discussion of factors affecting
correlation coefficients.

strong) within a certain restricted range of the variables, but be quite different overall, or within other restricted ranges of the same variables. Unless we are particularly interested in some restricted range of values of a variable, it is good practice to be sure that the full range of possible scores is considered. Grouping variables (or truncating) could also have the effect of distorting a measure of association. This was illustrated in Table 6.1, but it applies to this chapter as well.

The form of the distribution of the variables being correlated can have profound effects upon the magnitude that a correlation coefficient may take. For example, the maximum values that may be attained by Pearson's r depend on the relative forms of the distributions of the independent and dependent variables. Pearson's r only has a potential range from -1.00 to $+1.00$ when both variables X and Y are symmetric and have the same form. They do not necessarily have to be normally distributed; however, whatever form the symmetry of their distributions assumes, it must be the same for both. If the distributions of X and Y are both skewed in the same direction and both have the same degree of skewness, then r can reach a value of $+1.00$, but it cannot reach a value of -1.00. If the distributions of X and Y are skewed in opposite directions and both have the same degree of skewness, then r can reach a value of -1.00 but it cannot reach a value of $+1.00$ (Edwards, 1984:54–56).

Finally, the reliability of the measurements used for X and Y has an influence on the magnitude a correlation will reach. In general, unreliable measurements will depress the magnitude of a correlation coefficient. Consequently, if a correlation is weak, it may be because of poor measurement instruments rather than a lack of relationship between the variables being studied.

7.8 SUMMARY

In this chapter we have developed a series of measures of degree of association for nominal, ordinal, and interval variables. In each case our concern was with developing a measure that could be interpreted as an indication of the proportionate reduction in predictive errors to be made by using information about an independent variable, as opposed to using only information about the dependent variable.

We could predict modes, either overall or within categories, for an independent variable. We could predict category placement of cases. We could predict rank orderings, either by chance, or by using a rule based on the direction of association. We could develop an equation to compute a predicted score and compare that prediction with the overall mean of the dependent variable. In each case the association measure amounted to a version of a ratio that expresses the reduction in errors made possible by improved information, as against the total of possible errors that could be expected without that information.

These measures do not exhaust the range of possible coefficients, and you may well be able to devise a new coefficient yourself. The measures do have utility in sociology, however, because they permit comparisons between tables that would otherwise be difficult to make. By virtue of proper norming and proper selection of measures of association, an investigator is able to raise questions about how well theory permits an explanation and prediction of some phenomenon of interest.

In addition to the theme of the ratio as a means by which many different kinds of valid comparisons are possible, there has been another theme in this chapter. It points back to the original ideas an investigator has in mind in examining data. Many of the differences in the array of coefficients we have presented are the result of the different kinds of information a collection of data may contain, which may be relevant to some problems and not to others. The difference between asymmetric and symmetric measures, for example, reflects the difference in interest in how the two variables go together, versus how well one variable permits us to predict or explain variation in a dependent variable. We would select between measures according to the nature of the association and, in fact, according to what we want to treat as a perfect association. Clear formulation of a research problem will usually permit a clear selection of the most appropriate coefficient.

Up to this point we have dealt with univariate distributions one by one, and then we compared central tendency and variation, for example, between univariate distributions. Then we packaged a series of these univariate distributions into a single table and developed an overall series of measures on the table itself as a whole. We are about to do this same type of thing again. We have discussed single bivariate distributions. Now we will begin to examine sets of bivariate distributions—one for each special condition or value of a further set of control variables. Then we will again ask whether there is some way to put all of these separate tables together into some overall comparison. At each step, we selectively ignore the more irrelevant information contained in data, and we focus an investigative light on those features that contain the comparative information that gives us the answers to the questions we ask.

CONCEPTS TO KNOW AND UNDERSTAND

PRE measures
regression equation
   intercept
   slope
   b-coefficient
   standardized regression equation
   scatter diagram
linear relationship
curvilinear relationship
nature of association
least-squares criterion
standard error of the estimate
explained variation
unexplained variation
total variation
lambda
asymmetric, symmetric relationships

concordant, discordant pairs
Kendall's tau-b
Somers' d_{yx}
gamma
Pearson's product–moment
   correlation r
eta
Spearman's rho
correlation matrix
selecting appropriate measures
cautions
ecological fallacy
causation versus association
built-in correlations
homoscedasticity
coefficient of determination
uncertainty coefficient

PROBLEMS AND QUESTIONS

1. Compute lambda and the uncertainty coefficient for Table 6.6 (p. 170), using "income" as the dependent variable and "ethnic background" as the independent variable. What kinds of different information do each of these coefficients provide? Under what conditions would they be quite different from each other?

2. Develop an analytic organization of measures of association or a "decision tree" showing how to decide on a measure. As a start you might use the guides given in the text and consider the points mentioned in the section on selecting measures of association.

3. Find an interesting pair of scores measured on about 70 cases and compute correlation and regression coefficients. Create a scatter diagram, draw in the regression line, and interpret the statistics you have computed.

4. Using tables A and B from Problem 5 of Chapter 6, compute gamma, tau-b, and Somers' d_{yx}. Compare the three measures. To what do you attribute the differences among them? How was each influenced by the skewed marginal distribution of the tables? Why were they influenced to different extents? Do these measures of association alter in any way the interpretation of the tables which you based on percentages in Problem 5 of Chapter 6?

5. The table below is from a longitudinal study of white males. The men in this sample were asked when they expected to retire. Later during the same year, after they had actually left the labor force, the reason for their leaving was determined. Using the appropriate measure of association, determine whether there is a relationship between their expected retirement age and the reason for leaving the labor force. What is your conclusion?

	Expected Retirement Age	
Reason for Leaving Job	Less Than 65	65 and Over
Involuntary	10	22
Health	19	51
Retirement	41	38
Other voluntary reason	4	18

Source: U.S. Department of Labor (1975:182, Table 15.13).

6. Use an alternate measure of association to measure the relationship between the variables in Problem 5. How does this measure compare with the first one you used? How do you explain the difference?

7. Compute Spearman's rho (r_s) and Pearson's product–moment r for the following data. How do the two measures compare? To what can the difference be attributed?

TOTAL HOSPITAL BEDS AND TOTAL ELDERLY
POPULATION IN NEBRASKA, 1970

Planning and Service Areas	Population 65 and Over (X)	Hospital Beds (Y)
A	38,675	3,658
B	16,737	1,069
C	10,695	177
D	10,998	310
E	13,482	381
F	9,957	389
G	17,421	643
H	7,481	381
I	16,014	623
J	6,351	382
K	6,925	275
L	11,902	664
M	4,763	170
N	5,832	198
O	6,323	266

Source: U.S. Bureau of the Census (1975a, Table 11).

8. The data below from 52 Standard Metropolitan Statistical Areas are percentages of occupied dwelling units with 1.01 or more persons per room and murder and manslaughter rates (per 100,000 population). Compute r, b_{yx}, and a_{yx} for these data. Construct a scatter diagram for the data and plot the regression line. How strong is the correlation? How may it be interpreted?

DENSITY OF DWELLING UNITS AND MURDER AND MANSLAUGHTER RATES FOR STANDARD METROPOLITAN STATISTICAL AREAS WITH 200,000 POPULATION OR MORE IN 1970

Occupied Units with 1.01 or More Persons per Room (Percentages)	Murder & Nonnegligent Manslaughter Rates (per 100,000)
5.2	8.3
4.3	3.2
11.5	9.7
4.0	3.8
6.1	3.7
6.1	7.3
8.1	1.4
8.3	20.8
11.0	16.5
9.7	11.1

(Table continues.)

DENSITY OF DWELLING UNITS AND MURDER AND MANSLAUGHTER RATES FOR STANDARD METROPOLITAN STATISTICAL AREAS WITH 200,000 POPULATION OR MORE IN 1970

Occupied Units with 1.01 or More Persons per Room (Percentages)	Murder & Nonnegligent Manslaughter Rates (per 100,000)
11.3	11.4
7.1	16.8
12.7	13.9
10.1	12.9
5.1	2.0
9.6	17.7
5.8	5.6
5.8	3.3
5.3	6.0
6.2	5.3
7.7	5.0
11.7	15.8
9.0	18.0
8.9	16.2
8.2	15.9
9.3	7.8
5.5	17.2
6.6	5.6
9.2	19.5
11.4	21.4
6.0	8.0
17.1	13.5
8.5	14.5
6.9	3.0
6.1	13.6
5.4	7.4
6.1	4.4
7.8	20.2
7.5	3.5
18.6	5.5
6.0	3.0
5.9	3.0
8.8	8.2
11.3	15.1
8.7	13.6
7.1	15.6
6.3	4.8
11.1	12.5
12.1	20.7
6.2	3.8
8.1	9.6
9.4	16.3

Source: U.S. Bureau of the Census (1975b:886ff).

GENERAL REFERENCES

Blalock, Hubert M., Jr., *Social Statistics*, rev. ed. (New York, McGraw-Hill), 1979.

Coleman, James S., *Introduction to Mathematical Sociology* (New York, The Free Press of Glencoe), 1964.
    See especially Chapters 6 and 7.

Costner, Herbert L. (ed.), *Sociological Methodology: 1971* (San Francisco, Jossey-Bass), 1971.
    See especially Chapter 10 by Robert K. Leik and Walter R. Gove, "Integrated Approach to Measuring Association," which presents a more detailed treatment of the subject.

Freeman, Linton C., *Elementary Applied Statistics for Students in Behavioral Science* (New York, Wiley), 1965.
    See especially Section C.

Liebetrau, Albert M., *Measures of Association* (Beverly Hills, Calif., Sage), 1983.

Mueller, John H., Karl F. Schuessler, and Herbert L. Costner, *Statistical Reasoning in Sociology*, third ed. (Boston, Houghton Mifflin), 1977.
    See especially Chapters 9, 10, and 11.

Schroeder, L. D., D. L. Sjoquist, and P. E. Stephan, *Understanding Regression Analysis* (Beverly Hills, Calif., Sage), 1986.

LITERATURE CITED

Costner, Herbert L. "Criteria for Measures of Association," *American Sociological Review*, 30 (June 1965), pp. 341–353.

Edwards, Allen L., *An Introduction to Linear Regression and Correlation*, second ed. (San Francisco: W. H. Freeman), 1984.

Furstenberg, Frank F., Jr., "The Transmission of Mobility Orientation in the Family," *Social Forces*, 49 (June 1971), p. 598.

Hammond, Kenneth R., and James E. Householder, *Introduction to the Statistical Method* (New York, Knopf), 1962, pp. 212–214.

Haug, Marie R., and Marvin B. Sussman, "The Indiscriminate State of Social Class Measurement," *Social Forces*, 49 (June 1971), p. 559.

Hunter, Albert, "The Ecology of Chicago: Persistence and Change, 1930–1960," *American Journal of Sociology*, 77 (November 1971), pp. 425–444.

Kim, Jae-On, "Predictive Measures of Ordinal Association," *American Journal of Sociology*, 76 (March 1971), pp. 891–907.

Marsh, John F., Jr., and Frank P. Stafford, "The Effects of Values on Pecuniary Behavior: The Case of Academicians," *American Sociological Review*, 32 (October 1967), p. 743.

Riley, Matilda White, *Sociological Research: A Case Approach*, vol. 1 (New York, Harcourt, Brace and World), 1963.

Robinson, W. S., "Ecological Correlation and Behavior of Individuals," *American Sociological Review*, 15 (1950), pp. 351–357.

Rossi, Alice, "Status of Women in Graduate Departments of Sociology, 1968–1969," *American Sociologist*, 5 (February 1970), p. 4.

Senders, Virginia L., *Measurement and Statistics* (New York, Oxford University Press), 1958.

Somers, Robert H., "A New Asymmetric Measure of Association for Ordinal Variables." *American Sociological Review*, 27 (December 1962), pp. 799–811.

U.S. Bureau of the Census, "Household Income in 1970" and "Selected Social and Economic Characteristics of Households," *Current Population Reports*, Series P-60, No. 79 (Washington, D.C.), 1971. (a)

————, "Income in 1970 of Families and Persons in the United States," *Current Population Reports*, Series P-60, No. 80 (Washington, D.C., U.S. Government Printing Office) 1971. (b)

————, *Social Statistics for the Elderly: State Level System Users Manual* (Washington, D.C., U.S. Government Printing Office), 1975. (a)

————, *Statistical Abstract of the United States, 1975* (Washington, D.C., U.S. Government Printing Office), 1975. (b)

U.S. Department of Health, Education and Welfare, Public Health Service, "Children and Youth: Selected Health Characteristics," Series 10, No. 62 (Washington, D.C., U.S. Government Printing Office), 1971.

U.S. Department of Labor, *The Pre-Retirement Years*, vol. 4, Manpower R&D Monograph 15 (Washington, D.C., U.S. Government Printing Office), 1975.

Yoels, William C., "Destiny or Dynasty: Doctoral Origins and Appointment Patterns of Editors of the *American Sociological Review*, 1948–1968," *American Sociologist*, 6 (May 1971), p. 135.

DIAGRAM 7.1 Ordinal Measures of Association

THE PROBLEM

We want to describe the association between two ordinal variables in this example, mobility-orientation of 10–19-year-old children in 466 households and the mobility orientation held for them by their parents (see Furstenberg, 1971:598).* Data were collected from a representative sample of households on the Lower East Side of New York in 1960. The two mobility orientation scales included items about educational and occupational values, goals, and attitudes about achievement.

The study was concerned with the extent to which attitudes about mobility are transmitted from parent to child as indicated by the similarity of parent and child views. A strong association between the child's view and the child's parents' view would suggest that mobility orientation is indeed transmitted between parent and child. The author's hypothesis was that there was little such transmission, and that the association would thus be rather low. We will use these data to show how to compute the various types of pairs of cases (in this example, a "case" is a child, measured on the child's and the parent's mobility orientation for the child). Several measures of ordinal association may be computed from these types of pairs and these are discussed in this chapter. In actuality, only one of these measures of association would be computed; the criteria for deciding on which one are also discussed in Chapter 7. The author reported tau-*b* as the appropriate measure in this instance.

(Y) Child's Own Mobility Orientation	(X) Mobility Orientation a Child's Parents Hold for the Child			
	Low	Medium	High	Total
High	ᵈ 29	55	68	152
Medium	59	53	48	160
Low	ₛ 71	37	30	138
Total	159	145	146	450

COMPUTING NUMBER OF PAIRS OF DIFFERENT KINDS

In practice, only those types of pairs needed for the selected measure of association would be computed. Here all five types of pairs will be computed for illustration.

Step 1: Examine the table and determine which diagonal is the "positive" diagonal, that is, the one that extends from the "high-high" cell to the "low-low" cell on both variables. In this example, that diagonal is from the lower left to the upper right on the table. Label one end of this

* Sixteen cases of "don't know" or "no answer" were excluded from the table in Furstenberg's study.

DIAGRAM 7.1 *(Continued)*

diagonal s and label one end of the "negative" diagonal d. This step assures that N_s and N_d pairings will be computed properly and, thus, that the coefficient's sign will accurately reflect the direction of association in a table.

Step 2: Compute types of pairs. T = total number of unique pairings of cases:

$$T = \frac{N(N-1)}{2}$$

$$T = \frac{450(450-1)}{2} = 101,025$$

N_s = number of "concordant" pairs. This is computed by locating the cell in the "s" corner of the table as indicated in Step 1, above. This is the first "target cell" and its frequency is multiplied by the sum of all cell frequencies above and to the right of the target cell (in this instance, since "s" is in the lower left corner of the table). This is illustrated in the schematic diagram at the left below, where the darkened, single cell is the target cell that is multiplied by the sum of cell frequencies in cells indicated by shading. To this first product are added similar products formed by taking each additional cell in the table (that has cells above and to the right) as successive "target" cells. In the table above, there are four such products, which are summed as follows:

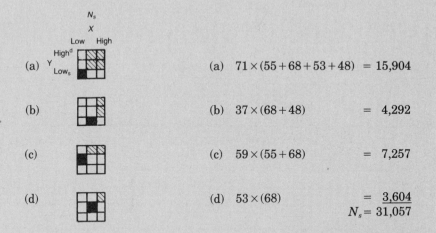

(a) $71 \times (55 + 68 + 53 + 48) = 15,904$

(b) $37 \times (68 + 48) = 4,292$

(c) $59 \times (55 + 68) = 7,257$

(d) $53 \times (68) = \underline{3,604}$
$N_s = 31,057$

It is helpful to notice the logic involved in these computations. The 71 cases in the first "target" cell have $55 + 68 + 53 + 48$ cases, which are ranked differently and also ranked higher on both variables than the target cell's 71 cases. The number of pairs that could be created would be equal to 71 times the total of $55 + 68 + 53 + 48$, which is 15,904. Likewise, for each target cell, the total pairings of this concordant kind can be

DIAGRAM 7.1 *(Continued)*

computed for it and the sum of these computations equals the total number of unique pairs that are concordant.

N_d = number of "discordant" pairs. This is computed in exactly the same fashion as N_s, *except* that target cells start in the "d" corner and work down the negative diagonal. In this case, target cells are multiplied by the sum of cell frequencies for cells that are below and to their right. These computations yield the following sum:

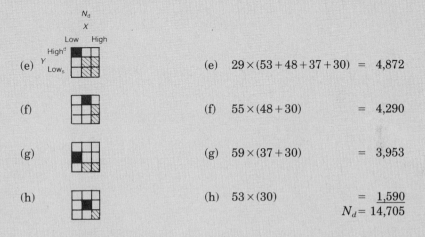

(e)	$29 \times (53 + 48 + 37 + 30) = 4{,}872$
(f)	$55 \times (48 + 30) = 4{,}290$
(g)	$59 \times (37 + 30) = 3{,}953$
(h)	$53 \times (30) = \underline{1{,}590}$
	$N_d = 14{,}705$

It is immediately apparent that there are more concordant pairs in these data, and thus that the association is a positive association. This will result in a plus sign on any of the association measures we eventually compute.

T_x = pairs tied on the independent (X) variable but not on the dependent (Y) variable. These pairs are those formed within the same category of the X variable (*i.e.*, tied on X) as indicated in the graphic illustration below at the left. Starting with a target cell at the top of a column, this is multiplied by the sum of cell frequencies for cells immediately below the target cell, etc. The computations are given below:

T_y = pairs tied on the Y variable but not on the X variable. These are computed exactly as T_x pairs are, *except* that products are formed within categories of the Y variable only. In this case, target cells are multiplied

DIAGRAM 7.1 *(Continued)*

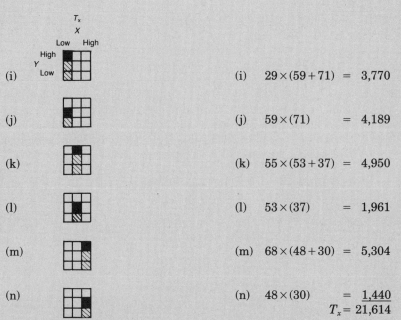

(i)	(i) $29 \times (59 + 71) = 3{,}770$
(j)	(j) $59 \times (71) = 4{,}189$
(k)	(k) $55 \times (53 + 37) = 4{,}950$
(l)	(l) $53 \times (37) = 1{,}961$
(m)	(m) $68 \times (48 + 30) = 5{,}304$
(n)	(n) $48 \times (30) = \underline{1{,}440}$
	$T_x = 21{,}614$

only by the sum of cell frequencies to the right, within rows, as illustrated in the diagram at the left. The computations are as follows:

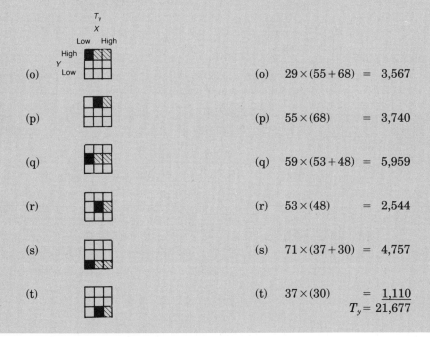

(o)	(o) $29 \times (55 + 68) = 3{,}567$
(p)	(p) $55 \times (68) = 3{,}740$
(q)	(q) $59 \times (53 + 48) = 5{,}959$
(r)	(r) $53 \times (48) = 2{,}544$
(s)	(s) $71 \times (37 + 30) = 4{,}757$
(t)	(t) $37 \times (30) = \underline{1{,}110}$
	$T_y = 21{,}677$

DIAGRAM 7.1 *(Continued)*

T_{xy} = pairs tied on both the X and the Y variables. These consist of the sum of pairs that can be formed out of cases that fall in the same cell (*i.e.*, have the identical value on X and also an identical value on Y). These are computed for each cell as follows:

$$\frac{f(f-1)}{2}$$

where f is the cell frequency for a given cell. These computations on each cell are then summed over all cells to equal T_{xy}. The computations on this table are as follows:

$$
\begin{aligned}
29(29-1)/2 &= 406 \\
55(55-1)/2 &= 1{,}485 \\
68(68-1)/2 &= 2{,}278 \\
59(59-1)/2 &= 1{,}711 \\
53(53-1)/2 &= 1{,}378 \\
48(48-1)/2 &= 1{,}128 \\
71(71-1)/2 &= 2{,}485 \\
37(37-1)/2 &= 666 \\
30(30-1)/2 &= 435 \\
T_{xy} &= 11{,}972
\end{aligned}
$$

As a check, you should notice that the sum of these five kinds of pairs equals the total number of possible, unique pairings:

$$
\begin{aligned}
N_s &= 31{,}057 \\
N_d &= 14{,}705 \\
T_x &= 21{,}614 \\
T_y &= 21{,}677 \\
T_{xy} &= 11{,}972 \\
T &= 101{,}025
\end{aligned}
$$

Step 3: Compute the appropriate ordinal measure of association: tau-*b*, gamma, or Somers' d_{yx}, which are discussed in the body of Chapter 7.

DIAGRAM 7.2   Regression and Correlation Coefficients

To illustrate the computation of regression coefficients and the Pearsonian correlation coefficient, we will use the illustrative data from Figure 7.3. These are raw data, and $N = 5$.

(X) Education (in Years)	(Y) Income (in $1000's)	X^2	Y^2	$X \cdot Y$
1	2	1	4	2
5	7	25	49	35
3	3	9	9	9
2	5	4	25	10
4	8	16	64	32
$\Sigma X = 15$	$\Sigma Y = 25$	$\Sigma X^2 = 55$	$\Sigma Y^2 = 151$	$\Sigma X \cdot Y = 88$
$\bar{X} = 3$	$\bar{Y} = 5$			

THE PROBLEM

Compute coefficients a_{yx} and b_{yx} in the following regression equation: $Y' = a_{yx} + b_{yx}(X)$, and compute the correlation coefficient r. There are a number of different approaches, including using a computer program (which is probably most common and useful), using definitional formulas given in this chapter, or using the formula that uses data converted to z-score form prior to computations. Among the hand-computation procedures, the raw data formulas used below seem most useful.

Step 1:   Examine a scatter diagram of the data to determine whether or not a linear regression line is appropriate. Often computer programs provide scatter plots. If a correlation coefficient is low, one reason may be that the linear model does not fit the data very well (see Figure 7.3).

Step 2:   Find the following sums from the raw data: ΣX, ΣY, ΣXY, ΣX^2, ΣY^2. If data are in a grouped frequency distribution form, then X will refer to category midpoints that have to be multiplied by category frequencies to form sums indicated by: ΣfX, ΣfY, ΣfXY, ΣfX^2, ΣfY^2.

COMPUTING b_{yx}, THE SLOPE COEFFICIENT

The slope coefficient or b-coefficient, of the regression equation expresses the number of units of change expected in the dependent variable, Y, given a single unit increase along the scale of the independent variable. Expressed differently, it is the ratio of the covariation between X and Y, over the variation of the independent variable (see Formula 7.14). Variation is expressed in terms of variances, thus the b-coefficient can be expressed as:

$$b_{yx} = \frac{\text{Covariance of } X \text{ and } Y}{\text{Variance of } X}$$

DIAGRAM 7.2 *(Continued)*

The variance of X may be computed by this familiar formula:

$$s_x^2 = \frac{\Sigma X^2 - (\Sigma X)^2/N}{N}$$

The covariance of X and Y may be computed by the following formula where XY is used instead of XX, or X^2, and $(\Sigma X)(\Sigma Y)$ is used in place of $(\Sigma X)(\Sigma X)$, or $(\Sigma X)^2$. N is, of course, the number of cases (or pairs of scores).

$$s_{xy}^2 = \frac{\Sigma(XY) - (\Sigma X)(\Sigma Y)/N}{N}$$

Thus, b_{yx} could be computed from raw data by this process:

(D.1)
$$b_{yx} = \left[\frac{\Sigma(XY) - (\Sigma X)(\Sigma Y)/N}{N} \right] \Bigg/ \left[\frac{\Sigma X^2 - (\Sigma X)^2/N}{N} \right]$$

$$= \frac{N(\Sigma XY) - (\Sigma X)(\Sigma Y)}{N(\Sigma X^2) - (\Sigma X)^2}$$

Computing for this example, the process would be as follows:

$$b_{yx} = \frac{5(88) - (15)(25)}{5(55) - (15)^2} = \frac{440 - 375}{275 - 225} = \frac{65}{50} = +1.3$$

COMPUTING a_{yx}, THE Y-INTERCEPT CONSTANT

Since a regression line goes through the point that is at the intersection of the mean of X and the mean of Y, we could substitute these mean values into the regression equation and solve for a_{yx}, as follows:

(D.2)
$$a_{yx} = \bar{Y} - b_{yx}(\bar{X})$$

or,

$$a_{yx} = \frac{\Sigma Y - b_{yx}(\Sigma X)}{N} = \frac{25 - 1.3(15)}{5} = \frac{25 - 19.5}{5} = 1.1$$

Thus,

$$Y' = 1.1 + 1.3(X)$$

COMPUTING THE PEARSONIAN CORRELATION COEFFICIENT, r

The correlation coefficient is defined by the following formula as the ratio of the covariance of X and Y over the product of the standard deviations of X and of Y.

(D.3)
$$r = \frac{\left(\Sigma XY - \dfrac{(\Sigma X)(\Sigma Y)}{N} \right)}{N} \Bigg/ \sqrt{\frac{\Sigma X^2 - (\Sigma X)^2/N}{N} \ \frac{\Sigma Y^2 - (\Sigma Y)^2/N}{N}}$$

DIAGRAM 7.2 *(Continued)*

Simplified, this can be expressed as:

$$r = \frac{N\Sigma XY - (\Sigma X)(\Sigma Y)}{\sqrt{[N\Sigma X^2 - (\Sigma X)^2][N\Sigma Y^2 - (\Sigma Y)^2]}}$$

Computing for this example:

$$r = \frac{5(88) - (15)(25)}{\sqrt{[5(55) - (15)^2][5(151) - (25)^2]}} = \frac{440 - 375}{\sqrt{(275 - 225)(755 - 625)}}$$

$$= \frac{65}{\sqrt{6500}} = +.81$$

Pearson's r can be interpreted in a PRE sense by squaring its value; thus, $r^2 = .65$, meaning that 65% of the variation is "explained" by the linear relationship between these two variables.

DIAGRAM 7.3   Organization of Measures of Degree and Nature of an Association

	Measures of Degree of Association		Nature of Association
Level of Measurement	Symmetric	Asymmetric	
Nominal	$\|Q\|$ $\|\phi\|$ *	ϵ λ_{yx} U_{yx}	Percentaged table
Ordinal	Q G ϕ r_s t_b	d_{yx}	Percentaged table
Interval ratio	r	E_{yx}	$Y' = a_{yx} + b_{yx}(X)$ or some other, specified form.

* Other delta-based measures would go here. Since C, T, and V are not properly normed and easily interpreted, they are excluded here (see footnote to Section 6.5.1).

Part IV

DESCRIPTIVE STATISTICS: THREE OR MORE VARIABLES

CHAPTER 8

Elaborating the Relationship between Two Variables

8.1 EXAMINING RELATIONSHIPS BETWEEN VARIABLES

Up to this point we have been concerned with the relationship between two variables, often an independent and a dependent variable. The last chapter dealt with a variety of measures of the *nature* and *strength* of their relationship, and these measures were developed so that comparisons could be made either between bivariate associations, in studies involving different subgroups, or between associations in studies involving different variables. In this chapter we will begin to introduce more systematic differences between tables whose coefficients we want to compare. In fact, we will begin to define subpopulations for which the basic relationship between the same independent and dependent variables can be examined from subpopulation to subpopulation. Careful comparison will permit us to draw conclusions about the effects of other variables on the *relationship* between the original pair of variables. Let us begin with an example.

8.1.1 Morale and Interaction—An Illustration of Elaboration

What happens to people's involvement in society as they age? Do they continue to be engaged pretty much the way they always have in the round of social activities, perhaps substituting more leisurely for more strenuous pursuits; or do people typically disengage from society in the normal sequence of their aging? One theory holds that people disengage from society as they enter old age, and that this is a mutually satisfactory process both for the individual and for society. Such decreases in social interaction would have a positive or at least a minimal impact on their levels of morale or satisfaction. An alternative viewpoint is that individuals attempt to continue to be engaged as they were in middle age, although the form of social engagement may shift. In this view, declines in social interaction have the effect of decreasing the level of morale and satisfaction of an individual. Both lines of reasoning make predictions for the relationship between morale and degree of social interaction for people moving toward the end of life, and data supporting each viewpoint have been presented.

Mark Messer addressed himself to this problem and tried to examine the relationship between morale and interaction in somewhat greater detail (Messer, 1967). He reasoned on the basis of other research that the age concentration of the person's environment is a crucial factor that also has to be considered (*i.e.*, the extent to which older people live only with other older people or with people of a mix of different ages). There is reason to believe, for example, that social interaction rates are higher when a person is in an age-homogeneous environment. This may be because people prefer to interact with age peers. It has been argued that age grouping (especially in the socialization process of young people) helps support transitions from one role to another by insulating people from some of their other roles, which may be inconsistent with the new role, until the problems and strains of their role transition are settled. If this is the case, Messer argues, high morale among those in an age-concentrated environment should be a function of the normative system that this concentration permits rather than a function of the social interaction of an individual. In a mixed-age environment, on the other hand, morale should depend upon a higher rate of social interaction.

To examine his ideas, Messer contrasted 88 older people living in an age-concentrated public housing project for the elderly in Chicago against 155 elderly people living in a mixed-age public housing project. In each setting, however, he was interested in the relationship between morale and the degree of social interaction, and he expected a stronger positive association in the mixed-age setting. Table 8.1 presents his results.

Interestingly, it does appear that age concentration of the environment has an effect upon the relationship between morale and interaction. In fact, the percentaged tables show a very low and negative relationship between morale and interaction in the age-concentrated environment (Table 8.1B), but the two variables are positively associated in the mixed-age environment. His hypothesis about the effect of age-concentration seems to be confirmed.* The test of his ideas involved an examination of a relationship between two variables within the categories of a third variable. This leads to an important modification and possible linking of the two initial theories.

This kind of examination is called **elaboration** because a basic relationship of interest is examined under a variety of different conditions. In this instance, introduction of the third variable into the analysis helped *specify* conditions under which the relationship would be strong or weak. This kind of examination is often

---

* Messer's hypothesis was that social interaction is causally prior to morale, and he treats his data in these terms. It is possible, however, that morale may have some influences upon future interaction. The importance of theoretical ordering of variables for the way one conducts research will be emphasized later in this chapter in Section 8.2.3.

TABLE 8.1 PERCENTAGE DISTRIBUTION OF MORALE FOR ELDERLY IN PUBLIC HOUSING

A. *Percentage Distribution of Morale by Social Interaction*

(Y) Morale	Interaction (X) Low	Interaction (X) High	Total
High	14.3	28.2	21.4
Medium	58.0	50.0	53.9
Low	27.7	21.8	24.7
Total	100.0	100.0	100.0
	(119)	(124)	(243)

$$d_{yx} = .15$$

B. *Percentage Distribution of Morale by Social Interaction and Age Environment*

(Y) Morale	(T₁) Age-concentrated Environment Interaction (X) Low	High	Total	(Y) Morale	(T₂) Mixed-age Environment Interaction (X) Low	High	Total
High	26.7	25.9	26.1	High	10.1	30.3	18.7
Medium	60.0	50.0	53.4	Medium	57.3	50.0	54.2
Low	13.3	24.1	20.5	Low	32.6	19.7	27.1
Total	100.0	100.0	100.0		100.0	100.0	100.0
	(30)	(58)	(88)		(89)	(66)	(155)

$$d_{yx} = -.09 \qquad\qquad d_{yx} = +.25$$

Source: Messer (1967). Used by permission of the publisher.

used in sociology, and the purpose of this chapter is to show some of these patterns of relationship among three or more variables, to look at some of the ways they might be summarized, and to investigate how the different patterns might be interpreted. Let us start with some of the basic terminology.

8.1.2   Total Tables and Conditional Tables

The overall association between two variables is called a *total association*, or a **zero-order association**. Table 8.1A is an example of such a total association. Total tables include, as *N*, all the available cases to be examined in terms of the pair of variables. As in the example above, however, analysis of an overall bivariate relationship is usually aided if the cases are divided into subgroups corresponding to categories of an additional variable. Tables showing the associa- tion between two variables within categories of other variables are called **condi-**

BOX 8.1   BASIC IDEAS

Before you get into this chapter very far, you should check yourself on the following ideas.

Association of two variables (Section 6.4)
Independent, dependent, and control variables (Section 2.1.4)
Measures of association (Sections 7.2–7.7)
The idea of setting up important contrasts (Sections 3.1 and 3.2)

This chapter has two main parts. The first deals with comparisons of contrasting associations, that is, associations in several bivariate tables. The second concerns ways to make a statistical summary over a set of bivariate tables. In between, topics on experimental design and theory are introduced. These topics are central to an understanding of where and why we would want to use elaboration, standardization, or partial correlation techniques in examining the relationship between three or more variables.

---

tional tables, or conditional associations.* Table 8.1B shows two conditional tables. The additional variable that is introduced into the analysis as a basic split in the group of cases is called a *control* variable or *test* variable. We are interested in examining the basic relationship between two variables (X and Y) within categories of control variable(s) (T). Each conditional table includes only cases that have the same value (or range of values within a category) on that control variable, and there is a different conditional table for each category of the control variable(s). This procedure "controls" or "takes out" variation in that additional variable, permitting a clear focus on the relationship between the original independent and dependent variables within any given table.

The *terminology* is as follows:

zero-order association	*No* control variables (*i.e.*, no test factors). The basic relationship between two variables is examined *overall*.
first-order conditional tables	*One* control variable. The basic X-Y relationship is examined within each category of the control variable.
second-order conditional tables	*Two* control variables. The basic X-Y relationship is examined within categories created by all possible combinations of the categories of the *two* control variables.

---

* In some texts the "conditional" table is called a "partial" table and the process is called "partialing." We will reserve the term "partial" for the summary coefficients used later in this chapter, which express the relationship between X and Y variables where the effect of a control variable (or variables) has been statistically removed.

third-order conditional tables *Three* control variables. The basic *X-Y* relationship is examined within categories created by all possible combinations of the categories of the *three* control variables.

etc.

A second-order conditional table would be one in which, for example, morale and interaction (the basic relationship of interest in Table 8.1) were examined in each category of age concentration of environment and in public versus privately owned categories of housing. The basic table would then be examined separately for these four categories: (a) age-concentrated public housing, (b) age-concentrated private housing, (c) mixed-age public housing, and (d) mixed-age private housing.* Thus, there would be only four tables to compare, since there are only two categories of each control variable.

In general, a statistical analysis of the relationship between two variables proceeds by introducing control variables one at a time, starting from the original zero-order table and proceeding to first-order conditional tables and then to second-order conditional tables, etc. All first-order tables are examined, then second-order, etc.

As Table 8.2 illustrates, a number of different total and conditional associations can be created from the basic information given in higher-order conditional tables (but one cannot create higher-order tables out of lower-order tables). For example, from the first-order conditionals shown in Table 8.1B and reproduced in Table 8.2 as a frequency distribution, we could create three total associations: (a) one between morale and the degree of social interaction, (b) another between age concentration of environment and morale, and (c) a third between age concentration of environment and degree of social interaction. Notice that these total tables are created out of the marginals of the first-order conditional tables or, in the case of Table 8.2B, by adding together the two conditional tables cell by cell.

Table 8.2C is created by simply taking the row marginals for the age-concentrated and for the age-mixed environments in Table 8.2A and using these as the frequencies in columns of Table 8.2C. All 243 cases are classified in this total table. Table 8.2D is made up from the column marginals of Table 8.2A and again, all 243 cases appear in this total table.

Table 8.2E is another rearrangement of Table 8.2A. In this case, the control

---

* The four conditional tables in this example could be increased if the control variables had more categories than two each. If one has three categories and the other has four, then there would be $3 \times 4 = 12$ conditional tables to examine, and all 12 tables would be second-order conditional tables because they resulted from a combination of two control variables. In general, the number of conditional tables equals the product of the number of categories of all control variables. Obviously, the higher the order of the conditional tables, the more exacting and refined the analysis, and also the more cases one needs to include in a study to "fill up" the conditional tables. It is a general practice not to compute percentages where the base of the percentage is less than, say, 15 or 20, because such percentages would ordinarily be too unreliable (a topic for consideration in inferential statistics). Where many control variables are used, alternative ways to handle the problem that require fewer cases are usually used. Some of these will be discussed later in this chapter and in the next.

TABLE 8.2 OTHER TABLES MADE FROM THE FIRST-ORDER CONDITIONAL TABLES SHOWN IN TABLE 8.1

A. *Frequency Distribution of Morale by Interaction and Age Environment (from Table 8.1, above)*

	Age-concentrated Environment				Mixed-age Environment		
	Interaction (X)				Interaction (X)		
(Y) Morale	Low	High	Total	(Y) Morale	Low	High	Total
High	8	15	23	High	9	20	29
Medium	18	29	47	Medium	51	33	84
Low	4	14	18	Low	29	13	42
Total	30	58	88	Totals	89	66	155
	$d_{yx} = -.09$				$d_{yx} = +.25$		

B. *Frequency Distribution of Morale by Interaction*

	Interaction (X)		
(Y) Morale	Low	High	Total
High	17	35	52
Medium	69	62	131
Low	33	27	60
Total	119	124	243

$$d_{yx} = +.15$$

C. *Frequency Distribution of Morale by Age Environment*

	Age Environment (T)		
(Y) Morale	Concentrated	Mixed	Total
High	23	29	52
Medium	47	84	131
Low	18	42	60
Total	88	155	243

$$d_{yt} = +.12$$

D. *Frequency Distribution of Interaction by Age Environment*

	Age Environment (T)		
(X) Interaction	Concentrated	Mixed	Total
High	58	66	124
Low	30	89	119
Total	88	155	243

$$d_{xt} = +.23$$

continues

TABLE 8.2 *(Continued)*

E. *Frequency Distribution of Morale by Age Environment Controlling for Interaction*

	Interaction (X)			
	Low Age Environment (T)		High Age Environment (T)	
(Y) Morale	Concentrated	Mixed	Concentrated	Mixed
High	8	9	15	20
Medium	18	51	29	33
Low	4	29	14	13
Total	30	89	58	66
	$d_{yt} = +.28$		$d_{yt} = -.07$	

F. *Frequency Distribution of Interaction by Age Environment Controlling for Morale*

	Morale (Y)					
	Low Age Environment (T)		Medium Age Environment (T)		High Age Environment (T)	
(X) Interaction	Concentrated	Mixed	Concentrated	Mixed	Concentrated	Mixed
High	14	13	29	33	15	20
Low	4	29	18	51	8	9
Total	18	42	47	84	23	29
	$d_{xt} = +.40$		$d_{xt} = +.21$		$d_{xt} = -.04$	

Source: Data from Table 8.1 (Messer, 1967).

variable is social interaction and the basic relationship under examination is morale and age concentration. In Table 8.2F, the basic relationship of interaction and age homogeneity of environment is examined within categories of the morale variable. Although Tables 8.2E and 8.2F could be created from these same data, they would not be of central interest to Messer because they do not permit him to focus readily upon the relationship of morale and social interaction, the two key variables in his study. These examples do begin to illustrate, however, the variety of ways a set of data may be examined and, it is hoped, the importance of being clear about what it is that we are interested in studying.

8.1.3 The Role of Theory

What other variables may help explain the relationship between morale and interaction? Probably income, gender, cultural background, age, experience in other kinds of interaction settings, whether one is married or not, etc. The list could be rather long. Just which variables to pick is a serious problem and largely unsolvable without a heavy reliance on prior theoretical work and the results of research aimed at explanation of phenomena in a certain area of interest. Clearly, "toenail length" would be ruled out immediately as a general explanation of the relationship of morale and interaction. The role of some of the other variables mentioned above is not as clear-cut, and careful theoretical work would point the way toward finding significant variables. In Messer's research he was able to draw upon three kinds of previous theory, the disengagement theory, the activity theory, and ideas that relate age homogeneity to satisfaction and morale. An understanding of these ideas played a critical role in specifying important variables to control and the way results should be interpreted. Without this kind of starting point, an investigator would hardly be able to pick a relevant independent variable, let alone select those that are relevant from an infinity of variables that could be measured and used as control conditions. Even in very small studies, the possible combinations of variables becomes very large indeed, and as purely a practical matter, theoretical guidance in selection of relationships and control conditions becomes important.*

Research that is able to contrast two theories purporting to explain the same phenomenon is particularly fruitful for building scientific knowledge. Usually, however, research leads to an extension of a theory, to an addition of some condition under which the theory is found to lead to somewhat different predictions, or to the refinement of measurements and of "scope conditions" of a theory (*i.e.*, the class of phenomena to which the theory pertains). Much research is of an exploratory nature (as sharply opposed to "sloppy" or unplanned investigation), especially when explicit theories have not been developed.† But whether theories are explicit or not, investigators make use of theoretical knowledge in selecting variables and planning their analysis. The more clearly set out these ideas are, the more fruitful the analysis.

The **theoretical order** of variables is important in interpreting the outcome of the process of elaboration. An independent variable, if it is thought of theoretically as a causal variable, comes before the dependent variable (*i.e.*, it changes first) in any time sequence. The "test factor" or control variable, however, may have its effect at different points in time relative to the independent and dependent variables. If a test factor has its effect *before* both independent *and* dependent variables, it is called an **antecedent** test factor, as in this example:

SEX	SOCIAL INTEGRATION	SUICIDE
(T) $\longrightarrow$	(X) $\longrightarrow$	(Y)
Antecedent Variable	Independent Variable	Dependent Variable

* For example, in a modest study with only five variables, each having five categories, there are 10 zero-order tables, 150 first-order conditional tables, 1500 second-order tables, etc. Needless to say, typical research involves many more variables than five, and the problem quickly becomes one of avoiding irrelevant tables and selecting relevant ones.

† There are a number of excellent discussions on theory, on theory construction, and on the role of research. See especially Reynolds (1971), Blalock (1969; General References), Stinchcombe (1968), Chafetz (1978), Hage (1972), and Freese and Sell (1980).

If it has its effect *after* the independent *and* dependent variables, it is called a **consequent** test factor, as in this example:

NO. OF DELINQUENT FRIENDS	NO. OF DELINQUENT ACTS SELF-REPORTED	NUMBER OF TIMES ARRESTED
(X) $\longrightarrow$	(Y) $\longrightarrow$	(T)
Independent Variable	Dependent Variable	Consequent Variable

If the test factor has its effect *after* the independent variable but *before* the dependent variable, it is called an **intervening variable** or intervening test factor, as below:

NO. OF DELINQUENT FRIENDS	PARENTAL SUPERVISION	SELF-REPORTED NO. OF DELINQUENT ACTS
(X) $\longrightarrow$	(T) $\longrightarrow$	(Y)
Independent Variable	Intervening Variable	Dependent Variable

The basic dependent variable of interest will be referred to as variable Y in this chapter. Variable X will be the main independent variable of interest, and variable T will be the control or "test" variable. Where it is important to specify the category of the test variable, we will use subscripts such as T_1, T_2, etc. The relationship between Y and X within category T_1, then, may be symbolized as $YX:T_1$.

Drawing an **arrow diagram** of the relationships between variables included in a study is often a useful device for clarifying both the theory and the kinds of statistical analysis needed to answer a research question. For example, in a study of the relationship of organizational involvement and political participation, Erbe (1964) surveyed the available research literature, which suggested that two other variables were relevant. The theoretical relationships between variables used to guide his study were as follows:

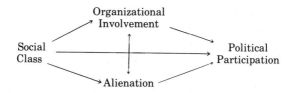

The above model immediately indicates the ordering of the variables, shows which are intervening, and suggests the apparent necessity of taking account of both social class and alienation in examining the "involvement–political participation" relationship.

As we shall see, the theoretical ordering of variables also has a key role to play in the interpretation of results. Virtually identical statistical results may be interpreted as evidence that a hypothesized causal link is spurious, that there is evidence of a causal sequence, or that there is evidence of independent influence, depending on the ordering of variables.

In many cases, of course, the time order of variables is established simply by the logic of the measurement or the type of variable. If there is any causal relationship between ethnic group and attitudes, the group membership probably is the causal factor. On the other hand, many variables are not clearly time ordered, especially when the investigator takes measurements of all variables at the same point in time. Where this confusion of time order is likely to occur, investigators would normally attempt to gather specific information about time order for the effects of the variables they want to study.

Time order could be established by gathering data at two or more points in time. In voting studies, for example, an investigator may create a "panel," or sample of individuals, who are questioned each month for 2 or 3 months before an election. Measures of attitudes, for example, at "time-point one" could be related to attitudes at "time-point two" and the time order would be clear. This is called a **panel study** design. Other **longitudinal** designs are used, for example, in the study of families or aging individuals over time. Sometimes it is sufficient to ask individuals to report on past events or to use records to establish the time order of variables.*

The centrality of theory as a guide to analysis (and research as a guide to development of theory) will become more evident as we proceed through this examination of statistical elaboration.[†]

8.1.4   Analysis of Conditional and Total Associations

The statistical analysis of *each* conditional table does not differ in any important respect from the analysis one might do on a total association between two variables. There is the full range of measures of association discussed in Chapter 7 from which to choose, and selection depends on the character of the variables and the problem at hand. In addition to strength and direction of association, one should examine the *nature* of each association. The only difference in dealing with conditional tables is that comparison between conditional tables takes on a more systematic character than it might otherwise. Tables shown above could be examined in terms of, say, Somers' d_{yx} and percentages. In Table 8.1B the association in the two conditional tables is $d_{yx} = -.09$ in the age-concentrated environment and $d_{yx} = +.25$ in the mixed-age environment. The stronger association between these two variables exists in only one of the two conditions— namely the age-heterogeneous condition—an outcome concerning the relationship between morale and interaction that Messer's logic had predicted.

The relationship in the conditional tables, however, is only part of the story. The more complete story would include the overall relationship in the three total associations (Tables 8.2B, C, and D) as a further basis for interpreting these data. Notice, for example, that the original zero-order association between

---

* Additional information on panel studies and ways they may be statistically examined may be found in Riley (1963: 559; see General References), Zeisel (1968), and Lazarsfeld and Rosenberg (1955).
[†] A classic statement on the bearing of theory on research and vice versa is made by Merton (1968).

interaction and morale (Table 8.2B) is positive but relatively weak $(d_{yx} = +.15)$. From a look at the conditional tables (Table 8.2A), it is clear that part of the reason for this overall weakness is that a weak negative association in one conditional table (*i.e.*, the age-concentrated group) is thrown together with a stronger positive association in the other conditional table (*i.e.*, the mixed-age group), and they tend to cancel each other out (in Table 8.2B) to some extent.

Notice, too, that the control variable, T, is positively related to each of the main variables, morale and interaction. The Somers' d for the table of age environment and morale (Table 8.2C) is $d_{yx} = +.12$, and that for the association of age environment and interaction (Table 8.2D) is $d_{yx} = +.23$. Part of the association observed between morale and interaction is due to the correlation of each of these two variables with the test factor. These outcomes are organized below.

$d_{yx} = +.15$ *Original total association* between morale and interaction is a result of these:

    Conditional associations:

$d_{yx} = -.09$ a) The association of morale and interaction in the age-concentrated environment

$d_{yx} = +.25$ b) The association of morale and interaction in the mixed-age environment

    Total associations with the test factor:

$d_{yx} = +.12$ c) The association of morale with the test factor, age environment

$d_{yx} = +.23$ d) The association of interaction with the test factor, age environment

8.1.4a The Lazarsfeld Accounting Formula

The relationship between a total association, a set of conditionals, and another set of total associations has been pointed out in the sociological literature by Lazarsfeld, and it is sometimes summarized in what is called the **Lazarsfeld accounting formula**. This equation states that a total association between X and Y can be accounted for by the conditional associations of X and Y within categories of T (test variables), plus total associations of X and Y separately with the test variables. For the three-variable situation where the variables are all dichotomies and where the degree of association is expressed in terms of delta (Δ),* the Lazarsfeld accounting formula is as follows:

$$\Delta XY = \Delta XY{:}T_1 + \Delta XY{:}T_2 + \frac{N_1 + N_2}{(N_1)(N_2)}(\Delta XT)(\Delta YT)$$

* Recall from Chapter 6 that delta (Δ) is simply the difference between observed and expected cell frequencies. In a 2 by 2 table the size of Δ is the same, regardless of which cells are contrasted (although the sign may be different). It is traditional (consistency is necessary for the Lazarsfeld equation to work out) to compute Δ on a 2 by 2 table by subtracting the expected from the observed frequency for only *one* cell, the high-X, high-Y cell in a table. Delta is not a normed measure of association, so it is generally not used except to illustrate how Lazarsfeld's formula works out numerically.

If the morale variable in Table 8.2 were grouped into only two categories (by putting the "high" and "medium" categories together, for example) and the expected cell frequencies were computed as discussed in Chapter 6, the resulting Δ values would be as follows:

$$+3.62 = -2.13 + 4.88 + \frac{243}{(88)(155)}(3.73)(13.09)$$

$$3.62 = -2.13 + 4.88 + .87$$

Using Δ on a fourfold table, this situation works out nicely: the total association at the left equals the sum of terms expressed in the equation at the right. In general, however, the weighting factor is not resolved, either for other measures of association or for tables larger than 2 by 2 tables.* The usefulness of Lazarsfeld's formula holds, however; associations between two variables can be accounted for by a set of conditional and total associations.

An examination of the accounting formula quickly yields a rich set of possible outcomes.† An association may or may not exist in the total table, which is cross-classifying an independent and a dependent variable. If no association exists, then the right side of the Lazarsfeld formula would have to "balance out" to 0. This might happen if the conditional tables yielded a negative and positive association that, added together in a total table, canceled each other out. Looking at the right side of the formula, an association may be shown in the total tables, in the conditional tables, or in both. Furthermore, the conditional tables may show the same strength and direction of association, or they may differ among themselves, with some showing a stronger association than other conditional tables, or some showing associations with a different direction or nature than others.

8.1.4b   Odds Ratios and Hierarchical Analysis

A system for the analysis of cross-classifications of three or more nominal or ordinal variables has been developed by Goodman (1965, 1969, 1972).‡ The complete system is involved and comprehensive and we will not treat it in detail here. We will, however, consider that part that deals with hierarchical analysis. This type of

---

* A slightly different approach using Yule's Q is described in Davis (1971: 188); see General References).

† In addition to the outcomes that are discussed in the chapter, there are other outcomes that should be noted because they are important, although of a different type than those discussed here. First, it is possible that the observed association between two variables is simply due to *chance*. This would be a possible outcome where the data are gathered by sampling procedures from some population. There is some chance that the sample will misrepresent the population from which it is drawn and show an association between two variables when, in fact, in the population there is none. Second, an observed association may be due to the research process itself. If observations are related rather than independent, the observed association could be accounted for largely by whatever research steps led to this relatedness. Generally, an investigator tries to design research so that observations are, in fact, independent observations.

‡ Davis (1974) has an excellent translation of the work of Goodman into terms that do not require mathematical sophistication.

analysis is very useful for sorting out the relationships between dependent, independent, and one or more test variables. A key concept in the system is the concept of an **odds ratio**. An odds ratio is the ratio between the frequencies of two categories of a variable. For example, consider the following frequency table consisting of fictitious data for a dependent (Y) and an independent (X) variable.

	Variable X		
Variable Y	Low	High	Totals
High	33	47	80
Low	52	23	75
Totals	85	70	155

Overall, of the 155 cases for which data are available, 80 scored high on variable Y and 75 scored low. Consequently, the odds ratio for variable Y is $80:75$ or $80/75 = 1.07$. In this case, the frequency for the high scores on Y was divided by the frequency for the low scores. It would have been equally proper to divide the low-score frequency by the high (*i.e.*, $75/80 = 0.94$). To be consistent, whenever the variable is ordinal, we will always divide the frequency for the *higher-score* category *by* that for the *lower*. If the variable is nominal, of course, the choice of which category frequency to put in the numerator and which in the denominator is a matter of personal preference. Once the choice is made, however, it should be used consistently to help clarify the analysis.

An overall odds ratio can also be computed for the X variable as follows: $70:85$ or $70/85 = 0.82$. Using Davis's (1974) system of notation, we will symbolize the

odds ratio for Y: $\begin{pmatrix} Hi \\ Y \\ Lo \end{pmatrix}$ and the odds ratio for X: $\begin{pmatrix} Hi \\ X \\ Lo \end{pmatrix}$. Thus,

$$\begin{pmatrix} Hi \\ Y \\ Lo \end{pmatrix} = 1.07 \quad \text{and} \quad \begin{pmatrix} Hi \\ X \\ Lo \end{pmatrix} = 0.82$$

When the frequencies of the two categories that constitute an odds ratio are equal the odds ratio will equal 1.00. When the frequency in the numerator exceeds that in the denominator, the odds ratio will exceed 1.00 The minimum odds ratio is 0, but there is no defined upper limit to odds ratios. An odds ratio such as $50/0$ is undefined since it is not permissible to divide by 0. If such an odds ratio came up in the analysis of a set of data (because of 0 cases in some category), an adjustment would need to be made.*

Because $\begin{pmatrix} Hi \\ Y \\ Lo \end{pmatrix}$ and $\begin{pmatrix} Hi \\ X \\ Lo \end{pmatrix}$ are odds ratios based on distributions of frequencies for *single* variables, they are referred to as *first-order odds ratios*. If an odds ratio for variable Y is computed taking a particular value of X into consideration, then it is referred to as a first-order *conditional* odds ratio of Y. For

---

* Davis (1972) suggests adding 1 to each cell of a table in order to circumvent this problem.

example, in the table there are two columns of values for variable X, one column in which the value of X is high and another in which the value of X is low. There are two frequencies for variable Y in each of these columns for variable X. When X is high, the frequency of high values on variable Y is 47 and the frequency of low values is 23. The odds ratio for these frequencies of Y is $47/23 = 2.04$. This odds ratio is a first-order conditional odds ratio and is symbolized as follows:

$$\left(\begin{matrix} Hi \\ Y \\ Lo \end{matrix} \; \middle| \; X = Hi \right) = 2.04$$

The vertical line in the expression means "given that," so the part of the expression after the vertical line reads, "given that the value of X is high."

There is a corresponding conditional first-order ratio for Y when the value of X is low:

$$\left(\begin{matrix} Hi \\ Y \\ Lo \end{matrix} \; \middle| \; X = Lo \right) = 33/52 = 0.63$$

If they are of interest to us, we can also compute two first-order conditional odds ratios for X, one for high values of Y and another for low values.* They would be as follows:

$$\left(\begin{matrix} Hi \\ X \\ Lo \end{matrix} \; \middle| \; Y = Hi \right) = 47/33 = 1.42$$

$$\left(\begin{matrix} Hi \\ X \\ Lo \end{matrix} \; \middle| \; Y = Lo \right) = 23/52 = 0.44$$

When an odds ratio is computed from two first-order conditional odds ratios, the resulting odds ratio is known as a *relative odds ratio*. For example, we could compute a relative odds ratio from the two first-order conditional odds ratios for Y as follows:

$$\frac{\left(\begin{matrix} Hi \\ Y \\ Lo \end{matrix} \; \middle| \; X = Hi \right)}{\left(\begin{matrix} Hi \\ Y \\ Lo \end{matrix} \; \middle| \; X = Lo \right)} := \frac{2.04}{0.63} = 3.24$$

This relative odds ratio is also called a *second-order odds ratio* and is symbolized as $\left(\begin{matrix} Hi & Hi \\ Y & X \\ Lo & Lo \end{matrix} \right)$.

---

* Whether we would be interested in these two conditional odds ratios would depend on the nature of the problem we were studying and the specific comparisons that were relevant to that problem.

A relative odds ratio (or second-order odds ratio), such as the one just computed, gives an indication of the amount of association that exists between two variables. If the second-order odds ratio is *less than 1.00*, the association between the two variables is *negative*. If the odds ratio is *greater than 1.00*, the association is *positive*. An odds ratio of 1.00 indicates that there is no association between the two variables—they are independent. Notice that the second-order odds ratio computed above is 3.24. This ratio indicates that variables X and Y are positively associated. A high value on X is associated with a high value on Y and a low value on X with a low value on Y.

It should also be noted that second-order odds ratios (as well as higher-order ones) are *symmetric*. That is, the second-order odds ratio computed from the two first-order conditional odds ratios for X will have the same value as the one just computed from the two first-order conditional odds ratios for Y (within errors of rounding). Thus,

$$\frac{\left(\begin{array}{c}Hi\\X\mid Y=Hi\\Lo\end{array}\right)}{\left(\begin{array}{c}Hi\\X\mid Y=Lo\\Lo\end{array}\right)}=\frac{1.42}{0.44}=3.23$$

If we were working with data that included a test variable (T), it would be possible to extend the notions we have been discussing and compute second-order conditional odds ratios such as the following:

$$\left(\begin{array}{c}Hi\ Hi\\Y\ \ X\mid T=Hi\\Lo\ Lo\end{array}\right)$$

and

$$\left(\begin{array}{c}Hi\ Hi\\Y\ \ X\mid T=Lo\\Lo\ Lo\end{array}\right)$$

Furthermore, from these two second-order conditional odds ratios a *third-order odds ratio* could be computed as follows:

$$\left(\begin{array}{c}Hi\ Hi\ Hi\\Y\ \ X\ \ T\\Lo\ Lo\ Lo\end{array}\right)=\frac{\left(\begin{array}{c}Hi\ Hi\\Y\ \ X\mid T=Hi\\Lo\ Lo\end{array}\right)}{\left(\begin{array}{c}Hi\ Hi\\Y\ \ X\mid T=Lo\\Lo\ Lo\end{array}\right)}$$

Given the availability of more variables, fourth-order, fifth-order, and higher order odds ratios could be computed.

Third- and higher-order odds ratios impart information about *interactions* between the variables being analyzed. A third-order odds ratio that equals 1.00 indicates that there is no interaction among the three variables being analyzed. Consequently, all the two-variable associations will be equal in magnitude for the different categories of the third variable. Thus, the association of X and Y would be the same regardless of whether T were high or low. Also, the association of Y and T would be the same regardless of whether X were high or low. Finally, the association of X and T would be the same regardless of whether Y were high or low. It would not mean, however, that all of the two-variable associations, YX, YT, and XT, would have the same strength or that any of them would necessarily depart from independence.

A third-order odds ratio that departs from 1.00 indicates that each of the three possible two-variable associations will differ in magnitude according to the value of the third variable. For example, the magnitudes of the relationships between variables Y and X will be different depending on the value of variable T. Thus, the relationship between Y and X might be positive when the value of T is high, but negative when the value of T is low; or the relationship between Y and X might be negative when the value of T is high and positive when it is low. It is also possible, however, for the relationship between Y and X to be positive whether T is high or low, but the magnitude of the relationship would be greater for one condition of T than for the other. Finally, the relationship between Y and X could be negative regardless of the value of T, but the magnitude of the relationship would vary depending on whether T was high or low.

Since third-order odds ratios, like second-order odds ratios, are *symmetric*, the relationship between variables Y and T would vary according to the value of X and the relationship between variables X and T would vary according to the value of Y.

An odds ratio might have more than one test or control variable. Furthermore, this is true for odds ratios at any level. Consequently, a first-order odds ratio could be conditional upon the values of two or more test variables, as could a second-order or a third-order ratio. A second-order odds ratio conditional upon the values of two test variables would be symbolized, for example, as follows:

$$\begin{pmatrix} Hi\ Hi \\ Y\ \ X \mid T = Hi\ \text{and}\ Z = Hi \\ Lo\ Lo \end{pmatrix}$$

There would be four such second-order conditional odds ratios with two test variables with dichotomous categories: the one presented above, one in which the value of T would be high and the value of Z would be low, one in which the value of T would be low and the value of Z would be high, and one in which the values of both T and Z would be low. Of course, if there were more than two categories for variables T and Z, there would be more possible odds ratios.

Comparison of these conditional odds ratios makes it possible to sort out the relationships between the dependent variable, the independent variable, and the test variables. Analysis of cross-classifications of nominal or ordinal variables through the use of odds ratios constitutes a *hierarchical type of analysis* because of the different levels or orders that odds ratios may take and because of the interpretations that can be made about the relationships between two or more variables from an examination of these various levels.

In general, only those odds ratios that are relevant for a specific research problem are computed. Nor is it necessary to hand-compute those that are relevant. Computer programs are available to do the job and they are especially useful when there are more than three variables involved. Proper conceptualization of a research problem will usually point to those odds ratios that are of particular interest. A model can be developed by specifying, in terms of odds ratios, which variables should be related and which should not be. Such a model would specify which of all possible odds ratios would be expected to depart from 1.00. All those not specified would be assumed to be 1.00. As a matter of fact, models can be delineated in which all odds ratios above a certain specified order can be set to 1.00. Goodman (1972) has provided techniques for specifying such models and then for testing them to see whether they fit empirical data. These techniques are discussed in Section 16.2.

8.2 PATTERNS OF ELABORATION

The interpretation of patterns of outcome revealed by the accounting formula and by hierarchical analysis will be illustrated and discussed in the sections that follow. Strategies for elaboration are listed in Diagram 8.1 (p. 325).

8.2.1 Specification

Under what conditions does a relationship hold up? Stated differently, under what situations can a given independent variable explain why a dependent variable is distributed as it seems to be? In Messer's study, discussed above, ideas from theory suggested a variable, age concentration of the environment of a person, that was then introduced into the analysis as a control or test factor. The result was that under one condition of the control variable the relationship remained strong and, in fact, increased, but under the other condition the relationship between morale and interaction changed (dropped) markedly. This is a pattern of **specification**. There is a *statistical interaction* between the control variable and the independent variable, and this specifies the level of the dependent variable. The outcome is different for different categories of the test variable and this has helped specify (at least one of the) conditions under which the relationship is maintained.

What other conditions govern the relationship of morale and interaction? We can think of several likely ones just on the basis of our general experience— physical health, whether the move to public housing was voluntary or not, general orientation to life in the past. Does the relationship hold up for younger people? For people in other cultures? For both men and women? These questions suggest a line of further investigation similar to that conducted by Messer. We could select further test factors and examine them for *differences* between the conditional tables.

8.2.2 The Causal Hypothesis

Several astute observers have noticed a curious relationship between dollar loss from fires and whether or not firemen happen to be at the fire. The association, as

shown below, is positive, rather strong, and seems to raise a number of em-
barrassing questions about the activities of those fire departments upon which
these data are based. As the hypothesis goes, firemen, rather than reducing the loss
from fire, are actually increasing that loss. Graphically, the hypothesis can be
shown thus:

(X)———————————→ (Y)   *Model 1*
Number of   Dollar Loss
Firemen at   from the Fire
the Fire

In these data 143 fires are examined in terms of the number of firemen present and
the ultimate fire loss in dollars.

(Y) Dollar Loss from the Fire	(X) Number of Firemen		
	None	One +	Total
Over $500	24	61	85
$500 or less	42	16	58
Total	66	77	143

$$Q = +.74$$

Do these data prove that firemen cause fire loss?

Although the data are sufficient to raise the suspicions of those who see
them, they are not, for a number of reasons, sufficient to settle the issue. First, there
are the usual questions about bias in collecting the data, sampling procedures,
adequacy of the measurement procedures, and computational accuracy. Second,
there is the issue of time ordering of the variables. Do we have evidence that the
firemen arrived in droves *before* the fire loss became severe? Third, the idea of
causal influence includes more than one association. Can the dependent variable
be altered by manipulating the independent variable? Is this possible under all
conditions? Finally, we would expect that the relationship shown above, if it
is indeed a causal one, would be maintained even if other *antecedent* variables
are controlled—even within the categories of other variables.* Is this the case
here?

---

* The notion of "cause" is a theoretical notion about the way some set of variables are
related. A causal relationship is said to exist if (a) the independent variable has its effect
before, in time, the dependent variable, (b) there is an association between the independent
variable and the dependent variable, and (c) if this relationship is maintained even when
antecedent test variables are controlled. It is often difficult to argue convincingly that all
potential test factors have been introduced, and sometimes the time order of variables is at
issue, particularly in survey research. Generally, too, there are conditions under which the
relationship holds and conditions where it does not. What seems well established today may
well be modified or explained away tomorrow, but this is the state of any explanatory theory
in any field.

8.2.2a   Testing for Spuriousness

We might suspect that, if we could somehow control for the initial size or threat of the fire, then we could explain away the apparent support for the causal hypothesis. If, in controlling for this test variable, the association between loss and number of firemen is *maintained*, then our causal interpretation would be still *not dis*confirmed, and we would continue to hold our initial causal model. On the other hand, we might expect that firemen are rarely called for small fires, and almost always called for larger ones, so that we would observe an overall association simply because firemen and loss are related to the initial threat from the fire (its initial size). We thus have a *competing model* to the causal one above. This model is also a causal model, but it relates each of the original variables to an antecedent variable, and not to each other, as below:

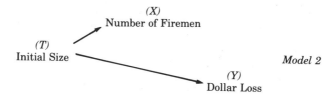

Under this model, we would expect to be able to show that, by controlling for the test factor, the association of X and Y would drop from its overall association of $Q = +.74$ down to no association, $Q = 0.0$. If this happens, we will say that our initial causal interpretation about the link between number of firemen and dollar loss is a **spurious interpretation**, and that Model 2 is a better explanation of the data than Model 1. Let us look at the data.

Table 8.3 shows the relationship of number of firemen and dollar loss separately for smaller and larger fires. This is clearly an antecedent variable, because it is measured as the state of the fire prior to any call for firemen and prior to any final determination of loss. The association in the conditional tables does drop to 0. Model 2 works better than Model 1 for these data.

What happened to the association that once was so strong? Lazarsfeld's accounting formula helps explain. If there is an initial association between X and Y and if the association does not exist in any of the conditional tables, then it must exist in the other total associations of each of the other variables and the test variable. This is, in fact, the case, as shown in Table 8.3. Computing delta for these data, the Lazarsfeld equation is as follows:

$$15.2 = 0.0 + 0.0 + \frac{143}{(55)(88)}(27.7)(18.6)$$

Notice that the two conditional tables, added together, do add up to the total association originally observed, and the other total tables can be created from the marginals on the pair of conditional tables in Table 8.3.

This pattern of outcomes, when controlling for an antecedent variable, is called "spurious" because the initial causal interpretation of the X-Y relationship is not borne out in these data, and the conditional associations drop to 0. This is another common type of analysis—the search for test factors that might show

TABLE 8.3   CONDITIONAL AND TOTAL TABLES FOR THE FIREMAN EXAMPLE*

A. First-Order Conditional Tables

(T-1) Small Sized Fire

| (Y) Dollar Loss | (X) Number of Firemen | | |
	None	One+	Total
Over $500.	4	1	5
$500. or Less	40	10	50
Total	44	11	55

$$Q = 0.0$$

(T-2) Large Sized Fire

| (Y) Dollar Loss | (X) Number of Firemen | | |
	None	One+	Total
Over $500.	20	60	80
$500. or Less	2	6	8
Total	22	66	88

$$Q = 0.0$$

B. Total Tables

| (Y) Dollar Loss | (T) Fire Size | | |
	Small	Large	Total
Over $500.	5	80	85
$500. or Less	50	8	58
Total	55	88	143

$$Q = +.98$$

| (X) Number of Firemen | (T) Fire Size | | |
	Small	Large	Total
One or more	11	66	77
None	44	22	66
Total	55	88	143

$$Q = +.85$$

* Data are fictitious.

hypothesized causal relationships to be spurious. This is the second pattern that would lead an investigator to control for other variables in the process of examining the relationship between two variables (the first was specification).

How much does the association have to drop in the conditional associations before we would conclude that the causal interpretation is spurious? This is hard to say. Any general drop would mean that the test factor helps account for some of the observed, zero-order association between the independent and dependent variables. The more the association drops in the conditional associations, the more the association could be attributed to the test factor. Usually the results are somewhat mixed, and an investigator has to be satisfied with the conclusion that there are several contributing factors to an adequate explanation of the distribution of a dependent variable.

When have sufficient control variables been introduced to assure us that a causal interpretation is the best interpretation? There is never a guarantee that this process is at an end. Many of the advances in the theory of an area come about by introducing a further test factor that alters a favorite causal interpretation. With a reasonably well thought out theory, we can be more confident that all *relevant* test factors have been examined, but this is no ultimate proof or guarantee that controlling on some other variable will not show that a long-held causal hypothesis is, indeed, spurious.

8.2.3   A Causal Sequence of Influence

Why do people plan to vote the way they do? How can we explain their political behavior? This question was raised by Lazarsfeld and others in their research report, *The People's Choice* (1948:47).* They hypothesize that a person's placement in the social class structure of society explains voting intentions, but that this happens because higher social class individuals are more interested in politics, and this leads them to be more likely to plan to vote. It is a causal sequence from social class standing (as measured here by education level) to interest in politics to vote intentions.

$$(X) \longrightarrow (T) \longrightarrow (Y)$$

Social Class	Political	Voting	
(Education)	Interest	Intentions	Model 3

An alternative explanation might be that social class influences political interest, but that it also has an independent effect on voting intentions. This could be shown as follows:

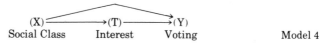

$$(X) \longrightarrow (T) \longrightarrow (Y)$$

| Social Class | Interest | Voting | Model 4 |

A further competing model might be that social class and political interest have independent effects on voting behavior.

---

* This example is cited in an excellent book on partialing and other procedures for handling survey data, by Rosenberg (1968:59; see General References).

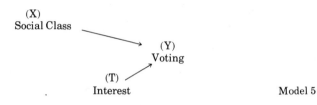

Model 5

Is it reasonable to think of these three variables as forming a causal sequence, or are the other two models more appropriate?

If there is a causal sequence of the kind in Model 3, we might be able to test it by controlling for the intervening variable, "interest," to see if the relationship between social class and voting drops to 0 in the conditional tables. This would happen because, if Model 3 is correct, removing the variation in the data in the "interest" variable would also have the effect of removing variation in social class, and thus would eliminate the association of social class and voting. On the other hand, if the association persists at some level, then Model 4 or 5 would more accurately represent the separate and direct contribution of social class to voting. The data are presented in Table 8.4.

TABLE 8.4   PERCENTAGE DISTRIBUTION OF VOTE PLANS

A. By Education

	(X)		
(Y) Vote Plans	Some High School	No High School	Total
Will not vote	8	14	11
Will vote	92	86	89
Totals	100	100	100
	(1613)	(1199)	(2812)

$$d_{yx} = .06$$

B. By Education Controlling for Political Interest

	(T) Political Interest					
	Great		Moderate		Low	
(Y) Vote Plans	Some High School	No High School	Some High School	No High School	Some High School	No High School
Will not vote	1	2	7	10	44	41
Will vote	99	98	93	90	56	59
Totals	100	100	100	100	100	100
	(495)	(285)	(986)	(669)	(132)	(245)

$$d_{yx} = .01 \qquad d_{yx} = .03 \qquad d_{yx} = -.03$$

Source: Rosenberg (1968: Tables 3.3 and 3.4). Copyright ©by Basic Books, Inc, Inc., Publishers, New York. Used by permission.

Notice that our expectation tends to be supported in the data. What association there was in the data in the original association of social class and voting plans is reduced by half or more in the conditional associations where political interest is controlled. Our conclusion, then, would be that these variables tend to form a causal sequence. Notice, too, that there was not much variation in voting plans in the first place and, furthermore, the independent variable as it is measured here does not help explain much of the variation that there is. This would suggest that there may be some separate influence of social class on voting plans, independent of political interest. Model 4 may more accurately reflect the weak streams of influence in this case. The coefficients (.01, .03 and $-.03$) represent the independent influence of social class (education as measured here) on voting plans when political interest is controlled. Let us examine the association of political interest, controlling for the antecedent variable, social class, to see how much independent influence political interest has for voting plans. These coefficients of association can be computed by using the data in Table 8.4. They are as follows:

$d_{yt} = .15$ The association of political interest (variable T) and voting plans (variable Y) for individuals with *some high school.*

$d_{yt} = .22$ The association of political interest and voting plans for individuals with *no high school.*

Our conclusion would be that these coefficients of association are much stronger than those for the independent effects of social class (education). The X-Y stream in Model 4 would be considerably weaker than the T-Y stream. Finally, looking at the conditionals for social class and interest, controlling for the *consequent* variable, voting intentions, we can see to what extent Model 5 would more accurately represent the data. If the conditional associations drop to 0 in this case, we would say that Model 5 would be a better fit. The data are as follows (again computed from information contained in Table 8.4, rearranged):

$d_{tx} = .14$ The association of political interest (here called variable T) and social class (variable X) for those with *no plans to vote.*

$d_{tx} = .12$ The association of political interest and social class among those *planning to vote.*

Comparing the coefficients for the three sets of conditional tables, then, the X-T and T-Y links appear to be much stronger than the X-Y link when the third variable is controlled. Since the association between X and Y drops when T is controlled, the conclusion is that the data fit Model 3 best. There is no reason to give up the idea of causal sequence; at the same time, the direct influence of social class on voting plans is considerably less than its indirect influence through political interest. Our hypothesis is confirmed.

8.2.3a   Independent Causation

The point to be emphasized is that we have contrasted our expectations for three competing models of the relationship between these variables (Models 3, 4, and 5). By controlling on an intervening variable, we are able to conclude that a causal sequence interpretation is *not disproven* if the association in the partials drops to 0. To the extent that it does not drop to 0, there is evidence of some *independent* effect of the independent variable on the dependent variable, irrespective of the test factor that intervenes. By controlling on the other two variables, we are able to examine the relative strengths of association. By controlling on X, for example, we can see whether T is related to Y and how strongly, independently of X. By controlling on Y, we can see to what extent X and T are related independently of Y. If the partial associations went to 0 in this latter test, we would have concluded that the two variables X and T independently influence the dependent variable Y and are related only by virtue of this common influence. Model 5 would have represented the outcome of **independent causation** in this case.

8.2.4   Suppressor Variables

In all the cases discussed thus far, we have started with some total association between the independent and dependent variables and introduced control variables that would help explain the original association further. Is it sufficient to elaborate only those relationships that are not 0 to start with? The answer is *no*: It is possible to have a third variable act to *suppress* the observed relationship between two variables and, in fact, to mask it completely as a **suppressor variable**. This sometimes happens with variables such as gender or race when a relationship in one conditional association is equally strong but opposite in direction to the association in another conditional. The result is that one relationship could cancel the other out, so that the overall association may be 0. Without some sort of expectation (*i.e.*, a theory) it is, of course, difficult to separate a situation where there is a zero-order association when one would not expect any association to appear no matter what controls were introduced from a situation where there is a zero-order association that is definitely not expected given previous research and theorizing. This latter case will serve as an illustration of a suppressor effect.

Reiss hypothesized, on the basis of prior research and theory, that there would be a negative relationship between social class and permissiveness attitudes, with attitudes toward sexual permissiveness measured by a series of scales (Reiss, 1967:59, 61). For the purposes of this study, he dichotomized permissiveness, although the puzzling findings were checked by including more categories of permissiveness attitudes and several different measures of social class. The data were gathered on about 800 students from five schools in the East. Table 8.5 presents these data.

Using gamma (G) as a measure of association, he found that there was no association between permissiveness attitudes and social class. This finding occurred in each of the five schools, with a dozen different social class measures and with a variety of other controls as well. Prior research by Kinsey indicated that religion was an important explanation of variation in sexual relations, so

TABLE 8.5   Percentage Distribution of Permissiveness Attitudes for the Student Sample

A. By Social Class

Sexual Permissiveness Attitudes	Social Class			Total
	Low	Medium	High	
High	49	46	50	49
Low	51	54	50	51
Total	100	100	100	100
	(383)	(189)	(225)	(797)

Gamma = .01

B. By Social Class and Church Attendance

High Church Attendance

Sexual Permissiveness Attitudes	Social Class			Total
	Low	Medium	High	
High	42	26	23	34
Low	58	74	77	66
Total	100	100	100	100
	(262)	(98)	(102)	(462)

Gamma = −.35

Low Church Attendance

Sexual Permissiveness Attitudes	Social Class			Total
	Low	Medium	High	
High	64	67	72	68
Low	36	33	28	32
Total	100	100	100	100
	(113)	(89)	(119)	(321)

Gamma = +.14

Source: Data from Reiss (1967:Table 4.1 and 4.2). Copyright © 1967 by Holt, Rinehart and Winston, Inc. Reprinted by permission of Holt, Rinehart and Winston.

frequency of church attendance was introduced as a control variable, as shown in Table 8.5B. Among the high church attenders, the social class–permissiveness association was −.35 (gamma), and it was +.14 among the low church attenders. Thus, the higher the social class, the lower the level of permissiveness among high church attenders, and the opposite was true of low church attenders. Further investigation indicated that the same kind of outcome occurred whenever the student group was divided into categories of generally "conservative" and generally "liberal" on a number of other variables such as political preference, beliefs about integration of schools, and civil rights activity. This and related analyses eventually led Reiss to a general proposition, namely: "The stronger the amount of general liberality in a group, the greater the likelihood that social forces will maintain high levels of sexual permissiveness" (Reiss, 1967:73). Social class standing is one of those forces.

Had Reiss stopped his analysis upon finding no association between social class and permissiveness attitudes, he would have missed an important series of findings about variables that masked the relationship.

8.2.5   Social System Analysis

For many sociological problems, investigators are interested in two levels (or more) of analysis, the *individual* level and the *group* level. They may argue, for example, that differences between groups stem from the different kinds of individuals who compose the group (called *structural analysis*), or they may feel that individuals behave in certain ways not only because of their own characteristics but because of the character of the group itself of which they may be members (called *contextual analysis*). In addition to these two types of analysis are one-level analyses using groups as the unit and one-level analyses using individuals as the unit.*

In a social system analysis involving three variables and two levels, each individual case is characterized by three measured variables. Usually, two of these variables, the independent and dependent variables, are individual characteristics. and the third variable is a group-level variable that is used to characterize the kind of group context the individual is in. We could consider creating one table cross-classifying the independent and dependent variable for all individuals in a given *kind* of group. One such table would be created for each of the different kinds of groups included in the study, and systematic comparisons would be made within and between groups.

8.2.5a   Structural Analysis

A **structural analysis** might involve comparison of a specific cell (*e.g.*, percentage psychotic among foreign-born) across *groups* that differ in some interesting respect (*e.g.*, degree of social disorganization of the area) to see what effects the differences between groups might have when the structure of the group is held constant (*e.g.*, only foreign-born are compared or only native-born). Riley points out another kind of structural analysis, used by Durkheim, in which group differences are controlled by examining the association of two variables (*e.g.*, religion and suicide) within separate groups (*e.g.*, within different types of countries).

---

* In this chapter we shall illustrate some of the possibilities that involve conditional tables and an examination for interaction between group and individual variables, one of the patterns of relationship between three variables. An excellent discussion of social system analysis with some examples from the field of sociology is found in Riley (1963; see General References).

8.2.5b   Contextual Analysis

A **contextual analysis** combines information from all three variables, examining the way the relationship between two variables (*e.g.*, two individual variables, or perhaps an individual and a context variable) may change systematically, for *individuals*, across groups that are set up to differ in a systematic way on a group-level variable. Here the analysis will focus on individuals as the unit being investigated, and it will measure individuals on variables that reflect individual properties as well as some aspect of the kind of group context within which an individual is located. Let us turn to an example at this point.*

In a study of job satisfaction, previous research has supported the notion that older workers are more likely to be satisfied than are younger workers. It has also been argued that the "age" of the company or department of a company has an influence on job satisfaction (and on the creativity and productivity of the organization), but the data are conflicting and generally show little difference by group age. Is it possible that job satisfaction, for example, is influenced not only by individual age of the worker, but also by the organizational age of the group context within which the worker is situated?

One of the authors examined this question using data on some 235 individuals who worked in 35 branches or departments of a large research and development organization. Individuals were asked their own ages and a series of questions designed to measure their degree of job satisfaction. One of these job satisfaction questions was whether or not individuals felt their department was a highly supportive work environment. In addition, the departments themselves were "measured" on organizational age, and this was done by computing the arithmetic mean age of individuals in that department. Individuals then, were characterized by their own ages, the ages of the departments they worked in, and their job satisfaction.

Both individuals and departments were classified on age into "young," "middle age," and "old" in this analysis. There were, of course, a number of different departments in each age bracket, but rather than examine each organization separately, as we might do in one kind of structural analysis, we put the data together in three summary tables, one for each kind of group (*i.e.*, younger, middle-aged, and older groups). These data are presented in Table 8.6.

Overall, 48.1% of the workers felt their work context to be "highly supportive." Older individuals were more likely to agree that their context was highly supportive, as is indicated in Table 8.7. The pattern by group age is reversed in that the percentage is lower for the older age group than for younger age groups.

Data in the three summary tables (Table 8.6) show an interesting added pattern. Young or middle-aged individuals in middle-aged or older group contexts are less likely to feel their department is a highly supportive work context. Compare these figures to, for example, older individuals in young groups, older

---

* An excellent example of contextual analysis that makes use of some of the procedures to be discussed in the next chapter is given in McDill *et al.* (1967). At issue is whether or not the socioeconomic context of the school influences individual achievement. Data came from 20 public high schools.

TABLE 8.6   PERCENTAGE FEELING THEIR DEPARTMENT IS A HIGHLY SUPPORTIVE WORK ATMOSPHERE, BY INDIVIDUAL AGE AND ORGANIZATIONAL AGE

Younger Departments

Feelings about Support of Work Context	Individual Age			
	Young	Middle	Older	Total
Highly Supportive	51.4	48.6	64.3	52.2
Not Highly Supp.	48.6	51.4	35.7	47.8
Total	100.0	100.0	100.0	100.0
	(35)	(37)	(14)	(86)

Middle-aged Departments

Feelings about Support of Work Context	Individual Age			
	Young	Middle	Older	Total
Highly Supportive	30.4	45.9	58.6	46.1
Not Highly Supp.	69.6	54.1	41.4	53.9
Total	100.0	100.0	100.0	100.0
	(23)	(37)	(29)	(89)

Older Departments

Feelings about Support of Work Context	Individual Age			
	Young	Middle	Older	Total
Highly Supportive	(40.0)	37.5	48.7	45.0
Not Highly Supp.	(60.0)	62.5	51.3	55.0
Total	100.0	100.0	100.0	100.0
	(5)	(16)	(39)	(60)

Source: Data are from 235 individuals in 35 branches of a research and development organization, collected in 1966 by Robert Biller, who kindly made these data available for the author's study of individual and organizational age. Branch heads and managers have been eliminated in the above data.

TABLE 8.7   PERCENTAGE OF WORKERS FEELING THEIR BRANCH PROVIDES A HIGHLY SUPPORTIVE WORK ATMOSPHERE, BY GROUP AND BY INDIVIDUAL AGE

By Group Age		N(100%)		By Individual Age		N(100%)
31–35 yrs	52.2%	(86)	Young	22–32 yrs	42.9%	(63)
36–39 yrs	46.1%	(89)	Middle	33–39 yrs	45.6%	(90)
40–55 yrs	45.0%	(60)	Older	40–64 yrs	54.9%	(82)

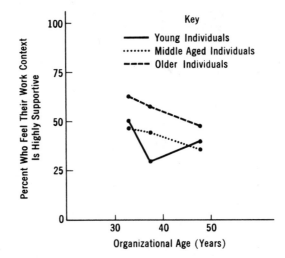

Source: Data from Table 8.6.

FIGURE 8.1 Percentage Distribution of a Job Statisfaction Measure by Age of Individuals and Their Organization Context for Workers in Departments of a Research and Development Organization.

individuals in middle-aged groups, or the young and old individuals in their own age groups. The differences in percentage feeling their department work context is supportive in Table 8.6 show a pattern of *statistical interaction* between individual and group age. Not only are there differences by individual age or by organizational age, separately, but there are much more impressive differences when specific combinations of these two age measures are examined. This effect can be shown clearly in the graph in Figure 8.1.

In these graphs, each of the dots indicates one of the combinations of individual age and group context age. There are nine such combinations here and, thus, nine dots. Each dot is placed at a point that corresponds to (a) the percentage of individuals who feel that the work context is supportive, and (b) the average age of the department within which they are located. The dots representing young individuals are connected and, separately, those dots for the middle-aged individuals and for the older individuals. There are three lines on the graph, one for each category of individual age and the lines reflect changes in the dependent variable, percentage feeling the context is supportive, across the differently aged departmental contexts. These lines converge or cross each other, which means that there is statistical interaction; the level of the dependent variable depends not only on the addition of a group effect to an individual age effect, but depends on the specific combination of those two variables.

Figure 8.2 illustrates some of the other outcome possibilities one might find in two-level analysis of this sort. In Figure 8.2a, for example, the two lines that represent the different categories of one of the independent variables are identical. This means that there is no individual effect. The fact that the lines have a slope indicates, however, that there is a group effect on the dependent variable. Here the

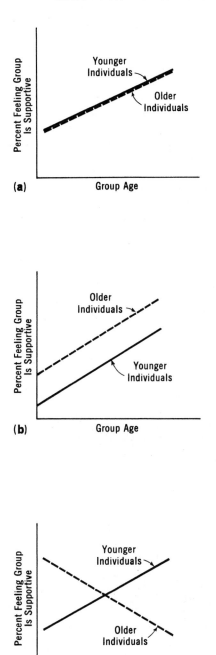

(a) *Group Effect Only.* Here there is a difference in the dependent variable for contexts of different kinds, but there is no difference within any of the contexts in terms of the other individual characteristic. If the lines were together as they are here, but parallel with the *X*-axis, there would be no group or individual difference in the dependent variable.

(b) *Separate Individual and Group Effects.* In this graph, the lines are parallel, sloped, and different. This indicates that there is a change in the dependent variable for different kinds of contexts and that the effect is the same on each kind of individual within these contexts. There is also an individual effect since the two lines are separated. Within each kind of context, there is an individual difference on the dependent variable. Note that the dotted line might be above or below the solid line. That is, the correlation of the group variable may differ in sign from the correlation of the individual variable.

(c) *Interaction Effects.* Here the lines are different, both sloped, and they are not parallel. This indicates that the dependent variable depends upon the specific combination of individual and group independent variables. It is not a simple case of adding together separate group and individual effects as in situation (b), above. Here, because the context examined is older, older individuals tend to decline on the dependent variable, but younger individuals tend to increase.

FIGURE 8.2 SOME POSSIBLE PATTERNS OF GROUP AND INDIVIDUAL EFFECTS [for a discussion of some of these patterns, see General References: Riley (1963) and Davis *et al.* (1961).]

BOX 8.2 STATISTICAL INTERACTION

If two variables, *A* and *B*, each explain some of the variation in a dependent variable, their combined explanatory power may be simply a sum of their separate effects. That is, they each contribute a part to their combined effect on a dependent variable. They have an *additive effect*.

Contrasted with this additive effect are *interaction effects* in which specific combinations of categories of two or more independent variables (or independent and control variables) explain more of the variation in a dependent variable than would be expected from a simple additive combination of their separate effects. The combination of certain categories of individual age and group age, for example, seems to result in differences in satisfaction beyond what would be expected from a sum of their separate effects.

Statistical interaction is an important kind of outcome of the relationship of two or more independent variables to a dependent variable, not only because it helps determine the kind of analysis to use, but also because our theories often lead us to expect interesting and unusual effects of particular combinations of values of two or more independent variables.

---

dependent variable within a given context for a certain kind of people is measured in terms of percentages, although it could be measured in terms of medians or means, etc., depending on the kind of measurement or feature of interest.

In Figure 8.2B, there is an added effect due to individual differences in the independent variable. Notice that the individual differences are the same within the different categories of group context, and thus both the group and the individual variables make separate, additive contributions to the explanation (or prediction) of the dependent variable. To say it somewhat differently, the individual difference occurs regardless of the particular group context one is talking about (and vice versa). In Figure 8.2c, however, the level of the individual-age effect on the dependent variable depends on the group context in important ways. Older individuals will have a higher level of the dependent variable than younger individuals only if the group context happens to be younger. The reverse is true in this example, where group contexts are older and there is a type of middle-age group context for which there are no individual differences. This is an example of statistical interaction, since the relationship between two variables (independent and dependent) depends critically on the value of the other independent variable. In this case, since one of the variables is group level and the other is individual level, the analysis permits us to draw conclusions about group and individual effects. Interaction could also occur, of course, between variables both at the same level of analysis (*e.g.*, both group or both individual), and this was the case in the example of specification discussed earlier in this chapter (Section 8.2.1).

8.2.6   Summarizing Patterns of Elaboration

Some of the more frequent and interesting patterns of outcome are of the type referred to as *specification* or *statistical interaction* (Section 8.2.1), a condition in which the conditional tables differ substantially from each other. Patterns where there is an original association but the conditional associations are near 0 are interpreted as evidence that a causal hypothesis is *spurious* (8.2.2a), that a *causal sequence* hypothesis is supported (8.2.3), or that there is *independent causation* (8.2.3a), depending on the antecedent, intervening, or consequent theoretical ordering of a test factor. Finally, there is the pattern in which a *suppressor* variable (8.2.4) operates to mask out an original association, and this is discovered only after examining other conditional and total tables. The *Lazarsfeld accounting formula* (8.1.4a) is useful in organizing patterns of outcome when total and conditional associations are examined. In general, the procedure used is one of (a) examining lower-order tables first and higher-order tables later in an analytic sequence, and (b) controlling on other relevant variables in the examination of an *X-Y* relationship of interest. What is controlled and how depends critically on the ideas brought to the research by the investigator. *Hierarchical analysis* (8.1.4b), which utilizes odds ratios, is also useful for sorting out all of the patterns of statistical interaction to be found in total and conditional tables.

8.3   THE AVERAGE PARTIAL TABLE

The lines of elaboration we have been discussing direct the investigator to examine (a) any difference there may be in the conditional tables, or (b) the overall change in degree of association in the conditionals. In situation (a), the only analysis procedure is elaboration (or some form of dummy-variable regression analysis, as discussed toward the end of the next chapter), and elaboration involves the examination of the different conditional tables. In situation (b), however, where we are merely interested in whether the association in the conditionals remained the same, decreased, or increased, there are some other types of statistical summary that could be carried out to simplify the analysis. Two of these procedures will be discussed in this section. One is *standardization* and the other is *partial association*. In both instances the result is a kind of "average" of the outcome in a set of conditionals. In the first instance an average partial table is the result, and in the second it is an average of the measures of association that might be computed on separate conditionals.

In either case, these further descriptions are used in the same way they were used earlier. One would compare the original association of two variables with the association after a test factor has been "partialed out" or controlled. If, for example, the test factor were an antecedent variable, and if the partial associations dropped to 0 when the test factor was controlled, the conclusion would be as it was when individual conditional tables were examined—any causal hypothesis would then be a *spurious* interpretation of the relationship between the main independent and dependent variables. Let us discuss the idea of standardization first and then turn to partial association coefficients.

BOX 8.3   CONDITIONAL TABLES AND PARTIAL ASSOCIATION

One of the organizing themes in this book has been the search for ever more powerful summary statistics. This search led to the step from a univariate distribution to summary index numbers such as the mean and standard deviation. Again, at the point where several univariate distributions were being contrasted, we stepped up to an overall treatment of these separate distributions as columns in a bivariate table, and this led to the use of measures of strength and nature of association. In the first part of this chapter we began by looking at a basic table relating two interesting variables within categories of other control variables. Now, in this section, we are again looking for a more powerful summary procedure that will express in one table or in one number what we would otherwise get from a potentially large number of conditional tables.

Although terminology is not altogether consistent in statistics, we will here observe a distinction between the examination of *conditional* tables and their associated *conditional association* measures (the last section) and *partial association* measures, which summarize a set of conditional relationships and express the part of the total association that exists after the effects of control variables have been statistically removed. It should be noted, however, that some texts refer to the examination of conditional tables as "partialing."

8.3.1   Standardization

Standardization is a procedure that statistically removes fluctuations in the data due to some control variable, thus permitting an investigator to examine and compare tables without this contaminating effect. Two procedures will be discussed, both of which are called *direct* standardization. The first procedure applies to figures, such as rates, which we may want to compare, and the second applies to whole tables. We will use the first to explain the idea of standardization. Computational methods for the second procedure are presented in Diagram 8.2 (p. 326).

8.3.1a   Standardizing Rates

Suppose that we are interested in studying the rate at which women of different educational backgrounds produce children. Examining U.S. Census reports, we discover that among those women in the child-bearing ages (*i.e.*, 15–49 years) there are 302 children under 5 years old per 1,000 women who had 8 years of elementary school education, and that the rate increases to 393 per 1,000 for women with 4 years of high school, and it drops again to 383 per 1,000 for those with 4 or more years of college (U.S. Bureau of the Census, 1970). There appears to be a difference in number of children under 5 by education.

At this point we might be somewhat suspicious of our comparison because, after all, women in their 20s are more likely to have young children than women in their 40s. If there is a higher percentage of young women in the high school educated group, we would expect a higher rate. Some of the difference in rates may be due to differences in age distribution of elementary, high school, and college trained women. The problem, then, is to create three rates that somehow take out the effects of any differences there may be in the distribution of age.

To do this, we will choose some standard age distribution as our **standard population**. The choice is arbitrary, but the rates will be more interesting if the standard population is realistic. For this purpose we will use the actual overall age distribution of women in the United States in 1969. This distribution is as follows:

"STANDARD POPULATION" NUMBER OF WOMEN BY AGE BETWEEN AGES 15 AND 49, UNITED STATES, MARCH, 1969 DATA

Age of Woman	Number (in 1000's)
15 to 24	17,109
25 to 29	6,608
30 to 39	11,388
40 to 49	12,472
Total	47,577

Source: U.S. Bureau of the Census (1970).

Now we need detailed "children-under-5" rate figures for women of different ages and for the three different education categories. These rates are given below.

NUMBER OF CHILDREN UNDER 5 YEARS OLD PER 1,000 WOMEN, BY AGE OF WOMAN AND NUMBER OF YEARS OF SCHOOL COMPLETED, UNITED STATES, 1969

	Schooling Completed		
Age of Woman	Elementary	High School	College
15 to 24	251/1000	413/1000	110/1000
25 to 29	916/1000	879/1000	673/1000
30 to 39	456/1000	407/1000	572/1000
40 to 49	96/1000	76/1000	88/1000

Source: U.S. Bureau of the Census (1970).

The standardization procedure involves multiplying each of the *rates by the number of women in the standard population who are in that age bracket*. This produces an "expected" number of children under 5 if there were as many women in that age group as there are in the standard population. The expected number of children under 5 can then be summed separately for each of the levels of schooling, divided by the number of women in the standard population, and expressed as a rate per 1,000 women. Thus 17,109 (1,000's) women aged 15 to 24 would be *expected* to have 4,294,359 children if they all had completed only 8 years of elementary school [*i.e.*, multiply the rate (251, from the table above for those in the 15–24 age bracket with elementary schooling) by the 17,109 women in that age group in the standard population]. On the other hand, one would expect 7,066,017 children under 5 years of age ($413 \times 17,109 = 7,066,017$) if all the women of this age in the standard population had a high school education. Finally, one would expect fewer children under 5 if this same group had a college background ($110 \times 17,109 = 1,881,990$). These computations are shown below:

COMPUTATION USING RATES FOR WOMEN WITH *ELEMENTARY SCHOOL* EDUCATION

Age of Woman	Rate per 1,000		Standard Population (1,000's)		Expected Number
15–24	251	×	17,109	=	4,294,359
25–29	916	×	6,608	=	6,052,928
30–39	456	×	11,388	=	5,192,928
40–49	96	×	12,472	=	1,197,312
Totals			47,577		16,737,527

$$\text{Standardized Rate} = (1000)\frac{16,737,527}{47,577,000} = 352$$

COMPUTATION USING RATES FOR WOMEN WITH A *HIGH SCHOOL* EDUCATION

Age of Women	Rate per 1,000		Standard Population (1000's)		Expected Number
15–24	413	×	17,109	=	7,066,017
25–29	879	×	6,608	=	5,808,432
30–39	407	×	11,388	=	4,634,916
40–49	76	×	12,472	=	947,872
Totals			47,577		18,457,237

$$\text{Standardized Rate} = (1000)\frac{18,457,237}{47,577,000} = 388$$

COMPUTATION USING RATES FOR WOMEN WITH *COLLEGE* EDUCATION

Age of Woman	Rate per 1,000		Standard Population (1,000's)		Expected Number
15–24	110	×	17,109	=	1,881,990
25–29	673	×	6,608	=	4,447,184
30–39	572	×	11,388	=	6,513,936
40–49	88	×	12,472	=	1,097,536
Total			47,577		13,940,646

$$\text{Standardized Rate} = (1000)\frac{13,940,646}{47,577,000} = 293$$

As a result of applying the different rates to the same standard population, we can compute rates that can be *compared:* 352, 388, and 293 per 1,000 women. The following table summarizes the rate of children under 5 years old per 1,000 women aged 15–49 both before standardization on age and after standardization.

RATE OF CHILDREN UNDER 5 YEARS OLD PER 1,000 WOMEN AGED 15 TO 49, UNITED STATES, 1969, BEFORE AND AFTER STANDARDIZATION ON AGE

Education Completed	Not Standardized	Standardized On Age
Elementary	302	352
High School	393	388
College	383	293

The age structure of the population upon which the rates are computed is the *same* for each education group, so the overall differences between age-standardized rates must be due to the different rates that apply to differently educated women. These are called age-standardized rates because the effect of age structure of the population is removed (by making it the same). There are a number of uses of this standardization procedure in demography in comparisons of death rates, birth rates, or disease rates for populations of different makeup.

8.3.1b   Using the Consumer's Price Index to Standardize Current Dollars

A common example of standardization in the United States is the use of the Consumer Price Index to take out the effects of inflation in order to make a comparison of dollars over time more comparable. There are several indices that are used for this purpose, but the U.S. Bureau of Labor Statistics' Consumer Price Index is widely used. The index is created by combining current retail prices for a standard "shopping basket" of goods and services purchased by city wage earners and clerical workers. Table 8.8 provides the consumer price index for a number of years, using an average of 1982–1984 as the base that is set to 100. One can use the index to adjust current dollars from different years into "standard" 1982–1984 dollars. The relative size of the index in different years reflects differences in the purchasing power of a dollar.

Suppose that a family has an annual income of $15,000 in 1967 when the CPI was 33.4. What would that income be in terms of 1982–1984 base-period dollars? We could use the CPI indices in Table 8.8 to find the equivalent income in 1982–1984. Since the 1982–1984 CPI is set to 100, or nearly three times what it was in 1967, the ratio of these indices (100.0/33.4 = 2.99) could be used to adjust the 1967 income to the equivalent base-period purchasing power. The $15,000 income in 1967 would be equivalent in 1982–1984 purchasing power to 2.99 ($15,000) = $44,850. Using the same approach, the $15,000 income in 1967 would have the same purchasing power as $8,760 in 1946 (19.5/33.4 = .584 and .584 × $15,000 = $8,760), or as $55,650 in 1989 (124.0/33.4 = 3.71 and 3.71 × $15,000 = $55,650). A

TABLE 8.8 CONSUMER PRICE INDEX (1982–1984 = 100)

1946	19.5	1961	29.9	1976	56.9
1947	22.3	1962	30.2	1977	60.6
1948	24.1	1963	30.6	1978	65.2
1949	23.8	1964	31.0	1979	72.6
1950	24.1	1965	31.5	1980	82.4
1951	26.0	1966	32.4	1981	90.9
1952	26.5	1967	33.4	1982	96.5
1953	26.7	1968	34.8	1983	99.6
1954	26.9	1969	36.7	1984	103.9
1955	26.8	1970	38.8	1985	107.6
1956	27.2	1971	40.5	1986	109.6
1957	28.1	1972	41.8	1987	113.6
1958	28.9	1973	44.4	1988	118.3
1959	29.1	1974	49.3	1989	124.0
1960	29.6	1975	53.8	1990	135.9

Source: 1990 President's Economic Report, Table C-60. Data beginning 1978 are for all urban consumers; earlier data are for urban wage earners and clerical workers.

1946 income of $8,760, and a $15,000 income in 1967, and a $55,650 income in 1989 would be comparable to a 1982–1984 base period of $44,850 in terms of purchasing power. Using the consumers price index permits one to adjust dollars in different years to dollars in a standard base period, and this standardization permits comparison of dollars across time without confounding by changes in the purchasing power of the dollar.

8.3.1c Standardized Tables

The second application of standardization is called *test factor standardization* (see Rosenberg, 1962; Davis, 1984), and it too statistically removes the effect of control variables so that the relationship between independent and dependent variables can be examined without this source of contamination. This permits an investigator to compare an original total association of two variables and the same association where the effects of some test factor have been statistically removed. This would permit us to make some kind of judgment about what is happening to the strength of association in the conditionals in general. Test factor standardization is illustrated in Diagram 8.2 (p. 326). To explain the logic behind standardization, consider the two conditionals from Table 8.3. They have been percentaged and reproduced in Diagram 8.2.

In these conditional tables there is no association between X and Y, although, when cell frequencies are added together, they produce a rather striking association ($Q = .74$) in the total table. Looking at the differences between the total and conditional tables, we can see that the effect of the test factor has been to "rearrange" cases within each conditional table, and thus to rearrange the distribution of column and row totals, so that each cell frequency in a given conditional contributes a *differently weighted amount* to the total table. In the T-1 partial, for example, the "4" in the upper left cell contributes 17% of the 24 in that cell of the total table, while the "1" contributes only 2% to its combined-cell total, and the cell with 40 cases contributes 95% of the cases in its cell in the total table. This differential weighting of cells does not reflect the relative number of cases in the distribution in a column of the conditional table and, thus, it does not reflect the pattern in the conditionals.

To handle this differential weighting, we can take two steps. *First*, percentage each conditional table so that the effect of different column (or row) totals is removed. If Y is the dependent variable, percentaging would be in the direction of the independent variable. In effect, this gives proper "rates" of appearance of the different values of Y within the separate categories of the independent and control variables. Then, all the cells in a conditional table (*i.e.*, all cells in one category of the control variable) are weighted equally (*i.e.*, multiplied by the same weight), and the weighted percentages can then be summed over all conditional tables to create a standardized percentage table.

$$\Sigma(\omega_i P_i) = \text{standardized percentage for a cell}$$

where ω is the weight and P is the cell percentage, summed over all the conditional tables.

There is no mathematical reason for picking any particular set of weights. The important point is that, whatever they are, they should be *uniform* within any one conditional table. There are some subsidiary reasons for selecting certain weights, however. If the sum of weights used in the different partial tables is equal to 1.00, then the resulting standardized table will be a percentaged table with column (row) totals equal to 100%. If weights total to more than 1.00, then column (row) totals in the standardized table will be larger than 100% and would have to be represented, following the usual procedures but being careful to percentage in the same direction that the conditionals were percentaged. It is also useful to select a weighting factor for a given conditional table that is equal to the proportion of all of the cases that fall in that category of the control variable. This, in effect, gives more weight to the conditional table that has the greater number of cases (and may, therefore, be more stable) and gives less weight to conditionals with very few cases (which may therefore be less stable). This is the procedure described in Diagram 8.2 (p. 313), although, again, the important point is that weights within a partial not be changed from cell to cell. Computations for a 2 by 2 table are shown in Diagram 8.2, although test factor standardization is a technique that applies equally well to any size table and any number of categories of control variables.

The standardized percentage table may be converted back to frequencies, using the total table marginals as the base of the corresponding 100% values. Appropriate measures of strength of association could be computed on the standardized frequency table in the usual way, and the original total association could be compared with the association computed on the standardized table.

Standardization was used as an analysis technique by McAllister in a study of residential mobility among blacks and whites in the United States (McAllister *et al.*, 1971). The literature, they note, suggests that blacks move more often than whites and that blacks' mobility is more likely to be local. Using data from a national survey, the investigators examined moving behavior between 1966 and

TABLE 8.9 MOBILITY DIFFERENCES BETWEEN RACES: STANDARDIZED AND TOTAL

Total Association

Mobility Behavior	Ethnicity		Total
	Black	White	
Stayed	48.7	58.6	56.9
Moved	51.3	41.4	43.1
Total	100.0	100.0	100.0
	(263)	(1226)	(1489)

$$|Q| = .20$$

Total Association Standardized on Owner/Renters

Mobility Behavior	Ethnicity		Total
	Black	White	
Stayed	58.4	55.9	56.3
Moved	41.6	44.1	43.7
Total	100.0	100.0	100.0
	(263)	(1226)	(1489)

$$|Q| = .11$$

Source: McAllister *et al.* (1971). Used by permission. Reprinted from McAllister, Ronald, Edward Kaiser, and Edgar Butler, "Residential Mobility of Blacks and Whites: A National Longitudinal Survey," "*American Journal of Sociology,* 77 (November, 1971), p 445–456 by permission of the University of Chicago Press. Copyright 1971 by the University of Chicago and published by the University of Chicago Press.

1969 for blacks and whites. The data are given in Table 8.9. The absolute value of $Q(|Q|)$ equals .20, meaning that whites were more likely to have stayed than is true of blacks. The comparison, as they point out, is not an appropriate one, however, because blacks are more likely to be renters, and renters of any ethnic status tend to move more frequently than owners. The standardized table in Table 8.9 controls for the effects of the owner–renter variable. You will notice that the association drops to a $|Q|$ of .11, indicating that there are minor ethnic differences in moving behavior and, if anything, blacks tend to be the stayers. Note, too, that the nature of association shifted as well.

Test factor standardization is useful whenever we are interested in what happens in a set of conditional tables in general. It does not permit us to examine *differences* between conditional tables (*i.e.*, interaction effects).

8.3.2 Measures of Partial Association and Correlation

The third approach to an analysis of three or more variables to be discussed in this chapter is the computation of some type of "average" over the measures of association for each of the conditional tables. These coefficients are called

coefficients of partial association and, like the standardized table, they are compared with the original total association to determine what happened *in general* to the strength of association when one or more test factors were introduced. Like the standardized table, partial association coefficients do not provide information about any pattern of *differences* between separate conditional tables over which they are computed. Two procedures for computing partial coefficients will be introduced here, one for ordinal data and one for interval data.

8.3.2a   Ordinal Partial Association Coefficients

A relatively direct approach to the creation of ordinal partial association coefficients involves combining computations based on each conditional table into a single, "average" coefficient. For those ordinal measures of association that make use of counts of pairs (*i.e.*, gamma, Somers's d_{yx}, etc.), the combination is accomplished by simply adding together the count of a certain type of pair computed on each of the conditional tables. Thus, a total of concordant pairs (N_s) could be arrived at by adding up the N_s computations on each of the conditional tables. A similar summing could be made for other types of pairs, such as the discordant pairs (N_d), pairs tied on X but not on $Y(T_x)$, pairs tied on Y but not on $X(T_y)$, and pairs tied on both X and $Y(T_{xy})$. The appropriate totals would then be substituted into the formula for the ordinal coefficient one wants. For example, ΣN_s could be used instead of N_s, ΣN_d could be used instead of N_d, and so on.

The formula for *partial gamma* (G_p) then becomes

(8.1)
$$G_p = \frac{\Sigma N_s - \Sigma N_d}{\Sigma N_s + \Sigma N_d}$$

where the N_s and N_d sums are taken over the N_s and N_d components computed on each conditional table separately. This formulation of partial coefficients for ordinal measures would be appropriate for Somers's d_{yx} and other ordinal measures as well.

The creation of a partial G can be illustrated with data from an article by Ransford, "Skin Color, Life Chances, and Anti-White Attitudes" (1970), in which some 312 black males were interviewed shortly after a mid-1960s race riot in the Watts area of Los Angeles. His hypothesis was that skin color itself has an influence on the structure of opportunity, even when such variables as educational experience are taken into account. Table 8.10 presents the total and conditional associations of occupation and skin color, controlling for three categories of formal education.

The original, zero-order G was $-.26$, indicating that the lighter the skin color, the higher the occupational standing. The N_s and N_d counts are indicated in Table 8.10 for the total association and for each of the three conditional tables, where education is controlled. What is the effect of educational controls? The partial G can be computed to answer this question by summing N_s and N_d counts over the three tables as follows:

$$\Sigma N_s = 412 + 754 + 833 \ = 1999$$

$$\Sigma N_d = 714 + 526 + 1695 = 2935$$

TABLE 8.10 PERCENTAGE DISTRIBUTION OF OCCUPATIONAL LEVEL BY SKIN COLOR AND EDUCATION FOR BLACK MALES INTERVIEWED IN LOS ANGELES IN THE MID-1960s

	Total			Education								
				Less than H.S.			H.S. Grad.			Some College		
	Skin Color			Skin Color			Skin Color			Skin Color		
Occupation	Lt	Md	Dk	Lt	Md	Dk	Lt	Md	Dk	Lt	Md	Dk
	%	%	%	%	%	%	%	%	%	%	%	%
White Coll.	52	40	29	15	05	13	32	17	9	83	70	58
Blue Collar	39	48	50	69	70	45	50	67	76	17	26	36
Unemployed	9	12	21	15	24	42	18	15	14	0	4	6
Total	100	100	100%	99	99	100%	100	99	99%	100	100	100%
	(64)	(159)	(85)	(13)	(37)	(31)	(22)	(46)	(21)	(29)	(76)	(33)

$$N_s = 6640 \qquad N_s = 412 \qquad N_s = 754 \qquad N_s = 833$$
$$N_d = 11239 \qquad N_d = 714 \qquad N_d = 526 \qquad N_d = 1695$$
$$G = -.26 \qquad G = -.27 \qquad G = +.18 \qquad G = -.34$$

Source: Based on rearranged data from Ransford (1970:171, Table 1).

The partial G, then, is:

$$G_p = \frac{1999 - 2935}{1999 + 2935} = \frac{-936}{4934} = -.19$$

Comparing the original G of $-.26$ with the G_p of $-.19$ indicates that indeed the association does drop when education is taken into account. It is also clear, however, that the association does not drop to 0, thus partially supporting the author's initial expectation.

Again it should be pointed out that the partial G, like partial correlation coefficients, answers only the question of what happens in the conditional tables *in general*. Notice in Table 8.10 that there are interesting differences in the strength of association between the G coefficients computed on the separate tables. The G_p indicates the proportionate reduction in error in predicting rank on one variable from rank on the other variable after the effects of education have been taken out.*

---

* A more precise statement of the meaning of gamma (G) is found in Section 7.3.2b. It is the proportionate reduction in error resulting, in this case, from using the "opposite-rank order" rule to predict the rank of pairs on one variable from a knowledge of the rank order on the other variable, rather than making a random guess of rank order (the minimum guessing rule), among pairs that are ranked differently on both variables (*i.e.*, the $N_s + N_d$ pairs), after the effects of education have been taken into account. It is relevant to note that education was taken into account to the extent that three categories were used. Somewhat different results may occur if more or fewer categories of the control (and other variables, for that matter) are used.

8.3.2b   Partial Correlation Coefficient for Interval Variables

A frequently used partial correlation coefficient for interval variables is $r_{yx \cdot z}$, which is computed from Pearson's r (discussed in Chapter 7). Although r is symmetric, subscripts are used to indicate which variables the correlation is between (the first two subscripts in front of the dot), and the variable(s) used as controls (those listed after the dot).

The partial correlation coefficient that statistically controls the effects of one control variable is called a **first-order partial correlation**, and it is computed from zero-order correlation coefficients as follows:

$$(8.2) \qquad r_{yx \cdot t} = \frac{r_{yx} - (r_{yt})(r_{xt})}{\sqrt{(1 - r_{xt}^2)(1 - r_{yt}^2)}}$$

Higher-order partial correlation coefficients can be computed in a similar manner, using the next lower partial correlation coefficient in the general formula above. Thus, for two control variables, the second-order partial becomes

$$(8.3) \qquad r_{12 \cdot 34} = \frac{r_{12 \cdot 3} - (r_{14 \cdot 3})(r_{24 \cdot 3})}{\sqrt{(1 - r_{14 \cdot 3}^2)(1 - r_{24 \cdot 3}^2)}}$$

Higher-order partials may be formed in similar ways.

Like the total correlation, r, the partial correlation varies from -1.00 to $+1.00$. Its square expresses the proportion of the variation in Y (or X) explained by its linear association with the other variable, X (or Y), after the linear effects of the control variables have been taken into account (statistically removed).*

An example of the use of the partial correlation coefficient is provided by Lightfield (1971), who studied factors influencing the recognition of sociologists by their peers. His data are based upon responses of 200 university sociologists, and as independent variables he used (a) status of the department from which the rated sociologists received their Ph.D., (b) quality of publications, and (c) quantity of publications. The dependent variable was a peer-recognition score. To guide his analysis, Lightfield presented an arrow diagram, shown in Figure 8.3.

The numbers in parentheses on the diagram are partial correlation coefficients, controlling on prior variables that influence the relationship between a given independent and dependent variable. Thus the .35 on the line between quality of publications and recognition by peers is the partial correlation of these two variables, with quantity of publications and status of the Ph.D. department

---

* Another way to express the meaning of the partial correlation coefficient is in terms of a correlation between two sets of residuals. Think of two linear regression equations, $Y' = a_{yt} + b_{yt}T$ and $X' = a_{xt} + b_{xt}T$, where T is the control variable whose effects are to be partialled out, statistically, and X and Y are the main independent and dependent variables whose relationship is of interest. If T is related to X and to Y, then the two regression equations will be able to improve the prediction of Y and X somewhat, but there will likely be some residual in both Y and X yet to be explained. That is, for many scores, there will be both a $Y - Y'$ and an $X - X'$ difference. The partial correlation coefficient is the Pearsonian correlation, r, between these two sets of residuals that remain after the linear effect of T has been statistically removed. After T has explained all it can in terms of a simple linear regression, the remaining correlation between X and Y is the partial correlation coefficient.

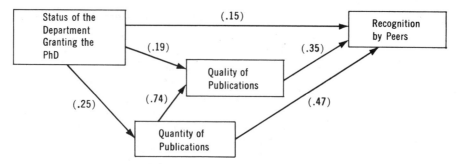

Source: Based on data from Lightfield (1971).

FIGURE 8.3 DIAGRAM OF THE EXPECTED RELATIONSHIPS BETWEEN THREE INDEPENDENT VARIABLES AND PEER RECOGNITION OF SOCIOLOGISTS (partial correlation coefficients are entered on the diagram).

controlled. The .47 between quantity and recognition is the partial correlation controlling for department status, etc. From this set of partial correlation coefficients, Lightfield was able to examine the strength of association of publication variables, without the "contamination" of the general linear effects of departmental status. The relationship between recognition and either of the publication variables is positive and relatively strong even when departmental status is taken into account.

These partial correlation coefficients can also be expressed in terms of the proportion of variance in one variable explained by its linear association with the other variable by simply squaring the coefficient. Here, quality contributes about 12% to the explanation of recognition and quantity contributes 22%.

It should be recalled that the correlation coefficient r indicates how well a straight line explains the relationship between two variables. One reason for a low correlation (or low partial correlation) may be that the relationships are not very close to linear, either in general or within categories of test variables.

8.4 STATISTICAL CONSIDERATIONS IN RESEARCH DESIGN

Statistical procedures for controlling on test factors are closely related to more general issues of research design. Courses and texts in research methods deal with these issues, and although we will not pursue these topics here, we would like to reemphasize their importance. Experimental designs approach the problem of handling other relevant variables somewhat differently, often through the use of random assignment of cases to control and experimental groups and through manipulation of an independent variable. Some of the advantages and possibilities of this approach are discussed in a classic book by the sociologist Samuel Stouffer (1962; see General References).

The explanation of a dependent variable usually leads an investigator to consider not only other measured independent variables, but also the representativeness of the dependent variable (avoiding, for example, stratification on a dependent variable), independence of observations, the amount of variation on dependent and independent variables, and a number of other possible explanations of outcomes (such as chance or history, etc.), which are nicely described by Campbell and Stanley (1963).

Since the dependent variable generally is critical to a study, investigators often spend extra effort on its measurement to assure themselves that they are measuring the kind of phenomenon they wish to measure. Sometimes this care takes the form of including several separate measures of the dependent variable so that parallel analyses can be conducted using the different measures. In fact, there are recommended styles of analysis that specifically call for research designed to include different measurements and different methods to get at the same research problem (Campbell and Fiske, 1969).

There is a very close relationship between research design and statistical analysis. This book is designed to introduce you to some of the possibilities and requirements of statistics for sociological research.

8.5   SUMMARY

In this chapter we have been primarily concerned with the problem of simultaneously handling more than two variables. Our general concern is with conditions under which a relationship between an independent and dependent variable holds or varies. The problem, ideas, or theory of an investigator become particularly important in selecting relationships to examine and control variables to introduce into an analysis.

Three general approaches to the analysis of relationships between three or more variables were introduced. The first approach was *elaboration* of a relationship of interest (summarized in Section 8.2.6). This involved creation of conditional tables showing the relationship of key independent and dependent variables within categories of control variables. Lazarsfeld's accounting formula and hierarchial analysis helped summarize possible patterns of elaboration involving both the idea of statistical interaction or differences in strength and direction of association among conditional tables, and the idea of a general decrease, constancy, or increase in the strength of association in conditional tables.

When an investigator is interested in the general effect of introducing a control or test variable upon an original relationship, two other approaches may prove especially useful. The first of these is standardization. Test factor standardization, for example, is a procedure for creating a single table showing the relationship between an independent and dependent variable with the effects of control variables statistically removed. This permits a direct comparison of the original total association and a standardized table. Standardization provides for a more compact analysis of the general effect of introducing a test factor, but it does not permit an examination of differences in conditional tables. Likewise, the third approach, partial correlation, permits an investigator to contrast two summary numbers. One number is an original coefficient of association and the other is a kind of average of the associations in conditional tables or, to put it differently, a measure of association between two variables with the effects of control variables statistically removed. Again, as in the case of standardization, partial association permits us to examine what happens in general when control variables are introduced into the analysis of a relationship.

The next chapter directs attention toward another way to put together information from a series of independent variables in order to improve predictions of a dependent variable and to examine theoretically proposed models more fully. Our movement to this point has meant a selective dropping of detail to achieve more comprehensive and pertinent descriptions of interesting segments of the social world. The pursuits described in the next chapter follow this general progression as well.

CONCEPTS TO KNOW AND UNDERSTAND

conditional tables
control variables, test factors
zero-, first-, second-order associations
role of theory
theoretical ordering of variables
antecedent variables
intervening variables
consequent variables
hierarchical analysis
odds ratio
Lazarsfeld accounting formula
specification
causal model
testing for spuriousness

causal sequence
suppressor variables
independent causation
statistical interaction
social system analysis
structural analysis
contextual analysis
standardization
standardizing rates
standard population
standardized tables
partial correlation coefficients
partial association coefficients

QUESTIONS AND PROBLEMS

1. Using data from Table 8.10, create a standardized table and compare it with the original association of occupation and skin color. Discuss the advantages and disadvantages of using a test factor standardization rather than a partial association measure to test the investigator's ideas. It may help to look up the article.

2. Select a theoretical model that has been drawn out as an arrow diagram (such as that in the middle of Section 8.1.3). Explain how one might use conditional tables, standardization, or partial correlation coefficients to test the model. What relationships would the model lead us to expect of data?

3. Does the high school environment have an effect on student performance? Select some "environment" variable, such as social class of the high school, an individual characteristic such as I.Q., and student performance, and then draw out a graph such as those shown in Figure 8.2 to illustrate (a) only an individual effect on performance, (b) only a group effect on performance, and (c) both group and individual effects. Carefully label each graph and describe what the

graphs show. It might help to read further about this type of analysis in Riley (1963; see General References) and the example of contextual analysis discussed by McDill *et al.* (1967).

4. The data from Tables 8.6 and 8.7 are presented below in frequency form. Analyze these data by using the techniques of partial association discussed in this chapter. Do the results agree with those reported in Section 8.2.5b? Now compute the relevant odds ratios for these data. Do these results agree with those reported in Section 8.2.5b? How do they compare with the analysis you did using techniques of partial association?

FREQUENCY TABLES FOR EMPLOYEES FEELING THEIR DEPARTMENT IS A HIGHLY SUPPORTIVE WORK ATMOSPHERE BY INDIVIDUAL AGE AND ORGANIZATIONAL AGE

Table A

| Feelings about Support of Work Contexts | Individual Age | | | |
	Young	Middle	Older	Total
Highly Supportive	27	41	45	113
Not Highly Supp.	36	49	37	122
Total	63	90	82	235

Table B

Younger Departments

| Feelings about Support of Work Contexts | Individual Age | | | |
	Young	Middle	Older	Total
Highly Supportive	18	18	9	45
Not High Supp.	17	19	5	41
Total	35	37	14	86

Middle-age Departments

| Feelings about Support of Work Contexts | Individual Age | | | |
	Young	Middle	Older	Total
Highly Supportive	7	17	17	41
Not Highly Supp.	16	20	12	48
Total	23	37	29	89

Older Departments

| Feelings about Support of Work Contexts | Individual Age | | | |
	Young	Middle	Older	Total
Highly Supportive	2	6	19	27
Not Highly Supp.	3	10	20	33
Total	5	16	39	60

5. The following tables are from a study of retired English men. They compare the type of work engaged in before retirement (blue-collar versus white-collar) with a measure of adjustment to retirement for different levels of self-perceived health status. Analyze these data using the techniques discussed in this chapter. Discuss the results of your analysis and explain why you used the specific techniques that you did.

Health Status: Poor

Level of Adjustment to Retirement	Type of Work	
	Blue-Collar	White-Collar
Poor	8	2
Fair	9	2
Good	6	6

Health Status: Fair

Level of Adjustment to Retirement	Type of Work	
	Blue-Collar	White-Collar
Poor	6	2
Fair	13	3
Good	28	8

Health Status: Good

Level of Adjustment to Retirement	Type of Work	
	Blue-Collar	White-Collar
Poor	9	1
Fair	8	3
Good	54	59

Total Table

Level of Adjustment to Retirement	Type of Work	
	Blue-Collar	White-Collar
Poor	23	5
Fair	30	8
Good	88	73

6. Use the data from Messer's study (Table 8.2) to compute odds ratios that illustrate the analysis of the impact of age environment.

GENERAL REFERENCES

Blalock, Hubert M., Jr., *An Introduction to Social Research* (Englewood Cliffs, N.J., Prentice Hall), 1970.
 Chapter 4 on "Explanation and Theory" is especially relevant to this chapter.
Davis, James A., *Elementary Survey Analysis* (Englewood Cliffs, N.J., Prentice Hall), 1971.
———, Joe L. Spaeth, and Carolyn Huson, "A Technique for Analyzing the Effects of Group Composition," *American Sociological Review*, 26 (April 1961), pp. 215–225.
Hyman, Herbert, *Survey Design and Analysis* (Glencoe, Ill., The Free Press), 1955.
Riley, Matilda White, *Sociological Research: A Case Approach*, vol. 1 (New York, Harcourt Brace and World), 1963.
 See especially the commentary after Unit 12, pages 700–738, which discusses social system analysis and structural and contextual analysis.
Rosenberg, Morris, *The Logic of Survey Analysis* (New York, Basic Books), 1968.
Stinchcombe, Arthur L., *Construction of Social Theories* (New York, Harcourt, Brace and World), 1968.
Stouffer, Samuel A., *Social Research to Test Ideas* (New York, The Free Press of Glencoe), 1962.
 See especially the 15th chapter, "Some Observations on Study Design."
Zeisel, Hans, *Say It with Figures*, Rev. 5th ed., (New York, Harper & Row), 1968, Chapter 10.

LITERATURE CITED

Blalock, Hubert M., Jr., *Theory Construction: From Verbal to Mathematical Formulations* (Englewood Cliffs, N.J.: Prentice Hall), 1969.
Bureau of Labor Statistics, "Using the Consumer Price Index for Escalation," U.S. Department of Labor, Report 761, January, 1989.
Campbell, Donald T., and Donald W. Fiske, "Convergent and Discriminant Validation by the Multitrait-Multimethod Matrix," *Psychological Bulletin*, 56 (March 1969), pp. 81–105.
Campbell, Donald T., and Julian C. Stanley, *Experimental and Quasi-Experimental Designs for Research* (Chicago, Rand McNally), 1963.
Chafetz, Janet Saltzman, *A Primer on the Construction and Testing of Theories in Sociology* (Itasca, Ill.: Peacock), 1978.
Davis, James A., "The Goodman Log Linear System for Assessing Effects in Multivariate Tables" (Chicago, National Opinion Research Center), 1972. (litho).
———, "Hierarchical Models for Significance Tests in Multivariate Contingency Tables: An Exegesis of Goodman's Recent Papers, "in Herbert L. Costner, editor, *Sociological Methodology 1973–1974* (San Francisco, Jossey Bass), 1974, pp. 189–231.
———, "Extending Rosenberg's Technique for Standardizing Percentage Tables," *Social Forces*, 62 (March 1984), pp. 679–708.
Erbe, William, "Social Involvement and Political Activity: A Replication and Elaboration," *American Sociological Review*, 29 (April 1964), pp. 198–215.
Economic Report of the President (Washington, D.C., U.S. Government Printing Office), 1990.
Freese, Lee, and Jane Sell, *Theoretical Methods in Sociology: Seven Essays* (Pittsburgh: University of Pittsburgh Press), 1980.
Goodman, Leo A., "On the Multivariate Analysis of Three Dichotomous Variables," *The American Journal of Sociology*, 71 (November 1965), pp. 290–301.

————, "How to Ransack Social Mobility Tables and Other Kinds of Cross-Classification Tables," *The American Journal of Sociology*, 75 (July 1969), pp. 1–40.

————, "A Modified Multiple Regression Approach to the Analysis of Dichotomous Variables," *American Sociological Review*, 37 (February 1972), pp. 28–46.

Hage, Jerald, *Techniques and Problems of Theory Construction in Sociology* (New York, Wiley), 1972.

Lazarsfeld, Paul F., Bernard Berelson, and Hazel Gaudet, *The People's Choice* (New York, Columbia University Press), 1948.

Lazarsfeld, Paul F., and Morris Rosenberg (editors), *The Language of Social Research* (New York, The Free Press of Glencoe), 1955, p. 231.

Lightfield, E. Timothy, "Output and Recognition of Sociologists," *American Sociologist*, 6 (May 1971), pp. 128–133.

McAllister, Ronald, Edward Kaiser, and Edgar Butler, "Residential Mobility of Blacks and Whites: A National Longitudinal Survey," *American Journal of Sociology*, 77 (November 1971), pp. 445–456.

McDill, Edward, L., Edmund D. Meyers, and Leo C. Rigsby, "Institutional Effects on the Academic Behavior of High School Students," *Sociology of Education*, 40 (Summer 1967), pp. 181–199.

Merton, Robert K., *Social Theory and Social Structure*, enlarged ed. (New York, The Free Press), 1968, Part I.

Messer, Mark, "The Possibility of an Age-Concentrated Environment Becoming a Normative System," *Gerontologist*, 7 (December 1967), pp. 247–251.

1990 President's Economic Report, Table C-60.

Ransford, H. Edward, "Skin Color, Life Chances, and Anti-White Attitudes," *Social Problems*, 18 (Fall 1970), pp. 164–179.

Reiss, Ira L., *The Social Context of Premarital Sexual Permissiveness* (New York, Holt, Rinehart, and Winston), 1967, pp. 59 and 60.

Rosenberg, Morris, "Test Factor Standardization as a Method of Interpretation," *Social Forces*, 41 (October 1962), pp. 53–61.

U.S. Bureau of the Census, "Population Characteristics," *Current Population Reports*, Series P-20, No. 205 (Washington, D.C., U.S. Government Printing Office), 1970.

DIAGRAM 8.1 Some Strategies of Elaboration

1. Be sure to keep a clear focus on which variable is the *dependent* variable.
2. Contrast explanatory models, if possible, using theory to develop alternatives, and guide the selection of control and independent variables. A useful procedure is to draw out expected relationships in the form of an arrow diagram.
3. Test factors used should be those related to the dependent variable. Take account of the ordering of variables in selecting control variables and interpreting outcomes of analysis.
4. If you are expecting or looking for statistical interaction, examine the conditional tables themselves.
5. Generally, start with a zero-order association and work toward more controls, examining first-order associations, then second-order, etc.
6. If you are looking for a general increase or decrease (or no change) in the conditional tables, consider using odds ratios, standardization, or the use of an appropriate measure of partial correlation. (See Section 8.3.)
7. In interpreting outcomes, remember where the data came from, the source of measurements, number of cases, reasonableness of contrasts, quality of the data, whether or not multiple measurements of variables were made, and independence of observations.

DIAGRAM 8.2 Test Factor Standardization

CONDITIONAL TABLES

		T-1		Total			T-2		Total
		X					X		
FREQUENCY		4	1	5			20	60	80
	Y	40	10	50		Y	2	6	8
		44	11	55			22	66	88
PERCENTAGE		9%	9%	9%			91%	91%	91%
		91%	91%	91%			9%	9%	9%
		100%	100%	100%			100%	100%	100%

		Total		
		X		Total
FREQUENCY		24	61	85
	Y	42	16	58
		66	77	143
PERCENTAGE		36%	79%	59%
		64%	21%	41%
		100%	100%	100%

WEIGHTS

   T-1 conditional contains $55/143 = .385$ of the total number of cases.
   T-2 conditional contains $88/143 = .615$ of the total number of cases.
          sum of weights = $\overline{1.000}$

COMPUTATION

T-1	9	9	9	each cell		3.5	3.5	3.5	
	91	91	91	multiplied →		35.0	35.0	35.0	
	100	100	100	by .385		38.5	38.5	38.5	summed, cell by
	(44)	(11)	(55)						cell, to form the
T-2	91	91	91	each cell		56.0	56.0	56.0	standardized
	9	9	9	multiplied →		5.5	5.5	5.5	table below
	100	100	100	by .615		61.5	61.5	61.5	
	(22)	(66)	(88)						

	Standardized Percentage Table				Standardized Frequency Table		
						X	
	59.5	59.5	59.5		39	46	85
	40.5	40.5	40.5	Y	27	31	58
	100%	100%	100%		66	77	143
	(66)	(77)	(143)				

Source: Data from Table 8.3.

DIAGRAM 8.2 *(Continued)*

SUMMARY OF STEPS IN CREATING A STANDARDIZED TABLE

1. Create conditional tables, controlling on desired test factors. The table may be of any size ($r \times c$).*
2. Properly percentage all conditional tables.
3. Select a weight to apply to all cells of a given conditional table. These weights might be the proportion of cases that are in a given conditional table, as in the example above. Weights should sum to 1.00 over all tables.
4. Multiply each cell percentage by the weighting factor for each conditional table.
5. Sum these weighted percentages across all conditional tables, cell by cell. The resulting table is the standardized percentage table.
6. Using the marginal frequencies from the total table, convert the standardized table percentages back into frequency form.
7. Compute measures of association on the standardized frequency table for comparison with measures computed on the total table.
8. Compare total and standardized tables (and association measures) to assess the effect of the test factor(s).

* Note that the conditional tables must have some non-0 frequency in each category of the independent variable (*i.e.*, the base of percentages) so that percentages in each category of the independent variable in the standardized table will total to 100.

CHAPTER 9

Path Analysis and Multiple Regression

There are several analysis techniques used in sociology that fit under the general label of multiple regression analysis. They are appropriate where an investigator is interested in predicting or analyzing scores on one dependent variable by combining the predictive power of several independent variables by means of an equation that is called a **multiple regression equation**. How well this equation is able to predict scores on the dependent variable is indicated by the **multiple correlation coefficient**, R. In many ways this procedure is simply an extension of correlation and regression procedures we have discussed where only one independent variable is used to predict one dependent variable.

This chapter is an introduction to some of the multiple regression techniques you are likely to encounter in reading reports of sociological research. Topics we will discuss include multiple correlation and multiple regression, path analysis, the use of dummy variables, and stepwise multiple regression. These are all quite easily understood at an interpretative-theoretical level, even though various computational alternatives and their derivation are details we will reserve for a separate course in regression analysis.*

---

* Clerical mathematics needed to compute some of the coefficients will generally be left to existing computer programs, as done by most investigators. Your instructor will explain how to use your computer and the programs that are available for multiple regression. These are generally quite easy to use in problems you may be assigned.

It is well to use some caution in accepting the output of "canned" computer programs that are not known to have been checked out by someone with sufficient statistical and mathematical insight to detect possible errors. Some checking procedures are provided in this chapter, but our focus will be on the theory necessary for an interpretative understanding of these techniques. With an adequate computer program you will be able to make use of the procedures yourself. An interesting commentary on the accuracy of some existing multiple regression programs is provided in Wampler (1970).

It is quite worthwhile and interesting to work through a more mathematical treatment of these topics in other books and courses. This would be especially helpful in interpreting results of more unusual applications of these techniques. See, for example, Draper and Smith (1966; General References).

9.1   MULTIPLE REGRESSION AND MULTIPLE CORRELATION

9.1.1   Some Assumptions of Multiple Regression

Multiple regression techniques are among the more interesting and useful because they help handle the kind of complexity that begins to reflect theoretical notions sociologists have about the social world they are trying to explain. Even so, a number of simplifications are involved in multiple regression analysis, and these may mean that the regression model as described here simply is not appropriate or useful for many research problems.*

One basic assumption of multiple regression analysis is that variables are related to the dependent variable in a simple *linear* fashion, and it is usually a good procedure to construct scatter diagrams to check this assumption. Sometimes simple transformations (such as logarithmic transformation) can be used to uncover a linear relationship. Another assumption is that effects of variables can be *added* together to form a prediction of the dependent variable. "Statistical interaction" cannot be handled unless it is "coded" and included as a separate variable in the regression equation. Furthermore, all the variables included in the multiple regression equation should be interpretable as *interval* measures. Finally, there is an assumption that an independent variable is not a function of another independent variable (*i.e.*, collinear). Usually, it is preferred that independent variables not be highly correlated, so that the effects of each variable on the dependent variable, with others taken into account, can be reliably computed.

BOX 9.1   REVIEWING CORRELATION AND THE NATURE OF ASSOCIATION

At this point you may want to check back on the idea of *correlation* (Chapter 7) and particularly the *nature* of association and possibilities for linear and curvilinear association (Section 7.4.4). Sometimes some function of the scores (such as logarithms), rather than the scores themselves, will have a simple linear relationship with another variable and thus satisfy a presumption of the Pearson correlation coefficient and linear regression analysis. The transformation of scores to produce this effect is something that requires experience and usually involves double-checking through the use of scatter diagrams both before and after transformations.

---

* There are a number of other multivariate procedures not discussed here that may be of interest to you. Here, multiple regression uses one interval dependent variable with several interval independent variables. *Canonical correlation* uses several interval dependent and several interval independent variables, finding weights among each such that the correlation between the two sets is maximized. *Discriminant function analysis* uses interval independent variables to create one or more functions of them to predict the (nominal) category of a dependent variable. *Factor analysis* seeks one or more linear combinations of interval variables that, together, explain as much of the covariation among a set of variables as possible. *Logit* and *Probit analyses* are multiple-regression procedures designed to predict dichotomous dependent variables. *LISREL* is a procedure used to test more complex models, including those where several indicators or scale items measure one or more underlying properties. Finally, there are test factor standardization techniques designed to examine multivariate models of nominal/ordinal variables.

Basic assumptions:

1. Independent variables are related in a linear fashion with the dependent variable and among themselves.
2. Effects of independent variables can be added together to yield a prediction of the dependent variable.
3. Independent variables are not highly correlated.*
4. All variables are interval variables.

Added assumptions if one is interested in running statistical tests of hypotheses about a population from random sample data:

5. The dependent variable is normally distributed within categories of independent variables, singly and in combination.
6. The variance in the dependent variable is equal across categories of the independent variables.

Added assumptions if multiple regression analysis is to be applied to testing causal models of the relationship between variables:

7. The theoretical ordering of independent and dependent variables should be known and be such that independent variables change first and dependent variables later.
8. The set of independent and dependent variables should be inclusive of all (major) variables influencing the dependent variables. That is, it should be a closed system.
9. Measures should have high (demonstrated) reliability and validity.
10. Disturbance terms (error terms) should be uncorrelated with each other or with independent variables directly connected to the same variable.

---

* For a discussion of correlated independent variables, see Berry and Feldman (1985) or Blalock (1972) (see General References).

FIGURE 9.1 Assumption of Multiple Regression Analysis

Some other assumptions are involved in specific applications of multiple regression (see Figure 9.1) and these will be discussed when introduced later in this chapter.*

9.1.2   One Independent Variable: A Review

In Chapter 7, "simple" (one independent and one dependent variable) linear regression equations were expressed in the following form where the *b*-value represents the slope or amount of change in the dependent variable for each unit

---

* Assumptions should be taken seriously, and consequences of departures from one assumption upon the importance of other assumptions is the subject of debate and inquiry in the field. The interval measurement assumption, for example, can be handled in some cases for even nominal variables by using what are called "dummy variables." The "not perfect-collinearity among independent variables" assumption is usually interpreted to mean "low intercorrelation" (low multicollinearity). Sometimes highly related variables are combined into a single scale. It may also be the case that the theoretical idea of variables related in a simple, linear fashion is intrinsically interesting as a model (or first model) of some aspect of the social world. In this case, the results of a regression analysis would be interesting, even though all the regression assumptions may not fit the world particularly well.

change in the independent variable and a_{yx} represents the Y-intercept where the regression line crosses the Y-axis.

(9.1)
$$Y' = a_{yx} + b_{yx}X$$

If the scores for both the Y and X variables were expressed not as raw measures but as standard scores (z-scores), the regression equation could be expressed as follows.* Note that we are changing the notation slightly so that all variables including the dependent variable are referred to by the same letter with a different subscript.[†]

(9.2)
$$z'_1 = b^*_1 + b^*_{12}z_2$$

The b^*_1 (b-star or beta-weight) coefficient in (9.2) corresponds to a_{yx} in Formula (9.1) and in fact the two equations are identical, term for term, except for the notation and the use of z-scores rather than raw Y and X scores in Formula 9.2.

Since the mean of a distribution of z-scores is 0, and since the regression "line" passes through $\bar{X}_1$, $\bar{X}_2$, then the value of b^*_1 will always be 0, and the regression equation in Formula 9.2 can be simplified to

(9.3)
$$z'_1 = b^*_{12}z_2$$

The correlation coefficient, r, can be seen as a correlation between the score on the dependent variable *predicted* by using the regression equation and the *actual* score of the dependent variable. The closer the prediction comes to the actual score, the higher the degree of linear association and the closer r comes to $+1.00$ or -1.00.

The **coefficient of determination** (the square of the correlation coefficient), r^2_{12}, indicates the proportion of variation in one variable (for example the dependent variable) that is explained by its linear association with the other variable expressed in the right side of the linear regression equation. Unexplained variation is expressed as $1 - r^2_{12}$ and is referred to as the *coefficient of nondetermination*.

---

* Recall that z-scores are simply the difference between a score and the mean of its distribution expressed in terms of the number of standard deviation units it is from the mean. The formula for z-scores is $z = (X_i - \bar{X})/s$. It is discussed in Section 5.4.7.

[†] *Note on symbolism:* Dependent variables are referred to in a number of different ways, for example, as Y or X_0 or X_1, etc. To indicate that this is a predicted value, some texts use a prime (X') and some use a "hat" $(\hat{X})$. We will use subscripts starting at 1, generally, to refer to variables and explain in context whether each variable is to be treated as an independent or dependent variable. A predicted dependent variable will be indicated by a prime, X'_1. Weights in the regression equation will be written in two ways, b for weights in an equation that uses raw scores, and b^* (b-star) for weights in an equation where standard scores (z-scores) are used. Appropriate subscripts will also be added: the first indicating the dependent variable; the second indicating the independent variable (and other variables in the equation will be listed following a dot in the subscript). Some texts also introduce Greek letters for the a and b coefficients where the population value (rather than sample value) is used. To simplify symbolism at this point, sample notation is used. Later (in inferential statistics), where the distinction between population and sample values is of concern, further symbolism may be introduced and explained.

9.1.3 Two or More Independent Variables

When more than one independent variable is used to explain variation in a single dependent variable, the basic regression equation in Formula 9.3 can be extended as in Formula 9.4 to include terms that combine a beta-weight (b^*) coefficient and the z-score for each of k independent variables. Thus we have the **multiple regression equation**:

$$(9.4) \qquad z'_1 = b^*_{12 \cdot 34 \ldots k} z_2 + b^*_{13 \cdot 24 \ldots k} z_3 + b^*_{14 \cdot 23 \ldots k} z_4 + b^*_{1k \cdot 234 \ldots} z_k$$

Here the standard scores for each independent variable are weighted according to the contribution that variable makes to the overall predicted sum, z'_1. Each b^* value represents the relative amount of contribution of that variable, after contributions of the other variables included in the regression equation are taken into account. In that sense, the b^* values are like a partial correlation coefficient in "holding constant" or "correcting" for the contribution of other included variables. Often this controlling effect is indicated in the subscripts for the b^* coefficients, as illustrated in the three-variable regression equation below. The dependent and independent variables to which the b^* value refers are written in front of the dot in the subscript, and the other variables included in the regression equation are listed after the dot. Those after the dot are "held constant" (*i.e.*, their contribution is partialled out). As usual, the first subscript is the dependent variable and the second is the independent variable for that particular b^* coefficient. Using two independent variables, Formula 9.4 becomes this:

$$(9.5) \qquad z'_1 = b^*_{12 \cdot 3} z_2 + b^*_{13 \cdot 2} z_3$$

Beta-weights (b^*) are computed in a way that minimizes the sum of squared deviations between predicted and actual scores on the dependent variable. This is called the **least-squares criterion**.

$$(9.6) \qquad \Sigma(z_1 - z'_1)^2 = \text{a minimum}$$

Using only two independent variables and substituting the right side of Formula 9.5 for z'_1, the quantity to be minimized is

$$(9.7) \qquad \Sigma(z_1 - b^*_{12 \cdot 3} z_2 - b^*_{13 \cdot 2} z_3)^2 = \text{a minimum}$$

This results in a maximum linear correlation between predicted and actual scores on the dependent variable.

To carry out the computations implied by the least-squares criterion, a set of normal equations (*not* related to the idea of a normal curve in any way), predicting correlations between each variable and the dependent variable, is created and solved for unknown b^*s. For the three-variable problem, these normal equations would be as follows:

$$(9.8) \qquad \begin{aligned} b^*_{12 \cdot 3} + r_{23} b^*_{13 \cdot 2} &= r_{12} \\ r_{23} b^*_{12 \cdot 3} + \phantom{r_{23}} b^*_{13 \cdot 2} &= r_{13} \end{aligned}$$

Note that solution for the b^*'s is possible because there are as many equations as unknowns. The intercorrelation of variables can be computed directly from the data in the study being used in the multiple-regression analysis. Thus, in a three-variable regression equation with two independent variables, the two normal equations in Formula 9.8 could be solved for the two betas, $b^*_{12 \cdot 3}$ and $b^*_{13 \cdot 2}$.

Usually, b^* coefficients are computed by means of computer programs for multiple regression. The beta coefficients can also be computed from a correlation matrix of the interrelationship between all pairs of variables to be included in the regression equation, but this procedure becomes tedious for more than one or two independent variables.

The b^* coefficients provide an investigator with a basis for comparing the relative contribution of one variable to a prediction of the dependent variable with the contribution of other variables in the equation. If $b^*_{12 \cdot 3}$ were larger than $b^*_{13 \cdot 2}$, then one would be able to conclude that a given amount of change in z_2 would result in more change in the dependent variable than the same amount of change in z_3. Thus, z_2 is a more potent influence than z_3. If $b^*_{12 \cdot 3}$ is twice the size of $b^*_{13 \cdot 2}$, then a given change in z_2 has twice the effect of the same change in z_3. Variables with b^* coefficients that are very nearly 0 have very little separate influence on the dependent variable, at least in the linear, additive way described by the multiple-regression equation. We might, on this basis, argue that such variables could well be eliminated from the prediction equation, and the beta coefficients could be recomputed.

At this point we are only talking about the way changes in some variables are related to changes in other variables, and not about influence in a "cause–effect" sense, although clearly, given other kinds of information, such as the theoretical ordering of the influence of variables, we might use evidence from a multiple-regression equation in evaluating a theoretical cause–effect argument.

9.1.4   Regression Weights and Partial Correlation Coefficients

The standardized beta weight (b^*) and the partial correlation coefficient share some characteristics. They both reflect the effect of an independent variable on a dependent variable when the effects of other included independent variables are taken into account, statistically. The dependent variable scores to be predicted are adjusted by subtracting out predictions that would be made by linear regression equations that omit a given independent variable. Thus, the b^* for a given independent variable is computed so that it best predicts (in the least-squares sense) the adjusted scores for the dependent variable. The partial correlation coefficient expresses the relationship between such predicted scores and the adjusted scores on the dependent variable.

The two coefficients provide different information, since the b^* indicates the *amount of change in the dependent* variable that is associated with a unit change on the independent variable (when other independent variables are taken into account.) Therefore, it is an asymmetric measure. The partial correlation coefficient is a symmetric measure that indicates the closeness of relationship, overall, between a dependent and independent variable when scores of the dependent variable have been adjusted to take out the effect of variation in other independent variables implied by their linear relationship with the dependent variable. The partial correlation coefficient provides a measure of the accuracy of prediction, and the beta coefficient provides a measure of the contribution of a variable to the prediction. By squaring the partial correlation coefficient, one can measure the proportion of the variation in the dependent variable that is explained by the direct contribution of an independent variable when effects of other included variables are taken into account. Both coefficients provide useful but different information.

9.1.5 Multiple Correlation

Multiple correlation $(R_{1 \cdot 23})$, like the simple product–moment correlation coefficient r_{yx}, is simply the correlation between the actual scores on the dependent variable and the scores on the dependent variable predicted by use of the multiple-regression equation. The beta weights (b^*), in fact, are computed so that this correlation is as high as possible for a given set of data. The multiple-correlation coefficient is symbolized by the capital letter R, and it varies on a scale from 0 to $+1.00$. The smaller the coefficient the weaker the correlation is, and the larger the coefficient, the stronger the correlation.

Like the coefficient r, the multiple-correlation coefficient can be interpreted more usefully by squaring it. R^2 has an interpretation quite akin to r^2: the proportion of the variation in the dependent variable that is explained by the regression equation.

$$R^2 = \frac{\text{Explained variation in } X_1}{\text{Total variation in } X_1}$$

(9.9)
$$R^2 = \frac{\Sigma(X_1' - \bar{X}_1)^2}{\Sigma(X_1 - \bar{X}_1)^2}$$

R^2 is called the coefficient of multiple determination; $1 - R^2$ is the proportion of variation in the dependent variable left unexplained by the multiple regression equation.

If all the intercorrelations of independent variables were 0, then the square of the multiple-correlation coefficient would simply be the sum of squared correlations between each independent variable and the dependent variable as in Formula 9.10.

(9.10)
$$R_{1 \cdot 23}^2 = r_{12}^2 + r_{13}^2$$

If, as is usually the case, some independent variables are related to each other, the overlap in contribution to the explanation of the dependent variable would have to be taken into account and eliminated from Formula 9.10 in order to arrive at R^2. This is done simply by adjusting each r by multiplying it times the related beta weight, b^*, as in Formula 9.11.

(9.11)
$$R_{1 \cdot 23}^2 = r_{12}b_{12 \cdot 3}^* + r_{13}b_{13 \cdot 2}^*$$

Both Formulas 9.10 and 9.11 could be extended to include more than two independent variables by simply adding additional terms.

It is apparent from Formula 9.10 that R will be 0 if all the correlations between dependent and independent variables are 0, and R cannot be less than the highest r relating any independent and dependent variable. R exceeds the highest r relating any independent and dependent variable by the largest amount when independent variables are independent of each other (*i.e.*, have 0 intercorrelations), since each variable contributes an added, separate amount to the prediction of the dependent variable.*

9.1.5a   Corrected R

The multiple-correlation coefficient computed from a sample tends to overestimate the coefficient for the population (the parameter). The extent to which the sample coefficient will be inflated depends upon the size of the sample and the number of independent variables in the regression equation. This bias in the direction of a higher R for sample data is due to the fact that the multiple-regression equation is tailored to the sample data to produce the highest R possible. This operation takes advantage of any chance difference between the sample and population distribution of scores to find a higher R by tailoring the b^*'s to the data. Since a small sample is likely to have a larger sampling error than a larger sample, an R computed from a small sample is more likely to deviate from the parameter it is meant to estimate. Furthermore, as the number of independent variables in the regression equation is increased, the job of tailoring the R to the sample data will be more efficient, thus exaggerating the bias. This bias can be corrected by using the following formula:

(9.12)
$$R_c^2 = 1 - \left(\frac{N-1}{N-k}\right)(1 - R^2)$$

where R^2 is the uncorrected coefficient, R_c^2 is the corrected coefficient, N is the sample size, and k is the number of independent variables included in the multiple-regression equation.

Generally, investigators are interested in finding sets of independent variables that will provide the best prediction of the dependent variable, the highest value of R. If they preselect independent variables by looking for variables that are most highly correlated with the dependent variable in their sample data and then introduce these into a multiple regression equation, the inflation of R is even greater than that eliminated by the correction in Formula 9.12. An R recomputed on a larger sample or on the population would be less than that found by preselecting independent variables. For this reason, investigators replicate their study on other new data and use relatively large samples as the basis of computing multiple regression coefficients and R.

Notice in Formula 9.12 that, when the sample size, N, and the number of independent variables, k, are equal, R^2 and R will always be equal to 1. This is

* Your attention should be called to another variation on the multiple-correlation coefficient. It is a *multiple partial correlation coefficient*, and it indicates the multiple correlation between a set of independent variables and a dependent variable when other independent variables are statistically controlled. Thus it is useful in examining the explanatory power of one set of independent variables on a dependent variable, controlling for effects of another set of independent variables. The multiple partial coefficient is introduced by Blalock (1979: 488; see General References).

because b^*'s would be calculated to perfectly fit each individual score, and there would be no difference between predicted and observed scores. The numerator in Formula 9.9 would be equal to the denominator simply because of this, and R^2 would equal 1. To compute the multiple-regression and multiple-correlation coefficients, then, the number of cases must exceed the number of variables (a rule of thumb is that the number of cases should be at least 10 times the number of independent variables).

9.1.6   An Illustration

Koslin and a group of investigators at Princeton University studied a group of 29 11- to 13-year-old boys at a boys' camp in Canada (Koslin *et al.*, 1968). In this study they developed a regression equation to predict the standing of a boy in a group, using several measures of the expectations group members had for each other's performance on several tasks. They argue that individuals within groups develop ways of evaluating the performance of their associates and that expected performance is related to status in the group. Higher-status individuals are expected to perform better than lower-status individuals on tasks of central concern to the group, regardless of actual performance. If this is the case, then one should be able to predict status within a group by knowing whether an individual's performance is overrated or underrated when compared with his actual performance.

Four different tests were devised to measure over/underrating: (a) a rifle task, (b) a canoe task, (c) a sociometric preference test, and (d) a height perception test. The rifle test, for example, involved having each boy in turn shoot four shots at a target while other group members watched. The target disappeared immediately after a shot was fired and the boy who shot as well as others in the group was asked to record how close to the bull's-eye the shot hit. The score on this task for each group member was the average over- or underestimate of his shooting accuracy (compared with where the actual shot hit) as made by group members who watched him shoot. The canoe task involved group estimates of the time a boy took to go a given distance. The sociometric questionnaire asked for teammate preferences, and the height test involved comparisons of an individuals' height in relation to other individuals in the group. In each case, the test was conducted in an ambiguous situation (*i.e.*, participant members had no watches to measure time, people were too far apart to allow actual comparisons of height, etc.). The dependent variable, group status, was determined by the investigators through careful observation of the boys over the course of the experiment. Data from the study are given in Table 9.1.

The regression equation computed by Koslin using standardized beta weights was as follows:

(9.13)
$$z_5' = .37z_1 + .37z_2 + .25z_3 - .07z_4$$

where z_1 is the rifle test, z_2 is the canoe task, z_3 is the sociometric test, z_4 is the height guessing test, and z_5 is the investigator's rating of individual social status in the group on the basis of careful observation data.

TABLE 9.1   CORRELATION MATRIX FOR THE PREDICTOR AND CRITERION VARIABLES

	(2) Canoe Test	(3) Sociometric Test	(4) Height Guess Test	(5) Observed Social Status
1. Rifle test	.52	.58	.48	.67
2. Canoe test		.61	.65	.67
3. Sociometric			.66	.64
4. Height guessing				.51

Source: Data from Koslin et al. (1968).

It is immediately evident that the rifle and canoe test are equally good predictors of observed status; the sociometric test is somewhat poorer but still a positive contributor to the prediction of observational status independently of the other three variables. Only height-guessing makes a very minor contribution to the prediction, and that contribution is a negative one. Thus their expectations were supported; over/underratings on relevant tasks could be used to predict observed status and over/underratings on irrelevant tasks contributed little in addition. Notice that this conclusion is not immediately apparent in Table 9.1, where zero-order correlation coefficients are all relatively high. In fact, from the b^* coefficients, one could say that a unit change in the rifle and canoe tests results in the same amount of change in observed status; but the sociometric test contributes only two-thirds as much as either of these variables ($.25/.37 = .676$). The impact of height is only 19% of that of the rifle or the canoe tests and only 28% of the impact of sociometric choice.

Partial-correlation coefficients provide somewhat different but consistent information. The partial-correlation coefficient .42 in Table 9.2 is the correlation between the rifle test scores and observed status when effects of the canoe test, sociometric choice, and height-guessing were taken into account. The square of these coefficients provides a way of judging the proportion of the variation in the dependent variable accounted for by a given independent variable after other

TABLE 9.2   MULTIPLE CORRELATION USING OBSERVED SOCIAL STATUS AS THE CRITERION (DEPENDENT) VARIABLE

Predictor	b^*	Partial r
1. Rifle test	.37	.42
2. Canoe test	.37	.39
3. Sociometric	.25	.26
4. Height guess	−.07	−.08
Multiple R = .79		

Source: Koslin et al. (1968). Used by permission.

effects are taken into account. Thus the rifle test explained 18%, the canoe test 15%, the sociometric test 7%, and the height guessing test about 0.5%.

Overall, the multiple-regression equation yielded a multiple correlation between predicted and observed social status scores of .79. Squaring the coefficient, it becomes evident that 62% of the variation in the dependent variable is accounted for by the multiple-regression equation and 38% (the difference between 62% and 100%) is left unexplained. Since there were four independent variables and a sample size of only 29, a better estimate of a population multiple correlation coefficient would be $R = .76$, a corrected value computed from Formula 9.12.

9.2 PATH ANALYSIS: AN APPLICATION OF MULTIPLE REGRESSION TO PROBLEMS OF THEORY

9.2.1 Path Analysis

The use of standardized multiple-regression equations in examining theoretical models is called **path analysis**. The objective is to compare a model of the direct and indirect relationships that are presumed to hold between several variables and observed data in a study in order to examine the fit of the model to the data. If the fit is close, the model is retained and used or further tested. If the fit is not close, a new model may be devised, or, more likely, the old one will be modified to fit the data better and then be subjected to further tests on new data.

This is precisely the interest we have if we want to build theoretical explanations of social phenomena. Where the underlying assumptions of path analysis are reasonably met, it provides a very pertinent way to relate theory and data when many variables are to be handled simultaneously. Correlation analysis alone, although helpful, does not provide as useful a measure of the impact of variables directly and indirectly on others.

In addition to the basic assumptions of multiple regression, path analysis includes two others (see Figure 9.1). First, the variables included in a model must be known to fall in some specific theoretical order in their effect on individuals; the independent variables must fall in causal order *before* the dependent variable. This is important and not always obvious.* Second, the model must be treated as a closed system in the sense that all important variables are explicitly included in the model. To some extent one can assess the extent to which this is true in the process of conducting the path analysis itself. Finally, for the purposes of this presentation, it is assumed that the influence of one variable on another is a one-way influence, or, in other words, the model is *recursive*; there is no feedback.

Path analysis models are generally illustrated, as in Figure 9.2, by means of one-headed arrows connecting some or all of the variables included in the model.

---

* At this point you may wish to refer back to discussions in Chapter 8 on the ordering of variables and the use of diagrams to make explicit the relationships between variables implied by a theory.

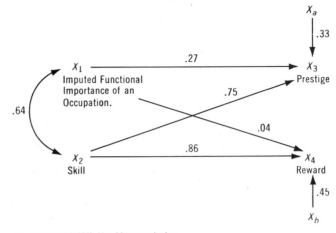

Source: Land (1970). Used by permission.

FIGURE 9.2   Davis–Moore Model of Social Stratification

Variables are distributed from left to right, depending upon their theoretical ordering, with the first independent variables at the extreme left. Intercorrelations (zero order) between variables not influenced by other variables in the model are given as numbers on curved, two-headed arrows. These are correlations between *exogenous* variables, those influenced only by variables prior to and outside the model. Exogenous variables provide a link between a model under consideration and outside variables.

Numbers entered on the direct paths reflect the amount of direct contribution of a given variable on another variable when effects of other related variables are taken into account. A path coefficient is symbolized by the letter p, with subscripts indicating the two variables it connects. It turns out that path coefficients are identical to the b^* coefficients in the standardized multiple-regression equation discussed earlier, where the regression equation reflects the structure of the model one is testing. These are computed as described earlier, from scores ultimately measured on a group of subjects. Thus the path model and the path coefficients provide a picture of the bit of the social world an investigator is interested in explaining, together with the coefficients describing the impact of independent variables. The impact is in terms of the amount of change in the dependent variable that is associated with a unit change in a given independent variable, over and above the contribution of other variables to the dependent variable.

The basic theorem of path analysis and the way by which a specific structural model is reflected in the computation of **path coefficients** is as follows:

(9.14)
$$r_{ij} = \sum_k (p_{ik} r_{kj})$$

where k includes each of the variables connected directly to the dependent variable i and prior to i in theoretical ordering shown in a path diagram. Let us develop these ideas in terms of a particular model and some actual data.

9.2.2 An Example of Path Analysis in Sociology

The Davis–Moore theory of social stratification argues that some type of social stratification will be found in every society because some types of duties must be performed for the society to continue as a functioning unit, and by some means individuals must be motivated to accept and perform these duties (see Land, 1970). The stratification system is one means of assuring the placement and motivation of individuals in the social structure. Rewards used in the process include income, leisure, and prestige. These are distributed among occupations, for example, in terms of (a) the importance of that occupation in the society and (b) the scarcity of appropriately trained individuals needed for the occupation. Figure 9.2 describes this model, except that income and leisure are combined as "other rewards," distinguished from "prestige." Note that two other "variables" (X_a and X_b) are included to reflect variables external to the model and measurement errors that may influence the dependent variables.

Using X rather than z to stand for the *standard score* of a given variable (since this is the convention here), one could express the graphic model in terms of a path model as follows:

(9.15a)
$$X_3 = p_{31}X_1 + p_{32}X_2 + p_{3a}X_a$$

(9.15b)
$$X_4 = p_{41}X_1 + p_{42}X_2 + p_{4b}X_b$$

This is a fully recursive model, since all the possible one-way arrows are drawn between the four explicit variables included in the model. These structural equations correspond to the multiple-regression equation in Formula 9.5, except for the addition of the residual variables X_a and X_b.* These equations can be rewritten in terms of path coefficients and zero-order correlation coefficients in a fashion parallel to that in Formula 9.8, remembering that path coefficients are standardized b*'s. In fact, in this example, one could use a computer program for a multiple-regression equation with standardized data using X_1 and X_2 to predict X_3, and X_1 and X_2 to predict X_4.[†] From the information provided, all the path coefficients would be readily available.

Using path coefficients and correlation coefficients, one could write a series of equations that would predict the correlation of independent and dependent variables. The basic theorem of path analysis, Formula 9.14, provides the model for constructing these path-estimating equations.

(9.16a)
$$r_{31} = p_{31}r_{11} + p_{32}r_{21} \quad \text{or (since } r_{11} = 1.0) \quad p_{31} + p_{32}r_{21}$$

(9.16b)
$$r_{32} = p_{31}r_{12} + p_{32}r_{22} \quad \text{or (since } r_{22} = 1.0) \quad p_{31}r_{12} + p_{32}$$

(9.16c)
$$r_{41} = p_{41}r_{11} + p_{42}r_{21} \quad \text{or (since } r_{11} = 1.0) \quad p_{41} + p_{42}r_{21}$$

(9.16d)
$$r_{42} = p_{41}r_{12} + p_{42}r_{22} \quad \text{or (since } r_{22} = 1.0) \quad p_{41}r_{12} + p_{42}$$

---

*· The X_a and X_b terms are called residual variables or, sometimes, the "random shock" or "error" terms.
† A helpful discussion of path analysis and a program for computing path coefficients is provided by Nygreen (1971; see General References).

Taking Formula 9.16a as an example and referring to the path diagram in Figure 9.2, notice that the correlation between variables 3 and 1 can be written in terms of the path between variables 3 and 1, *times* the correlation of variable 1 with itself (which is taken as 1.00, and so r_{11} and r_{22} could be dropped from these equations), *plus* the path from 3 to 2 times the correlation of variables 1 and 2. Here, k in the path theorem, Formula 9.14, has taken on two values, 1 and 2, reflecting the fact that variables 1 and 2 are directly connected and prior to variable 3. Four such equations are possible to predict each of the four correlations between independent and dependent variables. Since these four equations have only four unknown path coefficients, they can be solved for the path coefficients, and there are computer programs that carry out the calculations. All we need at this point are data from a study that measure variables included in the model. Land (1970) examined this model by using data gathered from 185 junior and senior high school students in Nevada and Massachusetts in 1962, where they made judgments of the functional importance, skill, prestige and rewards of 24 occupations (Lopreato and Lewis, 1963).

The zero-order intercorrelation (r) of variables is as follows:

	Variables		
Variables	(2)	(3)	(4)
(1) Imputed functional importance of an occupation	.64	.75	.59
(2) Skill	—	.92	.89
(3) Prestige	—	—	.87
(4) Reward	—	—	—

From these data and the path-estimating equations in Formula 9.16, path coefficients can be computed for the model. A multiple-correlation coefficient could be obtained for the regression equation predicting each dependent variable. Both path coefficients and the multiple-correlation coefficients are generally provided by standard computer regression programs.

With computed coefficients shown in Figure 9.2, one can turn to the crux of the problem, the evaluation of the fit of the model to the data. Goodness of fit of the model can be examined in a number of ways. Land (1970) cited three general approaches:

1. One could examine the amount of *variation* in dependent variables that is *explained* by variables linked as specified in the model.
2. One could examine the *size of path coefficients* to see whether they are large enough to warrant the inclusion of a variable or path in the model.
3. One could evaluate the ability of the model to *predict correlation* coefficients that were not used in computation of the path coefficients themselves.

In each of these cases, investigators usually contrast the usefulness of their model in these three respects with alternative models, and this is the heart of explanatory progress in any science.

Taking the goodness of fit criteria one at a time and applying them to the example, we see the following. **First,** the two multiple-correlation coefficients, squared, provide an estimate of the proportion of variation in the dependent variable explained by the linear combination of specific independent variables. In this case:

	Explained	Unexplained
$R_{3 \cdot 12} = .94$	$R_{3 \cdot 12}^2 = .89$	$1 - R_{3 \cdot 12}^2 = .11$
$R_{4 \cdot 12} = .89$	$R_{4 \cdot 12}^2 = .80$	$1 - R_{4 \cdot 12}^2 = .20$

Explaining 80% and 89% of the variance in a dependent variable is quite high, although clearly not perfect. The model leaves unexplained only 11% of the variation in variable 3 and 20% in variable 4. In this respect, the model is relatively satisfactory. The unexplained variation is due to variables or measurement error not included in the model and, for the sake of completeness, the square roots of these $(1 - R^2)$ values are ascribed to the residual variables a and b in Figure 9.2.

(9.17)
$$p_{3a} = \sqrt{1 - R_{3 \cdot 12}^2}, \qquad p_{3a} = .33$$
$$p_{4b} = \sqrt{1 - R_{4 \cdot 12}^2}, \qquad p_{4b} = .45$$

If these residual paths become large, then investigators would begin to search for an alternative model that included other independent variables and/or they would examine their measurement process for measurement error.

In terms of the **second** criterion of goodness of fit, the model shows some weaknesses. Most of the path coefficients are moderately high except the path from variable 1 to 4. One might consider eliminating this path from the model for this reason and recompute path coefficients. The indirect effect of functional importance on rewards can be found by multiplying path coefficients (or correlation coefficients on curved lines in the model) times each other, along each route connecting the two variables of interest, and then by summing over all connecting paths. In this case, $(.64)(.86) = .54$, which represents the indirect effect of variable 1 on variable 4. This is considerably stronger than the direct effect of 1 on 4.

Third, one could examine the fit between observed correlation coefficients not previously used in formulas for calculating path coefficients and predictions of correlation coefficients that would be derived from the model. If the fit is good, the model is supported; if not, some modification is perhaps needed. In this instance, the correlation between variables 3 and 4 was not used to estimate path coefficients in equations shown in Formula 9.16.

Since by the basic theorem of path analysis (Formula 9.14), a correlation equals the sum of products of coefficients along all connecting paths, the model would lead to the following prediction of r_{34}:

(9.18)
$$r'_{34} = p_{41}p_{31} + p_{41}r_{12}p_{32} + p_{42}r_{12}p_{31} + p_{42}p_{32}$$

or

$$
\begin{aligned}
(.04)(.27) \quad &= .011 \\
(.04)(.64)(.75) &= .019 \\
(.86)(.64)(.27) &= .149 \\
(.86)(.75) \quad &= \underline{.645} \\
r'_{34} &= .824
\end{aligned}
$$

This corresponds quite closely with the observed correlation of prestige and reward of .87.*

9.2.3   Testing an Alternative Model

Land (1970) goes on to contrast the goodness of fit of the Davis–Moore model with that of the Parsonian model, postulating an added variable, *values*, an unmeasured, underlying variable that is thought to account for the other four variables: skill, importance, prestige, and reward. This model and the path estimation equations are presented in Figure 9.3. Land concludes that neither the model in Figure 9.2 nor the model in Figure 9.3 can be rejected on the basis of the goodness of fit criteria, but that the Parsonian model somewhat more consistently fits these data. Other theoretical work suggests that the Davis–Moore theory of social stratification would be a better representation in social systems where there is *high* interdependence of work activities, and the Parsonian model would be better under conditions of *low* interdependence—theoretical ideas that suggest greater refinement and the possibility for further productive research.

Path-analytic models may be extended to include more variables, as is illustrated in Figure 9.4, in which Sewell *et al.* (1970) revised and extended earlier work aimed at explaining the occupational attainment of Wisconsin high school seniors. The data upon which path coefficients in this model are based are from 4,388 high school seniors first interviewed in 1957 and then reinterviewed in 1964.

9.3   MULTIPLE REGRESSION USING UNSTANDARDIZED SCORES

Up to this point, we have been converting raw scores on each variable to z-scores in the process of computing coefficients for a multiple-regression equation. This transformation sets the mean of each variable equal to 0 and the standard deviation equal to 1. Because of this transformation, we could legitimately compare beta weights (b^*) or path coefficients (p) to determine the relative contribution of each variable to the explanation of variation in a dependent variable. In this sense we could determine which of several independent variables is "most important."

---

* What is "quite closely" and what is not depends to a large extent upon the investigator's experience and judgment. Differences of .05 or .10 are likely to be considered small unless a meaningful alternative model is able to make closer predictions.

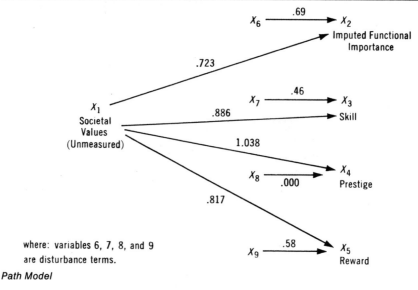

where: variables 6, 7, 8, and 9 are disturbance terms.

Path Model

$$X_2 = p_{21}X_1 + p_{26}X_6 \qquad X_4 = p_{41}X_1 + p_{48}X_8$$
$$X_3 = p_{31}X_1 + p_{37}X_7 \qquad X_5 = p_{51}X_1 + p_{59}X_9$$

Path Estimation Equations

$$r_{23} = p_{21}p_{31} \qquad\qquad p_{26} = \sqrt{1 - p_{21}^2}$$
$$r_{24} = p_{21}p_{41} \qquad\qquad p_{37} = \sqrt{1 - p_{31}^2}$$
$$r_{25} = p_{21}p_{51} \qquad\qquad p_{48} = \sqrt{1 - p_{41}^2}$$
$$r_{34} = p_{31}p_{41} \qquad\qquad p_{59} = \sqrt{1 - p_{51}^2}$$
$$r_{35} = p_{31}p_{51}{}^*$$
$$r_{45} = p_{41}p_{51}{}^*$$

Correlation Matrix (From Lopreato-Lewis matrix above, following Formula 9.16)

	(3)	(4)	(5)
2. Imputed Functional Importance	.64	.75	.59
3. Skill		.92	.89
4. Prestige			.87
5. Reward			

Goodness of Fit of Model in Figure 9.3

1. Predictability of dependent variables.
   Unexplained Variation in:
   Functional Importance 48%
   Skill   21%
   Rewards   25%
   Prestige   00%
2. Path coefficients large. The unmeasured, hypothetical variable, "values," is virtually identical to prestige.
3. Predicted $r_{35} = p_{31}p_{51} = (.886)(.817) = .72$ (Actual $r_{35} = .89$)
   $r_{45} = p_{41}p_{51} = (1.038)(.817) = .85$ (Actual $r_{45} = .87$)

---

* These equations, not needed to compute path coefficients, are used to check model's fit.
Source: Land (1970). Used by permission.

FIGURE 9.3 PARSONIAN MODEL OF SOCIAL STRATIFICATION

Intercorrelation of Variables (Pearsonian *r*):

	(2)	(3)	(4)	(5)	(6)	(7)	(8)
				Variables			
(1) 1964 Occupational Attainment	.618	.483	.463	.438	.384	.331	.363
(2) Education		.632	.696	.609	.535	.417	.486
(3) Level of Occupational Aspiration			.771	.565	.470	.366	.445
(4) Level of Educational Aspiration				.611	.459	.380	.418
(5) Influence of Significant Others					.473	.359	.438
(6) Academic Performance						.194	.589
(7) Socioeconomic Status							.288
(8) Mental Ability							

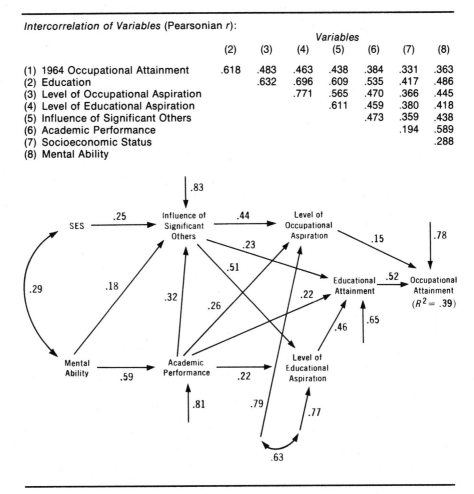

Source: Sewell et al. (1970). Used by permission.

FIGURE 9.4   THE SEWELL MODEL OF OCCUPATIONAL ATTAINMENT

Often, however, one is more interested in *predicting* the value of a dependent variable rather than analyzing a theoretic model or comparing the importance of independent variables. Using *z*-scores as data in a multiple-regression equation yields a prediction of the *z*-score of the dependent variable, rather than a predicted raw score expressed in units such as miles or feet, pounds or ounces, number of births, or number of friendship choices. The raw scores of variables used as independent variables in the prediction equation also have units of measurement, and it may be useful *not* to convert them to standard scores in making predictions. This means that the beta coefficients (b^*) would have to be altered somewhat to take account of the particular units of measurement used in the regression equation.

This would occur most frequently in an applied situation in which the multiple-regression equation is used to make some actual prediction of the value of a dependent variable. Such predictions might then be used to guide actions such as turning on air-conditioners, adding more of some ingredient, hiring more truck drivers, building more schools, deciding on closer supervision of new parolees, or admitting a college applicant. Regression weights (b) for predicting height in inches, using age in years and weight in pounds, would be quite different from what they would be if the prediction were to be made in feet rather than inches or if the weights were to be expressed in ounces rather than pounds.

Currently, much of the research in sociology is oriented toward testing general ideas about which variables are most important to include in a theory, rather than toward predicting actual score outcomes, although there is an indication that this situation may be changing. Voting studies, for example, predict the actual percentage vote for a given candidate. Population studies predict the actual birth rate or the size of a population, and those involved in controlling crime predict the likelihood that new parolees will commit the same crimes again (recidivate).

The following equation is written in a form appropriate for raw data. Notice that a constant term, $a_{1\cdot234}$, has been added to adjust for the location of scores on the dependent variable. This corresponds to the Y-intercept term in a simple regression equation discussed earlier in this book, $Y' = a_{yx} + b_{yx}X$ (see Section 9.1.2). The regression coefficients are now written without an asterisk (*) to indicate that we are *not* using standard score input data, but rather the raw scores themselves in their original measurement units.

(9.19)
$$X'_1 = a_{1\cdot234} + b_{12\cdot34}X_2 + b_{13\cdot24}X_3 + b_{14\cdot23}X_4$$

Again, the regression coefficients are *partial* regression weights, in the sense that they reflect the additional contribution of a variable beyond the contributions of other variables in the specific equation.

To change from a standardized beta weight (b^*) to an unstandardized one (b), the relative amount of variation in the dependent variable and independent variables must be taken into account. This is accomplished by multiplying the standardized beta by the *ratio* of the standard deviation of the dependent variable to the standard deviation of the specific independent variable to which the standardized beta weight refers.

$$b_{12} = b^*_{12}\frac{s_1}{s_2}$$

where s_1 is the standard deviation of the dependent variable, and s_2 is that for the independent variable. In general, the conversion can be expressed as follows for a four-variable problem:

$$b_{ij\cdot kl} = b^*_{ij\cdot kl}\frac{s_i}{s_j}$$

where the subscript i refers to the dependent variable, and subscript j to the specific independent variable in a regression equation with variables k and l. The intercept coefficient can be computed as follows, again for the four-variable problem. These procedures can be generalized for more variables simply by including comparable terms for each of the variables.

(9.20)
$$a_{i \cdot jkl} = \bar{X}_i - b_{ij \cdot kl}\bar{X}_j - b_{ik \cdot jl}\bar{X}_k - b_{il \cdot jk}\bar{X}_l$$

Computation of these coefficients would usually be left to a checked out multiple-regression computer program, where raw scores are specified as input. It is helpful and highly desirable, however, to be able to double-check some of the computations and make simple conversions by hand.

9.3.1   An Example of Multiple Regression Using Raw Scores

A very nice illustration of raw-score multiple regression procedures comes from the social stratification literature, where these procedures were applied to the problem of predicting an occupational prestige score that could then be used in further research involving a social class variable.

You have probably read about the NORC–North–Hatt scale of occupational prestige (Reiss, 1961), since it is widely used in research, and it is frequently quoted in introductory books in social sciences. Cecil North and Paul K. Hatt directed a study by the National Opinion Research Center (NORC) in March 1947 in which some 2,920 respondents were interviewed. Respondents were asked to give their "personal opinions of the general standing" of some 88 different occupations. Responses on a five-point scale (excellent, good, average, somewhat below average, and poor) were summarized to create scores that could vary from 20 to 100 (actually ranging from 33 for shoe shiner to 96 for U.S. Supreme Court justice), and these scores on 88 occupations are referred to as the North–Hatt scale.

Unfortunately for later investigators desiring to use these prestige scores for occupations in their own research, the 88 occupations did not cover the range of occupations that would come up in their data, so they were faced with the problem of guessing an approximate score or of predicting missing scores by using some prediction formula based on other available information. Otis Dudley Duncan helped solve the problem by multiple-regression techniques, using income and education as predictor variables (Reiss *et al.*, 1961). Forty-five of the 88 North–Hatt occupations matched sufficiently well with one of the occupational categories of the inclusive U.S. Census classification of all jobs into 425 categories (U.S. Bureau of the Census, 1960). For each of these categories, information was available on (a) the proportion of men in the occupation with 1949 incomes of $3,500 or more, and (b) the proportion of men in the occupation with four years of high school or more formal education. These variables were then used to predict the percentage of respondents who judged an occupation to be of "excellent" or "good" standing in the North–Hatt data.* The regression equation developed on these 45 occupations is as follows:

$$X_1 = -6.0 + .59X_2 + .55X_3$$

where X_1 is the occupation score, X_2 is income, and X_3 is education.

---

* Because, overall, income and education are related in a curvilinear fashion with age—the young and old are low on both—these two variables were standardized on age, and the age-adjusted proportions were used in the multiple-regression equation expressed here. Standardization is discussed in Chapter 8.

Using these weights in the multiple-regression equation, Duncan could create social class scores for the rest of the 425 U.S. Census occupational categories for which income and education data were available in the U.S. Census publications. Truck drivers, for example, are in an occupation in which 21% of the males in 1950 had 1949 incomes of $3,500 or more, and 15% had graduated from high school or had further education (both standardized for age).

Using the regression equation above, one could predict that the percentage of people who would rate that occupation as "excellent" or "good" would be 15%; this then becomes Duncan's Socioeconomic Index Score. The original NORC prestige rating was actually 13%, so the prediction missed the mark by only 2 percentage points. The coefficient of multiple determination, R^2, using the data on the 45 occupations, was .83; that is, for the 45 occupations, 83% of the variance in occupational prestige was accounted for, statistically, by this particular linear combination of indicators of income and education for these occupations.

Prior to using the multiple-regression procedure, scatter plots of X_1 by X_2, X_2 by X_3, and X_1 by X_3 were inspected, and the relationships were found to be essentially linear. One could compute the regression weights and R^2 from the following information using formulas discussed earlier in this chapter:

$$r_{12} = .84 \quad r_{12 \cdot 3} = .61$$
$$r_{13} = .85 \quad r_{13 \cdot 2} = .65$$
$$r_{23} = .72$$

It is interesting to note that a correlation of .99 was found between prestige scores for the 90 occupations in the 1947 North–Hatt study and a 1963 replication reported by Hodge and others (1964). Regression analysis was also used in this study to examine shifts over the 16-year period. One of the findings was that there had been an upward shift of scores for blue-collar occupations as compared to professionals and white-collar workers.

9.4 DUMMY VARIABLE MULTIPLE REGRESSION

Sometimes it happens that an independent variable we would like to use in a multiple-regression analysis is a nominal variable, rather than interval as the model assumes. It is possible to include these variables in the analysis by creating what is called a *dummy variable*. These are dichotomous variables indicating the presence (scored 1) or absence (scored 0) of a certain characteristic for each individual respondent.

If five categories for marital status were useful as an independent variable, one could represent the same information by four dummy variables as follows:

Marital Status	Dummy Variables
1 Currently married	1 Currently married (1 = Yes) (0 = No)
2 Never married	2 Never married (1 = Yes) (0 = No)
3 Widowed	3 Widowed (1 = Yes) (0 = No)
4 Separated	4 Separated (1 = Yes) (0 = No)
5 Divorced	

Using this scheme, if a person were widowed, rather than have a score of 3 on the marital status variable, that person would have four scores, one on each of the four dummy variables: 0, 0, 1, 0. That is, that person is 0 on currently married, 0 on never married, 1 on widowed, and 0 on separated dummy variables. The 0, 0, 0, 0 pattern would indicate a person who is *none* of the four explicitly mentioned marital statuses—so that person would have to be "Divorced," the only other coded possibility. In general, one creates one dummy variable fewer than there are categories of the nominal-level variable one is interested in including.* These dummy variables are all included in the usual multiple-regression analysis. If an individual has a score of 0 on one of the dummy variables, then that regression weight and term in the regression become 0. Otherwise, the regression weight represents the value of that particular marital status—that is, the amount that is to be added in the regression equation to represent the value of being of some particular marital status. Dummy variables can be used as independent variables, whether one is expressing scores in standard-score form or raw-score form, and they can be used in path analysis problems as well. Dummy variables provide a useful way to score specific combinations of values of variables in order to handle expected statistical interaction in multiple-regression equations where otherwise the model would assume an additive relationship among variables.

Duncan used dummy variables in predicting the percentage of respondents rating occupations as "excellent" or "good" as part of the study discussed in the previous section (Blau and Duncan, 1967). Instead of the education variable, he used the dichotomy "white-collar" versus "manual workers." Since there were only two categories in this variable, only one dummy variable is needed, with $Z = 1$ if the occupation is white-collar and $Z = 0$ if the occupation is a manual occupation. Again using the data on 45 occupations from the North–Hatt study for which U.S. Census data on the independent variables were available, Duncan derived the following regression equation:

$$X'_1 = 3.9 + .79X_2 + 19.8Z$$

$$R^2_{1 \cdot 2Z} = .76$$

This means that 19.8 points are added if the occupation is a type where $Z = 1$, that is, "white-collar." If $Z = 0$, then the 19.8 points would not be added, since 19.8 times 0 is 0.

The correlation between social class scores for occupations using the first formula and the dummy-variable formula above is .96. That is, education and the "white-collar/manual worker" variables produced very nearly the same ordering of scores.

When the dummy variable in a multiple-regression problem is the dependent variable rather than an independent variable, the analysis becomes what is known as a *discriminant function analysis*, that is, one in which the values of a number of continuous variables are used to predict categories of a nominal variable. Consideration of discriminant function analysis will not be included here since it is usually treated in specialized courses in multivariate analysis. If you wish to pursue the subject further on your own, see Van de Geer (1971) or Pedhazur (1982).

* For an explanation of dummy variables, see Suits (1957; General References).

9.5 STEPWISE REGRESSION PROCEDURES

If investigators are interested in searching their data for the best set of predictor variables, they may utilize a procedure called **stepwise multiple regression**. Usually carried out with the help of a computer, the stepwise regression procedure examines a larger number of potential predictors, starting with a single independent variable that is the best predictor of the dependent variable. Then a further variable is added, and this added variable is one that explains as much of the remaining variation in the dependent variable as possible. Then the "next best" variable is added, and so forth, each time adding a term to the multiple-regression equation. The purpose is to find a small set of independent variables that produces the highest R^2 possible. An investigator may drop variables out of the multiple-regression equation whenever their addition produces little increase in the coefficient of multiple determination, R^2.

It sometimes occurs that the most highly predictive set of variables is not the one "discovered" by starting with the "most predictive" variable in the procedure outlined above. Some computer programs solve this situation by examining essentially *all* the possible combinations of different numbers of independent variables chosen from the original data set.* This procedure permits an investigator to pick the best set of predictor variables, in the sense of explaining the maximum amount of variation in the dependent variable with a minimum number of linearly related independent variables.

Regardless of which approach is used, stepwise regression analysis is a procedure that capitalizes on any unusual predictive variation in the data at hand. The results are, so to speak, tailored to the existing data, and because of this the regression equation may not be applicable to data beyond that included in the study. Whenever theory suggests how the addition of particular variables affects prediction, comparison of steps helps test theory. It is a helpful search procedure, but as in the case of results from any search, we would want to replicate the study to see if variables included are still the best predictors beyond the data on which they were derived.[†]

9.5.1 An Example of Stepwise Multiple Regression

One example of the use of stepwise multiple regression procedures is in a study of factors predictive of the number of demonstrations among 104 selected, state-supported, nontechnical, nonspecialized colleges and universities during 1964 and 1965 (Scott and El-Assal, 1969). Representatives of 69 of the 104 schools responded with a variety of information including size of school. It was hypothesized that "multiversities" (those more complex in terms of number of degrees granted, size of departments offering various degrees, the ratio of dormitory to nondormitory

---

* Some of these computational alternatives are presented in the early part of Chapter 6 in Draper and Smith (1966; see General References).
[†] Sometimes investigators split their data set into two subparts (randomly), using one-half to explore and the remaining half to test these outcomes.

TABLE 9.3 Stepwise Multiple Regression Analysis, Predicting Number of Student Demonstrations at Selected, State, Nontechnical, Nonspecialized Colleges and Universities During 1964–1965

Independent Variables	R	R^2
School Size	.580	.336
School Size & Complexity	.591	.349
School Size, Complexity & School Quality	.593	.352
School Size, Complexity, School Quality & Community Size	.594	.353

Source: Scott and El-Assal (1969).

students, professor–student ratio, etc.) ferment more student unrest than other types of schools.

A stepwise multiple-regression analysis among size, complexity (the multiversity measure), quality, and community size was conducted, with the dependent variable being number of student demonstrations. Table 9.3 shows the multiple correlation coefficient and the coefficient of multiple determination, R^2, at each step as new variables were added to the multiple-regression equation. It is evident in these data that, school size is the best predictor of incidence of student protest demonstrations. Other variables added little additional explanatory power to the regression equation.

The authors note that the multiple-regression model is not a very good fit, since some of the variables are not "interval level of measurement," and, in fact, statistical interaction may be expected, rather than additive relationships between variables. In spite of this, they argue, the multiple-regression analysis helped pinpoint school size as a central variable in explaining the incidence of demonstrations.

9.6   CURVILINEAR ASSOCIATION IN MULTIVARIATE ANALYSIS

The basic assumption underlying the discussion of linear regression in this chapter is that variables are related in a simple straight-line fashion. This, of course, does not exhaust the possibilities and, in fact, a fair number of relationships in the field of sociology are either known to be or suspected to be curvilinear in nature. The relationship between income and age is an example. Where relationships do not come very close to the linear expectation, the multiple-regression equation will not be a very good fit, will not explain very much of the variation in a dependent variable, and, at worst, may be misleading to an investigator who is interested in the relative importance of variables for theory construction purposes.

Figuring out exactly the most appropriate form of relationship for a given set of variables is beyond this book, although two kinds of comments may help to provide a basis for further inquiry. First, it is an excellent practice to examine the *nature* of relationship between all pairs of variables in an analysis using either (a) a scatter plot of pairs of variables or (b) a careful comparison of the Pearsonian product–moment correlation coefficient r with eta (E), as discussed in Chapter 7. Computer programs are generally available that will print out scatter plots (or r's and E's). If the relationship is not linear between some of the variables, we might attempt a transformation of the data, such as the logarithmic transformation used by Winsborough (Duncan, 1966:8). This may straighten the relationships and permit us to use the multiple-regression procedures we have discussed. It may also be possible to standardize or "adjust" the raw data, to take account of curved patterns in the data, before the regression analysis is attempted. Duncan (Reiss *et al.*, 1961), for example, adjusted income and education scores for age differences prior to predicting social status scores. His regression equation included only education and income scores after these scores had the effects of age "standardized out," so to speak.

Second, we might attempt to deduce or compute an equation that describes the nonlinear relationship between variables. Some computer programs are available that search data and fit a more complex equation.

9.7 SUMMARY

In this chapter we discussed the general topic of multiple regression and multiple correlation, in which interval variables are related in a linear, additive fashion. The problem is to explain as much as possible of the variation in a single dependent variable by the linear association of two or more independent variables. R^2, the coefficient of multiple determination, provides a measure of the proportion of the variation in the dependent variable that has been explained by the multiple-regression equation.

One use of the multiple-regression equation is based on the use of standard scores (z-scores) as the input data. The beta weights (b^*) that result can be used to compare the relative importance of independent variables in the regression equation. With some notion of how variables are linked theoretically, we could use the procedure called *path analysis* to examne the fit between actual data and our theoretical model.

A second use of multiple-regression analysis involves predicting the actual value, rather than standard score, of the dependent variable. Computation procedures for finding nonstandardized regression weights (b) were discussed.

Finally, we discussed some of the procedures encountered when multiple-regression analysis is used. *Dummy variables* help handle categorical independent variables or anticipated interaction patterns. *Stepwise multiple regression* is a procedure for selectively adding more variables to a regression equation in a search for a maximally predictive set of independent variables. Although the multiple regression equations discussed assume a linear relationship between variables, it is possible to deal with curvilinear relationships by removing them, either through some type of transformation of the data prior to regression analysis or through the use of more complex regression equations. Computer programs become exceedingly useful in handling the clerical complexity of multiple-regression mathematics.

The multiple-correlation coefficient, R, like the bivariate measure, r, expresses how well a regression equation permits one to predict scores on a dependent variable. Like r^2, the square of the multiple-correlation coefficient, R^2, can be interpreted as the proportion of the variation in the dependent variable that is explained by the regression equation, rather than merely the mean of the dependent variable.

CONCEPTS TO KNOW AND UNDERSTAND

multiple-regression equation
assumptions of multiple regression
regression weights and partial
    correlation coefficients
standardized regression (beta)
    weight (b^*)
unstandardized regression weight (b)
multiple correlation
coefficient of multiple determination
    (R^2)

path analysis
corrected R
path coefficient
basic theorem of path analysis
path diagram
stepwise multiple regression
dummy variables
discriminant function

PROBLEMS AND QUESTIONS

1. Find an example of the use of multiple regression for either predictive or analytic purposes and describe its use for the problem the author describes. Are assumptions met? If not, what consequences may there be as far as you can tell? Is multiple regression the most appropriate technique for the problem? What alternatives might there be? Express clearly in your own words what it is that the multiple-regression equation, the regression weights, and the multiple correlation coefficients mean.

2. Blau and Duncan (1967:170) present a correlation matrix showing the zero-order relationships between five variables, as follows:

Zero-Order Correlations of Five Status Variables				
	(2)	*(3)*	*(4)*	*(5)*
(1) 1962 Occupational Status	.541	.596	.405	.322
(2) First job status		.538	.417	.332
(3) Education			.438	.453
(4) Father's Occupational Status				.516
(5) Father's Education				

Using the 1962 occupational status of the respondent as a dependent variable, create a path diagram showing the linkages between variables you would hypothesize to exist. Compare this with the Blau and Duncan model. Now, using information in this chapter about path analysis, create the equations needed to compute path coefficients for your model from the zero-order correlation matrix, above. If you have access to computer facilities, compute the path coefficients and appropriately label your path diagram. Again, compare these coefficients to those computed by Blau and Duncan. Finally, examine the model to reach a conclusion about how well the model fits the data (Blau and Duncan's data are from a national survey of some 20,700 men aged 20 to 64 years old, gathered in 1962).

3. Explain the difference between standardized beta weights, unstandardized regression weights, partial-correlation coefficients, and multiple-correlation coefficients.

4. The data in the accompanying correlation matrix are from a study known as PROJECT TALENT (Flanagan et al., 1964; Lohnes, 1966) based on a stratified random sample of senior males from American high schools and a follow-up of them 5 years after graduation. The correlations in the matrix are from the 14,891 whites who participated in the follow-up. The variables of the study are the following: (1) Occupational Attainment—the Duncan Socioeconomic Status Index of the occupation of the respondent at the time of the follow-up; (2) Household Head's Occupation—the Duncan Socioeconomic Status Index of the head of household in which the boy lived as a 12th-grade student; (3) Intelligence—composite score for respondent from several academic aptitude instruments; (4) Grades—self-reported grade point average of respondent for five academic subjects; (5) Occupational Aspiration—Duncan Socioeconomic Status Index of occupation 12th-grade boy said he would most like to enter; and (6) Educational Attainment—highest level of education reported as completed by respondent at time of follow-up.

   a. Using occupational attainment as the dependent variable, select two or three independent variables and draw a path diagram that predicts how they relate to each other. Do the path analysis and test the model using the three criteria suggested by Land (1970) and listed on p. 341.

   b. Compute the multiple R for the dependent variable and the independent variables you have selected. Compute a residual path coefficient from this R.

CORRELATION MATRIX FOR SIX VARIABLES FROM PROJECT TALENT DATA

	1	2	3	4	5	6
1. Occupational Attainment	—	.197	.337	.236	.262	.489
2. Household Head's Occup.		—	.290	.122	.209	.303
3. Intelligence			—	.341	.323	.526
4. Grades				—	.237	.377
5. Occupational Aspiration					—	.337
6. Educational Attainment						—

Source: Porter (1974).

5. The data in the accompanying table are from selected U.S. cities with populations of 100,000 or more in 1988. The HEALTH and POLICE data are annual expenditures for 1987 for those services (in millions of dollars); MURDER refers to homicides per 100,000 population; POPCHNGE is the percentage increase or decrease in population between 1980 and 1988; BLACK is the percentage of the population identified as black; and HISPANIC is the percentage of the population identified as Hispanic. Note that the 9's in the HEALTH column represent missing data.

a. Select one of the variables as a dependent variable and construct a path diagram to show the relationship between it and three or more of the other variables.

b. Compute path coefficients and test the model using three criteria suggested by Land (1970) and listed on p. 341. If the model does not fit well, try an alternative model.

c. Compute the multiple R and use it to compute the residual path coefficient.

d. Compute the standardized regression equation from these data; then convert the standardized beta weights into unstandardized b's. Use the unstandardized regression equation to predict two or three values of the dependent variable. How well does it work? How do you explain the magnitude of the prediction error?

DATA FROM SELECTED CITIES OF 100,000 OR MORE POPULATION IN 1988

Case	Health	Police	Murder	Popchnge	Black	Hispanic
1	2516.6	1423.6	25.8	4.0	25.2	19.9
2	5.6	512.5	21.6	12.9	17.0	27.5
3	65.8	529.4	22.0	−.9	39.8	14.0
4	29.2	202.2	25.5	6.5	27.6	17.6
5	180.8	277.5	22.4	−2.4	37.8	3.8
6	73.7	243.9	57.9	−13.9	63.1	2.4
7	1.0	108.0	13.4	22.2	8.9	14.9
8	10.3	114.7	36.0	9.1	29.4	12.3
9	28.8	66.2	15.3	19.7	7.3	53.7
10	.6	110.6	11.1	17.0	4.8	14.8
11	5.2	76.4	3.3	2.2	1.2	5.2
12	41.6	134.6	30.6	−4.5	54.8	1.0
13	313.8	122.0	12.2	7.8	12.7	12.3
14	115.2	64.0	16.3	3.8	21.8	.9
15	9999.9	74.1	5.1	17.3	4.6	22.3
16	12.8	55.8	26.0	−.2	47.6	.8
17	301.9	231.7	59.5	−3.4	70.3	2.8
18	12.1	92.3	13.3	−5.8	23.1	4.1
19	132.1	105.3	16.0	2.6	22.4	6.4
20	16.2	74.6	11.7	.8	22.1	.8
21	6.9	51.5	42.4	−4.7	55.3	3.4
22	13.7	102.0	25.2	−9.1	43.8	3.1
23	104.1	86.8	43.4	−.1	12.0	18.8
24	9.8	29.5	6.2	20.2	3.2	62.5

continues

DATA FROM SELECTED CITIES OF 100,000 OR MORE POPULATION IN 1988

Case	Health	Police	Murder	Popchnge	Black	Hispanic
25	6.6	64.2	11.1	1.1	9.5	2.6
26	56.5	42.5	15.7	5.6	23.3	.8
27	93.0	50.7	9.2	34.3	12.2	18.7
28	40.1	30.7	13.6	7.5	14.6	2.8
29	29.8	66.4	29.9	−2.0	27.4	3.3
30	7.0	40.0	22.4	10.8	22.8	12.6
31	46.8	76.7	32.9	−10.9	45.6	1.2
32	.1	42.1	48.8	−1.1	66.6	1.4
33	8.3	71.6	14.0	14.8	11.3	14.0
34	12.9	58.9	11.6	13.7	7.6	2.1
35	7.1	41.6	4.6	−11.5	24.0	.8
36	4.3	34.7	9.4	2.1	11.8	1.7
37	25.7	51.2	12.4	−3.9	33.8	.8
38	4.1	40.1	13.0	13.9	2.5	33.8
39	9999.9	41.7	8.1	16.7	3.7	24.9
40	.5	50.3	30.5	5.2	46.9	9.6
41	7.4	40.0	15.1	−7.1	7.7	1.3
42	1.4	28.2	12.4	16.6	31.0	1.1
43	.4	27.5	7.4	12.4	12.0	2.3
44	9.4	29.3	7.6	39.3	10.0	2.0
45	1.0	33.0	13.3	−12.4	26.6	2.7
46	9999.9	45.8	20.0	22.7	13.4	14.2
47	6.1	45.0	36.0	−4.7	58.2	18.6

Source: U.S. Bureau of the Census (1990).

GENERAL REFERENCES

Berry, William D., and Stanley Feldman, *Multiple Regression in Practice* (Beverly Hills, Calif., Sage), 1985.

Blalock, Hubert M., Jr., *Social Statistics,* rev. ed. (New York, McGraw-Hill), 1979.
  See especially Chapter 19 and Section 19.6.

Costner, Herbert L., editor, *Sociological Methodology, 1971* (San Francisco, Jossey-Bass), 1971.
  See especially Chapter 5 by George Bohrnstedt and T. Michael Carter for a discussion of "Robustness in Regression Analysis"; also Chapter 6 by Morgan Lyons, on "Techniques for Using Ordinal Measures in Regression and Path Analysis."

Davis, James A., *The Logic of Causal Order*, (Beverly Hills, Calif., Sage) 1985.

Draper, Norman R., and Harry Smith, *Applied Regression Analysis* (New York, Wiley), 1966.

Duncan, Otis Dudley, "Path Analysis: Sociological Examples," *American Journal of Sociology*, 72 (July 1966), pp. 1–16.

Edwards, Allen L., *An Introduction to Linear Regression and Correlation,* 2nd ed. (San Francisco, W. H. Freeman), 1984.

Land, Kenneth C., "Principles of Path Analysis," Chapter 1 in Edgar F. Borgatta, editor, *Sociological Methodology, 1969* (San Francisco, Jossey-Bass), 1969.

Nygreen, G. T., "Interactive Path Analysis," *American Sociologist*, 6 (February 1971), pp. 37–43.

Suits, Daniel, "The Use of Dummy Variables in Regression Equations," *Journal of the American Statistical Association*, 52 (1957), pp. 548–551.

Wonnacott, Thomas H., and Ronald J. Wonnacott, *Introductory Statistics* (New York, Wiley), 1969, Chapters 13 and 14.

LITERATURE CITED

Blau, Peter M., and Otis Dudley Duncan, *The American Occupational Structure* (New York, Wiley), 1967, p. 125.

————, "A Socioeconomic Index for All Occupations," Chapter 6 in Albert J. Reiss, *Occupations and Social Status* (New York, Free Press), 1961.

Flanagan, John C., *et al.*, *The American High School Student*, U.S. Office of Education, Co-operative Research Project No. 635 (Pittsburgh, Project TALENT Office, University of Pittsburgh), 1964.

Hodge, Robert W., Paul M. Siegel, and Peter H. Rossi, "Occupational Prestige in the United States, 1925–1963," *American Journal of Sociology*, 70 (November 1964), pp. 286–302.

Koslin, Bertram L., *et al.*, "Predicting Group Status from Members' Cognitions," *Sociometry*, 31 (March 1968), pp. 64–75.

Land, Kenneth C., "Path Models of Functional Theories of Social Stratification as Representations of Cultural Beliefs on Stratification," *Sociological Quarterly*, 11 (Fall 1970), pp. 474–484.

Lohnes, Paul R., *Measuring Adolescent Personality*, PROJECT TALENT Five-Year Follow-up Studies, Interim Report I (Pittsburgh, University of Pittsburgh), 1966.

Lopreato, J., and L. S. Lewis, "An Analysis of Variables in the Functional Theory of Stratification," *Sociological Quarterly*, 4 (Fall 1963), pp. 301–310.

Pedhazur, Elazar, J., *Multiple Regression in Behavioral Research*, 2nd ed. (New York, Holt, Rinehart and Winston), 1982.

Porter, James N., "Race, Socialization and Mobility in Educational and Early Occupational Attainment," *American Sociological Review*, 39 (June 1974), p. 307.

Reiss, Albert J., *et al. Occupations and Social Status* (New York: Free Press), 1961.

Scott, Joseph W., and Mohamed El-Assal, "Multiversity, University Size, University Quality, and Student Protest: An Empirical Study," *American Sociological Review*, 34 (October 1969), pp. 702–709.

Sewell, William H., Archibald O. Haller, and George W. Ohlendorf, "The Educational and Early Occupational Status Attainment Process: Replication and Revision," *American Sociological Review*, 35 (December 1970), pp. 1014–1027.

U.S. Bureau of the Census, *1960 Census of Population, Alphabetical Index of Occupations and Industries* (rev. ed.) (Washington, D.C., U.S. Government Printing Office), 1960.

————, *Statistical Abstract of the United States*, 1990. (Washington, D.C., U.S. Government Printing Office), 1990.

Van de Geer, John P., *Introduction to Multivariate Analysis for the Social Sciences* (San Francisco, W. H. Freeman), 1971.

Wampler, Roy H., "A Report on the Accuracy of Some Widely Used Least Squares Computer Programs," *Journal of the American Statistical Association*, 65 (June 1970), pp. 549–565.

Part V
INFERENTIAL STATISTICS:
SAMPLING AND PROBABILITY

CHAPTER 10

Probability

10.1   BACKGROUND TO INFERENTIAL STATISTICS

So far in this book, we have been dealing with the problem of *describing* whole distributions of a single variable or the relationship between two or more variables. There are several aspects of a distribution that need to be described, and there are many different techniques designed to help in that description. An investigator faced with this range of possibilities needs to know how to select the statistical tools that provide the most appropriate description.

The next part of this text deals with an inferential problem. Once the data at hand have been appropriately described, can one go further and generalize these descriptions to any larger population of cases? The short answer is yes; if a sample is drawn in a special way, then descriptions of that sample can be generalized to the larger population from which it was drawn. The risk of making an error in that "inferential leap" can be rather accurately described. *Inferential statistics is a set of tools used to help an investigator make inferences about a population from a description of data in a sample.*

Investigators are usually interested in making generalizations about a broader population of cases based on a sample of cases from that population. As we shall see, it is usually unnecessary (and often less accurate as well as more costly) to attempt to gather data from an entire population when a carefully drawn sample is available. The rest of this book focuses on how to make inferences about a broader population based on a description of a sample from that population.

You should be pleased to know that all the basic tools for handling this leap have been covered in past chapters. First, for example, the idea of statistical description has already been discussed. Here, the inferential leap is between two statistical descriptions. One is a description of a sample that one has in hand. The other is the statistical description of the population one wants to say something about. Thus, the arithmetic mean of a sample can give us information about the arithmetic mean of a population. You already know that descriptive measure from earlier chapters. The same can be said for other descriptive measures, such as correlations and regression coefficients, that we covered earlier. As we shall

BOX 10.1   IDEAS TO REMEMBER

The next chapters will draw upon ideas already covered in earlier chapters. In particular, you may want to refresh your memory on the following concepts.

Population, sample—refer to all relevant cases and to some smaller subset of these cases. "Population" and "sample" were first introduced in Section 1.1. How to define a population and how to draw different kinds of samples will be discussed in detail in Chapter 11.

Univariate description—how central tendency, variation, and form are described (Chapter 5). In particular, you should be familiar with the arithmetic mean (Section 5.3.3) and the standard deviation (Section 5.4.4) since these measures will be used repeatedly in the chapters to follow.

z-scores and the standard normal curve—for reasons explained below, you will find that the normal curve (Section 5.5.2) and standard (z) scores (Section 5.4.7) are frequently used. You will need to use tables of areas under a normal curve (Appendix, Table B) to find the area between the mean and a given z-score (Section 5.5.2b explains this).

Statistical descriptions—finally, you will notice that the following chapters are concerned with the inferential leap between a description of a sample and a description of a population. The descriptions involved are exactly those you have already covered—descriptions of the distribution of a single variable (Chapter 5), descriptions of the relationship between two variables (Chapter 7), and descriptions of the relationship between three or more variables (Chapters 8 and 9). In the chapters to follow, we will only make use of certain of these descriptions to illustrate the logic of inferential statistics. Nevertheless, you will want to refer back to earlier sections from time to time to refresh yourself on the characteristics of a particular descriptive statistic. Cross-references later in this book will help you make these connections.

discuss later, it is often helpful to introduce more symbols for clarity; the mean of a sample is referred to by familiar symbols (like X-bar, $\bar{X}$), but a new symbol is used for the mean when it refers to a population description (the Greek letter mu, μ).

There is a second way in which your knowledge of descriptive statistics will prove helpful. A key idea in inferential statistics is just another univariate distribution (called a *sampling distribution* because it describes how whole samples behave; the case has changed, but the ideas are the same). This distribution itself needs to be described (its central tendency, variability, and form, for example), just like any other univariate distribution. You already know how to describe such distributions. Again, while the ideas are the same, the labels for these ideas will differ somewhat in order to help make it clear when the description applies to this special distribution, the sampling distribution. For example, the old measure of variability for interval variables, standard deviation, will be called by a new term, *standard error*, when it applies to a sampling distribution. But the mean of a sampling distribution will simply be called the *mean of a sampling distribution*. We will point out these terminological changes along the way.

In an important sense, then, you have already covered most of the ideas in inferential statistics. The key change will be in how these ideas are used—how they help in the logic of making inferences about a population based on a sample.

10.1a   An Example of Induction

Recent work suggests that certain groups are superior to single individuals at problem-solving tasks. This superiority, however, is inversely related to how much hierarchical difference there is in the group, because hierarchically differentiated groups may exhibit less risk-taking behavior and be less efficient and less productive than groups that are relatively undifferentiated (Blau and Scott, 1963; Bridges, 1968).

To examine these ideas, we could use high schools. They fit the problem because high schools are organized groups that vary in degree of hierarchical differentiation. We could draw a probability sample (an important type of sample discussed in Chapter 11) of public high schools from the population of all public high schools in the country. For each high school in our sample, we could collect data on such variables as size of student body, size of faculty, number of levels of administration, amount of risk taking evident in seeking solutions to problems or making changes in the organization, percent of graduates entering college, and the like. Using data from the sample, we could test ideas about hierarchical differentiation and then generalize these findings to the broader population of high schools. Generalizing from sample to population would use probability theory to gauge the likelihood of error in those generalizations.

An introduction to probability is important to an understanding of how samples can be drawn and how inferences can be made from samples to broader populations.

To make scientific generalizations, then, sociologists employ sampling theory and probability theory. They take a sample of data relevant to theoretical ideas about which they wish to generalize, study the characteristics of the sample, and use probability theory as a basis for extrapolating the findings to the population the sample represents. This process is *induction*, which is central to the scientific method.*

Both probability theory and sampling theory are basic to the inductive process. Furthermore, they are so intricately interrelated that it is difficult to divorce one from the other. To present the ideas underlying these theories as simply as possible, however, we will treat the two topics separately.

This first chapter on *probability* will introduce some of the basic ideas used in dealing with risk, errors, outcomes, and chance. Building upon these ideas, Chapter 11 discusses ways samples may be drawn. This is particularly important

---

* There are some exceedingly interesting and important issues involved in induction in the social sciences, and these are usually discussed in philosophy of science or research methods courses. For a start, you might want to explore a general book in this area (see Brown, 1963).

where one is interested in drawing conclusions about populations from samples of them. Then Chapter 12 examines what we know about certain outcomes that result from drawing samples in certain ways. It is here that the key concept in inferential statistics—sampling distribution—is discussed and described. Later chapters make use of these notions in the process of making inferences from descriptions of samples.

10.2   AN INTRODUCTION TO PROBABILITY

Probability is a concept intuitively familiar to all of us. In our everyday experiences we are frequently concerned with the probability that an event of a certain kind will or will not happen. What is the probability that it will rain today? How likely is there to be a traffic jam? What is the chance of winning some money in Las Vegas?

Sociologists, too, frequently ask probability questions: What is the probability that children from broken homes will become delinquent? What is the probability that convicted murderers will be successful on parole? What is the probability of successful marriages for couples with very diverse religious backgrounds? What is the probability of survival of organizations that serve a single function? If sociologists can establish probability figures as answers to questions such as these, then they can add to their store of scientific generalizations and use the knowledge in predicting future outcomes.

In popular usage the concept of probability takes on many shades of meaning. However, in applied statistics, probability is formally defined in a very specific way.

10.2.1   Probability as an *a Priori* Concept

Probability refers to the *theoretical relative frequency of occurrence of a certain kind of event in the long run.* This is an ***a priori*** concept of probability, theoretically rather than empirically based.*

Let us analyze this definition of probability. Think of an **activity or process** that may have different *outcomes.* Some examples are giving birth to a boy or girl, drawing a sample of workers with different occupations, tossing a coin that has two sides, conducting an experiment in which one class is shown a film about older people and another class is not but both classes are asked to answer a questionnaire indicating whether they do or do not have favorable attitudes toward older people, recording the class standing (freshman, sophomore, junior, senior, other) of people walking into a college cafeteria, and noting whether or not a person who has been released from prison has returned to prison within a year.

* There is another concept of probability, known as Baysian probability, that makes use of subjective probability estimates (Mosteller *et al.*, 1961 or Oakes, 1986; see General References).

In each of these examples, we have identified a set of possible *outcomes* of the activity or experiment. The set of all possible outcomes of an activity or experiment can be taken to define a **universal set**—100% of the possible outcomes for that activity or experiment. The outcome of interest to us we call an **event**. We may be interested, for example, in the outcome "girl" of a birth, the outcome "blue collar" in the sampling of workers, the outcome "heads" in a toss of a coin, the outcome "favors older people" in the film experiment, the outcome "freshman" in the cafeteria observation activity, or the outcome "returned" in the prison release problem. We might, of course, be interested in any of the other possible outcomes.

Turning next to the phrase **theoretical relative frequency of occurrence**, imagine repeating the activity or experiment an infinite number of times, each time recording the outcome. The theoretical relative frequency of occurrence is the proportion of events (outcomes of interest to us) out of the total of all outcomes of that repeated activity or experiment. The probability is theoretical in the sense that it is derived from some theory of how the activity or experiment should turn out if the process were repeated an infinite number of times. For example, there appear to be biological reasons why there is a slightly higher probability that a birth will result in a boy. If only chance were operating in the selection of a sample of workers, one might expect somewhat higher likelihood of drawing a blue-collar worker simply because there are usually more such workers in industrialized societies. If a coin is "fair," we assign the theoretical relative frequency of occurrence of one-half (.50) to the event "heads" (that is what we define fairness of coins to mean). If 40% of the college population were freshmen, we might expect that chance would result in 40% freshmen among those visiting the cafeteria, and so on.*

Another important phrase in the definition of probability is the final one, **in the long run**. This phrase implies that probability does not apply to a single or a few repetitions of the activity or experiment. It applies to a large number of such instances. In the student body example, we would expect approximately 40% of the persons entering the cafeteria to be freshmen. Does this mean that of the next 10 persons entering the cafeteria, 4 will be freshmen? Not necessarily! Although that is a possibility, 10 observations are insufficient to constitute an adequate test. Probability refers to the relative frequency of occurrence of an event *in the long run*. The more persons we observe entering the cafeteria, the more closely the proportion of freshmen should approach 0.4, if indeed that theoretical probability applies.

Prediction in terms of probabilities is not prediction of a single outcome; rather, it refers to a large number of outcomes. This principle of prediction is that upon which the insurance business is based. When you take out a life insurance policy, you pay a premium based upon the probability that a certain number of people similar to you will die during each year. The insurance company is unable to predict individual deaths except in certain obvious cases—the uninsurable

---

* With more information, we may discover that we can actually make more accurate predictions than that implied in the use of the theoretical (chance) probability used in this situation.

people who are already dying. Prediction of death rates for large categories of people is, however, relatively accurate. The insurance company considers you as an element of a set of a large number of people with similar characteristics. For example, your set may be composed of white males between the ages of 20 and 24 (such sets are known as *cohorts*). Year-by-year studies of death rates for specific sets of people provide the insurance company with estimates of how many people in a particular set are likely to die during a given year. The amount of your premium is determined, in part, by the predicted death rate for your cohort. If the death rate for your cohort is high, the cost of your insurance will be high.

In a study that has become a classic in sociology, the French sociologist Emile Durkheim demonstrated that the suicide rates in certain parts of Europe prior to the turn of the century were consistent from year to year and could be predicted with impressive accuracy (Durkheim, 1897/1951). Durkheim could not predict which particular individuals would commit suicide, but he could predict the number of suicides that would occur during a given year in a given area.

10.2.2   Experiential Probability

Perhaps you have noticed that the concept of probability used in these two examples does not jibe with our original definition of probability. Our original definition was based upon theoretical relative frequencies—an ***a priori*** notion. Our theoretical notions of how a "fair" coin behaves when tossed provides an obvious example of *a priori*, theoretical probability. The death rates used by the insurance company and the suicide rates used by Durkheim come from empirical observations. The probabilities assigned to these events, therefore, are based upon **experience** rather than theory. In most instances, estimates of theoretical probabilities are derived from experience.

10.2.3   Some Probability Notation

Many aspects of probability can be illustrated using elementary concepts of **sets** and **Venn diagrams**.* The *set* of all possible outcomes that can occur in an activity or experiment is referred to as a *universal set*, which is defined as equal to 1.00 and symbolized U. The universal set is then divided up among possible alternatives. For example, suppose that we consider all students in a given high school as a universal set that is divided up into 40% who are freshmen and the remaining 60% who are nonfreshmen. The Venn diagram in Figure 10.1, for example, is made up of two sets: the .40 freshmen (probability is expressed as parts of 1.00) and .60 nonfreshmen, totaling to the universal set: 1.00.

The freshmen and nonfreshmen sets are called **disjoint sets** because they do not overlap; they are mutually exclusive of each other. Many sets are overlapping, a feature that is recognized in the general addition rule of probability,

---

* These diagrams illustrating logical relationships are named after John Venn (1834–1923), an English logician. They are also sometimes referred to as *Euler diagrams*, stemming from their use by Euler in his explanations in *Letters to a German Princess* (1772).

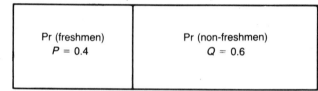

FIGURE 10.1 UNIVERSAL SET FOR STUDENT BODY DIVIDED 40%–60% BETWEEN FRESH-MEN AND NONFRESHMEN

discussed in the next section. *Pr* is often used to mean *probability*. Thus the probability of a student being a freshman in this high school is $Pr = .40$. Where there are two sets (a dichotomy) that together are *exhaustive* of all possibilities in a universal set and where these two sets are *mutually exclusive*, they are conventionally labeled P for one of the probabilities (your choice of which alternative to call P) and Q for alternatives not included in P. One of the principles of probability, then, is that $Q = 1 - P$ and $P + Q = 1.00$, the universal set. Where P is the probability of a student being a freshman in this high school, then $P = .40$, and .40 plus the probability of being a nonfreshman ($Q = .60$) sum to the universal set, $U = 1.00$.

Box 10.2 summarizes the notation for probabilities that we will use here and in future chapters. Symbols for the intersection of sets (the overlap of two sets), the union of sets (the total of a combination of two or more sets), conditional probability, and negation will be discussed and used below.

BOX 10.2   SYMBOLS USED IN TALKING ABOUT PROBABILITY

These will be defined and used in the next few sections.

U	Universal set
Pr	Meaning "the probability of"
∩	Intersection of sets
∪	Union of sets
|	Conditional "given that"
~	Not, negation

10.3   SOME PROPERTIES OF PROBABILITY

There are some rules of probability that are of particular interest to sociologists because they express basic properties of probability. These include the **general addition rule** (Section 10.3.1), the **special addition rule** (Section 10.3.2), the rule of **conditional probability** (Section 10.3.3), the **general multiplication rule** (Section 10.3.4), and the **special multiplication rule** (Section 10.3.5). We will examine each of these in turn. We will also take a brief look at stochastic processes, since they flow naturally from a discussion of the general multiplication rule and also since they are becoming increasingly important in sociology.

10.3.1 The General Addition Rule

The general addition rule states that for events A and B the probability that the event A *or* B will occur is equal to the probability that A will occur plus the probability that B will occur minus the probability that the event "both A and B" will occur. Symbolically, the rule is stated as follows:

(10.1)
$$Pr(A \cup B) = Pr(A) + Pr(B) - Pr(A \cap B)$$

where Pr stands for *probability*, $A \cup B$ stands for *the union of A and B*, and $A \cap B$ stands for *the intersection of A and B*.

The union of events A and B is defined as that set of elements included in event A or event B. It is important to note that in mathematical logic the conjunction *or* takes a special meaning. A or B means A, B, or both A and B.* The union of A and B, therefore, includes elements that are only in A, elements that are only in B, and elements that are in both A and B.

The intersection of A and B, denoted by $A \cap B$, is defined as the set of elements common to both event A and event B.

The general addition rule, then, gives the combined probability that event A will occur exclusively, event B will occur exclusively, or both A and B will occur simultaneously. Figure 10.2 is a Venn diagram of this situation.

You may be surprised that we subtracted the probability of the intersection of A and B in the formula. If you compare the formula and the Venn diagram, the reason for the subtraction can be shown. The probability of occurrence of event A, $Pr(A)$, is represented by the whole circle bounding subset A and the probability of occurrence of event B, $Pr(B)$, is represented by the whole circle bounding subset B. Since subsets A and B are overlapping, the area in which they

U (Area of U = 1.00)

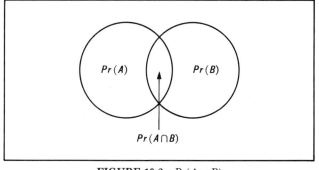

FIGURE 10.2 $Pr(A \cup B)$

---

* This is sometimes called the *inclusive* rather than the *exclusive* use of the logical connective *or*.

U (Area of U = 1.00)

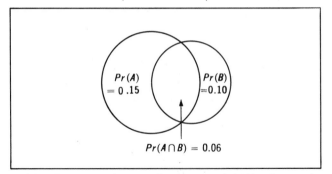

FIGURE 10.3 $Pr(\text{College Graduates} \cup \$50,000 + \text{Incomes})$

overlap, $Pr(A \cap B)$, has been added *twice* into the sum of $Pr(A)$ and $Pr(B)$; therefore, it is necessary to subtract this area to avoid duplication.

Let us apply this addition rule to an example in which we are interested in the probability of randomly drawing from the population of a city persons who are either college graduates or who have incomes in excess of \$50,000, or both. When we do not have theoretical bases for assigning probabilities of given outcomes of events, we can find the proportions that the given outcomes represent of all such events in the population of data available to us and use them as the desired probabilities. This **experiential** assignment of probabilities might be applied to the example as follows. Suppose that 15% of the population of the city are college graduates, 10% have incomes in excess of \$50,000, and 6% are both college graduates and have incomes in excess of \$50,000. Using these data, we can substitute into the formula for the general addition rule to find the probability of either college graduates, or those having incomes in excess of \$50,000, or both, as follows:

$$Pr(A \cup B) = Pr(A) + Pr(B) - Pr(A \cap B) = 0.15 + 0.10 - 0.06 = 0.19$$

The Venn diagram for this situation would be as in Figure 10.3.

If we eliminate the overlap in these probabilities, we can represent the probabilities of interest as in Figure 10.4.

U (Area of U = 1.00)

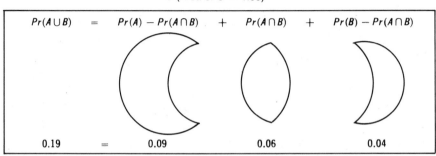

FIGURE 10.4 $Pr(\text{College Graduates} \cup \$50,00 + \text{Incomes})$ with Overlap Eliminated

Represented here are these elements: (a) the probability of graduates who do not have annual incomes in excess of \$50,000, *i.e.*, the part of circle A that excludes the overlap with circle B, or $Pr(A) - Pr(A \cap B)$; (b) the probability of noncollege graduates who have incomes in excess of \$50,000, *i.e.*, the part of circle B that excludes the overlap with circle A, or $Pr(B) - Pr(A \cap B)$; and (c) the probability of college graduates who also have incomes in excess of \$50,000, *i.e.*, the overlap between circles A and B, or $Pr(A \cap B)$. These three probabilities added together give us the computed probability of 0.19.

The addition rule can be extended to any number of events. For example, if three events were possible, A, B, and C, the addition rule would be.

(10.2)
$$Pr(A \cup B \cup C) = Pr(A) + Pr(B) + Pr(C) - Pr(A \cap B) - Pr(A \cap C)$$
$$- Pr(B \cap C) + Pr(A \cap B \cap C)$$

Notice that in this version of the formula the expression $Pr(A \cap B \cap C)$ is added back in, because subtracting out the expressions $Pr(A \cap B)$, $Pr(A \cap C)$, and $Pr(B \cap C)$ completely eliminated the intersection of A, B, and C.

10.3.2 The Special Addition Rule

In the case where the *events* that we are interested in are *mutually exclusive*, the formula for the addition rule can be simplified. For example, if A and B are mutually exclusive events (cannot occur together, they have zero overlap), the probability that either event A or event B will occur is given by the formula

(10.3)
$$Pr(A \cup B) = Pr(A) + Pr(B)$$

The probability of the intersection of A and B is 0, so it can be dropped from the formula.

Let $Pr(A)$ represent the probability voters register as Republicans and $Pr(B)$ represent the probability they register as Independents. Since it is not possible to be registered simultaneously both as a Republican and as an Independent, these two events are mutually exclusive. If $Pr(A) = 0.38$ and $Pr(B) = 0.15$, then the probability that randomly sampled voters will register either as Republicans or Independents is given by the following:

$$Pr(A \cup B) = 0.38 + 0.15 = 0.53$$

Literally, the special addition rule says *the probability of occurrence of any of a number of mutually exclusive events is equal to the sum of their separate probabilities.* In the case of three possible events, A, B, and C, the probability would be as follows:

(10.4)
$$Pr(A \cup B \cup C) = Pr(A) + Pr(B) + Pr(C)$$

This special case of the more general addition rule is of particular interest to sociologists because it fits many of the kinds of problems they are interested in and for which data are available. As we shall see later in Chapter 12, for example, the special addition rule plays an integral part in the generation of a binomial sampling distribution, and the binomial sampling distribution is useful for the analysis of small group data.

368 INFERENTIAL STATISTICS: SAMPLING AND PROBABILITY

10.3.3 The Rule of Conditional Probability

In the case of the addition rule, we were primarily interested in the occurrence of a composite event (although the general addition rule did include an adjustment for the possibility of the joint occurrence of events). In the example of college graduates and those with incomes in excess of $50,000, we might be interested in the proper technique for computing the probability that college graduates will have incomes in excess of $50,000.

It has been demonstrated that there is a relationship between level of education and annual income. Those persons who have completed higher levels of education tend to have higher annual incomes than those with less education. Because of this relationship, the probability of having an income in excess of $50,000 should be conditional upon level of education. The desired probability can be computed by using the rule of **conditional probability**. The formula for conditional probability is

(10.5)
$$Pr(B \mid A) = \frac{Pr(A \cap B)}{Pr(A)}$$

This means the probability that event B will occur, given the fact that event A has already occurred, is equal to the ratio of the probability of the joint occurrence of A and B, $Pr(A \cap B)$, to the probability of occurrence of event A, $Pr(A)$.

We are basically interested in a subset of the universal set. The question we wish to answer is: "If event A has happened, what is the probability that event B will happen?" Rather than focusing on the universal set, we look at that part of it representing A and ask ourselves how much of this subset also represents event B.

We found earlier that 15% of the population were college graduates, 10% had incomes in excess of $50,000, and 6% were college graduates with incomes in excess of $50,000. With these data, we can compute the probability that persons will have annual incomes in excess of $50,000 given the fact that they are college graduates. Let A represent college graduation and B having an income in excess of $50,000. Substituting the appropriate data into the conditional probability formula, we get

$$Pr(B \mid A) = \frac{0.06}{0.15} = 0.40$$

Thus, given the fact that persons sampled are college graduates, there is a 40% chance they will have incomes in excess of $50,000 per year. This probability figure differs considerably from the overall figure of 10% with incomes in excess of $50,000.

If we know that those we are considering do not have a college education, what can we say about the probability they will have incomes in excess of $50,000? Substituting the appropriate figures into the formula, we get

$$Pr(B \mid \sim A) = \frac{Pr(\sim A \cap B)}{Pr(\sim A)} = \frac{0.04}{0.85} = 0.047$$

where $Pr(\sim A)$ refers to the probability that persons are not college graduates and $Pr(\sim A \cap B)$ the probability that they are not college graduates but have annual incomes in excess of $50,000 (see Figure 10.4). Thus, given the fact that the persons we are considering do not have a college education, there is less than a 5% chance that they will have annual incomes in excess of $50,000.

10.3.4   The General Multiplication Rule

Through elementary algebra the formula for conditional probability may be rearranged to yield a formula for computing the probability of the joint occurrence of events. This formula is

(10.6)
$$Pr(A \cap B) = Pr(B \mid A) \, Pr(A)$$

or its equivalent

(10.7)
$$Pr(A \cap B) = Pr(A \mid B) \, Pr(B)$$

If we know the probability that event A will happen and the conditional probability that event B will happen given that A has already happened, we can compute the probability that both A and B will happen. This formula is known as the **general multiplication rule**.

Let us look at a possible application of this rule in sociology. Assume that 45% of married couples were engaged for less than 6 months. Further, assume that for those couples who were engaged less than 6 months, 20% of their marriages end in divorce within the first 5 years of marriage, whereas for those whose engagements were of 6 months or more duration the divorce rate is 10% during the first 5 years of marriage.

Given these data, we might ask, if we were to select couples who married 5 years ago, what is the probability that they were engaged less than 6 months (call this event A) and that their marriages were subsequently dissolved (call this event B)? The desired probability can be computed as follows:

$$Pr(A \cap B) = Pr(B \mid A) \, Pr(A) = (0.20)(0.45) = 0.09$$

We might also ask what is the probability of finding couples who had been engaged 6 months or more and whose marriages had been dissolved. Call the event being engaged 6 months or more $\sim A$, substitute the appropriate probabilities into the altered formula, and the resulting probability is

$$Pr(\sim A \cap B) = Pr(B \mid \sim A) \, Pr(\sim A) = (0.10)(0.55) = 0.055$$

This joint probability is somewhat lower than the first, indicating that short engagement is more likely to be followed by divorce than is longer engagement. Figure 10.5 is a tree diagram that includes the two joint probabilities we had just computed and the two remaining possibilities: those whose engagements were short and who are still married and those whose engagements were longer and who are still married. Notice that the four joint probabilities sum to 1.0. This is so because they represent all possible combinations of length of engagement and marital status. Notice, also, that each of the four joint probabilities is arrived at by multiplying the probabilities along the branches of the tree leading to it. Thus, the probability 0.495 is arrived at by multiplying 0.55 and 0.90: $(0.55)(0.90) = 0.495$.

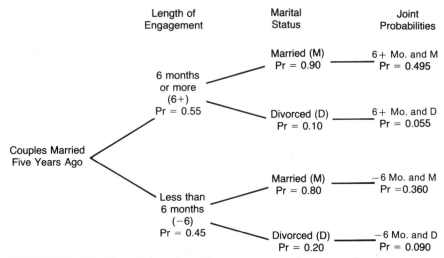

FIGURE 10.5 Tree Diagram of Joint Probabilities for Length of Engagement and Marital Status

The general multiplication rule also can be extended to more than two events. In the case of three events, A, B, and C, the formula for the joint occurrence of the three would be

(10.8)
$$Pr(A \cap B \cap C) = Pr(C \mid A \cap B)\, Pr(A \cap B)$$

or, since $Pr(A \cap B) = Pr(B \mid A)Pr(A)$,

$$Pr(A \cap B \cap C) = Pr(C \mid A \cap B)\, Pr(B \mid A)\, Pr(A)$$

Either expression may be used to compute the probability.

10.3.4a Stochastic Processes: An Application of the General Multiplication Rule

When we look at a sequence of events over time and compute the probabilities of their joint occurrence, we are dealing with **stochastic processes**. The general multiplication rule is appropriate for the computation of certain simple types of stochastic processes. The example we just considered involving the dissolution of marriage was, in effect, a two-step stochastic process. As sociologists move more toward the analysis of longitudinal data, the concept of the stochastic process will find its way increasingly into the literature.

10.3.5 The Special Multiplication Rule

The general multiplication rule applies to the probability of the joint occurrence of events regardless of whether the events are related or independent. If two events, A and B, are related, the probability of B is altered by the occurrence or nonoccurrence of A. When **two events are independent**, however, the occurrence or nonoccurrence of A will have no effect on the probability of occurrence of B. Furthermore, the occurrence or nonoccurrence of B will have no effect on the probability of occurrence of A. That is, $Pr(B|A) = Pr(B)$ and $Pr(A|B) = Pr(A)$.

Much of the data analysis engaged in by sociologists is aimed at determining whether variables (or events) are related or independent. In the field of descriptive statistics, considerable effort is devoted to an examination of techniques of analysis designed to determine whether variables are related or independent (see Chapters 6 through 9).

When sociologists ask a research question such as, "Are divorce rates higher among childless couples than among couples who have children?", they are asking whether the presence or absence of children in marriages alters the probability of divorce occurring. Essentially, this asks, "Is the presence or absence of children related to or *independent of* the probability of divorce?" If the events (having or not having children and getting or not getting a divorce) are independent, divorce rates for childless couples will be comparable to those for couples who do have children. Therefore, knowing that couples do or do not have children will not provide information useful for predicting whether they will or will not get a divorce.

It should be noted that lack of independence does not imply cause and effect. It merely implies that the occurrence or nonoccurrence of event A is associated with the occurrence or nonoccurrence of event B.

Given that events of interest are independent events, the general multiplication rule reduces to the special multiplication rule. This rule states that *the probability of the joint occurrence of independent events is equal to the product of their separate probabilities.* Therefore, if A and B are independent, the probability of their joint occurrence would be

(10.9)
$$Pr(A \cap B) = Pr(A)\,Pr(B)$$

This formula can be extended to include any number of independent events. The formula for the joint occurrence of the three mutually independent events, A, B, and C, for example, would be

$$Pr(A \cap B \cap C) = Pr(A)\,Pr(B)\,Pr(C)$$

As an illustration of the application of the special multiplication rule, assume we are interested in the probability that pregnant women will give birth to twins, will have red hair, and will be married to unemployed men. Presumably these three events are independent. At least there is no apparent reason to expect them to be related. If the probability of giving birth to twins is 0.01, the probability of having red hair is 0.15, and the probability of having an unemployed husband is 0.07, these figures can be substituted into the formula for the special multiplication rule to find the probability of the joint occurrence of the three events:

$$Pr(A \cap B \cap C) = (0.01)(0.15)(0.07)$$

$$= 0.000105$$

It is obvious that this combination of events is not very likely to occur by chance. As a matter of fact, given the individual probabilities we have used, such a combination would only be expected to occur by chance about 10.5 times out of 100,000 observations.

It should be noted that the events of concern do not need to take place simultaneously. The special multiplication rule applies equally well to events during a period of time, as long as the independence assumption is tenable.

In making use of the special multiplication rule, sociologists are often faced with a problem. In actual research data with which sociologists work, the independence of variables and observations being investigated must be carefully determined. This is frequently a difficult determination to make.

The difficulty of establishing independence of variables is illustrated nicely by an incident reported in *Time* magazine (April 26, 1968, p. 41). A witness reported that an elderly woman was mugged by a blonde girl with a ponytail who left the scene in a yellow car driven by a bearded black. An interracial couple fitting the description given by the witness was apprehended and tried for the crime. An expert witness was called by the prosecution to testify as to the probability of such a set of events occurring by chance. Using the special multiplication rule and what were considered conservative estimates of the probabilities involved (*e.g.*, that the probability of a car's being yellow was 0.1, that the probability of a couple in a car being interracial was 0.001, etc.), the expert estimated that there was about 1 chance in 12 million that another couple shared the same characteristics as the defendants. Largely on the basis of this testimony, the defendants were found guilty. However, the California Supreme Court overturned the conviction, holding that (a) there was not complete agreement among the witnesses on the characteristics in question, (b) that the probabilities assigned to the various characteristics were not shown to be even roughly accurate, and (c) that the special multiplication rule did not apply since the characteristics were not independent. On the latter question, the court pointed out that the probability of observing a black with a beard overlaps the probability that a bearded black might be part of an interracial couple.

BOX 10.3 RULES TO REMEMBER

You will find that the special addition rule (Section 10.3.2) and the special multiplication rule (Section 10.3.5) are the two rules of probability that will be most used in connection with the techniques discussed in the remainder of this volume. They are relatively elementary rules and should not be too hard to grasp. If they are not clear to you at this point, however, take time to review Sections 10.3.2 and 10.3.5. As you go over the discussion of the special multiplication rule, pay special attention to the concept of **independence** because this will be a key concept in later discussions of hypothesis testing.

In spite of the difficulties of establishing independence of variables, the special multiplication rule is often useful to sociologists in their research. Moreover, as we shall see later, the rule plays an important part in the test of certain statistical hypotheses (*e.g.*, see the discussion of chi-square (χ^2) in Chapter 16).

10.4 COMBINATORIAL RULES

In applications of probability theory it is sometimes necessary to determine the number of sequences in which elements of a set can be arranged. Is there a general way of determining how many possibilities there are? The answers to these questions are provided by one of a number of rules that constitute what is sometimes called combinatorial analysis. We will discuss only two of these combinatorial rules, the *permutation rule* and the *combination rule*, because these are the ones most relevant to our study of statistics.

10.4.1 The Permutation Rule

Permutations are ordered arrangements of elements. For example, assume that we have set up an experiment designed to test the reactions of subjects to different types of social pressures to conform. Furthermore, assume there are four experimental conditions and 10 subjects to be tested. We would wish to know in how many different orders 10 subjects might be assigned to four experimental conditions, and we might even plan to assign subjects to experimental conditions in every possible order. In Table 10.1 we have worked out a number of the different orders available to us. The letters A, B, C, and D represent four experimental conditions and the numbers represent subjects 1, 2, 3, and 4. The other subjects would be numbered 5, 6, 7, 8, 9, and 10.

Notice that in Table 10.1 we have used only the first four subjects and we already have 24 different orders in which they can be assigned to the experimental conditions. How many other arrangements are possible?

TABLE 10.1 SOME PERMUTATIONS OF TEN SUBJECTS ASSIGNED TO FOUR EXPERIMENTAL CONDITIONS

Trial No.	A	B	C	D	Trial No.	A	B	C	D
1	1	2	3	4	15	2	3	1	4
2	2	3	4	1	16	3	1	4	2
3	3	4	1	2	17	1	3	2	4
4	4	1	2	3	18	3	2	4	1
5	4	3	2	1	19	2	4	1	3
6	3	2	1	4	20	4	1	3	2
7	2	1	4	3	21	1	3	4	2
8	1	4	3	2	22	3	4	2	1
9	1	2	4	3	23	4	2	1	3
10	2	4	3	1	24	2	11	3	4
11	4	3	1	2	etc.				
12	3	1	2	4					
13	1	4	2	3					
14	4	2	3	1					

The permutation rule states that the number of ordered arrangements of M elements taken N at a time is given by the formula

(10.10)

$$P(M, N) = \frac{M!}{(M - N)!}$$

where $P(M, N)$ stands for the number of permutations of M elements taken N at a time, *i.e.*, M = population size, N = sample size, and $P(M, N)$ is the number of ordered samples of size N drawn from a population of size M.

The exclamation point behind each term in the formula stands for the factorial of the number specified by the term. Thus, $M!$ stands for the factorial of M. A **factorial** is the product of a sequence of consecutive integers down to 1. Therefore,

$$M! = M(M - 1)(M - 2)(M - 3)(M - 4)\ldots(3)(2)(1)$$

For example, if $M = 5$, the factorial of M will be

$$5! = (5)(4)(3)(2)(1) = 120$$

Take note of the fact that the factorial of 0 is defined as 1; that is, $0! = 1$.

Given the formula for permutations and an understanding of the meaning of a factorial, we can now substitute into Formula 10.10 and find the possible number of ordered arrangements of 10 elements taken 4 at a time:

$$P(10, 4) = \frac{10!}{(10 - 4)!} = \frac{10!}{6!}$$

$$= \frac{(10)(9)(8)(7)(6)(5)(4)(3)(2)(1)}{(6)(5)(4)(3)(2)(1)}$$

By canceling, we get

$$= \frac{(10)(9)(8)(7)\,(\cancel{6})\,(\cancel{5})\,(\cancel{4})\,(\cancel{3})\,(\cancel{2})\,(\cancel{1})}{(\cancel{6})\,(\cancel{5})\,(\cancel{4})\,(\cancel{3})\,(\cancel{2})\,(\cancel{1})}$$

$$= (10)(9)(8)(7)$$

$$= 5,040$$

BOX 10.4   SYMBOLS USED FOR COMBINATIONS AND PERMUTATIONS

M	Population size
N	Sample size
P	Number of permutations
$!$	Factorial operation $(n)\ldots(3)(2)(1)$
$P(M, N)$	Permutation of M elements taken N at a time (each order of N elements considered)
$\binom{M}{N}$	Combination of M elements taken N at a time (order of N elements is ignored)

Therefore, there are 5,040 permutations or ordered arrangements of 10 elements taken 4 at a time. It is obvious that it is impractical to include all these permutations in our experimental design and impossible without repeatedly measuring each element. What we would need to do instead is select some of the permutations to be used and ignore others.

10.4.2 The Combination Rule

Whereas a permutation is an ordered arrangement of elements, a combination is a selection of elements without regard to order. Thus, the 24 permutations in Table 10.1 would all be treated as one combination because they all contain the same elements (individuals 1, 2, 3, and 4), and only the order in which they appear differs.

The combination rule states that the number of combinations of M elements taken N at a time is given by the following formula:

(10.11)
$$\binom{M}{N} = \frac{M!}{(M - N)!\, N!}$$

Note that the expression

$$\binom{M}{N}$$

is not a fraction; it is merely the symbolic name for the combinations formula and is read, "The combination of M elements taken N at a time."

Suppose again that we have 10 subjects and wish to arrange them into as many groups of size 4 as are possible for purposes of studying their behavior in a problem-solving situation. How many different ways can they be arranged in groups of 4 without regard for order of arrangement? The answer is given by the following:

$$\binom{10}{4} = \frac{10!}{(10 - 4)!\, 4!} = \frac{10!}{6!\, 4!}$$

$$= \frac{(10)(9)(8)(7)(6)(5)(4)(3)(2)(1)}{(6)(5)(4)(3)(2)(1)\quad (4)(3)(2)(1)}$$

By canceling, we get

$$= \frac{(10)(9)(8)(7)(\cancel{6})(\cancel{5})(\cancel{4})(\cancel{3})(\cancel{2})(\cancel{1})}{(\cancel{6})(\cancel{5})(\cancel{4})(\cancel{3})(\cancel{2})(\cancel{1})\quad (4)(3)(2)(1)}$$

$$= \frac{(10)(9)(8)(7)}{(4)(3)(2)(1)} = \frac{5,040}{24} = 210$$

Therefore, there are 210 combinations of 10 elements taken 4 at a time as compared to 5,040 permutations. However, the 210 combinations are still probably too numerous to make use of in arranging groups of 4 to be tested. Again, we probably would select only certain combinations to test and ignore others.

By a slightly different interpretation of the meanings of M and N, we can use the combinations formula to find out, for example, how many different possible patterns of 4 wins and 1 loss there are in a baseball World Series that lasts five games. Let $M = 5$, the number of games played, and $N = 4$, the number of games won out of the 5. Substituting into the combinations formula with these figures, we get

$$\binom{5}{4} = \frac{5!}{(5-4)!\,4!} = \frac{5!}{1!\,4!}$$

$$= \frac{(5)(4)(3)(2)(1)}{(1)\quad(4)(3)(2)(1)}$$

By canceling, we get

$$= \frac{(5)(\cancel{4})(\cancel{3})(\cancel{2})(\cancel{1})}{(1)\quad(\cancel{4})(\cancel{3})(\cancel{2})(\cancel{1})}$$

$$= 5$$

Therefore, there are five different patterns of 4 wins and 1 loss that are possible in 5 games. The patterns are the following:

> win loss win win win
> win win loss win win
> win win win loss win
> loss win win win win
> win win win win loss

Of course the fifth pattern listed is not a possibility in a World Series because the first team to win 4 games wins the Series; therefore, this pattern reduces to a 4-game Series.

It may appear that order of occurrence was considered in this example, although we said earlier that order was not taken into account for combinations. The only distinction made here, however, was between wins and losses. We did not consider the four wins as four distinct events. If we had labeled the four wins as win A, win B, win C, and win D, and made distinctions in the order in which they occurred, then we would have permutations rather than combinations.

10.5 SUMMARY

In this chapter we have discussed and used two alternative concepts of probability (*a priori* and that based on data) and have presented a number of rules of probability useful to sociologists. We have presented examples of a sociological nature to suggest the kinds of applications that sociologists might have for these rules. We ended the chapter with two useful combinatorial rules.

In the following chapter we will see how probability theory plays an integral part in the sampling process, and in later chapters we will see that probability theory serves as the basis for the whole inductive process.

CONCEPTS TO KNOW AND UNDERSTAND

universal set	stochastic process
disjoint sets	conditional probability
overlapping sets	independent events
union	mutually exclusive events
intersection	general and special multiplication
a priori probabilities	rules
outcome	general and special addition rules
event	permutation
experiential probabilities	combination

QUESTIONS AND PROBLEMS

1. Give examples of finite and infinite universal sets that are of interest to sociologists. Read five article abstracts from a professional journal in sociology (such as those that appear in *The American Sociological Review*). Determine which universal sets the authors were interested in. Determine whether each of the universal sets is finite or infinite.

2. Explain what is meant by the statement that a probability is not a prediction of a single event, but of a large number of events. If 75% of the students who take a course in elementary statistics earn at least a C for a final grade, can you assume that you have a 75% chance of earning at least a C?

3. Give an example of two variables of interest to sociologists to which the general addition rule would apply (that is, two variables that are not mutually exclusive). Give an example of two variables of interest to sociologists to which the special addition rule would apply (that is, two variables that are mutually exclusive).

4. Write the formula for the general addition rule for $Pr(A \cup B \cup C \cup D)$.

5. Which of the following pairs of variables are independent of each other, and which of them are conditionally related? For those pairs that are conditionally related, specify which variable of each pair is logically prior to the other.
   a. religious affiliation and income
   b. size of family and income
   c. number of children under 18 and color of mother's eyes
   d. number of years of formal education and number of books read per year
   e. length of women's skirts and unemployment rates
   f. political affiliation and marital status
   g. age and country of birth

6. Think of two examples for which the permutation rule would be applicable and two examples for which the combinations rule would be applicable.

7. Rules of probability are often illustrated with coin tosses or throws of dice. Suppose that you have a "fair" pair of dice; what is the probability of the event:
   a. Getting two 6's?

    b. Getting a 6 and a 2?

    c. Getting either a 4 and a 2 or a 5 and a 1?

    d. Getting an outcome that adds to 6?

8. Your job is to set up tennis matches between pairs of people and you have a list of seven people, below, who have signed up to play. One person in each pair will be given the responsibility of picking up equipment for his or her match.

    Sally   Sue
    Joe   Steve
    Jim   Debby
    Laura

    a. How many different pairs can be created, where each person is assigned responsibility for the equipment for each pair?

    b. Ignoring who has responsibility for the equipment, how many different pairs can be created?

    c. How many pairs can be created where Sally is included on a team? Which rule of probability did you use to arrive at this conclusion?

    d. Now suppose that you want to create teams of four players from the above group of seven. How many different teams could be created (ignoring order of team selection)?

    e. How many different ways could the seven be arranged in order on a list?

GENERAL REFERENCES

Adler, Irving, *Probability and Statistics for Everyman: How to Understand and Use the Laws of Chance* (New York, New American Library, A Signet Science Library Book), 1963. This is a well written book that covers many of the ideas of probability theory in an understandable but solid fashion. Alder includes many interesting examples and puzzles.

Freund, John E., *Modern Elementary Statistics*, 3rd ed. (Englewood Cliffs, N.J., Prentice Hall), 1967. Part II is a lucid discussion of probability theory and probability distributions. Freund's approach to the subject of probability is more general than we have presented in this book.

Hays, William L., *Statistics*, 3rd ed. (New York, Holt, Rinehart and Winston), 1981. Chapter 1 introduces probability and set theory.

Kemeny, John G., J. Laurie Snell, and Gerald L. Thompson, *Introduction to Finite Mathematics* (Englewood Cliffs, N.J., Prentice Hall), 1956. Consult this book for a more detailed treatment of set theory. It is also a good source for illustrations of how mathematics is applied to the social sciences.

Kerlinger, Fred N., *Foundations of Behavioral Research*, 3rd ed. (New York, Holt, Rinehart and Winston), 1986. Chapter 4 briefly introduces the ideas of sets.

Mosteller, Frederick, Robert E. K. Rourke, and George B. Thomas, Jr., *Probability with Statistical Applications* (Reading, Mass., Addison-Wesley), 1961. This book was written as the text for a televised course in probability and statistics. Although it is mathematically sound, the approach is intuitive.

Oakes, Michael, *Statistical Inference: A Commentary for the Social and Behavioral Sciences*, (New York, Wiley), 1986.

LITERATURE CITED

Blau, Peter M., and W. Richard Scott, *Formal Organizations* (San Francisco, Chandler Publishing Company), 1963.

Bridges, Edwin M., *et al.*, "Effects of Hierarchical Differentiation on Group Productivity, Efficiency, and Risk Taking," *Administrative Science Quarterly*, 13 (September 1968), pp. 305–319.

Brown, Robert, *Explanation in Social Science* (Chicago, Aldine), 1963.

Durkheim, Emile, *Le Suicide* (Paris), 1897. See also the translation: Emile Durkheim, *Suicide* (Glencoe, Ill., The Free Press), 1951.

CHAPTER 11

Sampling Principles

Sociologists are engaged in the business of developing generalizations about social behavior. They attempt to generate explanatory schemes to predict the orderly operating process of social systems such as total societies (*e.g.*, the United States), communities (*e.g.*, Solvang, California), organizations (*e.g.*, the American Legion), and groups (*e.g.*, the corner gang). These explanatory schemes are generally tested through the analysis of limited bodies of data, called **samples**. The samples are not inherently interesting in themselves, however. They are only of interest to the extent that they represent larger, more significant bodies of data known as **populations**. In sociology, as in most other sciences, samples are used to represent populations because they are more economical to study and because selection of a sample is often feasible when complete enumeration of a population is not. Note that "sample" refers to all N cases, collectively. Individual cases in a sample are called "cases" or "elements" (not samples).

11.1 SAMPLE REPRESENTATIVENESS

Both for selecting a representative sample and for generalizing results to the population sampled, probability theory is crucial. In this chapter we will examine how probability theory operates in the selection of a representative sample, and in the following chapters we will examine the part probability theory plays in the generalization process itself.

To illustrate, it has been found that the level of academic performance of students is related to the degree of racial segregation of the school they attend. For example, students of predominantly black schools tend to score lower, on the average, on standardized reading tests than students of predominantly white schools. There are several possible explanations for this phenomenon. One possibility is that the quality of teaching is lower in black schools than in white schools. Another is that facilities in white schools are superior. A third possibility is that the typical educational program is geared to white children and is not relevant and meaningful to black children. A fourth possibility is that black children are not being offered sufficient motivation to learn to read and achieve. Numerous other factors also suggest themselves as possible explanations.

Suppose we wish to do a study of elementary schools with varying percentages of minority children to see how much their students vary in average reading ability, and to isolate those factors that best explain any differences we find. The public elementary schools of Los Angeles vary widely in percentage minority enrollment. Furthermore, since Los Angeles is a city with a heterogeneous population largely free of sectional peculiarities, it might be considered a good choice of a **population** to study.

Since there are 411 public elementary schools in Los Angeles (or were, as of 1985), a study of the whole population of schools would probably be beyond the limits of our resources in staff, dollars, and time. Therefore, we may decide to sample the schools and carry on an intensive analysis of those schools in the **sample**. We would naturally wish to get a **representative sample** of schools so that we would have a good spread in percentage of minority enrollment and a realistic distribution of those other variables relevant to differences in average reading ability of students. If the sample is representative, we can feel justified in generalizing the findings from the sample data to the population of all public elementary schools in the city. The basic sampling technique that is indicated for drawing a representative sample from a known population is called a **simple random sample**, or **SRS**.

It should be noted that not all simple random samples are representative of the population from which they are drawn. It is the case, however, that simple random samples are more likely to represent accurately characteristics of the population than not. Any differences between characteristics of a sample and the characteristics of the population it represents will be random differences for which we will learn later how to compensate.

11.2   THE SIMPLE RANDOM SAMPLE

A **simple random sample** is one in which the *elements* of the population are listed, and then N of them are randomly selected to be elements of the sample. The random selection is carried out in such a way that (a) **every element in the population has an equal probability of being included** in the sample, and (b) **every possible combination of N elements has an equal probability of constituting the sample**. Simple random selection *does not mean haphazard selection*; rather it means a selection process that gives *each element in the population an equal chance of appearing in the sample.* *

If there are M elements in a population, according to the first condition noted above, the probability of each individual element being included in the sample should be $1/M$. Therefore, in drawing a simple random sample of the 411 public elementary schools in Los Angeles, the selection process should be such that each and every school should have a probability of 1/411 of being included.

---

* In general, a random sample is one in which each element has a calculable probability of being selected. Simple random samples are those where this probability is equal for each population element.

Furthermore, the probability of each school being included in the sample should be **independent** of the probability of any other school being included. This is necessary to meet the second condition for a simple random sample: that every possible combination of N elements has an equal chance of constituting the sample.

11.2.1   Sampling with and without Replacement

Assume that we number the 411 schools in the population serially from 001 through 411, and then we randomly select 100 of these numbers to represent the elements of the sample. (We will explain below how the random selection is accomplished.) If we take as the elements of the sample the first 100 numbers between 001 and 411 that we select, even though some of the numbers may appear two or more times in the sample, we are said to be **sampling with replacement**. If in drawing the 100 numbers we insist that they be 100 unique numbers, then we are said to be **sampling without replacement**.

Both of these sampling procedures meet the first condition of the simple random sample; that is, that every element in the population have an equal probability of being included in the sample. When we sample *with replacement*, for example, the first element drawn in the sample has a probability of $1/411$ because there are 411 numbers to choose from; the second element drawn has a probability of $1/411$ because there are still the same 411 elements to choose from, etc.

When we sample *without replacement*, the first element drawn has a probability of $1/411$. The second element drawn is drawn from a pool of the remaining 410 elements (the first element drawn no longer being eligible to be drawn a second time), but its probability of being drawn on the second draw is *conditional* upon its *not* having been drawn on the first draw. The conditional probability of the second element being drawn, then, is $(410/411)(1/410)$, that is, the probability that it was not drawn on the first draw, times the probability of its being drawn on the second draw. Notice that this probability is $(410/411)(1/410) = 1/411$, which is the same probability that the first element had of being drawn on the first draw. Similarly, the third element in the sample will have a probability of $(410/411)(409/410)(1/409) = 1/411$ of being drawn third; and thus each succeeding element, according to the rule of conditional probability, will also have a probability of $1/411$ of being included in the sample.

Although the process of sampling *with replacement* also meets the second condition for a simple random sample, sampling *without replacement* does not. Some of the samples that are possible when sampling with replacement are not possible to achieve if a finite population such as this is sampled without replacement. For example, it would be impossible to achieve any of the 411 samples that include the same element drawn repeatedly 100 times. Nor is it possible to get any of the samples in which an element appears more than once. Thus the second condition for simple random samples, that all possible samples of N elements are equally probable, does not hold for sampling without replacement from finite populations.

The number of possible samples of sizes $N = 100$ that could be drawn from the population (M) of 411 schools where one samples **without replacement** (and order of drawing specific elements is not considered) is computed by the following formula:

(11.1)

$$\binom{M}{N} = \frac{M!}{(M - N)! \, N!}$$

which you will undoubtedly recognize as the formula for the number of combinations of M elements taken N at a time.

If we were to draw random samples of size 100 *without replacement* from the population of 411 elementary schools, we would have the following possible number of different samples:

$$\binom{411}{100} = \frac{411!}{(411 - 100)! \, 100!}$$

$$= \frac{411!}{311! \, 100!}$$

$$= 161{,}218 \times 10^{90}$$

As you can see, the number of samples possible when sampling without replacement is astronomical. The number possible when sampling with replacement is even larger.

Technically speaking, sampling with replacement is a necessary condition for a simple random sample, and we will reserve the term **simple random sample** for those random samples drawn *with* replacement. Those samples drawn without replacement we will call **random samples drawn without replacement**.

Many sociologists have qualms about sampling with replacement because of the possibility of drawing the same case more than once. They have real reservations about counting one person's questionnaire responses more than once, for example, or giving the same person more than one questionnaire to fill out. It is important to note, however, that sociologists are not interested in unique cases of a certain type (*e.g.*, Sam Smith who is tall), but only in *kinds of cases* defined in terms of variables of interest (*e.g.*, tall people).

Thus the problem with sampling without replacement is that the researchers may run out of certain *kinds*. That sort of problem can be solved by sampling with replacement or sampling from a very large population. The second solution, by the way, is more often adopted than the first by sociologists on the grounds that, if the size of the sample N is relatively small as compared to the size of the population M, it makes very little difference whether the sampling is done with or without replacement.

It is our position that **sampling with replacement is technically correct** and that it is entirely legitimate to count data from a population element more than once if that element appears in the sample more than once. If there is sound reason for sampling without replacement (*e.g.*, because of the nature of the problem being investigated or because of the quality and quantity of the data available), then that procedure is justifiable and should be followed. However, sampling without replacement should *not* be preferred merely on the grounds that sample elements should all be unique.

11.2.2   The Basic Nature of the Simple Random Sample

The simple random sample is the basic sampling technique for inferential statistics. It has served as the model from which other random sampling techniques were developed. Furthermore, it can be shown that the formulas employed to compute statistics such as means and standard deviations from sample data were derived using assumptions consistent with the logic of the simple random sample. This we will do here for the mean.

In deriving formulas to measure central tendency and variability, statisticians began with the assumption that they were working with a random variable X. Given a distribution of scores of the random variable X for which each value of X has a known probability, $Pr(X)$, and for which $\Sigma Pr(X) = 1.00$, the expected value of the variable X, $E(X)$ (also known as the expectation of X), is defined to be

(11.2)
$$E(X) = \Sigma [X Pr(X)]$$

Thus each value of X is multiplied by the probability of its occurrence, $Pr(X)$, and the products are summed over the X scores to arrive at the expected value of X. This expected value of X is, in fact, the mean of X (μ).*

As an example, suppose we had a distribution of X scores in which X took five different values and those values of X had the following probabilities of occurrence associated with them:

X Value	Pr(X)
5	.15
7	.10
9	.30
12	.20
16	.25

Then,

$$E(X) = 5(.15) + 7(.10) + 9(.30) + 12(.20) + 16(.25)$$
$$= 10.55$$

and $\mu = 10.55$. Note that $\Sigma Pr(X) = 1.00$ $(.15 + .10 + .30 + .20 + .25 = 1.00)$. If M is the total number of scores in a distribution, then each of the M scores can be assigned a probability, $Pr(X)$, of $1/M$, ignoring the possibility that two or more of those scores may have the same value. Thus, the formula for the mean of X (μ), can be written as

(11.3)
$$\mu = \Sigma [X(1/M)]$$

* Note that, in order to keep sample and population means separate, the Greek letter mu (μ) is used as the symbol for a population mean and X-bar $(\bar{X})$ is used to symbolize a sample mean. There will be other points at which different symbols are used to distinguish sample and population values.

Since $1/M$ is a constant and X is a variable, we can apply the principle that *the sum of a constant times a variable is equal to the constant times the summation of the variable.* We may thus rearrange the terms as follows:

(11.4)
$$\mu = \frac{1}{M}\Sigma X$$

which is equivalent to the familiar

(11.5)
$$\mu = \frac{\Sigma X}{M}$$

Thus, when Formula 11.5 is used to compute a mean the assumption is made that each score in the distribution of scores has an equal probability of occurrence. The assumption of equal probability of occurrence within the total population is carried over to the formula for computing the sample mean in accordance with the requirement that, in a simple random sample, each element has an equal probability of occurrence in the sample. Thus, the formula for the sample mean is given as

(11.6)
$$\bar{X} = \frac{1}{N}\Sigma X \quad \text{or} \quad \frac{\Sigma X}{N}$$

This same line of reasoning was followed in the derivation of the formulas for most statistical techniques. Therefore, *whenever you use the techniques covered in this book to make inferences to populations, you are, in effect, assuming that your data constitute a simple random sample.* Any use of these techniques to generalize from a sample that is *not* a simple random sample must be recognized as a violation of this assumption and must be taken into account in interpreting results.

11.2.3   Random Selection

Crucial to simple random sampling is the process by which elements are selected for inclusion in the sample. If the conditions of simple random sampling are to be met, the selection process must be random.

If we wished to draw a simple random sample of 100 schools from the population of 411, how would we go about doing it? It might occur to you that a good procedure would be to write the names of the 411 schools on slips of paper, put the slips in a container, mix the slips thoroughly, and draw slips from the container. To sample with replacement, we would replace each slip drawn after recording the name of the school on it (if we wished to draw a random sample without replacement, we would not replace the slips drawn). Although this procedure seems to be a logical way of drawing a simple random sample, it is not generally satisfactory because it is difficult to mix the slips thoroughly enough to approximate a random process.

When the first military draft lottery was conducted in December 1969, using birth dates encased in capsules and placed in a bowl, a statistician conducted an analysis of the order in which birth dates were selected and came to the conclusion that the process was not random. He found that the birth dates in the last

6 months of the year were more likely to be drawn early than were birth dates in the first 6 months (he estimated that there were only 5 chances in 1,000 that the results would turn out as they did if each date had an equal probability of being drawn). The statistician, John Ware, said:

> I would guess that they probably placed the capsules containing the numbers into the bowl in a chronological order.... In the course of shuffling the capsules, the months tended to stay together. The first six months may have been placed in the bowl first and they tended to stay together and be drawn last.... The cards were stacked against its being fair.... It wasn't the best procedure with our current knowledge of probability (Berman, 1969).

The best approximation to a random process is achieved through the use of random numbers. Random numbers are generated in such a way that all integers from 0 through 9 occur with about equal frequency. The digits appear in a random fashion; that is, each individual digit has an approximately equal probability of appearing in any position. Computers may be used to generate numbers that behave like random digits (called pseudo-random numbers). A table of random numbers is reproduced as Table 11.1. Table 11.1 was produced by a computer. The digits are grouped in columns to make them easier to read, but the groupings in the tables are without any other significance. Appendix A provides one computer program to create pseudo-random numbers. Other programs are available in statistical packages.

To illustrate the use of random numbers to draw a simple random sample of 100 schools from the Los Angeles population of 411 elementary schools, we will use Table 11.1. **First**, we take the list of schools and number them serially from 001

TABLE 11.1 RANDOM NUMBERS

93	71	61	68	94	32	88	65	97	80	92	05	24	62	15	95	81	90	68	31	39	51	03	59
05	27	69	90	64	94	92	96	26	17	73	10	27	41	22	02	75	86	72	07	17	85	78	34
76	19	35	07	53	39	49	56	62	33	44	42	36	40	98	32	32	57	62	05	26	06	07	39
93	74	08	28	82	53	57	93	12	84	38	25	90	83	82	45	26	92	63	01	19	89	01	88
32	58	08	51	49	36	47	33	31	12	36	91	86	01	97	37	72	75	85	85	13	03	25	52
45	81	95	29	79	61	95	87	71	00	17	26	77	09	43	10	06	16	88	29	35	20	83	33
74	35	66	35	29	72	01	91	82	83	16	70	07	11	47	36	11	13	30	75	86	78	13	86
65	59	27	48	24	54	76	88	65	12	25	96	74	84	39	34	13	28	59	72	04	05	74	02
28	46	17	65	74	11	40	14	90	71	22	67	69	08	81	64	74	49	16	43	59	15	29	26
65	59	08	02	41	32	64	43	44	96	24	04	36	42	03	74	28	38	73	51	97	23	78	67
54	84	65	47	59	65	13	00	48	60	88	61	81	91	61	71	29	92	38	53	27	95	45	89
09	80	86	30	05	14	33	56	46	07	80	90	89	97	57	34	78	03	87	02	67	55	98	66
64	85	87	53	90	88	23	16	81	86	03	11	52	52	75	80	21	56	12	71	92	55	09	97
33	34	40	32	30	75	75	46	10	51	82	16	15	83	38	98	73	74	91	87	07	61	50	27
12	46	70	18	95	37	50	58	71	20	71	45	32	95	04	61	89	75	53	31	22	30	84	20
40	18	90	11	94	75	04	46	60	32	28	46	66	87	95	77	76	22	07	91	55	70	34	48
83	14	65	17	78	55	48	94	97	23	06	94	54	13	74	08	07	55	35	89	72	65	34	46
74	15	52	84	33	85	31	08	00	74	54	49	34	28	07	86	43	93	71	62	81	85	64	60
32	14	17	49	84	71	41	44	72	88	84	36	39	72	43	23	80	77	98	38	78	52	02	92
45	95	74	96	44	03	12	65	11	69	85	50	03	58	65	40	15	03	64	04	71	36	69	94

to 411. **Second**, to minimize bias in entering the table of random numbers, we use some reasonably random process of deciding where to start. There are 48 vertical, single-digit columns and 20 horizontal rows of numbers on the page. We might use a random process to pick the column and row at which we will enter the table. For example, we might determine the number of the column by rolling one die to find the first digit of a two-digit number between 01 and 48. If a 1, 2, 3, or 4 comes up, we would use that as our first digit. If a 5 comes up, we could treat it as a 0 and use it as our first digit. If a 6 comes up, we could ignore it and roll the die again. The second digit of the column number could be obtained in a similar way, perhaps by summing two tosses, so the number could be 0 through 8. The appropriate starting row would be determined through a similar procedure. A roll of a die would determine the first and second digit in the row number.

If you examine Table 11.1, you will notice that the digits 486 are underlined. The 4 is in the seventeenth vertical, one-digit column and in the eleventh horizontal row. This is the starting point arrived at by rolling the die.

Since we need three-digit numbers (001 through 411), we will consider the 4 the first digit and also include the digits in columns 18 and 19. Starting at that point we read down the column (or in some other previously determined direction), reading three digits at a time, until we come to a three-digit number between 001 and 411 inclusive. The first such number happens to be 105. The school numbered 105, therefore, will be the first element of the sample. We continue reading down the column until we find another eligible number (the next one is 007) and the school bearing that number is the second element of the sample. We read down the three-digit column looking for more eligible numbers, then shift to the next columns of three digits (single-digit columns 20, 21, and 22), etc., reading as far along as necessary to draw the sample of 100 schools. Table 11.2 includes the actual list of 100 eligible random three-digit numbers drawn from the table. The schools bearing these numbers would constitute the sample of 100 elements.

If you examine the 100 numbers drawn, you will notice that there were eight numbers that were drawn twice. These were 052, 075, 083, 161, 189, 220, 325, and 352. The number 052, for example, was drawn sixteenth and seventy-nineth. Since we wished to draw a simple random sample, we drew the sample **with replacement**. The data collected from the eight schools selected twice would be counted twice in the analysis. Consequently, the sample size would be 100 although there are actually only 92 different schools in the sample.

If researchers could not bring themselves to sample with replacement and count cases more than once, they could use the table of random numbers to draw a random sample without replacement. **To sample without replacement**, the only alteration necessary in the procedure described would be to ignore a number whenever it was duplicated and continue selecting numbers from the table until 100 unique numbers had been drawn. If the sample were drawn without replacement, however, the researchers would be obligated to take account of that fact in interpreting the results of their data analysis.

For convenience in describing how random numbers are used in sampling, the table of random numbers (Table 11.1) is provided here. In actual practice, tables of random numbers are seldom used for sampling any more because of the availability of computers.

TABLE 11.2   ONE HUNDRED ELEMENTS DRAWN WITH REPLACEMENT. USING THE TABLE OF RANDOM NUMBERS

105	115	189	395	284
007	161	380	178	098
116	363	015	260	333
092	083	168	198	124
344	191	133	130	401
236	252	285	352	321
017	295	387	040	161
316	358	038	279	408
225	159	220	076	195
122	220	036	223	086
088	393	207	103	075
182	037	205	325	288
071	389	301	083	297
228	046	075	378	352
306	078	351	150	024
052	275	271	084	300
102	232	187	034	323
259	006	331	202	189
267	129	162	052	119
043	021	404	325	177

11.2.4   A Preliminary Word about Sampling Error

Sampling error is a term that refers to the variability that could occur from sample to sample if repeated random samples of the same size were drawn from the same population. The term will be discussed in detail later on, but a preliminary word is in order at this point. One of the attractive features of the simple random sample is that there is a straightforward way of measuring *sampling error* and gauging the degree of representativeness of the sample. Fortunately, because of the procedures used to sample, whatever sampling error there is will be random. In the next chapter (Sections 12.2.2 and 12.5.4 in particular) we will consider, in detail, the measurement of sampling error.

The magnitude of sampling error in a simple random sample is a function of the size of the sample. Interestingly enough, what is crucial is *not* the relative proportion of the population represented by the sample. Rather, as the number of cases in the sample is increased, sampling error decreases without regard to population size. It can generally be assumed that large samples will have small sampling errors. It is desirable, then, to take as large a sample as is feasible. It would *not* make sense, however, to draw a sample so large that it would be a major part of a population. It would be better to take a complete enumeration of the population and eliminate sampling error. Sampling is, in part, an economy measure, and there is no economy in taking a sample that includes most of the cases in the population.

Although sampling error tends to be minimized in large samples there are, unfortunately, increasing opportunities for **nonrandom errors** to creep into the data collection and data processing procedures. There are, for example, more chances for interviewers to make errors or introduce biases in the process of asking questions and recording answers. Furthermore, the chances that coders will make errors in processing questionnaire or interview-schedule data also increase as the volume of data being processed increases. These types of nonrandom errors, *unlike* random errors, cannot be handled through statistical analysis. Only through careful research design and execution can nonrandom error and bias be minimized. Careful research design and execution are central topics in courses in research methods.

11.2.5   Feasibility of Simple Random Samples

Although the simple random sample is fundamental to inferential statistics, most of the studies in the sociological literature are not based on data from simple random samples. One exception is a study by Portes (1969). Portes studied factors involved in the integration of Cuban refugee families who migrated to Milwaukee. His data were based on a simple random sample from the population of names of family heads found in lists provided by the Cuban Association of Wisconsin and other similar organizations. He drew a sample of 48 families from the population of 152 families. In this case, a simple random sample was feasible because the population was small and a list of population elements was available.

To draw a simple random sample, it is necessary to have a **list of population elements (cases)**. If such a list is not available or cannot be created, a simple random sample is not feasible.

If the units of study are social systems, it is more likely that a list can be obtained than if the units of study are individuals. Populations whose elements are social systems are generally more stable and permanent than populations whose elements are individuals. For example, we were able to obtain a list of the public elementary schools in Los Angeles; furthermore, we had some confidence that the list was accurate because schools generally exist over a period of many years. It would be a much more difficult task to compile a single list of all of the students enrolled in those 411 schools, and we would have much less confidence in the accuracy of such a list if it were compiled. Populations involving a large number of individual elements are generally fluid. That is, they are in a state of constant change. People are born, they die, and they move. Thus, publications such as city directories, telephone directories, and mailing lists are seldom accurate to begin with and they "decay" at the rate of about one-third per year.

When sociologists are interested in studying large populations of individuals (*e.g.*, the labor force of the United States), they are not likely to be able to get a list from which to draw a simple random sample. And, of course, those populations most significant from a theoretical standpoint are also those for which lists are least likely to be available. If random samples of such populations are desirable, but simple random samples are not feasible, then how may random samples be drawn? The answer is that sociologists must turn to some sampling design that is random, but a variation of the basic simple random sample. One such sampling design is the **cluster sample**.

11.3 THE CLUSTER SAMPLE

A **cluster sample** is usually a multistage random sample. It is drawn by taking a series of simple random samples. That is, the population is first divided into a set of clusters, and a simple random sample of the clusters is drawn. Then the clusters sampled at the first stage are each divided again into clusters, and a simple random sample of those clusters is drawn, etc. The final elements sampled may be either clusters or individuals.

The clusters from which the initial random sample is drawn are called **primary sampling units**; the subclusters sampled next are called **secondary sampling units**; etc. The final clusters or individuals sampled are called the **ultimate sampling units**.

Assume that we wish to draw a random sample of all registered voters in the United States. It might be drawn as follows. **First**, we could take a simple random sample of the 50 states and the District of Columbia, using a table of random numbers. The states selected would constitute our primary sampling units. **Second**, from each of the states drawn in the first stage we could draw simple random samples of counties or congressional districts. These would be the secondary sampling units. **Next**, we might divide the sample counties or districts into voting precincts and draw simple random samples of precincts. These would be tertiary (third-stage) sampling units. **Finally**, we could collect data from all registered voters in the sample precincts, or we could draw simple random samples of registered voters from the lists of voters in the sampled precincts. The registered voters in the sample precincts would constitute the ultimate sampling units. Table 11.3 outlines such a multistage random or cluster sample. Notice that lists of registered voters are required only for the sample precincts drawn and not for all precincts in the country.

The cluster sample has the virtues of (a) not requiring a single list of all of the elements of the population, and (b) having the elements in the sample geographically clustered.*

TABLE 11.3 EXAMPLE OF MULTISTAGE RANDOM SAMPLE OF REGISTERED VOTERS

Sampling Level	Subset or Element	Example
Primary sampling units	States	Missouri
Secondary sampling units	Counties	Howard County
Tertiary sampling units	Precincts	Precinct #10
Ultimate sampling units	Registered voters	Thomas Anderson

---

* Clusters need not be geographically identified, but this is common and has the advantage of reducing travel costs involved in personal interviewing.

11.3.1   Homogeneous and Heterogeneous Clustering

The cluster sample is most efficient when the clusters that form the primary sampling units are homogeneous with respect to relevant variables, and the ultimate sampling units are heterogeneous; that is, **when between-cluster variability is small relative to within-cluster variability**. This cuts the costs of a study because it is possible to take a smaller sample of primary sampling units where geographic dispersion is a problem, and to do complete enumerations or to select relatively large samples of ultimate sampling units that tend to be geographically clustered.

In the example just cited, states would be quite heterogeneous with respect to political party membership. Ideally, they should be homogeneous; therefore, it would be wise to include a sizable number of states in the sample in order to assure a more representative sample.* Ideally, the precincts sampled should be quite heterogeneous. Actually, some precincts would be heavily Republican, others heavily Democratic, and still others mixed. Thus, individual precincts would differ in their contributions to the representativeness of the total sample. From the standpoint of sampling efficiency, it would be desirable if all precincts that appeared in the final sample were of the mixed type.

Cluster samples generally have a larger sampling error than simple random samples of the equivalent size depending upon the magnitude of between-cluster and within-cluster variation. When the clusters that form primary sampling units are heterogeneous, and the ultimate sampling units are homogeneous within the final clusters sampled, then the sampling error of a cluster sample is maximized as compared to a simple random sample of the same size. As the clusters that form the primary sampling units become more homogeneous, and the ultimate sampling units within the final clusters sampled become more heterogeneous, the sampling error of the cluster sample becomes smaller and approaches that of a simple random sample (when between-cluster variation is 0).

The disadvantage—that sampling error tends to be greater in a cluster sample than in a simple random sample—must be weighed against the advantages accrued from the use of the technique. An important advantage is that the cost per case of collecting data from a cluster sample tends to be less than that for a simple random sample because of lower locating and travel-related costs. Another important advantage is that a cluster sample is often feasible when a simple random sample is not. Whereas the simple random sample requires a single list of the elements in the population, a cluster sample does not.

11.3.2   Further Examples of Cluster Sample Designs

We described how we might draw a cluster sample of all registered voters in the United States. If we wished to draw a national sample, but did not wish to restrict it to registered voters, how might we go about it? We might start as we did in our

---

* To increase the homogeneity of the states, they could be grouped into regions first and then sampled so that various regions would be represented. This procedure would introduce the notion of *stratification*, to be discussed in the following section of this chapter.

previous example, sampling states and then counties. Instead of sampling voting precincts, however, we might sample Bureau of the Census enumeration districts and residences within the enumeration districts.*

The residences to be used as the ultimate sampling units could be sampled using detailed maps of enumeration districts. Once the residences had been sampled, interviewers could be sent to them with instructions to interview the head of the household or some other previously determined inhabitant.

A cluster sample reported in the sociological literature was used in a study by Spaeth (1968). The sample design was a rather simple two-stage design used in a 4-year longitudinal study of June 1961 college graduates, conducted by the National Opinion Research Center. The first stage was a sample of 135 of the accredited colleges and universities in the United States that grant baccalaureate degrees. The second stage consisted of prospective June 1961 graduates of the 135 colleges and universities. All 135 schools cooperated in the study and 85% of the prospective graduates sampled returned completed questionnaires. The sample consisted of some 41,000 students.

11.3.3   Cluster Samples and Statistical Analysis

As we pointed out earlier in this chapter, statistical techniques generally are based on the assumption that the data to be analyzed constitute a simple random sample of some population of data. Therefore, when data come from a cluster sample rather than a simple random sample, caution must be exercised in interpreting results. Since sampling error is generally greater in cluster samples than in simple random samples, tests of statistical significance may be misleading. If the actual sampling error is underestimated, a result that appears to be statistically significant may not be significant at all. If adjusted formulas are available for use with cluster samples, these should be used in lieu of the formulas that appear in this volume.† For many techniques, however, no such adjusted formulas have been derived. When adjusted formulas are not available and unadjusted formulas are used, it is advisable to make very stringent tests of hypotheses and to be conservative in interpreting results.

11.4   THE STRATIFIED RANDOM SAMPLE

Another variation from the simple random sample is the stratified random sample. The stratified random sample serves different purposes than the cluster sample. It is used when it is desirable to make comparisons among subpopulations of a

---

* The Bureau of the Census has divided the United States into thousands of enumeration districts (EDs) that are sometimes as large as a county and sometimes as small as part of a residential block. In any case, each is small enough to be enumerated by a single field worker (see Scott, 1968).
† For a discussion of proper formulas for cluster samples, see Kish (1965: 151; see General References).

population or when it is desirable to reduce either sampling error or the costs of a study.

A **stratified random sample** *is one in which the population is divided into subpopulations* (**strata**) *and then simple random samples are drawn from each of the subpopulations.* When it is desirable to make comparisons among subpopulations, the subpopulations are usually differentiated on the basis of one or more of the **independent** variables of the study. For example, if sociologists were going to study comparative family planning practices of Protestants, Catholics, and Jews, religious affiliation could be used to divide the population into three strata—one consisting of Protestants, one of Catholics, and one of Jews (other religions would be excluded for this problem). Then a simple random sample would be drawn from each stratum.

As mentioned above, stratified random samples may also be used in lieu of simple random samples to **reduce sampling error or the costs of a study**. This is so because a properly executed stratified random sample (*i.e.*, one with small within-strata variation) will have a **smaller sampling error than a simple random sample of the same size**; or, alternatively, a stratified random sample with fewer cases, if properly drawn, will have a sampling error of the same magnitude as a larger simple random sample. Stratified sampling is also used as a **method of controlling extraneous variation**.

When the purpose of drawing a stratified random sample is to reduce sampling error (or sample size), the population is stratified using a variable or variables correlated with the independent and/or dependent variables of the study. For example, if researchers wished to conduct a study of the relationship between high school students' scholastic performance and their income aspirations 10 years hence, they might use a stratified random sample that is stratified on gender and socioeconomic status. Gender would be an appropriate variable on which to stratify because males and females differ to some degree both in their academic performance in high school and in their aspirations. Socioeconomic status would be appropriate to use as a basis for stratification since it is correlated with at least one of the other two variables (academic performance and aspiration) and probably both (see Sewell, 1971).

The result of dividing the population into subpopulations based on gender and socioeconomic status is to make of the population a number of subpopulations, each of which is more homogeneous in composition than the total population. If gender and socioeconomic status were used to stratify a population (where socioeconomic status was divided into the categories high, moderate, and low), six subpopulations would result, as follows:

High socioeconomic status males
High socioeconomic status females
Moderate socioeconomic status males
Moderate socioeconomic status females
Low socioeconomic status males
Low socioeconomic status females

Since both gender and socioeconomic status are correlated with the independent and dependent variables of this study, each of the six strata (subpopulations) should be more homogeneous on academic achievement in high school and/or on income aspirations than the total population. If such were the case, a composite sample of 200 drawn by taking six separate simple random samples from the strata would have a smaller sampling error than a simple random sample of 200 drawn from the total population. This is so because the size of the sampling error is affected by the degree of homogeneity in the population or subpopulation being sampled.

A stratified random sample might be preferred to a simple random sample as a means of reducing sampling error, or it might be preferred as a money saver as a result of the reduced sample size needed. If the sampling error of a simple random sample of 200 cases was small enough to satisfy the researchers, they might opt instead to reduce the size of their sample. If, for example, they could reduce sampling error 10% by using a stratified random sample in lieu of a simple random sample, then a stratified random sample of 180 cases would have a sampling error equal to a simple random sample of 200 cases.

The crucial factor in using the stratified random sample to reduce sampling error (or the size of the sample) is **to select variables on which to stratify that are correlated with the independent and/or dependent variables**. If uncorrelated variables are selected, the resulting strata will be no more homogeneous than the total population, and the stratification process will have been a wasted effort.* In the study of the relationship between high school academic performance and income aspirations, if we were to use hair color as a basis for stratifying and were to divide the population into three strata consisting of brunettes, redheads, and blondes, we would probably be stratifying to no avail. Hair color is probably not correlated either with high school academic performance or with income aspirations.

To execute a stratified random sample successfully, one must have **more information about the population** than is necessary for a simple random sample. Not only is a list of the population necessary, but it is also necessary to be able to divide the overall list into a number of "subpopulation lists" differentiated on some relevant variable or variables. As a matter of fact, one of the strengths of the simple random sample is that when one is ignorant of the relevant variables other than the independent and dependent variables, one can execute a simple random sample with some confidence that the unknown but relevant variables will be sampled in approximately the proportions in which they occur in the total population. At least, any difference between their proportions in the total population and their proportions in the sample will be a random difference, attributable to chance.

11.4a Proportionate and Disproportionate Stratified Random Samples

Stratified random samples may be either proportionate or disproportionate. In the **proportionate stratified random sample**, *the variables used as a basis for stratifying are represented in the sample in the same proportions as they are in the population.* For example, if the population from which we were drawing the

---

* Also, degrees of freedom will have been lost, another statistical consequence of sampling procedures that will be discussed later.

TABLE 11.4   DISTRIBUTION OF GENDER AND SOCIOECONOMIC STATUS IN A POPULATION OF HIGH SCHOOL SENIORS

	Gender		
Socioeconomic Status	*Males*	*Females*	*Total*
High	10,800 (12%)	11,700 (13%)	22,500 (25%)
Moderate	14,400 (16%)	18,000 (20%)	32,400 (36%)
Low	18,000 (20%)	17,100 (19%)	35,100 (39%)
Total	43,200 (48%)	46,800 (52%)	90,000 (100%)

stratified random sample of high school students included all the high school seniors in the state of Illinois, and if gender and socioeconomic status were distributed throughout that population as in Table 11.4, we would draw our sample so that the variables would be distributed throughout the sample in the same proportions.

Assuming that the population of high school seniors in Illinois numbers 90,000, let us consider how we would go about drawing a proportionate stratified random sample of 450 cases. **First**, the population would be stratified by gender and socioeconomic status into the six strata listed above. **Then** we would prepare six separate lists of population elements, one for each stratum. **From** each of these lists we would draw a simple random sample of 450/90,000 = .005 of the elements, using random numbers. Using a constant fraction (450/90,000 or .005) to sample from each of the six lists would result in an overall sample of 450 cases distributed as in Table 11.5.

TABLE 11.5   COMPOSITION OF PROPORTIONATE STRATIFIED RANDOM SAMPLE OF 450/90,000 HIGH SCHOOL SENIORS

	Gender		
Socioeconomic Status	*Males*	*Females*	*Total*
High	54 (12%)	58 (13%)	112 (25%)
Moderate	72 (16%)	90 (20%)	162 (36%)
Low	90 (20%)	86 (19%)	176 (39%)
Total	216 (48%)	234 (52%)	450 (100%)

Note that the 450 cases in the sample will be distributed by gender and socioeconomic status in almost exactly the same proportions as they are in the total population, if our sampling lists are accurate.

Once the proportionate stratified random sample has been drawn and data collected, the data can be *analyzed in a number of different ways.* First, the cases in the six strata could be compared separately on the relationship between their high school achievement and their income aspiration levels; second, socioeconomic status could be held constant by comparing the data for the 216 males with those for the 234 females; third, gender could be held constant by comparing the data for the 112 high socioeconomic status students, the 162 moderate socioeconomic status students, and the 176 low socioeconomic status students; or, finally, the whole sample of 450 cases could be analyzed as a unit, ignoring differences in gender and socioeconomic status. This can be done without distortion because the sampling fractions are the same for each stratum. Since the two variables used to stratify are correlated with the independent and dependent variables, we would expect each of the six strata to be more homogeneous than the overall population.

In this example the number of cases appearing in each of the six sample strata are similar enough to make comparisons among the strata. However, in the case of a population in which some categories of the stratifying variable have few cases and others many, it might facilitate analysis to take a **disproportionate stratified sample** in order to make the sample strata more equal in size.

For example, studies of equality in employment opportunities often report data separately by race. Because of the great differences in the number of people in different racial groupings, we might first stratify on race, using the U.S. Census racial divisions: (a) white, (b) black, (c) American Indian, Eskimo, or Aleut, (d) Asian or Pacific Islander, and (e) other. The 1990 Census estimates that 80.3% of the U.S. population is white, 12.1% black, 0.8% American Indian/Eskimo/Aleut, 2.9% Asian/Pacific Islander, and 3.9% other races (Census, 1991). A *proportionate* sample probably would not include enough in the American Indian or Asian or even Black categories to provide an adequate basis for comparison, and the size of the subsample of whites may be larger than necessary. To make such comparisons, it would be desirable to draw samples with *more than* a representative number from the American Indian and Asian categories and *less than* a representative number of Whites. Thus a **disproportionate stratified sample** would be appropriate where one might draw subsamples of the same size (*e.g.*, 200 cases from each of the five strata).

It should be reemphasized that stratified random samples are generally more difficult to execute than simple random samples because they require more information about the population. Not only is it necessary to have lists of the elements in the population, but information is also needed on the distribution of relevant variables in the population. When the necessary information is available, the stratified random sample is a useful one for the sociologist.

Marjorie Fiske Lowenthal and her associates engaged in a series of studies of adult socialization patterns and adaptations of elderly people in the San Francisco area (*e.g.*, Lowenthal and Haven, 1968). In their research they utilized stratified random sample techniques. For example, a stratified random sample of 600 persons aged 60 or older was drawn from 18 census tracts in San Francisco, and those in the sample were interviewed three times at approximately annual

BOX 11.1   DEALING WITH NONREPRESENTATIVE FACTORS

In Section 6.3.3, three rules for percentaging a table to make comparisons were presented. When there is a *nonrepresentative* variable, that is, one that does not reflect the actual distribution of a characteristic in the population, the rule is that percentaging must be in the direction of that unrepresentative factor. One of the sources of unrepresentative factors is disproportionate stratified random sampling. The sample is not representative on the stratifying variable; cases are disproportionately selected. This must be taken into account in percentaging tables (see Section 6.3.3a). It is important in using disproportionate stratified samples to stratify on independent variables (not on a dependent variable) so that percentaging can be done in a way that permits proper comparisons to be made.

The strata in disproportionate stratified random samples are usually kept distinct in statistical analysis. If strata from a disproportionate stratified random sample are to be combined rather than kept distinct in the analysis, each of them must be weighted before they are combined. Weighting is done by multiplying (data on variables from) each case by a weighting factor (often a new weighting variable created in a statistical program such as SPSS).

The weight for cases for a given stratum is the **inverse of its sampling fraction**. That is, if a stratum were oversampled at, say, twice its real proportion in the population, then data on each case in that stratum would be multiplied by 1/2 or .5 so that data from each case contributed only half-a-case worth in any overall summary statistics. Each case in underweighted strata would likewise count proportionately more than one. Overall, weights must average 1.0, so the total sample size is not affected.

intervals. This sample was stratified on gender, three age levels, and social living arrangements (alone or with others). The very elderly, males, and persons living alone were disproportionately represented in the sample to make comparisons possible. Lowenthal and her associates, in effect, used stratification as a means of controlling relevant variables.

It is important to note that adjustments may need to be made in the usual statistical formulas when analyzing data from stratified random samples (Kish, 1965:77; see General References). This is particularly important when the strata are sampled disproportionately. The formulas will need to be adjusted to take sizes of strata into account. If sampling is proportionate, the usual formulas may generally be used, but tests of hypotheses will generally underestimate the significance of results, because the sampling error is smaller for stratified random samples than for simple random samples.

Investigators who are designing complex samples so that they can be used for generalization, and yet be cost effective, usually consult with statistical experts. Brent Scott, and Spencer (1988) have developed a microcomputer program that incorporates much of the technical knowledge used in designing a sample.

The program asks questions about the planned research sample and suggests alternatives to an investigator as the sampling plan is worked through. This program, called EXSAMPLE (Brent, Scott, and Spencer, 1988), then prints out a suggested sampling plan and its consequences for various research decisions. This program is an example of an expert-system approach to integrating technical knowledge to help in reasoning through complex problems.

11.5   THE SAMPLING FRAME

Each of the random sampling techniques mentioned has required some sort of list or series of lists of the elements in the population to be sampled. The actual lists from which random samples are drawn are referred to as **sampling frames**. In the case of the simple random sample, a single list of population elements is required. In the case of the cluster sample, the sampling frame consists of a series of lists, each of which duplicates a set of primary, secondary, etc., population units (*e.g.*, such as states, counties, blocks, or families). So, too, for stratified random samples it is necessary to come up with a sampling frame of lists representing the various strata to be sampled (*e.g.*, married women, married men, etc.).

Ideally, these lists should be accurate, complete enumerations of population elements. In fact, however, it is seldom the case that such lists are available. Figure 11.1 illustrates this problem. Lists that are available *may not be totally accurate or complete*, although they may be the best available. For example, the local telephone book is usually out of date as a list of families in a community, because new families have moved in and others have moved away since the phone book was published. Then, too, there may be elements in the available lists that *logically do not belong there*. For example, if one were sampling residential blocks and then houses within those blocks in order to interview heads of households, vacant houses and businesses would not belong among those elements to be sampled. Therefore, it would be logical to remove those units from the lists from which the sample was to be drawn. You can imagine the difficulty of finding a suitable sampling frame of voters, or voters over 65 years of age, or people who are members of two or more groups, or houses used for illegal drug sales, etc.

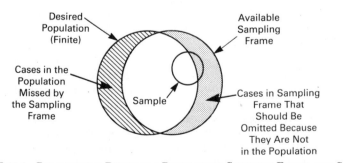

FIGURE 11.1   RELATIONSHIP BETWEEN A POPULATION, SAMPLING FRAME, AND SAMPLE

The problem of finding adequate sampling frames is a major topic in research methods courses. It is also one reason why alternatives to simple random samples are used. For example, multistage cluster sampling is useful because lists of broader elements (such as stages) are likely to be more stable and accurate. The problem of finding and checking a list of ultimate sampling units is more manageable when it can be focused on a smaller number of primary or secondary units that have been drawn in a sample. Up-to-date and special-purpose sampling frames can be purchased from organizations who make a business of developing and maintaining up-to-date lists. Researchers developing their own sampling frames sometimes cross-check lists from several sources or conduct a prestudy to develop a sampling frame. Questionnaires often include some screening questions to help screen out cases that should not be included in a sampling frame. Finally, researchers pay considerable attention to "nonresponses", because these represent another way in which cases that should be included in a sampling frame (and sample) can slip away, leaving a study that is potentially biased and not representative of the target sampling frame itself.

Intermediate, then, between the population and the actual sample is a sampling frame. To the extent that the sampling frame matches the population, generalization from the sample to the population is warranted. To the extent that the sampling frame differs from the population, generalization becomes problematic. At the end of this chapter we will discuss another methodological problem involved in drawing samples from finite sampling frames when one wants to make inferences to infinite populations.

11.6   NONPROBABILITY SAMPLES

In each of the sampling techniques discussed so far, the notion of randomness has been involved, and in each case it has been possible to estimate sampling error. It is crucial that we be able to estimate sampling error, because such estimates serve as bases for the inferential leap from sample to population data.

There are sampling techniques in use that are nonrandom (and, thus, nonprobability) in nature. These techniques are sometimes used because they involve lower data collection costs or because they avoid the problems involved in trying to draw random samples. They should generally be avoided because there is no legitimate basis for estimating sampling error from them.

We will mention several of these nonrandom techniques in passing because they do appear in the literature, and you should be aware of the fact that they are nonrandom.

Systematic Samples.   A systematic sample is one in which cases separated from each other by some set interval (such as every tenth case) are drawn from a list. The procedure is frequently used to sample from files, but it can also be used to sample houses on a street (such as every eighth house) or people waiting in a line. Such samples are obviously not random. If every tenth person in a line was selected, the tenth person, twentieth person, etc., would have a probability of 1.00 of being included in the sample, whereas persons not in those positions would have probabilities of 0.00 of being included.

If the starting point for a systematic sample is chosen randomly, it is possible to assign a probability of being chosen to each element in the population. For example, if one wished to sample every tenth name on a list, the starting point could be randomly selected from among the numbers between 1 and 10. Thus, if the first number selected randomly were 3, then the 3rd, 13th, 23rd, etc., cases would be included in the sample. This would have the effect of dividing the population into 10 sets of cases. Each set would have 1 chance in 10 of being selected in the sample. Likewise, each element in the population would have 1 chance in 10 of being included.

Such systematic samples with random starts are sometimes characterized as *pseudo-random samples* because, although probabilities of selection can be assigned to population elements, there is still no assurance that the sample will have the characteristics of a random sample. The crucial factor is the procedure used in constructing the list from which the systematic sample is drawn. If there is no biasing order to the list, the systematic sample with a random start may be a reasonable approximation of a random sample. If, on the other hand, the way the list is ordered does introduce bias, the resulting sample may be far from random.

Kish (1965:120; see General References) gives the following example of how the ordering of a list may bias the results of a study:

> Suppose that the 20,000 mortgages (loans) granted by a bank to home buyers were numbered in the order they were granted over 15 years. There is a tendency for these mortgages to increase from the lowest to the highest number on the list, because the cost of homes and the mortgages granted have been rising over the years. This tendency is much greater for the debt remaining unpaid. because the mortgages are reduced gradually by the monthly payment of the home buyers. Hence, for a study of mortgage debts, the mean of a systematic sample of 1 in 100 could depend greatly on the choice of a random number: a sample consisting of the first element in each zone (1, 101, 201, etc.) could have a much smaller mean than a sample of the last elements (100, 200, 300, etc.). The monotonic trend induces a variation among the 100 possible sample means.

Judgment Samples.   A judgment sample consists of those cases believed by the investigator to be "typical" of the population. What is considered typical may be nothing more than a matter of intuition or guessing. Even if the typical is based upon averages, it is still a matter of judgment rather than randomness.

Quota Samples.   Perhaps the most popular sampling technique utilized by market researchers and pollsters is a nonrandom technique known as a quota sample. It is an attractive technique because it is economical, convenient, and a fast way of getting data. A quota sample is drawn by specifying the characteristics desired in the subjects to be interviewed and then depending upon each of the interviewers to find and interview a quota of persons who possess the desired characteristics. Obviously, the procedures used by interviewers to fill their quotas are nonrandom.

Voluntary Samples.   The voluntary sample is composed of subjects who volunteer to participate in a study. It would be naive to assume that people who volunteer to participate do so randomly. Those who volunteer undoubtedly have some good reason for doing so.

As an example of a well-known study that was based on a nonrandom sampling technique, we will look briefly at the "Kinsey Report." Kinsey, Pomeroy, and Martin's study, *Sexual Behavior in the Human Male* (1948), was based on interviews with a voluntary sample of 5,300 white American males. Kinsey and his associates approached the members of 163 groups of various kinds (penal institutions, mental institutions, male prostitutes, junior high school students, speech-clinic groups, rooming-house groups, conscientious objectors, hitch-hikers, N.Y.A. workers, children's homes, etc.), lectured to them, and asked for volunteers. In those cases where "perhaps half or more" of the members of a group volunteered, the researchers attempted to persuade all the group members to submit to interviews (Kinsey, 1948:95). In some groups, 100% of the members were interviewed.

Sampling and statistical problems associated with the Kinsey Report formed the basis for a volume of critical essays sponsored by the American Statistical Association (Cochran *et al.*, 1954). Especially vulnerable to attack was the implication that the findings of the study could be generalized beyond the 5,300 men interviewed. It was pointed out, for example, that the responses of those who were in the 100% groups differed from the responses of those who were strictly volunteers. This difference was attributed to the presence of subjects in the 100% groups who had to be pressured into being interviewed and who submitted reluctantly (Cochran *et al.*, 1954:55).

One critic pointed out that there was evidence that many men volunteered to be interviewed because they had personal sex problems with which they needed help or because they had questions about sex that they wanted answered. Kinsey and his associates attempted to dismiss this as a possible source of bias on the grounds that the problems and questions that motivated these men to volunteer were "the everyday sexual problems of the average individual" (Cochran *et al.*, 1954:57).

The Kinsey study and the controversy that followed its publication offer a good object lesson about the hazards of nonrandom sampling techniques such as voluntary samples.

Although nonrandom sampling techniques are often more economical and convenient than random sampling techniques, **these advantages are outweighed by the disadvantage of being unable to estimate sampling error**. Therefore, the use of nonrandom techniques by sociologists should generally be discouraged.

11.7 METHODOLOGICAL PROBLEMS ASSOCIATED WITH INFERENTIAL LEAPS

As scientists sociologists are ultimately interested in developing generalizations about social behavior that are applicable to *infinite populations*, sometimes referred to as **general universes** (the terms *population* and *universe* may be used interchangeably). Because of the methodological problems involved, however, they generally do research on special (or finite) populations or samples of special populations (see Section 1.1). In fact, sampling theory, as it is practically applied, focuses on **special populations or universes** rather than on general universes.

By virtue of the fact that a simple random sample requires a list of population elements, researchers are generally restricted to working with finite populations. They are faced, then, with the methodological problem of how to make the inferential leap from the special universe with which they must work to the general universe of ultimate interest.

As Sjoberg and Nett (1968:130) have said:

> An essential step in conceptual clarification of the selection procedure is distinguishing between the special, or working, universe and the general one. The *special* (or *working*) *universe* is that specific, concrete system (or subsystem) from which one selects his units of study, notably his respondents. Statisticians refer to such a system as a *universe* or *population*, and usually they are content to work within its narrow boundaries. On the other hand, any theoretically oriented social scientist envisions still another kind of universe. If he studies a particular group or social system, he entertains the notion that his findings will, in part at least, hold for other groups or systems—not just in the United States but in other parts of the world as well. For his ultimate goal is establishing generalizations that extend beyond any time-bound social setting. We therefore define the *general universe* as that abstract universe to which the scientist assumes, however, tentatively, that his findings will apply. Put another way, every sample is a subsample of a broader type—mankind being, for purposes of generalization, the ultimate category.

The question of how the theoretical leap is made from the special to the general universe is not a simple one to answer.* As a matter of fact, it is a question that is frequently not faced by sociologists when they do research. In many cases sociologists select some special universe of interest to them, sample from it, and limit their generalizations to it. For example, Sewell and Shah (1968) conducted a study of the relationship between educational aspirations, intelligence, socioeconomic status, and parental encouragement. The data for their study came from a survey of graduating seniors in all public, private, and parochial schools in Wisconsin in 1957. They generalized their findings from the 10,308 seniors in their sample to all high school seniors in Wisconsin in 1957. They did not imply that their findings were generalizable beyond that population.

Other sociologists attempt to justify the leap to a broader universe (not necessarily a general universe) than the one from which they sampled. Barnett and Griffith (1970), for example, conducted a study of the relationship between anomia, achievement values, and attitudes toward population growth. They sampled women 18 years of age or older living in a public housing project in Flagstaff, Arizona. In a footnote (Barnett and Griffith, 1970:48–49) they say,

> It should be pointed out, in justification of the limited population represented in this survey, that a limited population contains fewer extraneous sources of variation which can affect a relationship under study than a broad population; hence, drawing from a limited population all the cases one can afford to draw increases the probability of making correct decisions about hypotheses stating causal relationships in that (limited) population (Herbert Hyman, *Survey*

---

* See Sjoberg and Nett (1968:129–159; see General References) for a discussion of the nuances of this problem.

Design and Analysis, New York: Free Press, 1955:81). And even though the findings were technically then applicable only to the limited population, it must nonetheless be kept in mind that "it is more probable that a hypothesis holds true outside the population on which it is confirmed than the contrary of the hypothesis holds true in the new population" (Hans Zetterberg, *On Theory and Verification in Sociology*. Third Edition, Totowa, New Jersey: Bedminster, 1965:128).*

Wallis (1949) offers an alternative rationale for making the inferential leap from sample data to a general universe. He says,

> On the other hand, any batch of data that we do get is a random sample from some population; that is to say, indefinitely many repetitions of the procedure that produced the sample will produce a population. We have two handles to manipulate in generalizing from human samples: one is to define our population as closely as we can and then attempt to approximate randomness in our sampling procedure; the other is to analyze our actual sampling procedure as well as we can and then attempt to describe the actual population to which it relates.

Wallis's statement may be adopted as a basis for making the **theoretical leap** from conclusions about a special universe to generalizations about a general universe, even when the data we examine are only a sample from a special universe. By studying the characteristics of this sample and making explicit the sampling design used to arrive at the sample, the sociologist can get some clues as to the nature of the general universe.

There are, then, at least two leaps typically involved in the process of drawing conclusions about general universes. One is the leap from sample to special universe from which the sample was drawn. The other is the leap from special to general universe. To complicate matters even further, to the extent that the sampling frame differs from the special universe, intermediate between the sample and the special universe is the sampling frame. In fact, then, there may be *three leaps involved in generalizing: sample to sampling frame, sampling frame to special universe, and special universe to general universe.*

The theoretical leap from sample to special universe is the subject of this volume, and the random sample is the technique designed to justify the leap. The leap from the special to the general universe is neither directly addressed nor justified by current theories and techniques of statistical inference. This leap is a serious methodological problem that sociologists have to face squarely as they develop general laws of social behavior. Figure 11.2 diagrams the relationship between sample, sampling frame, and general (or special) universe.

As much as possible the special universe to be sampled should be selected on the basis of theoretical considerations, that is, in terms of its logical connections with the general universe to which we ultimately wish to predict. Too often, unfortunately, the selection of a special universe is based instead upon such mundane factors as convenience, personnel resources, time, and money.

* Reprinted by permission of the publisher, Sage Publications, Inc.

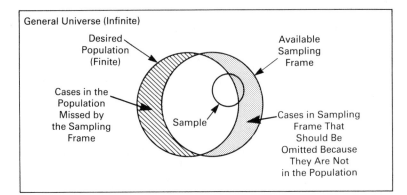

FIGURE 11.2 Relationship between General Universe, Special Universe, Sampling Frame, and Sample

11.8   SUMMARY

In this chapter we attempted to show that probability theory is the foundation upon which sampling theory is built. We pointed out that the simple random sample is the basic sampling model for the theory and techniques of inferential statistics. It was explained that other random sampling techniques are variations from the basic simple random sample design (see Diagram 11.1, p. 408, for an overview of random sampling).

We explained that the **simple random sample** requires that sampling be done **with replacement** and pointed out that the random sample without replacement does not satisfy all of the necessary conditions for a simple random sample. It was noted, however, that sociologists frequently use random samples without replacement because of reluctance to count data more than once (or convenience, or efficiency, or because it approximates a simple random sample in their case). The **procedure of selecting sample elements from random numbers** was described and, in this connection, the importance of being able to acquire a list of population elements was emphasized. It was pointed out that sociologists often study populations for which a single list of elements is not available. Because simple random samples are not feasible in such situations, the cluster sample may be used in lieu of the simple random sample.

The **cluster** sample was described as a multistage simple random sample, and the procedures involved in drawing a cluster sample were discussed. It was noted that a cluster sample generally has a larger sampling error than a simple random sample, but also generally costs less per case to collect data.

The **stratified random sample** was described and discussed as another possible alternative to the simple random sample. Its value for making subpopulation comparisons or for reducing sampling error and/or study costs was explained. It was pointed out that the stratified random sample generally has less sampling error than a simple random sample, but that it requires more information and is thus more difficult to execute.

Each random sampling technique depends upon the availability of a list or series of lists of population elements, clusters, or subpopulations. It was noted that the lists available do not always duplicate the desired population lists; consequently, the list or series of lists from which random samples are actually drawn are known as **sampling frames**. To the extent that these sampling frames do duplicate population lists, the samples drawn represent the populations of interest. To the extent that the sampling frames deviate from the population lists, generalization is problematic.

Various nonrandom sampling techniques were noted in passing, but the reader was advised against using these techniques because they do not allow estimation of sampling error, and estimation of sampling error is crucial to statistical inference.

Finally, we discussed the **methodological problem** involved in making the inferential leap from sample data and special universes to the general universes in which we are ultimately interested as scientists. We explained that this problem is not faced directly by statistical theory or sampling theory, so it must be attacked at a more abstract and more general methodological level.

We feel obligated to mention the fact that the sampling designs presented in this chapter were generally oversimplified. Sample designs actually used in research tend to be more complex. It has been our experience that a sampling design must be tailored to a particular study and a particular population. A detailed discussion of the fine points of sample design is beyond the purview of this volume.

In the next chapter we will introduce the concept of the **sampling distribution** and specific techniques for estimating sampling error. At that point we will have the tools necessary to launch into a discussion of parameter estimation and hypothesis testing.

CONCEPTS TO KNOW AND UNDERSTAND

sample
simple random sample
random sample without replacement
sampling error
random numbers
population list
cluster sample
homogeneous cluster
heterogeneous cluster
sampling frame

primary sampling units
ultimate sampling units
strata
proportionate stratified random
    sample
disproportionate stratified random
    sample
nonrandom sample
special universe (finite)
general universe (infinite)

QUESTIONS AND PROBLEMS

1. Make a list of the U.S. states and the District of Columbia (51 entries). Number the entries, and using a table of random numbers or the computer program in Appendix A, draw a simple random sample of 20 states. Draw a random sample

without replacement of 20 states. Are there any differences in the two samples? How great is the difference?

2. Describe how you would draw a simple random sample of the student body of your college or university?

3. Make a systematic comparison of the simple random sample, the cluster sample, and the stratified random sample. In what circumstances would a cluster sample be used instead of a simple random sample? In what circumstances would a stratified random sample be used instead of a simple random sample?

4. Explain how you would go about drawing a simple random sample of the names in the telephone book.

5. Read a research article in a sociology journal and analyze the sampling design used. Was it random or nonrandom? Was the sampling design well planned and well executed? How could the design be improved?

6. Why is a random sample generally preferable to a nonrandom sample?

7. Explain the relationship between a sample, a sampling frame, a special universe, and a general universe. Give an example that illustrates the relationship among the four.

GENERAL REFERENCES

Ackoff, Russell L., *The Design of Social Research* (Chicago, University of Chicago Press, 1953).
> Chapter 4 of this book, the chapter on sampling, is an excellent short summary of various sampling designs. A series of tables are included that make some important points about various techniques of sampling.

Kalton, Graham, *Introduction To Survey Sampling* (Beverly Hills, Calif., Sage), 1983.

Kish, Leslie, *Survey Sampling* (New York, Wiley, 1965).
> This book offers a good, technical discussion of sample designs and the practical problems facing the researcher in designing a sample. It features illustrations based upon survey samples of human populations.

Sjoberg, Gideon, and Roger Nett, *A Methodology for Social Research* (New York, Harper & Row, 1968).
> The authors spend some time discussing the methodological problem of generalizing from special to general universes. They include a consideration of the factors that enter into the selection of a special universe to be sampled.

Stephan, Frederick F., and Philip J. McCarthy, *Sampling Opinions* (New York, Wiley, 1958).
> This book is a good, basic source for those readers who wish to learn sampling procedures for large scale surveys. The second part of the book examines a number of empirical studies and analyzes their sampling designs.

LITERATURE CITED

Barnett, Larry D., and Jerry J. Griffith, "Anomia and Achievement Values and Attitudes toward Population Growth in the United States," *Pacific Sociological Review,* 13, 1 (Winter 1970), pp. 47–52.

Berman, Art, "Statistician Brands Draft Lottery Unfair," *Los Angeles Times,* December 3, 1969, part 1, p 13.

Brent, Edward E., James K. Scott, and John C. Spencer, *EXSAMPLE: An Expert System to Assist in Designing Sampling Plans* (Columbia, Mo.; Idea Works, Inc.), 1988.

Cochran, William G., Frederick Mosteller, and John W. Tukey, with the assistance of W. O. Jenkins, *Statistical Problems of the Kinsey Report on Sexual Behavior in the Human Male* (Washington, D.C., American Statistical Association, 1954).

Kinsey, Alfred C., Wardell B. Pomeroy, and Clyde E. Martin, *Sexual Behavior in the Human Male* (Philadelphia, W. B. Saunders, 1948)

Lowenthal, Marjorie Fiske, and Clayton Haven, "Interaction and Adaptation: Imtimacy as a Critical Variable," *American Sociological Review*, 33, 1 (February 1968), pp. 20–30.

Portes, Alejandro, "Dilemmas of a Golden Exile: Integration of Cuban Refugee Families in Milwaukee," *American Sociological Review*, 34, 4 (August 1969), pp. 505–518.

Scott, Ann Herbert, *Census U.S.A.* (New York, Seabury Press, 1968), pp. 128–129.

Sewell, William H., "Inequality of Opportunity for Higher Education," *American Sociological Review*, 36 (October 1971), pp. 793–809.

Sewell, William H., and Vimal P. Shah, "Social Class, Parental Encouragement, and Educational Aspirations," *American Journal of Sociology*, 73, 5 (March 1968), pp. 559–572.

Spaeth, Joe L., "Occupational Prestige Expectations among Male College Graduates," *American Journal of Sociology*, 73, 5 (March 1968), pp. 548–558.

U.S. Bureau of the Census, News Release, March 11, 1991 (CB91–100).

Wallis, W. A., "Statistics of the Kinsey Report," *Journal of the American Statistical Association*, 44 (1949), p. 468.

DIAGRAM 11.1   Random Sampling Techniques

THE SIMPLE RANDOM SAMPLE

Requirements

1. Every element in the population has a known and equal probability of being drawn.
2. All combinations of N elements are equally probable.

Steps in Drawing a Simple Random Sample

1. Find an accurate list of all population elements or a reasonably accurate sampling frame.
2. Sequentially number all of the elements.
3. Use a table of random digits (with a random start) or a computer random number program. Select a set of random digits within the range of sequential numbers assigned to the elements of the population. For each number drawn, include the element of the population bearing that number in the sample, and continue this process of selecting numbers until the desired sample size is achieved.
4. For a rigorous simple random sample (with replacement) use the random numbers to select N population case numbers (where N is the desired size of sample), regardless of whether any particular case number appears more than once. This technique satisfies both of the requirements listed above.
5. For a less rigorous random sample (without replacement) ignore any number that comes up a second, third, or greater number of times and go on to the next set of random digits in the range until an adequate number of unique population case numbers has been drawn. This technique meets the first of the two requirements listed above, but not the second.

Advantages and Disadvantages

1. Simple random samples tend to produce samples that are representative with respect to relevant characteristics of the population.
2. However, a sampling frame consisting of a single list is necessary to execute the technique.

THE CLUSTER SAMPLE

Requirements

1. Every element in the population has a known probability of being sampled.
2. At each stage of sampling the clusters have a known probability of being sampled.
3. The clusters from which the primary sampling units are drawn are homogeneous.
4. The clusters from which the ultimate sampling units are drawn are heterogeneous.

Steps in Cluster Sampling

1. Secure a list of clusters (*e.g.*, states) to serve as primary sampling units.

DIAGRAM 11.2 *(Continued)*

2. Draw a random sample of primary sampling units from that initial list.
3. Secure lists of clusters from each of the primary sampling units drawn.
4. Draw random samples of secondary sampling units from each of the lists of clusters secured in step 3.
5. Proceed to secure lists of clusters for each succeeding level to be sampled and randomly sample from each of those lists until the ultimate sampling units are finally drawn.

Advantages and Disadvantages

1. Cluster samples are feasible to execute when other random sampling techniques are not.
2. However, cluster samples generally have greater sampling error than other random sampling techniques.

THE STRATIFIED RANDOM SAMPLE

Requirements

1. Sufficient information about the population must be available to divide the population into subpopulations (strata) that have been stratified on relevant characteristics of the population (*e.g.*, gender, socioeconomic status).
2. The proportions of the total population that the subpopulations constitute must be available.
3. Lists of the elements in the subpopulations must be available.
4. The elements of the subpopulations each have a known probability of being sampled.

Steps in Stratified Random Sampling

1. Divide the population into subpopulations (strata) on the basis of characteristics that are relevant to what is being studied.
2. Secure lists of the elements in each of the subpopulations.
3. Draw random samples of elements from the subpopulations (either proportionate to their presence in the population or disproportionately, depending upon the type of data analysis planned).

Advantages and Disadvantages

1. If properly executed, a stratified random sample will tend to minimize sampling error.
2. However, execution of a stratified random sample requires knowledge of which population characteristics are relevant to the study and the ability to generate lists of elements in the subpopulations.

CHAPTER 12

Sampling Distribution:
A Pivotal Concept

Very often a social scientist wants to make a statement about a whole society, a large group, or frequently occurring behaviors. Sampling is an essential tool of such inquiry. *First*, and most importantly, study of a carefully drawn sample yields all the information that is needed—study of an entire population would not add more. *Second*, study of a large and unwieldy population may introduce more error by making the task of processing the information too complex and cumbersome or by using up resources that would better be focused on careful data collection and analysis. *Third*, the process of inquiry sometimes requires destruction of sampled items (perhaps in testing water glasses for strength) so that testing of all items would be self-defeating.

Our knowledge about what to expect of samples is based upon drawing samples in a way that permits us to utilize probability theory. As we have seen in the last chapter, there are several different ways of going about the business of drawing samples from populations. Some of these sampling plans permit us to apply our knowledge of probability and some prevent us from applying probability theory and thus from knowing what to expect of these samples. **Inferential statistics makes use of what we know about how samples behave.**

Most introductory statistics texts develop the logic of inferential statistics by *limiting their discussion to simple random samples drawn with replacement or from relatively large or infinite, defined populations.* We will also follow this practice, leaving discussions of alternative probability sampling procedures to more advanced texts.*

Our first step will be to illustrate how we can use probability theory to describe what we might expect from drawing simple random samples. Then we will show how you can describe in general the expected behavior of simple random samples. Finally, we will use this knowledge in a logical process of drawing conclusions from sampled data.

---

* For a discussion of how samples behave when they are drawn by other, more complicated sampling plans, see Kish, *Survey Sampling* (New York, Wiley), 1965.

12.1   SAMPLING: A CASE APPLICATION

Corporations represent one special way in which individuals form a social relationship. In our society, these relationships have something of the status of "artificial persons." They are legally acknowledged through articles of incorporation that explicitly formalize objectives of the relationship and rules by which those involved are to carry on their relationship over time. In 1980, according to the U.S. Census, there were 2,749,000 retail establishments, of which 515,000 or about 19% had the legal form of a corporation.

Suppose we decide to draw a sample of retail establishments—a simple random sample from this relatively large population of establishments—in order to study differences between retail establishments that have corporate versus other legal forms of organization. For purposes of illustration, let us draw a sample of size 4 ($N = 4$).

Can we describe how we would expect samples of this sort to behave? Because we are using a probability sampling procedure, the answer is yes. For example, one possible sample would be to draw three corporations (C) followed by a retail establishment that is not a corporation (N).

$$C \quad C \quad C \quad N$$

We can readily **compute the probability of drawing such a sample** because we know that:

1. The probability of drawing a corporation is .19 since there are known (in this instance) to be 19% corporations in the population of retail establishments. The probability of drawing a noncorporation would be $1.00 - .19 = .81$ (see Section 10.2.3).

2. The outcome of each draw is independent of the outcome of other draws if we follow simple random sampling procedures, and thus we can use the special multiplication rule (Section 10.3.5) to compute the probability of this series of four independent events. The probability of drawing such a sample is

$$(.19)(.19)(.19)(.81) = .0056$$

3. Each sample is mutually exclusive of other possible samples, which means, as we shall see, that we can make use of the special addition rule as well (see Section 10.3.2).

We could have drawn many different samples other than the C C C N pattern noted above. Specifically, we could have drawn 2^N (where N is the sample size and 2 is used because there are two outcomes on each draw, being a corporation or not) or 2^4 or 16 different samples (taking order in which each element was drawn into the sample into account.)

Our sample of size 4 could have been any one of the 16 different sequences of sample draws listed in Table 12.1. The likelihood of any one of these different samples is given at the right in Table 12.1. For example, the probability of drawing a sample with no corporations at all is .4305. This means that, in the long run, if we were to draw random samples of size four from this population, we should

TABLE 12.1 THE 16 POSSIBLE SAMPLES OF SIZE 4

	Sample Draw				Probability of This Sample
	1st	2nd	3rd	4th	
1.	C	C	C	C	$(.19)(.19)(.19)(.19) =$ .0013
2.	C	C	C	N	$(.19)(.19)(.19)(.81) =$ .0056
3.	C	C	N	C	$(.19)(.19)(.81)(.19) =$ .0056
4.	C	N	C	C	$(.19)(.81)(.19)(.19) =$ .0056
5.	N	C	C	C	$(.81)(.19)(.19)(.19) =$ .0056
6.	C	C	N	N	$(.19)(.19)(.81)(.81) =$ .0237
7.	C	N	C	N	$(.19)(.81)(.19)(.81) =$ .0237
8.	N	C	C	N	$(.81)(.19)(.19)(.81) =$ .0237
9.	C	N	N	C	$(.19)(.81)(.81)(.19) =$ .0237
10.	N	C	N	C	$(.81)(.19)(.81)(.19) =$ .0237
11.	N	N	C	C	$(.81)(.81)(.19)(.19) =$ .0237
12.	C	N	N	N	$(.19)(.81)(.81)(.81) =$ .1010
13.	N	C	N	N	$(.81)(.19)(.81)(.81) =$ .1010
14.	N	N	C	N	$(.81)(.81)(.19)(.81) =$ .1010
15.	N	N	N	C	$(.81)(.81)(.81)(.19) =$ .1010
16.	N	N	N	N	$(.81)(.81)(.81)(.81) =$ .4305
					Total = 1.0004*

* Does not add to 1.0000 because of errors in rounding.

expect to get a sample with 0 corporations about 43% of the time. **We know this because we can utilize probability rules and we can do that because of the way we carefully drew the samples.** The way samples are drawn is thus very important.

For most research purposes, we would have no interest in the order in which cases are drawn into our sample, whether corporations were drawn first or last. We would only be interested in the number (or percentage) of corporations our entire sample contains. Although there is only one way to draw a sample with 100% corporations and only one way to draw a sample with no corporations, there are several ways to draw a sample with only one corporation. Table 12.1 shows the various possibilities. There are four samples with one corporation, six samples with two corporations, and four samples with three corporations out of the four establishments in the sample.

In general, we could have computed **the number of different ways of getting a particular number of corporations in the sample by making use of the combinations formula** (Formula 10.11) discussed in Chapter 10. If we let N equal the total number of choices or the sample size, r equal the number of corporations, and $N - r$ the number of noncorporations, we can substitute into the combinations formula as follows to find out, say, the number of ways one could draw a sample with two corporations and two noncorporations. These are the samples listed as numbers 6 through 11 in Table 12.1.

(12.1)
$$\binom{N}{r} = \frac{N!}{(N-r)!\,r!}$$

$$\binom{4}{2} = \frac{4!}{(4-2)!\,2!} = \frac{4!}{2!\,2!} = \frac{(4)(3)(2)(1)}{(2)(1)(2)(1)} = \frac{24}{4} = 6$$

Using the **special addition rule**, we can sum the probabilities of drawing a sample with two corporations and two noncorporations for the six possible ways of drawing this kind of sample. Table 12.2 lists these summary results for each of the possible kinds of samples.

By the adoption of a simple notation system and the combined expression of the special multiplication rule and the special addition rule (as represented by the combinations formula), we can come up with a formula for computing the probability of the chance occurrence of any particular pattern of corporations and noncorporations in a sample of size N:

(12.2)
$$Pr(N, P, r) = \frac{N!}{(N-r)!\,r!}P^r Q^{N-r}$$

where N = the size of the sample, r = the number of corporations the sample contains, $N - r$ = the number of noncorporations in the sample, P = the probability of drawing a corporation from the population, and $Q = 1 - P =$ the probability of drawing a noncorporation.

To demonstrate how Formula 12.2 works, we will apply it to the situation of drawing one corporation and three noncorporations:

$$Pr(4, .19, 1) = \frac{4!}{3!\,1!}(.19)^1 (.81)^3$$

$$= \frac{(4)(3)(2)(1)}{(3)(2)(1)(1)}(.19)(.5314)$$

$$= \frac{24}{6}(.10097) = (4)(.10097) = .4039$$

TABLE 12.2 Distribution of Possible Sample Outcomes for Samples of Size 4 Drawn from a Population of Retail Establishments in Which the Probability of Randomly Drawing a Corporation Is .19 and the Probability of Drawing a Noncorporation Is .81

Possible Outcomes of Samples of $N = 4$		Computational Form	Probability of This Kind of Sample
4(100%)	corporations	$1(.19)^4(.81)^0$	.0013
3(75%)	corporations	$4(.19)^3(.81)^1$	.0222
2(50%)	corporations	$6(.19)^2(.81)^2$	.1421
1(25%)	corporations	$4(.19)^1(.81)^3$	.4039
0(0%)	corporations	$1(.19)^0(.81)^4$	.4305
		Total =	1.0000

BOX 12.1 REVIEWING RULES OF PROBABILITY

Your understanding of Formula 12.2 depends upon your grasp of three rules that were discussed in Chapter 10: the *special addition rule of probability* (Section 10.3.2), the *special multiplication rule of probability* (Section 10.3.5), and the *combination rule* (Section 10.4.2). If these three rules are still not clear to you, now would be an excellent time to review these sections; then reread the first pages of this chapter.

Thus there are four ways to achieve a sample with one corporation and three non-corporations. The probability of any one of these ways is .10097 and the probability of achieving a sample with one corporation from samples of size 4 is the product, or $4 \times .10097 = .4039$. This figure is given in Table 12.2 as well. The formula would work as well for figuring the probability of achieving a sample with no corporations (or all corporations), remembering that $0! = 1$ and any number raised to the power of 0 is equal to 1.

In Table 12.2 are listed the five kinds of sample outcomes, the computational forms for computing the probability of each of these outcomes, and their associated probabilities. Note that if the five probabilities are added, they sum to 1.0000 (except, perhaps, for rounding error in some instances). These are all the possible kinds of samples.

Table 12.2 shows that there is about thirteen-hundredths (.13%) chance of drawing a sample with 4 out of 4 corporations if indeed we are using random sampling procedures and are drawing a sample from a population in which there are 19% corporations. The sample of three corporations and one noncorporation we started with in this section we would expect 2.22% of the time. On the other hand, under these conditions we might expect to get one corporation in our sample about 40% of the time (if we were to draw repeated samples of size 4). This distribution of probabilities is called a **sampling distribution**.

It simply describes how often we can expect samples of this kind. This knowledge will be the basis upon which we build a logical argument that helps us make inferential statements about populations on the basis of samples we have drawn from them. For example, if we did draw one sample from this population and found that we drew all corporations, we might begin to (a) doubt whether we carefully applied **random** sampling procedures, (b) doubt whether the population we were actually drawing from really had only 19% corporations, or (c) decide that this was merely one of the more unusual but expectable chance samples we might get. Our doubts stem from a knowledge of how rare this sample outcome is under the circumstances. On the other hand, we would not be at all surprised to draw a sample with only one corporation.

Before we proceed with the use of sampling distributions in the logic of inferential statistics, we need to summarize more carefully what a sampling distribution is. The concept of the sampling distribution is one of the most important in inferential statistics. We will spend the rest of this chapter examining the concept in general and a number of specific sampling distributions that are commonly used by sociologists.

12.2   THE CONCEPT OF THE SAMPLING DISTRIBUTION

There are three distinct types of distributions of data: (a) one that characterizes the distribution of elements of a population (the **population** distribution), (b) one that characterizes the distribution of elements of a sample drawn from a population (a **sample** distribution), and (c) one that describes the expected behavior of a large number of simple random samples drawn from the same population (a **sampling** distribution). The third type of distribution differs from the other two in a number of ways that will be discussed below, but a paramount difference is that the units that are distributed are summary measures of whole samples of values rather than individual values of characteristics of single cases.

There are many different sampling distributions, depending upon (a) the population being sampled, (b) the size of the sample, N, (c) the method of sampling (we will consider only simple random samples here), and (d) the statistic (function of sample scores) in which we are interested.

A sampling distribution is *a theoretical probability distribution of sample statistics* (*e.g.*, sample means or sample proportions). A sampling distribution may be generated by taking all possible random samples (each with at least one element different) with a fixed N from a population, computing a statistic for each sample, and plotting the distribution of these statistics around the parameter that they estimate. For example, consider a population of 249,000,000 individuals in the United States in 1990. Assume that we drew a simple random sample of 1000 from that population and computed the mean age of the individuals in the sample.* Assume further that we repeated this procedure until we exhausted all the unique simple random samples of 1000 in the population. The *distribution of the means of all of these samples* would constitute a **sampling distribution of means**.

Of course, sociologists engaged in research do not actually generate a sampling distribution in this way except in cases of limited distributions of possible samples such as the four retail establishments discussed earlier. Usually, they draw one sample, compute a statistic from the sample, and use knowledge about the nature of its sampling distribution to generalize to the corresponding parameter of the population.

Every statistic, whether it be a frequency, a proportion, a mean, a variance, or whatever, has a sampling distribution. For a statistic to be useful in making inferences about parameters its sampling distribution must be known. One important job for the statistician who develops a new statistical technique is to specify its sampling distribution (or distributions). This has, in fact, been done for the commonly used statistics. These sampling distributions are usually presented in tabular form and are included in statistics books (as they are in this one; see Appendix Tables A through D). As an example, the normal distribution is the appropriate sampling distribution for several different statistical techniques when large samples are used (*e.g.*, means, proportions). The chi-square distribution is the appropriate sampling distribution for variances and the chi-square (χ^2) technique.

---

* Of course, this is a purely hypothetical example since it would not be feasible to draw a simple random sample of the United States population because of the lack of an accurate sampling frame.

Student's t-distribution is the appropriate sampling distribution for means from small samples when the population variance is unknown. The point is that there are many statistical techniques and many sampling distributions (although there are not as many sampling distributions as there are techniques). It is necessary to match the technique with the appropriate sampling distribution in order to generalize from sample data to population.

A sampling distribution is usually a **univariate distribution** and as such may be described in terms of its central tendency, variability, and form.* We will examine each of these characteristics of a sampling distribution in turn.

12.2.1   Central Tendency

In referring to the central tendency of a sampling distribution, it is customary to speak of the **expected value of a statistic**. The expected value is *the average value a statistic assumes for its sampling distribution* (e.g., the arithmetic mean of the sampling distribution of the statistic). That is, the expected value is the value that we would obtain if we added all of the statistics in the sampling distribution and divided by their number. An expected value is indicated by the capital letter E. The expected value of the sample mean, therefore, is designated $E(\bar{X})$. If the average or expected value of a statistic is, in fact, the parameter that it estimates, then the statistic is said to be an **unbiased estimator** of the parameter. We will establish later in this chapter that the sample mean is an unbiased estimator of the population mean [*i.e.*, $E(\bar{X}) = \mu$].

The fact that a statistic is an unbiased estimator does not give us any notion of just how closely a particular statistic from a single sample estimates its parameter. It merely assures us that *on the average* such statistics will equal the parameter. Still, the information that an estimator is unbiased is useful because it tells us that any difference that exists between a particular statistic, and its parameter is attributable to random error rather than systematic bias in the statistic itself. *Since scientific generalizations are based upon replications of a study rather than a single, isolated study, when we make repeated estimates of a parameter, it is important that our estimation errors not be systematic.*

12.2.2   Variability

A second important characteristic of a sampling distribution is the extent to which the sample statistics vary around their parameter. Variability of scores in population distributions or sample distributions can be measured by techniques such as ranges, variances, and standard deviations. When we measure the variability of sample statistics around their parameters, we refer to the summarizing measurement as a **standard error**.† There are standard errors for frequencies,

---

* There are multivariate sampling distributions of interest in more specialized research.
† The standard deviation computed on a sampling distribution of a statistic is called the *standard error* to distinguish it from other standard deviations.

proportions, means, medians, variances, correlation coefficients, and so forth. Standard errors, in general, measure the random variation of statistics around the parameter that they estimate. The size of the standard error is dependent, in part, upon the size of the sample from which the statistic is computed. In accordance with *the* **law of large numbers**, *as sample size increases, the standard error decreases.* The larger the N, the more closely statistics will cluster around their parameter.*

12.2.3   Form

The form of a sampling distribution refers to the shape of the particular curve that describes the distribution. It is important to determine, for instance, whether the distribution is symmetrical or skewed, leptokurtic, mesokurtic, or platykurtic. Is the distribution of the statistic normal or J-shaped? The particular form that a sampling distribution assumes is a significant factor to be considered in generalizing from statistics to parameter. For example, we will see later in this chapter that under certain conditions the sampling distribution of means is platykurtic. Ideally, it is possible to describe a sampling distribution with a mathematical formula (a density function) such as the formula for the normal distribution.

We will not attempt to discuss all of the known sampling distributions in this chapter. Rather, we will examine a few of the more commonly used ones in order to obtain a feeling for what sampling distributions are. Other sampling distributions will be discussed and described as they are needed in later chapters.

In this chapter we will direct our attention to the central tendency, variability, and form of sampling distributions. In the next chapter we will direct our attention to the question of how sampling distributions enter into the estimation of parameters. In the remaining chapters we will see how sampling distributions are used in tests of hypotheses.

12.3   THE BINOMIAL SAMPLING DISTRIBUTION FOR FREQUENCIES

The distribution of the number of corporations in samples of four retail enterprises that we examined earlier is a particular instance of a sampling distribution known as the **binomial probability distribution**. The binomial probability distribution is *the appropriate sampling distribution for dichotomized, nominal scale data for which the observations are independent.* For example, if the observations are independent, such measurement categories as male/female, success/failure, white/nonwhite, in-group/out-group have binomial sampling distributions. Binomial sampling distributions are expressed in terms of **frequencies of occurrence** of each of the dichotomous categories. Thus, the distribution of corporations/noncorporations was composed of the possible simple random samples of size 4 for which the number of corporations varied from 4 to 0.

---

* There are some exceptions that are not of interest in the usual research situation where a population mean could potentially be computed.

Alternatively, we can treat outcomes not as frequencies but as proportions or percentages. In this section we will use *frequency* as the way of describing sample outcomes where variables are dichotomous. In the following section we will use *proportions* (or percentages) as the way dichotomous sample outcomes are summarized.

Formula 12.2 allows us to compute directly the probability of occurrence of each of the terms of a **binomial sampling distribution**. Barring measurement error and other nonrandom errors, the probabilities computed from the formula are exact.

$$Pr(N, P, r) = \frac{N!}{(N - r)!r!} P^r Q^{N-r}$$

12.3.1 Central Tendency

Because the binomial sampling distribution is *an exact distribution*, the N and P upon which the distribution is based are parameters. Given these parameters, it is possible to compute a measure of central tendency that is also a parameter. The mean of a binomial distribution (μ_B) is given by

(12.3)
$$\mu_B = NP$$

where N is the size of the sample for each of the samples in the distribution, and P is the proportion of outcomes of interest (*e.g.*, the outcome of interest earlier was the proportion of corporations). The resulting mean is a frequency of outcomes of interest and may be interpreted as the number of such outcomes to be expected, on the average.

Using Formula 12.3, we may compute the mean of the sampling distribution of number of corporations in samples of four retail establishments.

$$\mu_B = NP = (4)(.19) = .76$$

Therefore, the central tendency of this sampling distribution is represented by a mean of .76 corporations. Such a low number of expected corporations is, of course, consistent with the results of our earlier examination of this sampling distribution.

12.3.2 Variability

It is also possible to compute a parameter as the measure of sampling error of the binomial sampling distribution. The formula for the **standard error** is

(12.4)
$$\sigma_B = \sqrt{NPQ}$$

where N and P have the same meanings as above and $Q = 1 - P$. This *standard error* measures the variability of sample frequencies of outcomes of interest, among themselves and around the mean of the sampling distribution.

Computing the standard error for the sampling distribution of the number of corporations in samples of four retail establishments with Formula 12.4 gives the following result:

$$\sigma_B = \sqrt{NPQ} = \sqrt{(4)(.19)(.81)} = \sqrt{.6156} = .7846$$

This, then, is a **measure of the variability of frequencies** of corporations from simple random samples of size 4 from a population of retail establishments composed of 19% corporations. A graph of this distribution is presented in Figure 12.1.

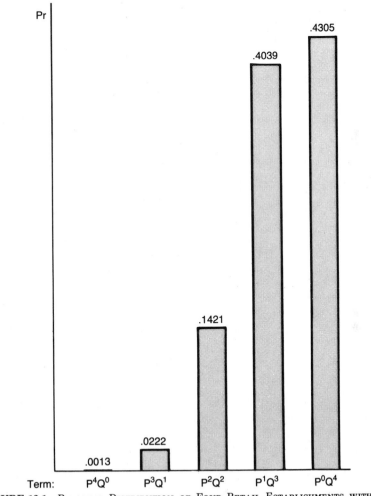

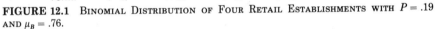

FIGURE 12.1   Binomial Distribution of Four Retail Establishments with $P = .19$ and $\mu_B = .76$.

12.3.3   Form of the Distribution

The form of the binomial sampling distribution is dependent upon N and P. When $P = 0.5$, the distribution will be symmetrical. As N approaches infinity, the binomial distribution **approaches the normal distribution**. Even when P is not equal to 0.5, as N becomes larger the binomial distribution begins to look more like the normal distribution. As a matter of fact, as long as both NP and NQ are equal to or greater than 5, the normal curve is a satisfactory approximation to the binomial distribution. Thus, if $P = Q = 0.5$, the normal approximation is satisfactory when N is as small as 10. If $P = 0.4$, N would have to be at least 13 to satisfy the rule of thumb. For a $P = 0.3$, N would have to be at least 17, and so forth. Figures 12.2A, 12.2B, and 12.2C portray the graphs of these three binomial distributions. Although the graphs of Figures 12.2B and 12.2C are not symmetrical, they do resemble the normal distribution.

When the normal distribution closely approximates the binomial distribu-

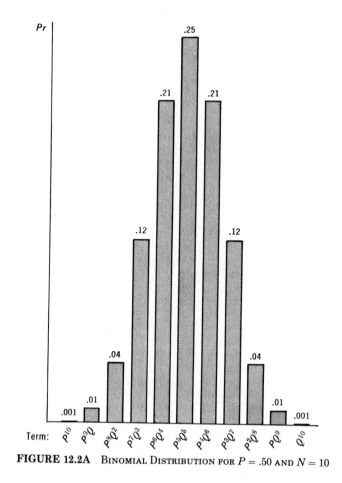

FIGURE 12.2A   Binomial Distribution for $P = .50$ and $N = 10$

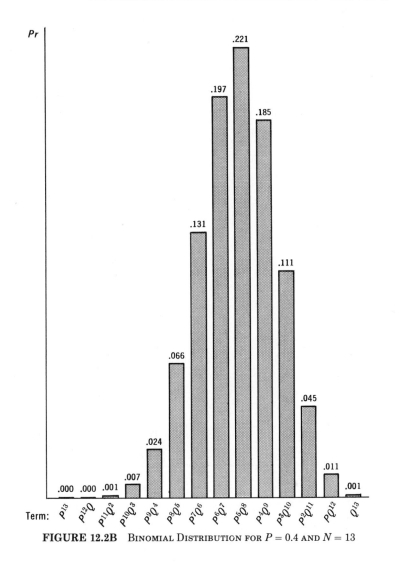

FIGURE 12.2B   BINOMIAL DISTRIBUTION FOR $P = 0.4$ AND $N = 13$

tion, the standard error of the binomial distribution takes on a meaning similar to that of the standard deviation (see Section 5.4.4). That is, approximately 68% of the sample frequencies can be expected to fall within 1 standard error of the mean of the binomial sampling distribution, approximately 95% within 2 standard errors, and just about all of the sample frequencies within 3 standard errors. When the binomial distribution departs from normality the standard error does not carry this interpretation.

It is obvious from an examination of Table 12.2 that the binomial sampling distribution in question has a definite skew toward sampling corporations. That is, the bulk of the samples in the distribution are heavily weighted in favor of

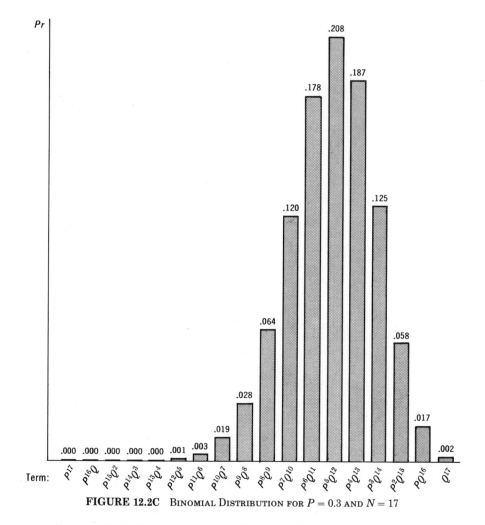

FIGURE 12.2C BINOMIAL DISTRIBUTION FOR $P = 0.3$ AND $N = 17$

sampling relatively few corporations. The size of the standard error for this distribution, .7846, reflects variability of sample frequencies larger than the mean since the mean itself is so low. In a skewed distribution such as this one, the interpretation of the standard error is complicated.

Whenever P deviates from 0.5, if N is not large, the binomial sampling distribution will be skewed. This is not a problem, however, since the binomial distribution yields **exact probabilities**.

12.3.4 Use of the Normal Approximation to the Binomial Distribution

As long as the sample N is small, it is advisable to use the binomial sampling distribution directly because it does give exact probabilities. As N gets larger, however, the binomial distribution becomes more and more cumbersome to work with. The advantage of being able to compute exact probabilities is soon outweighed by the amount of work involved in computing them. It is possible to program a computer to do the necessary computations.

An alternative is to use the normal curve as an approximation to the binomial sampling distribution. We will examine the normal approximation to the binomial distribution to determine how probabilities are computed and to compare those probabilities with the exact probabilities given by the binomial sampling distribution itself.

Although the sample size in our earlier example does not meet the rule of thumb for using the normal approximation to the binomial distribution, let us assume we had drawn a larger sample, say, a sample of size 17. First, we must find the mean and standard error of the binomial sampling distribution of retail establishments where the sample size is 17 and the probability of drawing a corporate form of social organization is .19, as before. Thus $N = 17$, $P = .19$, and $Q = (1.0 - .19) = .81$. Given these data, we can compute the mean and standard error as follows:

$$\mu_B = NP = (17)(.19) = 3.23$$
$$\sigma_B = \sqrt{NPQ} = \sqrt{(17)(.19)(.81)} = \sqrt{2.62} = 1.62$$

To illustrate the similarity of the normal curve approximation to the binomial distribution, let us look at the sample outcome of getting 4 corporations out of the sample of 17 retail establishments.

The four corporations is a **frequency** and frequencies are discrete values. The **normal curve**, on the other hand, is a *continuous distribution for which probabilities are represented by areas under the curve*, so it is not possible to use the normal curve to find a probability for a value like 4. What must be done is to **correct for continuity** by treating the integer 4 as though it were a continuous score with a lower boundary of 3.5 and an upper boundary of 4.5. It is possible to find an area under the normal curve between 3.5 and 4.5 to represent the desired probability. Using the following standard score formula, this is just what we shall do.

(12.5)
$$z = \frac{X - \mu_B}{\sigma_B}$$

BOX 12.2   REVIEWING NORMAL CURVE AND STANDARD SCORE

The discussion that follows draws upon your knowledge and understanding of the normal curve and the concept of a standard score. Both the normal curve and the standard score are extremely important for any discussion of inferential statistics. You will find that they will be referred to again and again throughout this volume. If you feel that you need to refresh your memory on these topics at this point, you should review Sections 5.4.7 and 5.5.2.

where X takes the value of first the lower boundary of the frequency and then the upper boundary.

Solving for the two z's, we get

$$z_1 = \frac{3.5 - 3.23}{1.62} = .17$$

and

$$z_2 = \frac{4.5 - 3.23}{1.62} = .78$$

To find the probability associated with the occurrence of a simple random sample with 4 corporations and 13 noncorporations, it is necessary to compute the area under the normal curve lying between these two z-scores. The area between a z-score of .17 and a z-score of .78 under the normal curve is approximately $.2823 - .0675 = .2148$ (see Appendix Table A); thus this represents our approximation to the desired binomial probability.

Using Formula 12.2 to compute the exact binomial probability for this kind of sample outcome (4 corporations and 13 noncorporations), we get

$$\frac{N!}{(N-r)!\,r!}\,P^r Q^{N-r} = \frac{17!}{(17-4)!\,4!}(.19)^4 (.81)^{17-4}$$

$$= 2380(.0013)(.0646) = .1999$$

The normal curve approximation, .2148, is slightly higher but close to the exact binomial probability, .1999 (the calculations for the normal approximation include rounding error). If the sample size had been larger and normal curve calculations carried out to more decimal places, the approximation would have been closer. When N is large, it is much more convenient to use the normal approximation to the binomial distribution than it is to find probabilities from the binomial itself. The approximated probabilities are generally close enough to those computed from the binomial to do the job.

We could use the normal curve approximation to find the probability of each possible sample outcome. Usually, however, only those outcomes of immediate interest are calculated. In this instance, we were interested in the probability of getting a sample with four corporations. Sometimes we may be interested in the probability of getting samples *as extreme or more extreme than 4 out of 17* (e.g., those with 4, 3, 2, 1, or no corporations).

12.4   SAMPLING DISTRIBUTIONS OF PROPORTIONS

If the data with which we are working are dichotomized, if simple random sampling is involved, and if the measurements are independent, then the binomial distribution is the appropriate sampling distribution to use, whether the data are in frequency, proportion, or percentage form.

When the sample size N is small, the easiest way to analyze data of this type is to work with frequencies rather than proportions and use the binomial sampling distribution directly. As we have just seen, when N is large and the proportion P is not too extreme so that the rule of thumb applies (both NP and NQ are equal to or greater than 5), then the normal curve may be used as an approximation to the binomial sampling distribution.* Alternatively, proportions can be computed, converted to standard scores, and evaluated on the normal sampling distribution. If we use the latter alternative, then we need to know how to compute the mean and standard error of proportions.

BOX 12.3 SYMBOLS FOR STATISTICS AND PARAMETERS

The distinction between samples and populations is central to inferential statistics. The symbols used generally reflect this distinction as well. In the text, different symbols are used for sample statistics or estimates and population values themselves. You should be alert to this distinction and extend this illustrative list for yourself.

Sample Statistic		Population Parameter
$\bar{X}$	arithmetic mean	μ
p	proportion p	P
$q = 1 - p$	proportion q	$Q = 1 - P$
s	standard deviation	σ
$s_{\bar{x}}$	standard error of the mean	$\sigma_{\bar{x}}$
s_p	standard error of proportion	σ_p

12.4.1 The Mean Proportion

It can be demonstrated that *a proportion is a mean for its distribution of scores.* Remember that the mean is merely the sum of scores of a distribution divided by the number of scores (*e.g.*, $\bar{X} = \Sigma X/N$).

In the General Social Survey of 1989[†] a national probability sample of respondents in the United States were asked whether they smoked. Of the 1,033 people who responded to the question, 309 said they did smoke and 724 said they didn't. Suppose that we assign a score of 1 to every person who said he did smoke and a 0 to each of those who said he didn't. If we add all of the 1 and 0 scores and divide by the number of respondents (1,033), we will get a mean. Since the 309 1's equal 309 and the 724 0's equal 0, the sum of the scores is 309 and the mean is $309/1033 = 0.299$. Therefore, the proportion 0.299 is the mean.

Earlier we found that the mean of a binomial distribution is simply N multiplied by P and this mean is for a frequency count. This could be changed to a

---

* If N is large but P is extreme (either very small or very large), the Poisson distribution may be used as the sampling distribution. Since this situation rarely occurs in sociology, the Poisson distribution has not received much attention (but see Pierce, 1970).

[†] See Box 3.5.

proportion simply by dividing NP (the average frequency of occurrence of events of interest) by N, yielding the mean of the sampling distribution of proportions. Substituting f for NP, we have

(12.6) $$\bar{P} = P = f/N$$

In the case of the General Social Survey data, the 0.299 is a sample proportion (mean), which is an unbiased estimator of the population proportion. That is, p estimates P, the mean of the sampling distribution of proportions.

12.4.2   Standard Error of a Proportion

If the population proportion (P) is known, the **standard error of the proportion** can be computed from the following formula:

(12.7) $$\sigma_p = \sqrt{\frac{PQ}{N}}$$

where σ_p is a parameter.

Formula 12.7 for the standard error of a proportion is directly related to the formula for the standard error of the binomial distribution (standard error of a frequency). The standard error of the binomial distribution is

(12.8) $$\sigma_B = \sqrt{NPQ}$$

Since a frequency is converted to a proportion by dividing by N, we can divide the standard error of a frequency by N to get the standard error of a proportion. Thus

$$\frac{\sigma_B}{N} = \frac{\sqrt{NPQ}}{N} = \sqrt{\frac{NPQ}{N^2}} = \sqrt{\frac{PQ}{N}} = \sigma_p$$

When it can be assumed that the normal curve is a good approximation to the sampling distribution of P, then σ_p can be interpreted like a standard deviation. That is, the statement can be made that approximately 68% of the sample proportions in the sampling distribution will be within 1 standard error of the population proportion.

If the population proportion is not known, our best estimator of it is the sample proportion (p). Since N has to be large in order to use the normal approximation to the binomial sampling distribution in the first place, the sample proportion (p) from such a sample should estimate the population proportion (P) fairly closely, and the error introduced by substituting p for P in the standard error formula should not be serious. We shall distinguish the standard error based on the sample proportion from the one based on the population proportion by calling the former s_p and the latter σ_p. The s_p is the statistic that estimates the parameter σ_p.

The formula for the standard error based on the sample proportion is

(12.9) $$s_p = \sqrt{\frac{pq}{n}}$$

In the case of the General Social Survey data, the proportion of respondents who admitted to smoking was 0.299 ($p = 0.299$) and $q = 1 - 0.299 = .701$. The sample is sufficiently large for the normal curve to be a close approximation to the binomial sampling distribution (Np and Nq are both larger than 5). Thus the sample estimate of the standard error of the sampling distribution of proportions is

$$s_p = \sqrt{\frac{(.299)(.701)}{1033}}$$

$$= \sqrt{\frac{.2096}{1033}}$$

$$= \sqrt{.0002029}$$

$$= .014$$

We would expect from the normal curve table (Appendix Table A) that approximately 68% of the sample proportions for samples of this size would be within 1.4% of the mean proportion (P) of smokers in the population.

12.5   SAMPLING DISTRIBUTIONS OF MEANS

Thus far we have used the binomial sampling distribution and have applied the normal approximation to it to evaluate various patterns of dichotomous events. Initially, the statistic we examined in the light of its sampling distribution was a *frequency* (or a *sum* of 0's and 1's). In addition, we examined sampling distributions of proportions or percentages.

Frequently, sociologists focus their attention on **means**. They might be interested in the mean income of a group of professionals, or the mean number of days of work lost because of strikes, or the mean size of high-socioeconomic-status families. *The mean is a much used statistic because it yields the greatest amount of information about the central tendency of a relatively symmetrical distribution of scores.*

Whenever sociologists draw a simple random sample and compute a mean, they do it in order to generalize about the central tendency of the population from which the sample comes. For example, if they drew a simple random sample of workers and computed the mean number of years those persons had worked, their purpose would be to make some statement about the mean number of years worked by all workers in the population from which the sample was drawn. To be able to do this, they must know the nature of the sampling distribution of means.

In this section our purpose is to determine the central tendency, variability, and form of the sampling distribution of means so that we will have the information necessary for making generalizations.

To begin with, we will explore further the potential of the sample mean as an unbiased estimator of the population mean. We will also examine sample variances and standard deviations as unbiased estimators since they play an integral part in the measurement of sampling error.

12.5.1 Unbiased Estimators: Sample Means

The sample mean is considered a good estimator of the population mean because it is an unbiased estimator. In Section 12.2.1, we have seen that unbiased estimators are those that, on the average, equal the parameter being estimated.

We will now use a simplified example to illustrate the fact that the sample mean is an unbiased estimator of the population mean. Assume that in a community there are 7 block clubs that have each existed for the following number of years: 2, 3, 5, 6, 11, 13, 16. If we consider these 7 block clubs as a *population*, then the population mean for years of existence is $56/7 = 8$.

Column 1 of Table 12.3 lists all of the possible simple random samples of size 2 that may be drawn from the population of 7 longevity scores; column 2 lists the mean for each sample. There are 49 such samples. If the sample mean is an unbiased estimator of the population mean, then the mean of the 49-sample means should equal the population mean. As Table 12.3 shows, the sum of the 49-sample means is 392, and the mean of the sample means $[E(\bar{X})]$ is $392/49 = 8$, which is, in fact, equal to the population mean.

The fact that an estimator is unbiased is no implied assurance that a single estimate (sample mean) will be equivalent to the population mean. As a matter of fact, examination of the sample means in Table 12.3 indicates that only 8.2% of them are equal to the population mean.

In general, unbiased estimators are more desirable even though there are instances in which biased estimators have known sampling distributions or can be corrected. Biased sampling procedures will result in biased estimations as well.*

TABLE 12.3 Sample Means, Variances and Standard Deviations of Simple Random Samples (with Replacement), $N = 2$, from a Population Consisting of 7 Block Clubs with Longevity Scores of 2, 3, 5, 6, 11, 13, 16

(1) Sample	(2) $\bar{X}$	(3) $\Sigma(X - \bar{X})^2$	(4) $\dfrac{\Sigma(X - \bar{X})^2}{N}$	(5) $\dfrac{\Sigma(X - \bar{X})^2}{N-1}$	(6) $\sqrt{\dfrac{\Sigma(X - \bar{X})^2}{N-1}}$
2, 2	2.0	.0	.00	.0	.00
2, 3	2.5	.5	.25	.5	.71
2, 5	3.5	4.5	2.25	4.5	2.12
2, 6	4.0	8.0	4.00	8.0	2.83
2, 11	6.5	40.5	20.25	40.5	6.36
2, 13	7.5	60.5	30.25	60.5	7.78
2, 16	9.0	98.0	49.00	98.0	9.90
3, 2	2.5	.5	.25	.5	.71

continues

* The impression that unbiased estimators are always "best" is not always true since there are sometimes conflicting criteria for what an investigator may wish to define as a "best" estimator.

TABLE 12.3 *(Continued)*

(1) Sample	(2) $\bar{X}$	(3) $\Sigma(X - \bar{X})^2$	(4) $\dfrac{\Sigma(X - \bar{X})^2}{N}$	(5) $\dfrac{\Sigma(X - \bar{X})^2}{N - 1}$	(6) $\sqrt{\dfrac{\Sigma(X - \bar{X})^2}{N - 1}}$
3, 3	3.0	.0	.00	.0	.00
3, 5	4.0	2.0	1.00	2.0	1.41
3, 6	4.5	4.5	2.25	4.5	2.12
3, 11	7.0	32.0	16.00	32.0	5.66
3, 13	8.0	50.0	25.00	50.0	7.07
3, 16	9.5	84.5	42.25	84.5	9.19
5, 2	3.5	4.5	2.25	4.5	2.12
5, 3	4.0	2.0	1.00	2.0	1.41
5, 5	5.0	.0	.00	.0	.00
5, 6	5.5	.5	.25	.5	.71
5, 11	8.0	18.0	9.00	18.0	4.24
5, 13	9.0	32.0	16.00	32.0	5.66
5, 16	10.5	60.5	30.25	60.5	7.78
6, 2	4.0	8.0	4.00	8.0	2.83
6, 3	4.5	4.5	2.25	4.5	2.12
6, 5	5.5	.5	.25	.5	.71
6, 6	6.0	.0	.00	.0	.00
6, 11	8.5	12.5	6.25	12.5	3.54
6, 13	9.5	24.5	12.25	24.5	4.95
6, 16	11.0	50.0	25.00	50.0	7.07
11, 2	6.5	40.5	20.25	40.5	6.36
11, 3	7.0	32.0	16.00	32.0	5.66
11, 5	8.0	18.0	9.00	18.0	4.24
11, 6	8.5	12.5	6.25	12.5	3.54
11, 11	11.0	.0	.00	.0	.00
11, 13	12.0	2.0	1.00	2.0	1.41
11, 16	13.5	12.5	6.25	12.5	3.54
13, 2	7.5	60.5	30.25	60.5	7.78
13, 3	8.0	50.0	25.00	50.0	7.07
13, 5	9.0	32.0	16.00	32.0	5.66
13, 6	9.5	24.5	12.25	24.5	4.95
13, 11	12.0	2.0	1.00	2.0	1.41
13, 13	13.0	.0	.00	.0	.00
13, 16	14.5	4.5	2.25	4.5	2.12
16, 2	9.0	98.0	49.00	98.0	9.90
16, 3	9.5	84.5	42.25	84.5	9.19
16, 5	10.5	60.5	30.25	60.5	7.78
16, 6	11.0	50.0	25.00	50.0	7.07
16, 11	13.5	12.5	6.25	12.5	3.54
16, 13	14.5	4.5	2.25	4.5	2.12
16, 16	16.0	.0	.00	.0	.00
Totals	392.0	1204.0	602.00	1204.0	192.33

Population mean = 8.00
Population variance = 24.57
Population standard deviation = 4.96

12.5.2 Unbiased Estimators: Sample Variances and Standard Deviations

The formula for the **population variance** is

(12.10)
$$\sigma^2 = \frac{\Sigma(X - \mu)^2}{M}$$

BOX 12.4 REVIEWING VARIANCE AND STANDARD DEVIATION

The material in this and in the following section requires that you be comfortable with the concepts of *variance* and *standard deviation*. Considerable stress is placed on them and on other measures of variability in the field of descriptive statistics. It is not unusual for students to become confused about the measures of variability, particularly concerning the variance and the standard deviation. If you have not quite mastered their meaning and use, it would be advisable for you to take time now to review some basics of descriptive statistics (see Chapter 5).

---

where X is a score, μ is the population mean, and M is the size of the population.

The **variance** is *an average squared deviation of the scores in a distribution from their mean*. In Formula 12.10, for the population variance, the population mean is used as the point of origin from which the deviations of the scores are measured. When only sample data are available, the population mean (μ) is unknown. We must, therefore, substitute our best estimate of the population mean for that parameter. Since the sample mean is an unbiased estimator of the population mean, it may be used in lieu of the population mean. If we replace the population M with the sample N in the denominator, we obtain the following formula for the **sample variance**:

(12.11)
$$(\text{biased})\, s^2 = \frac{\Sigma(X - \bar{X})^2}{N}$$

where X is a sample score, $\bar{X}$ is the sample mean, N is the sample size, and s^2 is the sample variance.

The problem that arises with Formula 12.11 is that it yields a *biased* estimator of the population variance [*i.e.,* $E(s^2) \neq \sigma^2$]. Using Formula 12.10 for the population variance (σ^2) on the longevity scores for the 7 block clubs, we get a variance of 24.57 and a standard deviation (σ) of 4.96.

In column 3 of Table 12.3 we have computed the sum of squared deviations of the sample scores from the sample mean [$\Sigma(X - \bar{X})^2$] for each of the 49 samples with $N = 2$, and in column 4 we have computed the sample variance, $s^2 = \Sigma(X - \bar{X})^2/N$, for each. If these sample variances were unbiased estimators of the population variance, we would be able to add them, divide by 49, and get a variance of $\sigma^2 = 24.57$. The total for column 4 is 602.0. Note, however, that dividing this sum by 49 does not give 24.57; it gives 12.29. In this example the average sample variance seriously *underestimates* the population variance.

These estimators are biased because we used the sample mean ($\bar{X}$) in place of the population mean (μ) in the formula for the sample variance. According to the second property of the mean, *the sum of squared deviations of a set of scores around their mean is minimal*. That is, the sum of squared deviations around the mean is smaller than the sum of squared deviations of the scores around any other value. The set of sample scores was used to compute the sample mean in each case and in the numerator of Formula 12.11 the sum of squared deviations of the sample scores was taken around the sample mean, so the second property of the mean applies. The expression $\Sigma(X - \bar{X})^2$ is minimal. If the sample mean is exactly equal to the population mean, then no harm is done. Whenever there is any difference between $\bar{X}$ and μ, $\Sigma(X - \bar{X})^2$ will be too small. It is not likely that $\bar{X}$ will be exactly equal to μ; it will usually differ somewhat (in 49 samples only 8.2% of the sample means were equal to the population mean). The sample variance computed from $s^2 = \Sigma(X - \bar{X})^2/N$ will tend, on the average, to *underestimate* the population variance.

Fortunately, it is possible to correct the formula for s^2 and come up with an unbiased estimator. Instead of increasing the size of the numerator (which tends to be too small), the correction can be made by decreasing the size of the denominator to $N - 1$. It has been found that the expected value of the uncorrected sample variance $[E(s^2)]$ is equal to $[(N-1)/N]\sigma^2$. That is, it is, on the average, too small by $(N-1)/N$. If each sample variance is divided by this factor so that

(12.12)
$$\text{(corrected)}\, s^2 = \frac{N}{N-1}\frac{\Sigma(X - \bar{X})^2}{N}$$

$$= \frac{\Sigma(X - \bar{X})^2}{N-1}$$

the estimator will be unbiased.

In column 5 of Table 12.3, we have recalculated the 49 sample variances using the correction factor. The sum of the 49 variances is 1204.0. Dividing the sum by 49, we obtain an average of 24.57, equivalent to the population variance.

Because it is an unbiased estimator, the corrected sample variance is the preferred measure of variability in inferential statistics. The corrected formula is used in this book.

Even with correction, the sample standard deviation (s) is a biased estimator of the population standard deviation. Remember that the population standard deviation for the 7 longevity scores is 4.96. In column 6 of Table 12.3, we have computed the 49 sample standard deviations using the corrected formula; that is,

(12.13)
$$\text{(corrected)}\, s = \sqrt{\frac{\Sigma(X - \bar{X})^2}{N-1}}$$

The sum of these 49 sample standard deviations is 192.33. When this sum is divided by 49, the average is 3.93, not 4.96. The sample standard deviation is, on the average, still too small.

Because it is a biased estimator, the sample standard deviation has been used less in inductive statistics than the variance. It is used, however, as we shall soon see. A number of statisticians have pointed out that there is a **further correction** that can be made to s to make it an unbiased estimator (Bolch, 1968; Cureton, 1968; Jarrett, 1968; Brugger, 1969). The **size of the correction factor depends upon the sample size**. Table 12.4 gives the correction factors for various N's between

TABLE 12.4   Correction Factors for Unbiased Estimation of the Standard Deviation[a]

N	a_1	a_2	N	a_1	a_2
2	1.77245	1.25331	34	1.02275	1.00760
3	1.38198	1.12838	35	1.02209	1.00738
4	1.25331	1.08540	36	1.02145	1.00717
5	1.18942	1.06385	37	1.02086	1.00697
6	1.15124	1.05094	38	1.02029	1.00678
7	1.12587	1.04235	39	1.01976	1.00660
8	1.10778	1.03624	40	1.01925	1.00643
9	1.09424	1.03166	41	1.01877	1.00627
10	1.08372	1.02811	42	1.01831	1.00612
11	1.07532	1.02527	43	1.01788	1.00597
12	1.06844	1.02296	44	1.01746	1.00583
13	1.06272	1.02103	45	1.01706	1.00570
14	1.05788	1.01940	46	1.01668	1.00557
15	1.05373	1.01800	47	1.01632	1.00545
16	1.05014	1.01679	48	1.01597	1.00533
17	1.04700	1.01574	49	1.01564	1.00522
18	1.04423	1.01481	50	1.01532	1.00511
19	1.04176	1.01398	60	1.01272	1.00425
20	1.03956	1.01324	70	1.01088	1.00363
21	1.03758	1.01257	80	1.00950	1.00317
22	1.03579	1.01197	90	1.00843	1.00281
23	1.03416	1.01142	100	1.00758	1.00253
24	1.03267	1.01093	110	1.00688	1.00230
25	1.03130	1.01047	120	1.00630	1.00210
26	1.03005	1.01005	130	1.00582	1.00194
27	1.02888	1.00965	140	1.00540	1.00180
28	1.02783	1.00931	150	1.00503	1.00168
29	1.02682	1.00897	160	1.00472	1.00158
30	1.02590	1.00866	170	1.00445	1.00149
31	1.02503	1.00836	180	1.00420	1.00141
32	1.02423	1.00810	190	1.00395	1.00130
33	1.02347	1.00784	200	1.00378	1.00127

Source: Bolch (1968:27).
[a] Multiply the correction factor for the given sample size by s. Use the factors in column a_1 if s is not corrected (i.e., has N in the denominator). Use the factors in column a_2 if s is corrected (i.e., has $N - 1$ in the denominator).

2 and 200. If the standard deviation has been computed using the $N - 1$ correction, the table entry under a_2 is multiplied by s to give the unbiased estimate. If the $N - 1$ correction factor has not been used, then the correction is $a_1 s$.

To use this correction on our sample standard deviations in column 6 of Table 12.3, we would multiply each s by $a_2 = 1.25331$. Since a_2 is a constant, we can merely multiply it by $E(s) = 3.93$ to get the corrected average standard deviation. Thus,

$$(1.25331)(3.93) = 4.93$$

We did not get exactly 4.96 using the correction factor, but that was due to the standard deviations in column 6 of Table 12.3 being decimal fractions rounded to two decimal places. The difference between 4.96 and 4.93 amounts to compounded errors in rounding.

12.5.3   A Note on Computational Formulas for s^2 and s

The corrected formulas for s^2 (12.12) and s(12.13) with which we have been working are useful definitional formulas because they express clearly the operations performed. However, they are not very desirable computationally. Altering the form of the numerator of each [usually called the "sum of squares" and symbolized by Σx^2, where $x = (X - \bar{X})$] results in formulas that are much more convenient for computing and minimizing rounding error. It can be shown algebraically that

(12.14)
$$\Sigma(X - \bar{X})^2 = \Sigma X^2 - \frac{(\Sigma X)^2}{N}$$

Substituting into Formulas 12.12 and 12.13, for the sample variance and standard deviation, we arrive at the following computational formulas:

(12.15)
$$s^2 = \frac{\Sigma X^2 - [(\Sigma X)^2/N]}{N - 1}$$

and

(12.16)
$$s = \sqrt{\frac{\Sigma X^2 - [(\Sigma X)^2/N]}{N - 1}}$$

where X is a sample score and N is the sample size.

12.5.4   Standard Error of the Mean

The measure of sampling error that indicates the magnitude of the deviations of sample statistics around their parameter is called the **standard error**. Column 2 of Table 12.3 constitutes a sampling distribution of sample means based on simple random samples of size 2 from the population of longevity scores of 7 block clubs. Figure 12.3 depicts a histogram of this distribution with all of the sample means rounded to the nearest even integer. Note that the sample means cluster around the population mean of 8. *The standard error of the mean is a measure* (the standard deviation) *of the variability of sample means around their population mean.*

If we treated the 49 sample means in this sampling distribution as though they were scores, we could devise a formula very similar to Formula 12.10 for the standard deviation to measure the variability of these 49 "scores" around the population mean. Such a measure of variability is called a "standard error" rather than a "standard deviation." **Whereas the standard deviation measures the variability of scores around their mean, the standard error of the mean measures the variability of sample means around the population mean.**

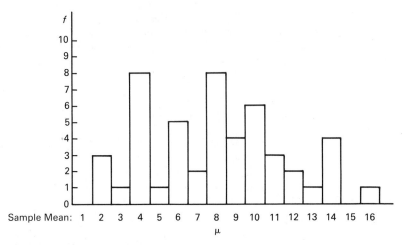

Source: Data from Table 13.4

FIGURE 12.3 HISTOGRAM OF THE SAMPLING DISTRIBUTION OF MEANS OF RANDOM SAMPLES OF SIZE 2 FROM THE POPULATION OF LONGEVITY SCORES OF 7 BLOCK CLUBS (SAMPLE MEANS ROUNDED TO NEAREST INTEGERS)

Taking the formula for the population standard deviation,

$$\sigma = \sqrt{\frac{\Sigma(X - \mu)^2}{M}}$$

we merely substitute $\bar{X}$ for X and n_s for M to convert it into a formula for the standard error of the mean:

(12.17)
$$\sigma_{\bar{X}} = \sqrt{\frac{\Sigma(\bar{X} - \mu)^2}{n_s}}$$

where $\bar{X}$ is a sample mean, μ is the population mean, n_s is the number of samples, and $\sigma_{\bar{X}}$ is the standard error of the mean (a parameter).

We will use this latter formula to compute the standard error of the mean for the sampling distribution of 49 sample means in Table 12.3. Table 12.5 reproduces the 49 sample means from Table 12.3 and shows the deviations and squared

TABLE 12.5 DEVIATIONS AND SQUARED DEVIATIONS OF 49 SAMPLE MEANS FROM A POPULATION MEAN (8.0) OF A POPULATION CONSISTING OF 7 BLOCK CLUBS

(1) Sample	(2) $\bar{X}$	(3) $\bar{X} - \mu$	(4) $(\bar{X} - \mu)^2$
2, 2	2.0	−6.0	36.00
2, 3	2.5	−5.5	30.25
2, 5	3.5	−4.5	20.25
2, 6	4.0	−4.0	16.00
2, 11	6.5	−1.5	2.25

continues

TABLE 12.5 *(Continued)*

(1) Sample	(2) $\bar{X}$	(3) $\bar{X} - \mu$	(4) $(\bar{X} - \mu)^2$
2, 13	7.5	−.5	.25
2, 16	9.0	1.0	1.00
3, 2	2.5	−5.5	30.25
3, 3	3.0	−5.0	25.00
3, 5	4.0	−4.0	16.00
3, 6	4.5	−3.5	12.25
3, 11	7.0	−1.0	1.00
3, 13	8.0	.0	.00
3, 16	9.5	1.5	2.25
5, 2	3.5	−4.5	20.25
5, 3	4.0	−4.0	16.00
5, 5	5.0	−3.0	9.00
5, 6	5.5	−2.5	6.25
5, 11	8.0	.0	.00
5, 13	9.0	1.0	1.00
5, 16	10.5	2.5	6.25
6, 2	4.0	−4.0	16.00
6, 3	4.5	−3.5	12.25
6, 5	5.5	−2.5	6.25
6, 6	6.0	−2.0	4.00
6, 11	8.5	.5	.25
6, 13	9.5	1.5	2.25
6, 16	11.0	3.0	9.00
11, 2	6.5	−1.5	2.25
11, 3	7.0	−1.0	1.00
11, 5	8.0	.0	.00
11, 6	8.5	.5	.25
11, 11	11.0	3.0	9.00
11, 13	12.0	4.0	16.00
11, 16	13.5	5.5	30.25
13, 2	7.5	−.5	.25
13, 3	8.0	.0	.00
13, 5	9.0	1.0	1.00
13, 6	9.5	1.5	2.25
13, 11	12.0	4.0	16.00
13, 13	13.0	5.0	25.00
13, 16	14.5	6.5	42.25
16, 2	9.0	1.0	1.00
16, 3	9.5	1.5	2.25
16, 5	10.5	2.5	6.25
16, 6	11.0	3.0	9.00
16, 11	13.5	5.5	30.25
16, 13	14.5	6.5	42.25
16, 16	16.0	8.0	64.00
Totals	392.0	0.0	602.00

Population mean = 8.00
Population variance = 24.57
Population standard deviation = 4.96

Data are from Table 12.3.

deviations of the sample means from the population mean (8), which are necessary to compute the standard error of the mean. Notice that the sum of the squared deviations of the sample means from the population mean, $\Sigma(\bar{X} - \mu)^2$, is 602. We can substitute this sum into the formula along with $n_s = 49$ and solve for the standard error of the mean:

$$\sigma_{\bar{x}} = \sqrt{602/49} = \sqrt{12.29} = 3.51$$

The standard error 3.51 can be interpreted in the same way as the standard deviation is interpreted (since it is a standard deviation), if the sampling distribution is normal or nearly normal. Given a normal sampling distribution, about 68% of the sample means in the sampling distribution can be expected to fall within 1 standard error of the population mean (8 $\pm$ 3.51).

We have illustrated for one sample that the sample mean is an unbiased estimator of the population mean and that the variability of sample means around the population mean can be assessed using the standard error of the mean. To generalize from a sample mean to a population mean, however, it is necessary to know more specifically the form of the sampling distribution. In this respect, we are aided by a powerful theorem known as the **central limit theorem** that provides us with information about the form of the sampling distribution of means.

12.5.5 The Central Limit Theorem

The central limit theorem states that *if repeated simple random samples of size N are drawn from a normally distributed population with mean μ and standard deviation σ, the means of such samples will be normally distributed with mean μ and standard error σ/√N.*

Moreover, *if the N of each sample drawn is large, regardless of the shape of the population distribution, the sample means will tend to distribute themselves normally with mean μ and standard error σ/√N.**

In effect, the central limit theorem tells us that the means of simple random samples drawn from a population that is itself normally distributed will result in a sampling distribution that is normal, even though the sample N is small.

The second part of the theorem adds that, as long as the sample N is large, even though the population is not normally distributed, we will still tend to get a normal sampling distribution of means. What is meant by "large" has not been firmly established. When N is as large as 30, the normal distribution may be a fairly good approximation to the sampling distribution of means. A more conservative estimate of a large N is 100 or more. Naturally, the larger the N, the better the sampling distribution is approximated by the normal curve.

Although the distribution of scores for the 7 block clubs is certainly not normally distributed and the sample size is only 2, we will see whether the normal distribution gives us a reasonable approximation to the sampling distribution of 49 sample means. We have already determined that the mean of the sampling

---

* This is the case if the population distribution has a finite mean and variance as it would have in most social science research applications.

distribution is 8 and the standard error is 3.51. To answer a question such as. "How many of the sample means will fall between 5 and 11, inclusive?" we can substitute these values into a standard score formula

(12.18)
$$z = \frac{\bar{X} - \mu}{\sigma_{\bar{X}}}$$

treating 5 and 11 as though they were continuous scores and using the lower and upper bounds of that interval, 4.5 and 11.5, in Formula 12.18 to obtain the following:

$$z_1 = \frac{4.5 - 8}{3.51} = -1.00$$

$$z_2 = \frac{11.5 - 8}{3.51} = +1.00$$

The area under the normal curve between $z = -1.00$ and $z = +1.00$ is .6826 (see Appendix Table A). Since there are 49 sample means, this indicates we would expect, if our sampling distribution was in fact normal, to find 33.45 or 33 means between 5 and 11, inclusive. If we actually count the number of sample means between 5 and 11 in Table 12.5, we find that there are 29. Thus, in this case at least, even with an N as small as 2, the normal approximation to the sampling distribution of means is fairly close. *Be aware, however, that this illustration does not prove that the normal approximation will work in other situations with such small N's!* It is our position that a normal sampling distribution should not be assumed when sampling from a nonnormal population unless N is at least 100.

Another important fact brought out by the central limit theorem is that the standard error of the sampling distribution is given by $\sigma/\sqrt{N}$. In our block club example we obtained the standard error of the mean directly from the sampling distribution using Formula 12.17. The resulting standard error was 3.51. Presumably, we should get the same standard error by using the following formula as well:

(12.19)
$$\sigma_{\bar{X}} = \frac{\sigma}{\sqrt{N}}$$

To show this, we need to go back to the longevity scores for the 7 block clubs and use the population standard deviation 3.95 that we had computed from them earlier. Dividing this standard deviation by the square root of N, we get

$$\sigma_{\bar{X}} = \frac{4.96}{\sqrt{2}} = \frac{4.96}{1.414} = 3.51$$

Thus, once we know the population standard deviation, we can find the standard error of the mean with Formula 12.19. *Since the sampling distribution is a theoretical distribution, we would not actually need to generate it as we did from the population of block clubs.* We only reproduced that sampling distribution for purposes of illustration. Ordinarily, we would not have the data available (nor is it necessary to go to the trouble of trying to collect them) to compute the standard error directly with Formula 12.17. *If the population standard deviation is known, the standard error should be computed using Formula 12.19.*

The problem, however, with using Formula 12.19 to compute the standard error of the mean is that we often do not know σ. In research typically conducted by sociologists, both the population mean and standard deviation are usually unknown. These frequently are the parameters that sociologists are interested in estimating from their sample data. If σ is unknown, the best available estimate of it would be the sample standard deviation (s). If s is corrected using Table 12.4, it will be an unbiased estimator. Substituting the corrected s for σ, the following formula is used:

(12.20)
$$s_{\bar{X}} = \frac{s}{\sqrt{N}}$$

BOX 12.5 STANDARD SCORE FORMULAS AND THE NORMAL DISTRIBUTION

If you examine carefully Formula 12.18, you will notice that it is merely another version of the basic standard score formula that is used in descriptive statistics (see Chapter 5). This is the second variation of the basic formula that we have had occasion to use. The first variation appeared as Formula 12.5 in Section 12.3.4. Still other versions will be presented in later sections of this volume. It should be progressively more clear to you that the normal distribution is *the most important* theoretical distribution for the field of statistics.

---

where $s_{\bar{X}}$ is a statistic that estimates $\sigma_{\bar{X}}$. This estimate is also known as the standard error of the mean, the difference being that $\sigma_{\bar{X}}$ is the parameter and $s_{\bar{X}}$ is the statistic.

12.5.6 Student's *t*-Distribution

When $\bar{X}$ is a random variable as it is in a sampling distribution (*i.e.*, sample means from random sampling), generalizations may be made by converting $\bar{X}$ to a standard score with Formula 12.18 and by evaluating the results on the normal curve. Since $\bar{X}$ is a random variable in Formula 12.18 for z, the numerator is also a random variable. The denominator, however, is a constant because both σ and N are constants. If σ is unknown, as it often is when we are working with sample data, and we substitute the corrected s for it in Formula 12.18, the formula becomes

(12.21)
$$t = \frac{\bar{X} - \mu}{s/\sqrt{N}}$$

In this case the denominator as well as the numerator is a random variable because s is a statistic rather than a parameter. Therefore, both the numerator and the denominator are sensitive to sampling error. If N is large (100 or more), t is approximately equal to z. This is so because the greater the N, the closer s estimates σ. With a large N, then, the normal distribution is a good approximation to the distribution of t.*

When N is small, the t-distribution may diverge greatly from the normal distribution.[†] Because both the numerator and denominator are variables in Formula 12.21 for t while only the numerator is a variable for z, the t-distribution tends to have greater dispersion than the normal distribution. Over a number of different samples, if $\bar{X}$ is the same, z will be the same because the denominator is a constant; but this is not the case with t. The **variability of the t-distribution is related to the size of N**. The smaller the N, the more variable the t-distribution. There is, then, a whole series of t-distributions, a different one for each N. The **family of t-distributions** is presented in **Appendix Table B**. In Appendix Table B the first column is labeled df, which stands for **degrees of freedom**. When the mean from one sample is evaluated using the t-distribution, $df = N - 1$ (we will have more to say about the concept of degrees of freedom in Section 12.5.7). Each row of the table represents values of a different t-distribution with $N - 1$ degrees of freedom.

These t-distribution values were derived based upon the assumption that the numerator of the t ratio is normally distributed and the denominator is independent of the numerator. Since the sample mean $(\bar{X})$ appears in the numerator and the sample standard deviation (s) appears in the denominator and these two statistics are usually related, it would seem unlikely that the denominator would be independent of the numerator. There is one instance, though, in which the denominator is independent of the numerator. This prevails when simple random samples are drawn from a **normally distributed population** of data. Such an assumption necessarily limits the instances when the t-distribution will be applicable to sociological research.

The t-distributions are symmetrical, but they are flatter (more platykurtic) than the normal curve and their tails approach the baseline at a slower rate. The smaller the N, the flatter the curve and the more slowly the tails approach the base line. If you look at the entries for the t-distributions in Appendix Table B, you will notice that, when $N = 2$ ($df = 1$), the standard scores beyond which a total of 5% of the area under the curve falls are -12.706 and $+12.706$. In the normal curve, 5% of the area lies beyond standard scores of -1.96 and $+1.96$. When N is increased to 3 ($df = 2$), the standard scores cutting off the 5% in the tails drop to -4.303 and $+4.303$; for an N of 4, they become -3.182 and $+3.182$; etc. Figure 12.4 presents graphs of these t-distributions and the t-distribution with $df = \infty$ (infinity). This later distribution is, in fact, the normal distribution since it is the limiting distribution for the Student's t. When N reaches 100, the standard scores cutting off 5% in the tails of the t-distribution become -1.98 and $+1.98$,

---

* Whenever the sample is large, we will make it a practice of designating Formula 12.21 as z and using the normal curve to evaluate it.

[†] The t-distribution was studied and introduced as a sampling distribution by W. S. Gossett (1876–1937), a mathematical consultant for the Guiness Brewery in Dublin, Ireland. He found that, with small N's, the use of the sample standard deviation (s) results in a sampling distribution of means that is not normal. He published his findings in 1908 under the pen name "Student" because of a previous agreement with Guiness.

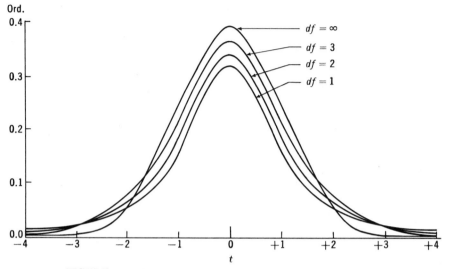

FIGURE 12.4 STUDENT'S *t*-DISTRIBUTIONS WITH 1, 2, 3, AND ∞ *df*

which are very close to the standard scores for the normal curve. Thus, **when N is equal to or greater than 100 the normal curve may be substituted** for the *t*-distribution, as a sampling distribution. In the following chapters we will demonstrate how the *t*-distribution can be used for estimating parameters and testing hypotheses.

We will digress for a moment from our discussion of sampling distributions now and examine in greater detail the concept of degrees of freedom.

12.5.7 Degrees of Freedom

The concept of **degrees of freedom** is one that arises frequently in connection with sampling distributions. It is relevant in particular to statistics with which families of sampling distributions are associated rather than a single sampling distribution. The standard normal curve is a single sampling distribution; thus the concept of degrees of freedom does not apply to it. Such sampling distributions as the *t*-distribution, the chi-square distribution, and the *F*-distribution are actually families of sampling distributions requiring the concept of degrees of freedom to determine which member of the family is applicable in a given situation.

Degrees of freedom, in general, relate to the number of restrictions (or parameters estimated) *put on data.* Initially, there are as many degrees of freedom as there are cases or categories of data. Each time a restriction is imposed upon the data, one degree of freedom is sacrificed.

For example, when we evaluate the ratio,

$$\frac{\bar{X} - \mu}{s/\sqrt{N}}$$

using the t-distribution, $df = N - 1$. For the t-distribution the number of cases (N) determines the number of degrees of freedom that are initially available. The one degree of freedom is sacrificed in the computation of the sample standard deviation (s). This degree is lost because of the restriction that the sum of deviations of scores around their mean must equal 0 [$i.e.$, $\Sigma(X - \bar{X}) = 0$]. Given this limitation, $N - 1$ of the scores in the sample may take any value, but once values have been assigned to $N - 1$ of them, the remaining score is determined. For example, if we had five scores whose mean was 10, four of the scores could take any values, but the fifth would have to have a value such that the sum of the deviations of all five from 10 would equal 0. Thus, values such as 1, 3, 14, and 20 might be assigned to the first four scores, but the fifth score would then have to have a value of 12. Or if the first four scores had values of 7, 11, 2, and 16 the fifth score would have to be 14.

When the appropriate number of degrees of freedom is established for a particular application of a statistical technique, then the specific sampling distribution that applies to that situation is also established.

Because **the formula for degrees of freedom varies from one sampling distribution to another and from one statistical technique to another**, it is difficult to devise definitive statements about finding degrees of freedom. In general, the explanation of the concept given here applies to each use, however. Throughout this book when degrees of freedom are relevant, the appropriate formula will be given.

12.5.8   Recapitulation

We have discussed how the **sampling distribution of means** is affected by normal and nonnormal population distributions, sample size, and knowledge (or lack thereof) about the population standard deviation. To bring the discussion into focus, in Diagram 12.1 (p. 449), we have recapitulated the essential points to be considered in choosing the appropriate sampling distribution and standard error.

12.6   SAMPLING DISTRIBUTIONS OF VARIANCES

By the nature of sociology there are sometimes situations in which variability may be as relevant or even more relevant than central tendency to the problem being investigated. For example, if a sociologist is engaged in the study of political activities of organizations, the variability of the characteristics of members may be an important factor in determining the amount of political activity in which the organization engages. It might be fruitful, for instance, to test the hypothesis that the greater the homogeneity of members, the more politically active the organization.

Blau has suggested that a number of empirical findings regarding formal organizations may be accounted for by two basic generalizations: "(1) increasing organizational size generates differentiation along various lines at decelerating rates; and (2) differentiation enlarges the administrative component in organizations to effect coordination" (Blau, 1970:201). These generalizations suggest the importance of examining measures of differentiation such as variances.

Blau contends that differentiation of responsibilities characteristic of larger organizations simultaneously leads to intraunit homogeneity and interunit heterogeneity (Blau, 1970:217). Although Blau primarily had job specialization in mind when he mentioned intraunit homogeneity and interunit heterogeneity, there is reason to believe that a number of variables that lend themselves to interval/ratio measurement (*e.g.*, age, income, years of education, etc.) might be significant indices as well.

At any rate, the point is that sociologists often need to examine variability of their data and make generalizations that are specifically addressed to the influence of variability on social situations. When focusing their attention on the variability of sample data, sociologists must have knowledge of the sampling distributions of measures such as variances.

It has been found that sample variances (s^2's) based upon simple random samples from a normally distributed population can be expressed as **chi-square scores** (χ^2's), which have a **chi-square sampling distribution** with $N-1$ degrees of freedom. The formula for converting sample variances to χ^2 scores is as follows:*

(12.22)
$$\chi^2 = \frac{(N-1)s^2}{\sigma^2}$$

where σ^2 is the population variance and s^2 is the sample variance.

12.6a   The Chi-Square Distribution

The chi-square sampling distribution is not one distribution but, like the *t*-distribution, a family of continuous distributions. The shape of a particular chi-square distribution depends upon the number of degrees of freedom involved. The chi-square table (Appendix Table C) is set up with degrees of freedom (*df*) in the first column and selected probabilities (areas in the tails of the distribution) in the other columns. χ^2 scores may vary from 0 to infinity.

Figure 12.5 illustrates curves of chi-square distributions for 1, 2, 3, 4, 5, 8, and 10 degrees of freedom. Notice that for one degree of freedom ($N = 2$) the chi-square distribution is essentially a single-tailed J-curve with the bulk of the χ^2 scores close to 0. Only 5% of the χ^2 scores in this particular distribution are as large as or larger than 3.841.

Another interesting fact about the chi-square distribution with one degree of freedom is that it is the same as that for z^2. Notice, for instance, that $(1.96)^2 = 3.841$. A *z*-score of 1.96 cuts 2.5% of the area off one tail of the normal distribution, while a χ^2 of 3.841 cuts 5% off of the right tail of the chi-square distribution.

---

* Note that the population variance (σ^2) is frequently unknown. We will show in Chapter 13 that a variation of this formula can be used even though the population variance is unknown.

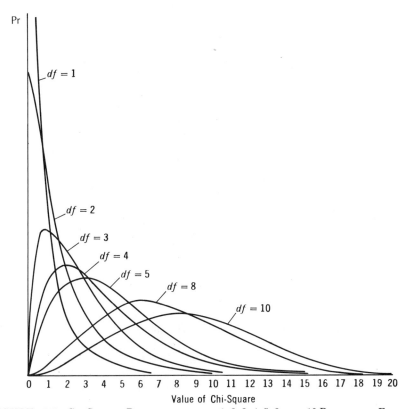

Pr

df = 1

df = 2

df = 3

df = 4

df = 5

df = 8

df = 10

0 1 2 3 4 5 6 7 8 9 10 11 12 13 14 15 16 17 18 19 20

Value of Chi-Square

FIGURE 12.5   CHI-SQUARE DISTRIBUTIONS FOR 1, 2, 3, 4, 5, 8, AND 10 DEGREES OF FREEDOM

Notice in Figure 12.5 that a chi-square distribution with two degrees of freedom ($N = 3$) also forms a J-curve, but it approaches the baseline more slowly than the curve for one degree of freedom. For the curve with three degrees of freedom, a tail falls on the left as well as on the right. The left tail starts at 0. Notice also that the mode of the curve for three degrees of freedom is at a χ^2 score of 1. It can be said, in general, that the mode of a χ^2 curve (with the exception of the curve with one degree of freedom) will be at $\chi^2 = df - 2$.

As the number of degrees of freedom (or sample size in the case of variances) increases, the chi-square distributions are less skewed. By the time degrees of freedom reach 10, the distribution is much less skewed, although it is not quite symmetrical. The chi-square table usually goes only to 30 degrees of freedom. For small samples, therefore, chi-square distributions can be used directly to evaluate sample variances. When the number of degrees of freedom exceeds 30, the chi-square distribution can be approximated by the standard normal distribution with a standard error of 1 and

(12.23)

$$z = \sqrt{2\chi^2} - \sqrt{2(df) - 1}$$

An alternative approach is possible with a large N because, as N gets large sample *standard deviations* tend to distribute themselves in an approximately normal distribution with a mean of σ and a standard error

(12.24)
$$\frac{\sigma}{\sqrt{2(N-1)}}$$

Since s tends to be close to σ when N is large, if σ is unknown, the unbiased s may be substituted for it without introducing serious error.

It will be explained in later chapters how these sampling distributions of variances and standard deviations can be used in (a) estimating parameters (Section 13.4) and (b) testing hypotheses (Section 15.6).

All we have said about sampling distributions of variances and standard deviations is based on the assumption that the population sampled is normally distributed. Unless N is extremely large, the error introduced when the population is not normally distributed may be considerable.

12.7 SUMMARY

In this chapter we examined the concept of the sampling distribution. We found that, in general, statistics distribute themselves around their parameters in predictable patterns with measureable degrees of variability. The predictable patterns are known as sampling distributions and the measures of variability are known as standard errors. We studied the central tendency, variability, and form of sampling distributions for frequencies, means, variances, and standard deviations.

The sampling distributions of these particular statistics by no means exhaust those of interest and importance to sociologists. The ones discussed in this chapter serve as typical examples of the kinds of sampling distributions with which sociologists work. The sampling distributions of other statistics will be discussed in chapters that follow (for example, the F-distribution in Section 15.3.2).

BOX 13.6 SOME IMPORTANT DISTINCTIONS

It is very important that you be able to distinguish among a **population** distribution, a **sample** distribution, and a **sampling distribution**. If this is not clear to you, go back and review all units under Section 12.2.

The **central limit theorem** was discussed in connection with the sampling distribution of the mean, but it has more general applicability than has been indicated thus far. Whenever the normal approximation to the sampling distribution of a statistic can be used with a large sample, it is because the central limit theorem is operative. We have found thus far that the normal approximation can be used with frequencies, variances, standard deviations, and means when N is large. In the following chapters we will find that the normal approximation applies to sampling distributions of other statistics as well.

The concept of the sampling distribution is really the pivotal concept for inferential statistics. It is at the very heart of the logic of statistical induction. In Chapter 13 we will see how sampling distributions are used in estimating parameters, and in the remaining chapters we will see how they are used to test statistical hypotheses.

BOX 12.7   FINITE POPULATION CORRECTION

You will note that the standard error formulas we have discussed make reference to **sample** size, N, but not population size, M. That is, the variability between probability samples depends upon the size of the sample, not the relationship of the sample to the population size. We should add one minor footnote, however. The only qualification is that, as a sample becomes a substantial part of a small finite population, sampling variability begins to decrease (and statistical decisions based on these samples would become somewhat more conservative). At the extreme, when a sample enumerates a whole finite population, sampling error no longer exists. If a sample is a large part of a small finite population, the standard error can be modified (reduced) by multiplying it by what is called a *finite population correction*:

$$\text{fpc} = \sqrt{1 - \frac{N}{M}}$$

Almost always, the reason for sampling is that enumeration is impossible, less accurate, or too costly. Thus, using a sample that is manageable and a minor part of a population is most often the practice. The finite population correction is therefore virtually never used. A sample of size 1000 has the same effective sampling variability whether it comes from a population of 50,000 or a population of 1 million! Researcher's typically draw national samples of 1500 individuals and rely on them for generalizations to the nation with exactly the same confidence that one would have using a sample of 1,500 from Des Moines to generalize to that city. It is the sample size (and, critically, how it is drawn) that matters most.

CONCEPTS TO KNOW AND UNDERSTAND

sampling distribution
expected value of a statistic
unbiased estimator
standard error
law of large numbers
binomial sampling distribution
normal sampling distribution
mean of a binomial distribution
standard error of a binomial
    distribution
sampling distribution of proportions

sampling distribution of means
Student's t-distribution
standard error of the mean
second property of the mean
central limit theorem
large N
degrees of freedom
sampling distribution of variances
chi-square distribution
standard error of a proportion

QUESTIONS AND PROBLEMS

1. Explain the relationship between a *sample* distribution and a *sampling* distribution.

2. What is the relationship between the standard error of the mean and the standard deviation?

3. Explain the meaning of the statement that the normal distribution is the limiting case of the binomial probability distribution.

4. Explain what is meant by the statement that the sample mean is an unbiased estimator of the population mean.

5. Why is the variance preferred to the standard deviation as a measure of variability in inferential statistics?

6. Prove algebraically that $\Sigma(X - \bar{X})^2 = \Sigma X^2 - (\Sigma X)^2/N$.

7. Why is $N - 1$ instead of N used as the denominator of the formula for the sample variance?

8. Of what significance for inferential statistics is the central limit theorem?

9. How does Student's t-distribution differ from the normal distribution? When may the normal distribution be used instead of the t-distribution?

10. Explain what is meant by degrees of freedom as applied to Student's t-distribution.

11. Draw a biased sample from the 7 block clubs described in Table 12.3. For example, draw samples with the restriction that the extreme scores (*e.g.*, 2 and 16) cannot appear in any sample. Then show that the mean of random samples of size 2 drawn in this biased way will be a biased estimate of the population mean. *Hint:* Identify all of the possible samples of size 2 under this biased sampling procedure, compute each sample's mean, and describe this distribution of sample means.

12. Create a sampling distribution. Select a dichotomous variable (*e.g.*, gender) and a small population (*e.g.*, members of your immediate family). Using the formula for computing binomial probabilities, compute the sampling distribution. For this problem, use sample size $N = 2$. Portray the sampling distribution in table form and also in the form of a histogram (carefully drawn). Now, create and portray the sampling distribution for the same situation where a larger sample size is used (but one less than the size of the population!). Comment on differences between these distributions.

13. Empirically set up a sampling distribution. You could do this in one of two ways. If you have access to a computer, identify a large data set (*e.g.*, some general social survey that is already stored in computer-readable form). Then identify a computer program that permits you to draw simple random samples of cases from this data set. Decide on a sample size (for example, size 30) and have the computer draw repeated random samples of this size (say, about 30 or so samples, if time permits). In each case, the computer should be instructed to

summarize statistically the data from the sample on the variable(s) in which you are interested. For example, if you were interested in age of individual, have the computer compute the mean and standard deviation of ages of the, say, 30 individuals included in each random sample. Then summarize statistically the results of your sampling. By making an overall description of all of the cases in the data set you would have, in effect, the population parameter. Compare the parameter with your sampling distribution summary. If you have resources, it would be helpful to repeat the problem using different sample sizes. If you do not have access to a computer, then draw and summarize small samples ($N = 10$, for example) from the data provided below. Complete a sampling distribution and describe that sampling distribution statistically. Then use the formulas given in this chapter to compute the mean and standard error of a sampling distribution of the kind you have empirically constructed. Comment on similarities between your empirical description and the computed one. Be sure that you use a table of random numbers to draw your simple random samples from the data below.

LISTING OF PERCENTAGE OF POPULATION LIVING IN URBAN PLACES OF 2,500 OR LARGER IN 1980 FOR STATES*

Line	State	% Urban	Line	State	% Urban	Line	State	% Urban
01	Maine	47.5	18	N. Dak	48.8	35	Ark	51.6
02	N.H.	52.2	19	S. Dak	46.4	36	La	68.6
03	Vt	33.8	20	Nebr	62.9	37	Okla	67.3
04	Mass	83.8	21	Kans	66.7	38	Tex	79.6
05	R.I.	87.0	22	Del	70.6	39	Mont	52.9
06	Conn	78.8	23	Md	80.3	40	Idaho	54.0
07	N.Y.	84.6	24	D.C.	100.0	41	Wyo	62.7
08	N.J.	89.0	25	Va	66.0	42	Colo	80.6
09	Pa	69.3	26	W. Va.	36.2	43	N. Mex	72.1
10	Ohio	73.3	27	N.C.	48.0	44	Ariz	83.8
11	Ind	64.9	28	S.C.	54.1	45	Utah	84.4
12	Ill	83.3	29	Ga	62.4	46	Nev	85.3
13	Mich	70.7	30	Fla	84.3	47	Wash	73.5
14	Wis	64.2	31	Ky	50.9	48	Oreg	67.9
15	Minn	66.9	32	Tenn	60.4	49	Calif	91.3
16	Iowa	58.6	33	Ala	60.0	50	Alaska	64.3
17	Mo	68.1	34	Miss	47.3	51	Hawaii	86.5

* *Source:* U.S. Bureau of the Census (1984:12, Table 12).

GENERAL REFERENCES

Games, Paul A., and George R. Klare. *Elementary Statistics: Data Analysis for the Behavioral Sciences* (New York, McGraw-Hill), 1967.

Chapter 8 is an unusually detailed discussion of sampling distributions. It stresses how to distinguish clearly between population distributions, sample distributions, and sampling distributions.

Hays, William L., and Robert L. Winkler. *Statistics: Probability, Inference, and Decision* (New York, Holt, Rinehart and Winston), 1971.

This book presents a more technical, more mathematically oriented discussion of sampling distributions. In Chapter 6 the normal distribution is shown to be the parent *distribution of the t*-distribution, the chi-square distribution, and the *F*-distribution.

LITERATURE CITED

Blau, Peter M. "A Formal Theory of Differentiation in Organizations," *American Sociological Review*, 35, 2 (April 1970), pp. 201–218.

Bolch, Ben W., "More on Unbiased Estimation of the Standard Deviation," *The American Statistician*, 22, 3 (1968), p. 27.

Brugger, R. M., "A Note on Unbiased Estimation of the Standard Deviation," *The American Statistician*, 23, 4 (1969), p. 32.

Cureton, E. E., "Unbiased Estimation of the Standard Deviation," *The American Statistician*, 22, 1 (1968), p. 22.

Jarrett, R. F., "A Minor Exercise in History," *The American Statistician*, 22, 3 (1968), p. 25.

Pierce, Albert, *Fundamentals of Nonparametric Statistics* (Belmont, Calif., Dickenson Publishing Co.), 1970.

U.S. Bureau of the Census, *Statistical Abstract of the United States: 1985*, 105th ed. (Washington, D.C., U.S. Government Printing Office), 1984.

DIAGRAM 12.1 Determining the Sampling Distribution and Standard Error of the Mean

1. If the population distribution is normal and the population standard deviation is *known*, regardless of whether N is large or small:
   Sampling distribution: normal curve
   Standard error, use Formula 12.19: $\sigma_{\bar{x}} = \sigma/\sqrt{N}$

2. If the population distribution is normal and the population standard deviation is *unknown*, when N is large (100 or more):
   Sampling distribution: normal curve
   Standard error, use Formula 12.20: $s_{\bar{x}} = s/\sqrt{N}$

3. If the population distribution is normal and the population standard deviation is *unknown*, when N is small (less than 100):
   Sampling distribution: Student's t with $df = N - 1$
   Standard error, use Formula 12.20: $s_{\bar{x}} = s/\sqrt{N}$

4. If the population distribution is not normal and the population standard deviation is *known*, when N is large:
   Sampling distribution: normal curve
   Standard error, use Formula 12.19: $\sigma_{\bar{x}} = \sigma/\sqrt{N}$

5. If the population distribution is not normal and the population standard deviation is *unknown*, when N is large:
   Sampling distribution: normal curve
   Standard error, use Formula 12.20: $s_{\bar{x}} = s/\sqrt{N}$

6. For all other situations: If the population distribution is not normal whether the population standard deviation is *known* or *unknown*, when N is small: neither the normal curve nor student's t-distribution applies.

CHAPTER 13

Point and Interval Estimates of Parameters

When the stock market crashed in 1929 and the Depression engulfed the country, the government found itself in the dark about the magnitude of the unemployment problem. At that time Census Bureau policy was to collect labor statistics using the "gainful worker" concept. Respondents were interviewed to determine the occupations at which they were usually employed. They could report occupations with which they identified, whether or not they were actually working at them. The use of the gainful worker concept thus made it possible for retired and institutionalized persons to report occupations. Although such information can be very useful to sociologists interested in the self-images of workers, potential workers, and ex-workers, it is not an effective way of determining how many people are in the labor force—how many are actually working, and how many are unemployed.

When the federal government took on an active role in regulating the economy, it realized that a more accurate measurement of employment and unemployment, taken at frequent intervals, was necessary. Accordingly, beginning with the 1940 census, the Census Bureau adopted the *labor force* concept. The labor force concept is based principally on each respondent's actual activity during the time surveyed. The respondent is categorized as working, looking for work, or doing something else.

Today, rather than collect labor force statistics only once every 10 years as part of the census, the Bureau of Labor Statistics surveys a sample of the population each month. During the week of the 12th day of the month, members of more than 70,000 households in 629 sample areas are interviewed to determine their work activities for the previous week. On the basis of these interviews, monthly estimates are made of the size of the labor force and the percentage of those in the labor force who are unemployed. These estimates are now considered to be so indicative of the state of the national economy that they get front-page coverage in the nation's newspapers and radio and television news bulletins.

Although it is not generally known, these much publicized figures are not parameters. They are estimates of parameters based on sample data. Nonetheless, the federal government takes these estimates seriously enough to use them as a basis for national policy. Certainly, these estimates are efficient enough to warrant the government's concern. The estimated unemployment rate has a standard error of 0.21%.

Parameter estimation is not merely of interest and importance to policy makers. It also plays an important role in science. Although labor force estimates are utilized primarily for policy making, they are also of interest to sociologists in the study of American society. After all, work is one of the primary sociological variables.

Sociologists not only make use of parameter estimates by agencies of the federal government; in the course of their research, they also frequently engage in estimation of parameters from sample data they have collected themselves. For example, Morse and Weiss conducted a study of a national sample of employed men to determine what meanings work had for them (Morse and Weiss, 1955). In response to a hypothetical question, "If by some chance you inherited enough money to live comfortably without working, do you think you would work anyway or not?" Eighty percent of the 393 men who responded said that they would continue to work (Morse and Weiss, 1955:191–92). Because Morse and Weiss were interested in the meanings of work for *all* employed men in the United States, their 80% figure can be looked on as a parameter estimate. Of course, in their study, they went on to ask other questions and to estimate other parameters. The point to consider is that they conducted a sociological study whose import depends primarily upon results of estimation procedures.

Sociologists have sometimes been prone to forget the important role played by estimation procedures in scientific research. Labovitz argued, for instance, that tests of significance are nonutilitarian for sociology (Labovitz, 1970; see also Morrison and Henkel, 1970). In the process of affirming his case against tests of significance he concluded,

> In a crude sense, it seems that many researchers have already rank ordered the importance of statistical techniques. Usually, a statistic is first developed and reported (*e.g.*, a rank correlation measure like d_{yx}), and then a claim is made that its sampling distribution will be developed later. The arguments in this paper would logically extend this rank order among techniques to the point of forgetting the development of the sampling distribution altogether. (Labovitz, 1970:147)

Such an extreme position ignores the fact that sampling distributions are necessary for estimating parameters as well as testing hypotheses. Although Labovitz recalled some important arguments against tests of statistical hypotheses, his arguments do not apply equally to estimation procedures.

13.1   POINT VERSUS INTERVAL ESTIMATES

When the Bureau of Labor Statistics estimate of the percentage unemployed is released each month, a single value is reported. *When a single value based upon sample data is used as an estimate of a parameter the estimate is known as a* **point estimate**. The daily newspaper may report that the unemployment rate for the month of July was 7%. What this actually means is that 7% of those surveyed by the Bureau of Labor Statistics interviewers reported themselves unemployed. Assuming that the survey sample is a random sample,* this statistic is used as a point estimate of the parameter, the percentage of the total labor force that is unemployed.

Point estimates are used in reports to the mass media because they are easiest for the general public to comprehend. Ideally, a point estimate should be based upon an unbiased estimator of the parameter (which this one is); error in estimation should be random; *and* the estimate should be an **efficient** one (*i.e.*, the standard error should be small so that the statistics cluster closely about the parameter they estimate). Unfortunately, a point estimate does not carry with it any indication of how closely it estimates the parameter—there is no probability figure attached to it, as such. Therefore, although it is a popular way of presenting an estimate, it leaves something to be desired.

There are real benefits to be derived from making an **interval** rather than a point estimate of a parameter. *An* **interval estimate** *is an estimate that consists of a range of values rather than a single value.* One advantage of the interval estimate is that the width of the interval tells something about the efficiency of the estimate. Another advantage is that we can attach a probability figure to the estimate. Because of these and other advantages to be pointed out later, interval estimates are generally preferred to point estimates for sociological research.

An interval used to estimate a parameter is known as a **confidence interval** and *the end points of the interval are known as* **confidence limits**. The use of the term *confidence* relates to the fact that probability figures may be associated with interval estimates, and the probabilities give us a notion of how much confidence we can have in our estimation procedure. We may select any probability level we wish to associate with our estimates. Two probability levels commonly used, however, are the 95% and 99% levels.

In this chapter we will discuss estimation procedures for several different statistics. Again, our discussion will be selective rather than exhaustive. We have chosen to examine statistics that are commonly used and that concisely illustrate estimation procedures. With estimation procedures for various statistics being generally similar, your ability to understand the procedures for a few techniques should enable you to apply them to other techniques as well.

13.2 ESTIMATING PROPORTIONS

Since we began this chapter with examples of the use of sample percentages as point estimates, we will turn first to a consideration of interval estimates for proportions (or percentages). Frequencies, proportions, and percentages are alternative ways of presenting the same data; therefore, they are interchangeable.

---

* Although the sampling plan is a probability sample, it is more complex than a simple random sample. It is, in fact, a variation of the cluster sample design.

13.2.1   Confidence Limits for Small Samples

When the sample size (N) is so small that the rule of thumb (NP and NQ are equal to or greater than 5) is not met, the binomial sampling distribution is appropriate and frequencies are used in Formula 12.2 (p. 413) to find the probability of each possible sample outcome. Figure 12.2a shows a binomial sampling distribution where $P = .5$ and $N = 10$. The possible outcomes and their associated probabilities are listed below.

BINOMIAL SAMPLING DISTRIBUTION FOR $P = .5$ AND $N = 10$

Number of Successes	Probability
0	.001
1	.01
2	.04
3	.12
4	.21
5	.25
6	.21
7	.12
8	.04
9	.01
10	.001

We could set up approximately the 90% confidence limits for this distribution by summing the probability of extreme outcomes, alternating ends of the distribution, until we reached the desired confidence level. Here we just exceed 10% in the tails of the distribution by summing probabilities for 0, 10, 1, 9, 2, and 8 successes. The sum is .102. Thus the 89.8% confidence limits would be 3 through 7 successes. Note that we could not set up the 95% confidence limits here. We could set up 97.8% confidence limits (2 through 8), however. It is also apparent that we do not need to compute probabilities for outcomes in the middle of the sampling distribution, only those in the tails sufficient to identify the critical region of interest. Even when a binomial sampling distribution is skewed, confidence limits are set up by alternating sides of the distribution in creating the needed sum.

13.2.2   Central Tendency, Variability and Form for Large Samples

When N is large and the proportion P is not too extreme so that the rule of thumb applies, then the normal curve may be used as an approximation to the binomial sampling distribution. As was explained in Section 12.4, for large samples, the sampling distribution of proportions will have a **mean proportion** equal to P, the population proportion, and a **standard error of the proportion** equal to

(13.1)
$$\sigma_p = \sqrt{\frac{PQ}{N}}$$

Furthermore, Section 12.4 explained that when the population proportion is unknown, the sample proportion (p) is used as an estimate of it and the standard error is estimated by

(13.2)
$$s_p = \sqrt{\frac{pq}{N}}$$

Given this knowledge of mean proportions and standard errors of proportions, it is now possible to compute confidence intervals.

13.2.3 Confidence Limits for Large Samples

Although the Bureau of Labor Statistics makes point estimates of the unemployment rates for the mass media, in their technical reports they give the information necessary to make interval estimates. As was mentioned earlier, the standard error of the proportion unemployed runs 0.21% (or .0021). Assume that in a given month the sample estimate of the proportion unemployed is 0.07. We will use this information to compute a 95% confidence interval.

Since N is large (over 70,000 households in the survey), we can expect the sampling distribution of the proportion unemployed to be approximated by the normal sampling distribution. The sample proportions will distribute themselves normally about their expected value [$E(p) = P$]. Moreover, 95% of the sample p's in the sampling distribution can be expected to fall under the normal curve between standard scores of -1.96 and $+1.96$ since these z-scores bound the middle 95% of the normal curve (see Appendix Table A).

Because the parameters P and σ_p are unknown, the **standard score** formula appropriate to this situation is

(13.3)
$$z = \frac{p - E(p)}{s_p}$$

where $E(p)$ is the expected value of p.

Since the sample proportion (p) is an unbiased estimator of P, $E(p) = P$. Of course, this is the very value that we wish to estimate. Rather than use p as a point estimate of this value, we will make an interval estimate.

Since 95% of the sample proportions in a sampling distribution can be expected to fall within ± 1.96 standard errors of P, if we knew P, we could construct a confidence interval around it and see whether our observed sample proportion is in the interval. Since we do not know P, what we shall do instead is construct the confidence interval about the observed p (an unbiased estimator of P).

If we take Formula 13.3 for z and manipulate it algebraically to solve for $E(p)$, we can come up with a formula for confidence limits that looks like the following:

$$E(p) = p \pm z(s_p)$$

We will substitute cl (confidence limits) for the unknown $E(p)$ in the formula because when we use the formula we will get lower and upper limits of a confidence interval rather than a single figure representing $E(p)$. The formula for finding the confidence limits of proportions will then be

(13.4)
$$cl = p \pm z(s_p)$$

To find the confidence limits, we substitute the appropriate values of p and s_p into the formula along with the appropriate z-scores. For example, to find the 95% confidence limits for proportion unemployed, we substitute $p = 0.07$ and $s_p = 0.0021$ and solve the equation with $z = -1.96$ and $z = +1.96$ in order to get the lower and upper limits of the interval:

$$cl_1 = 0.07 - 1.96(0.0021) = 0.07 - 0.004 = 0.066$$

$$cl_2 = 0.07 + 1.96(0.0021) = 0.07 + 0.004 = 0.074$$

Thus, the 95% confidence interval for the proportion unemployed has limits of 0.066 and 0.074. This gives one a somewhat different impression of the unemployment rate than the simple point estimate of 0.07.

13.2.3a   The Meaning of Confidence

What does the 95% probability figure mean? Does it mean that there is a 95% chance that the unemployment rate for the total population (the parameter) is between 0.066 and 0.074? This kind of interpretation of the probability associated with a confidence interval is a common error. Actually, the parameter (P) that we are trying to estimate is a fixed value. Its value at a given point in time does not fluctuate. Therefore, the probability that P is between 0.066 and 0.074 is 1 or 0. If it is, in fact, between these two limits, then the probability is 1; if it is not, the probability is 0. What fluctuates from sample to sample is p (the statistic). Since, by necessity, we constructed the interval around p instead of P, the position of the interval with regard to the parameter depends upon the location of the particular sample p we have available. Since 95% of the p's in the sampling distribution will be within ± 1.96 standard errors of the population proportion, and we used $\pm 1.96 s_p$ to construct the interval around p, any p within that range will result in a confidence interval that will include P. Any p not within the range of ± 1.96 standard errors of the parameter will result in a confidence interval that will not include the parameter. Figure 13.1 presents pictorially this situation to help clarify it. Because p_5 in Figure 13.1 was not in the range of ± 1.96 standard errors of P, its interval missed the parameter. Other p's, such as p_1 and p_2, were within the range, so their intervals do include P.

The 95% probability figure thus does not apply to a particular confidence interval (technically speaking, probability never applies to a particular event; it always applies in the long run). What it means is that if we made a very large number of interval estimates (such as those illustrated in Figure 13.1), each based on a sample p, 95% of the confidence intervals would include the parameter and only 5% of them would miss.

With respect to the confidence interval for proportion unemployed, the 95% probability figure means that if a great many independent interval estimates were made instead of just one 95% of those would include the parameter and 5% would not. The confidence we have in a single interval

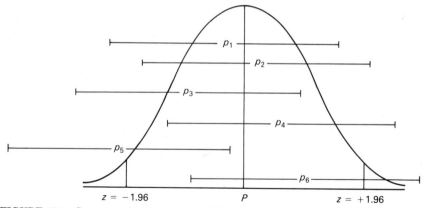

FIGURE 13.1 SAMPLING DISTRIBUTION OF PROPORTIONS ILLUSTRATING DISTRIBUTION OF CONFIDENCE INTERVALS OF SAMPLE p'S AROUND P

estimate comes from the fact that it is likely to be one of the 95% that include the parameter rather than one of the 5% that do not.

As we stated earlier (Section 13.1), the probability figures commonly used for finding confidence intervals are the 95% and 99% levels. The 95% level leaves 2.5% of the area of the sampling distribution in each tail of the curve, and the 99% level leaves 0.5% in each tail. The assumption is that a sample proportion (p) that deviates from the parameter (P) by so much that it falls into one of these tails represents an unusual occurrence. As we shall see later (in Chapter 14), a difference great enough to throw the sample proportion into one of the tails is considered a critical difference. Therefore, the tails of the sampling distribution beyond the 95% or 99% levels are usually labeled *critical regions*. The symbol α is sometimes used to refer to this probability, the proportion of the area under the curve that is in the tails of the distribution. We will have occasion to use this notation later in this chapter (Section 13.4.1).

13.2.3b Another Example

Although Morse and Weiss did not report interval estimates of parameters in their article, they did record the necessary data to make such estimates. We will take their finding that 80% of the respondents in their sample would continue to work and compute a confidence interval for it. Their $p = 0.80$ and $N = 393$ (Morse and Weiss, 1955:192, Table 1). Therefore, s_p may be computed as follows:

$$s_p = \sqrt{\frac{pq}{N}} = \sqrt{\frac{(0.80)(0.20)}{393}} = \sqrt{0.00041} = 0.02$$

The 95% confidence limits for $p = 0.80$ and $s_p = 0.02$ can then be computed as follows:

$$cl = p \pm 1.96 s_p$$

$$cl_1 = 0.80 - 1.96(0.02) = 0.80 - 0.0392 = 0.76$$

$$cl_2 = 0.80 + 1.96(0.02) = 0.80 + 0.0392 = 0.84$$

Thus, the 95% confidence interval has limits of 0.76 and 0.84. Notice that the width of this interval is 0.08, whereas the width of the interval for the unemployment rate was only 0.008. Although both are 95% confidence intervals, the standard error is much smaller for the employment data. Remember that the labor force survey is based on more than 70,000 interviews, whereas the Morse and Weiss data were based on only 393. The smaller the standard error, the narrower the confidence interval will be and the more efficiently it will estimate the parameter. Of course, the obvious way of decreasing the standard error is to increase the sample size. According to **the law of large numbers**, *the larger the sample, the smaller the standard error.*

Both examples we have used were 95% confidence intervals. Although there is nothing sacred about the 95% level, convention says that it is a reassuring level. Under some circumstances, the 95% level may not be considered adequate. It may be desirable to use the 99% level instead. In such a case it will be necessary to substitute $z = \pm 2.58$ for $z = \pm 1.96$ in the formula for confidence limits (see the normal curve table, Appendix Table A).*

We will compute 99% confidence limits for the unemployment data for the sake of comparison. Substituting into the formula, we get

$$cl_1 = 0.07 - 2.58(0.0021) = 0.07 - 0.0054 = 0.065$$

$$cl_2 = 0.07 + 2.58(0.0021) = 0.07 + 0.0054 = 0.075$$

Thus, the 99% confidence limits are 0.065 and 0.075 as compared to 0.066 and 0.074 for the 95% limits. The effect of increasing the probability from 95% to 99% was to increase the width of the confidence interval from 0.008 to 0.01. The increase was not drastic because the standard error is very small.

There are no hard and fast rules for determining a tolerable width for a confidence interval. It depends upon the research problem and the judgment of the researcher. For example, public opinion pollsters regularly make predictions of election outcomes. They draw samples of registered voters to ask them for whom they will vote. Although some pollsters use quota samples rather than random samples, they make estimates of what their sampling error would be if their samples were random. These rough estimates allow them to compute confidence intervals for their sample estimates, and they do surprisingly well with their predictions.

With two candidates in contention, it is important to predict which candidate will get the majority of the votes. During the 1960 election campaign when John F. Kennedy and Richard M. Nixon were the presidential candidates, the contest was so close that the lower limit of the confidence interval dipped below the 50% mark. Therefore, pollsters declared the election "too close to call." On the other hand, in the 1964 campaign when Lyndon B. Johnson and Barry M. Goldwater were the candidates, Johnson was so heavily favored that pollsters

---

* We should note that it is possible to have confidence limits that are not equally distant from the parameter estimate where the sampling distribution of some statistic is not symmetrical (and the excluded proportion under the curve is divided between the two tails of the distribution.) It is also possible to compute only one confidence limit depending upon the shape of the curve and the problem of interest.

predicted a Johnson victory without the slightest hesitation. The point estimate for a Johnson victory was about 64%; therefore, it was possible to use 99% confidence limits without fear of dipping below the 50% mark.

In the Nixon–Humphrey campaign of 1968, Humphrey closed the gap in the last days of the campaign, and the result was again in doubt. However, the Nixon–McGovern campaign of 1972 was similar to the 1964 campaign and pollsters predicted a Nixon victory with some degree of confidence (the predictions were generally within 2 or 3 percentage points of the final result, and one poll missed the percentage of the Nixon vote by less than 1%).

Ideally, a researcher would routinely use the 99% confidence interval and take a sample large enough to minimize sampling error. However, such mundane matters as availability of funds may compromise the design of a research project.

13.3 ESTIMATING MEANS

It has become increasingly obvious that illness and health are not solely physical or medical problems. They are only definable and meaningful within a social context. As a matter of fact, from the behavioral perspective, social definitions of illness and health are more predictive than the objective facts. Those who are recognized as being ill by others and see themselves as being ill are likely to play sick roles.* On the other hand, those who are physically ill but are not so defined by others nor by themselves are not likely to behave as though ill.

Sociological variables such as life styles, occupational patterns, and socioeconomic statuses are relevant for the study of the incidence of diseases. Also of relevance, sociologically, are studies of attitudes toward medical care and of behavior related to the treatment of illness. Sociologists interested in illness and health have available to them an excellent source of national health data for the United States because of the continuing surveys conducted by the U.S. Public Health Service. The data are from nationwide household interviews. The survey is a continuous one with a random sample of the civilian noninstitutionalized population of the United States being interviewed each week.

These data allow for an exploration of sociological and social psychological factors that influence attitudes and behavior towards illness and health. For example, it is known that older people and people from low-income families have more health problems than do younger or more affluent people (Lerner and Anderson, 1963; U.S. Senate, 1963). It is reasonable to expect that persons who are ill will seek help from physicians more often than persons who are not ill. Therefore, older persons and those from low-income families should have relatively high mean numbers of physician visits per year. We will look at the National Health Survey data for 1983 to see whether this is the case. In the process we will also learn the procedures used for estimating population means.

The random sample from which the physician visit data were collected included approximately 111,000 persons from 41,000 households. The population

---

* For a description of the set of behavior expectations attaching to the "sick role," see Parsons (1951:436).

covered by the survey was the civilian population of the United States living at the time of the interviews (U.S. Public Health Service, 1985).*

The Public Health Service report on physician visits is stated in terms of point estimates, but standard errors are also provided so we will use them to compute interval estimates of population means. A physician visit is defined as "consultation with a physician in person or by telephone, for examination, diagnosis, treatment, or advice."

Because of the nature of the data, it is convenient to compare physician visits of persons 65 years of age or older with physician visits of persons of all ages. The mean number of visits per year for persons of all ages was 5.0, whereas that for persons 65 or older was 7.6 (U.S. Public Health Service, 1985:82, Table 43). We will convert these point estimates to 95% interval estimates of their respective parameters. The appropriate sampling distribution when N is large is the normal distribution and the standard score formula, $z = (\bar{X} - \mu)/s_{\bar{x}}$, can be manipulated algebraically to come up with the following formula for confidence limits:

(13.5)
$$cl = \bar{X} \pm z(s_{\bar{x}})$$

where cl is a confidence limit (substituted for μ in the formula), $\bar{X}$ is the sample mean, $s_{\bar{x}}$ is the sample standard error of the mean (our best estimate of the population standard error and close when N is large), and z is the standard score. Note the similarity between this formula and the one for finding confidence limits for proportions (Formula 13.4).

The standard error of the mean number of physician visits for persons of all ages is approximately 0.05 and that for persons 65 and over is approximately 0.16 (U.S. Public Health Service, 1979:44–45). We can substitute the means and standard errors into the formula and solve for the confidence limits as follows:

For persons of all ages:

$$cl_1 = 5.0 - 1.96(0.05) = 5.0 - 0.098 = 4.90$$

$$cl_2 = 5.0 + 1.96(0.05) = 5.0 + 0.098 = 5.10$$

For persons 65 years of age or older:

$$cl_1 = 7.6 - 1.96(0.16) = 7.6 - 0.314 = 7.29$$

$$cl_2 = 7.6 + 1.96(0.16) = 7.6 + 0.314 = 7.91$$

The 95% confidence interval for persons of all ages, then, is 4.90 to 5.10, and that for persons 65 or older is 7.29 to 7.91. Notice that the two intervals do not overlap. This allows us to say, with some confidence, that the mean number of physician visits per year for persons 65 or older is higher than for persons of all ages in the United States. These data support the notion that those with higher rates of illness seek help from physicians more often.

The standard error for persons of all ages is smaller than that for persons 65 or older; consequently, the confidence interval is narrower. This is reasonable

---

* Note that these data do not include persons in hospitals, rest homes, etc. Therefore, there is a bias in favor of healthier people.

because the sample of persons of all ages is larger than the sample of persons 65 or older (the standard deviations may differ as well).

Now compare the mean number of physician visits per person per year of persons from low-income families (under $10,000 per year) with that for persons from higher income families ($20,000 or more per year).* The mean for physician visits from low-income families was 5.9 and from higher income families 5.2 (U.S. Public Health Service, 1985:82, Table 43). The standard errors of the means were approximately 0.218 and 0.114, respectively (U.S. Public Health Service, 1979:44–45). The 95% confidence limits may be found as follows:

For persons from families with incomes under $10,000:

$$cl_1 = 5.9 - 1.96(0.218) = 5.9 - 0.4273 = 5.47$$

$$cl_2 = 5.9 + 1.96(0.218) = 5.9 + 0.4273 = 6.33$$

For persons from families with incomes of $20,000 or more:

$$cl_1 = 5.2 - 1.96(0.114) = 5.2 - 0.2234 = 4.98$$

$$cl_2 = 5.2 + 1.96(0114) = 5.2 + 0.2234 = 5.42$$

Thus, the 95% confidence interval for persons from families with yearly incomes of $10,000 or less is 5.47 to 6.33, and that for persons from families with yearly incomes of $20,000 or more is 4.98 to 5.42. Again, notice that the two intervals do not overlap. This would give us some confidence that persons in the United States from lower-income families make more physician visits, on the average, than persons from higher-income families. This finding is consistent with our expectation. Persons from low-income families tend to be ill more than persons from high-income families and they make more physician visits.

Reviewing our analysis of physician visits, we find that the interval estimate for persons from families with incomes of $20,000 or more substantially overlaps that for persons of all ages (and incomes). The interval estimate for persons from families with incomes under $10,000, on the other hand, is higher than and does not overlap the interval for those in the $20,000 and over family income category. Furthermore, the interval estimate for persons 65 years of age or older is the highest of all and does not overlap that for persons from low-income families. Consequently, it may be concluded that those 65 and older visit physicians most often (on the average) followed by persons from families with low incomes. Behind the latter in physician visits are persons from families with incomes of $20,000 or more and persons of all ages and incomes.

Given that elderly people are more subject to health problems than those of other ages and that low-income people are more subject to health problems than higher income people, the findings of this analysis of physician visit data are consistent with what one would expect.

* Note that the comparison of mean physician visits for those 65 and over was with mean physician visits for people of all ages, including people 65 and over. Thus, we were comparing an estimate of a population mean with an estimate of the mean of a subpopulation of that same population. In this case we are working with estimates of means for two independent groups, those from low-income families and those from higher-income families.

13.3a   Estimates Based upon Small Samples

Thoits (1986) studied the relationship between multiple-role occupancy and psychological distress. Using panel data from surveys of 1,106 adult heads of households in Chicago and 720 adults in New Haven, Connecticut, she examined the influences of gender and marital status differences on this relationship. She was particularly interested in why women, especially married women, seem to have higher incidences of distress than men and hypothesized that the explanation was due, in part, to their differential possession of role identities.

She computed the mean number of role identities by gender and marital status for both the Chicago and New Haven samples. In general, she found clear gender differences in distress and role accumulation, but not among those who were never married (1986:270). Although Thoits did not make interval estimates of population means, we will use part of her data to illustrate how such estimates would be made.

Remember that when sample size is small and we use the sample standard deviation to estimate the population standard deviation, the sampling distribution to use is Student's t with $N-1$ degrees of freedom. Therefore, the formula for finding confidence limits is

(13.6)
$$cl = \bar{X} \pm t(s_{\bar{x}})$$

where t replaces z in signifying that Student's t-distribution will be used as the sampling distribution rather than the normal distribution.

The comparison we will make here is between the mean numbers of role identities possessed by widowed as compared to never-married men from the New Haven sample. The 15 widowed men in the New Haven sample had a mean of 2.1 role identities and a standard deviation of 1.1, whereas the 35 never-married men had a mean of 3.5 role identities and a standard deviation of 1.1. Substituting Thoits's data into Formula 13.6 to solve for the 95% confidence limits, we get the following:

For widowed men:

$$cl_1 = 2.1 - 2.145(1.1/\sqrt{15}) = 2.1 - 2.145(0.284) = 2.1 - 0.609 = 1.49$$

$$cl_2 = 2.1 + 2.145(0.284) = 2.1 + 0.609 = 2.71$$

where $t = 2.145$ with 14 df for the 95% confidence interval (see Appendix Table B).

For never-married men:

$$cl_1 = 3.5 - 2.042(1.1/\sqrt{35}) = 3.5 - 2.042(0.186) = 3.5 - 0.380 = 3.12$$

$$cl_2 = 3.5 + 2.042(0.186) = 3.5 + 0.380 = 3.88$$

where $t = 2.042$ with 34 df for the 95% confidence interval (see Appendix Table B).

Notice that the two confidence intervals do not overlap and that the one for the never-married men is considerably higher than the one for widowed men. This difference justifies the conclusion that the population parameter for never-married men is also higher than the one for widowed men.

It may be that widowhood among New Haven men is accompanied by a decrease in the number of roles occupied. This could be a side effect of the trauma resulting from the loss of a spouse. The survivors may withdraw from active participation in roles over which they have some control. On the other hand, the striking difference in the mean for widowed as compared to never-married men could be the result of differences in age distributions of the two groups. The widowed men may have fewer role identities simply because they are older or even elderly. Whether this is, in fact, the case is not possible to determine from Thoits' published account of her research.

BOX 13.1 RECONSIDERING KINDS OF SAMPLES

This is a good time to stop and remind yourself that the sampling distributions we have been using in this chapter to find confidence limits are generated based on the assumption of simple random sampling. The data from the examples we have used were not based upon simple random samples, unfortunately. It is difficult to find examples in sociological literature of confidence intervals computed from simple random samples. Random samples found in this material are more likely to be cluster samples than simple random samples. You should recall from Chapter 11 that cluster samples generally have larger standard errors than simple random samples; therefore, the confidence intervals we have computed in this chapter are probably somewhat narrower than they should be. This fact needs to be considered in interpreting results of the analysis of the data.

13.4 ESTIMATING VARIANCES

Since reading ability correlates with success in and out of school, educators, social scientists, and the public have been concerned about the effectiveness of reading programs in elementary schools. In California, reading tests are given periodically to elementary school students, and test results are made public to inform educators and taxpayers of the kind of job their schools are doing. Mean scores on reading tests are reported for each school for each grade tested.

For example, third and sixth grade students from elementary schools in the Los Angeles City School District were tested for reading ability in the spring of 1985 (*Los Angeles Times*, December 29, 1985, Part 2, p. 1 ff). Mean reading scores were computed for each of these grades for each school. Overall mean scores were also computed for third and sixth graders throughout the state of California. It was possible, therefore, to compare the means for individual schools in the Los Angeles District with the means for the state. Mean scores for both third and sixth graders in Los Angeles varied widely among the schools for which data were available. In some schools students in both grades tested were reading, on the average, well below the state means for their grades. In other schools students in both grades were reading well above the state means. In still other schools the results were mixed with students at one grade level reading above the state mean and students at the other level reading below.

These reading scores can provide answers to a number of significant questions that have both educational and sociological implications, such as: How do third graders in the Los Angeles Schools generally compare with third graders throughout the state of California? How do sixth graders compare? Do third graders deviate more from the state mean than sixth graders?

In addition to these questions about central tendencies, we might wish to raise questions about the variability of the mean reading scores from school to school. For instance, we might ask: Do mean reading scores for third graders vary more from school to school than those for sixth graders? The answer to this question can tell us something about the comparative effectiveness of the reading programs in the schools. An affirmative answer, for example, would indicate that there is a leveling process operating whereby the schools are teaching reading in such a way as to produce more uniformity in students, on the average, from school to school as they progress from third to sixth grade. A negative answer, on the other hand, would indicate that there is increasing differentiation among schools in the effectiveness of their reading programs. We will use these data on mean reading scores for Los Angeles Schools as a vehicle for learning how to make interval estimates of variances and standard deviations. And, in the process, we will see whether there are signs of decreasing or increasing differences between schools in the reading skills of students at the two different grade levels, third and sixth.

13.4.1   Small Sample Estimates

We will start by drawing a simple random sample of 32 schools from the 184 for which data were available in the spring of 1985, measuring the variability of mean reading scores for these schools and using these measurements to estimate parameters.* Table 13.1 presents the mean reading scores for third and sixth graders for the random sample of 32 schools. Notice that there were two schools without test data for third graders and three schools without test data for sixth graders.

Using the formula for the sample variance, Formula 12.15, we have computed the variances among sampled schools for the mean reading scores of third and sixth graders. These variances are 3460.92 and 2202.34, respectively. If we wished, we could use these as point estimates of the corresponding parameters. We will, however, make interval estimates instead.

It was stated in the last chapter (Section 12.6) that when the sample size is small, if sample variances are based upon simple random samples from a normally distributed population, they can be expressed as chi-square (χ^2) scores that have a chi-square sampling distribution with $N - 1$ degrees of freedom. The formula for

---

* Although mean reading scores were available for 184 elementary schools in the Los Angeles District, we will work here with data from samples as well as with population data in order to demonstrate how population variances and standard deviations may be estimated. The number 32 for the small sample resulted from the process used to sample population data within the Statistical Packages for the Social Sciences, since sample size was based upon a proportion of the total population of data.

TABLE 13.1   MEAN READING TEST SCORES FOR THIRD AND SIXTH GRADERS FROM A SAMPLE OF 32 LOS ANGELES ELEMENTARY SCHOOLS, SPRING, 1985

School Number	Third Grade	Sixth Grade
1	224	238
2	199	199
3	234	185
4	265	243
5	455	345
6	189	195
7	184	211
8	168	190
9	225	235
10	199	178
11	215	207
12	203	—
13	230	200
14	192	219
15	—	270
16	177	189
17	221	184
18	282	324
19	324	314
20	—	211
21	178	169
22	188	182
23	238	—
24	226	200
25	184	188
26	201	189
27	297	276
28	208	—
29	217	247
30	189	184
31	267	212
32	144	163

converting variances into χ^2 scores was given as follows:

$$\chi^2 = \frac{(N-1)s^2}{\sigma^2}$$

To find the lower and upper confidence limits, we will want the inequality

$$\chi^2_{\alpha/2} < \sigma^2 < \chi^2_{1-\alpha/2}$$

where α (alpha) refers to the area in the tails of the sampling distribution beyond the limits of the confidence interval. Thus, for the 95% confidence interval α would equal 0.05. Half of this, or 0.025, would be located in each tail of the sampling distribution, as the inequality suggests.

By elementary algebra, the inequality and the chi-square formula above can be converted into an expression for finding a confidence interval:

(13.7)

$$\frac{(N-1)s^2}{\chi_R^2} < \sigma^2 < \frac{(N-1)s^2}{\chi_L^2}$$

where the two χ^2 values in the denominator are values read from the table of areas under the chi-square distribution that define the area below which half of α would fall (e.g., χ_L^2) and above which half of α would fall (e.g. χ_R^2).

We will substitute the data for third graders into the inequality and find the 95% confidence limits. First, we must find the proper χ^2 values to substitute into Formula 13.7. If we are looking for 95% confidence limits, χ_R^2 will be the value that cuts 0.025 off the right tail of the chi-square distribution, with 29 degrees of freedom $(N-1)$. Remember that the number of cases for the third grade sample is 30 because of the absence of data for two of the schools sampled. Looking in the chi-square table (Appendix Table C), we find that this value for the right tail is 45.722. The other value, χ_L^2, will be the one that cuts 0.025 off the left tail of the distribution, with 29 degrees of freedom. This entry will be in the chi-square table under the column headed 0.975 (the area to the right of the given χ^2). The value is 16.047 (see Appendix Table C). We may now substitute these χ^2 values into the inequality along with $N-1=29$ and $s^2 = 3460.92$ as follows:

$$\frac{(29)(3460.92)}{45.722} < \sigma^2 < \frac{(29)(3460.92)}{16.047}$$

$$\frac{100366.68}{45.722} < \sigma^2 < \frac{100366.68}{16.047}$$

$$2195.15 < \sigma^2 < 6254.54$$

Thus, the 95% confidence limits for the variance of mean reading scores for third graders are 2195.15 and 6254.54. Notice that these limits are not symmetric around the sample variance of 3460.92. This is because the chi-square distribution with 29 degrees of freedom is not a symmetric distribution.

Now we will use the same procedure to find the 95% confidence limits for sixth graders from the sample of schools:

$$\frac{(28)(2202.34)}{44.461} < \sigma^2 < \frac{(28)(2202.34)}{15.308}$$

$$\frac{61665.52}{44.461} < \sigma^2 < \frac{61665.52}{15.308}$$

$$1386.96 < \sigma^2 < 4028.32$$

Degrees of freedom $(N-1)$ equal 28 because data are absent for three of the 32 schools in the sample. The chi-square values, 44.461 and 15.308, are the appropriate ones for 28 degrees of freedom, cutting 0.025 off each tail of the distribution.

In this case, the 95% confidence limits are 1386.96 and 4028.32. Later we will look at the actual parameters computed for all of the 184 schools for which test data were available, but, for now, what do the confidence limits for third graders as compared to those for sixth graders tell us? Note first that the sample variance of mean reading scores of third graders in the sampled schools is larger than the variance for sixth graders. At first glance, this would seem to indicate that there is less variability in the reading programs of the schools at the sixth grade level than there is at the third grade level. Notice, however, that the confidence limits for the sixth grade level substantially overlap those for the third grade level. What this indicates is that the population variances at the third grade and sixth grade levels may not be different. As a matter of fact, a population variance in the range from 2195.15 to 4028.32 would be a likely candidate to represent both the third and sixth grade levels. Consequently, this analysis does not give us clear-cut evidence that there is any more homogeniety in the results of reading programs that elementary schools in the Los Angeles District offer sixth graders than there is in the programs for third graders.

13.4.2 Large Sample Estimates

Now we will draw another simple random sample of schools from the population and make interval estimates using large sample procedures. Table 13.2 presents mean reading scores for a simple random sample of 101 schools.* Using the formula for the sample variance (Formula 12.15), we find that the variances for third and sixth graders are 2011.73 and 1475.60, respectively.

In the last chapter (end of Section 12.5a), we stated that when the size of the sample exceeds 30 the chi-square distribution can be approximated by a normal distribution with $z = \sqrt{2\chi^2} - \sqrt{2(df) - 1}$ and a standard error of 1. Using the facts that $\chi^2 = (N-1)s^2/\sigma^2$ and $df = N - 1$, we can substitute into the formula for z above and algebraically set up the following inequality for a confidence interval of the *standard deviation*, given that N is greater than 30:

(13.8)
$$\frac{s\sqrt{2N-2}}{\sqrt{2N-3}+z} < \sigma < \frac{s\sqrt{2N-2}}{\sqrt{2N-3}-z}$$

Note that this confidence interval is for the *standard deviation* rather than the variance. However, since the variance is the square of the standard deviation—if we wish to estimate the population variance rather than the standard deviation—once we have established the confidence limits for the standard deviation they may be squared to find the limits for the variance.

To find the 95% confidence limits using this inequality, we substitute $z = \pm 1.96$ into the formula and the appropriate s. Doing that for the third and sixth graders' mean reading scores from the sample of 101 schools results in

---

* The number 101 for the sample size resulted from the sampling procedure used by the Statistical Package for the Social Sciences. Sample size was based upon a proportion of the number of cases in the total population.

TABLE 13.2 MEAN READING TEST SCORES FOR THIRD AND SIXTH GRADERS FROM A SAMPLE OF 101 LOS ANGELES ELEMENTARY SCHOOLS, SPRING 1985

School Number	Third Grade	Sixth Grade	School Number	Third Grade	Sixth Grade
1	238	217	52	197	178
2	224	238	53	258	208
3	328	289	54	271	264
4	333	326	55	182	190
5	311	281	56	222	229
6	231	232	57	182	182
7	187	216	58	176	160
8	230	270	59	237	239
9	202	192	60	282	324
10	257	237	61	260	262
11	265	243	62	202	207
12	—	174	63	171	198
13	209	230	64	212	—
14	—	182	65	192	181
15	192	172	66	208	192
16	226	—	67	—	200
17	254	223	68	186	171
18	249	204	69	198	202
19	213	205	70	165	184
20	255	—	71	178	169
21	184	211	72	225	237
22	218	215	73	248	239
23	278	241	74	197	204
24	359	226	75	212	195
25	237	221	76	273	256
26	253	248	77	238	—
27	168	190	78	195	189
28	225	235	79	257	322
29	231	230	80	182	182
30	191	194	81	212	203
31	191	198	82	360	326
32	245	198	83	202	191
33	223	207	84	191	190
34	292	253	85	278	264
35	177	180	86	184	188
36	222	180	87	168	196
37	215	207	88	297	276
38	243	224	89	274	285
39	152	190	90	262	242
40	233	204	91	175	218
41	215	197	92	217	247
42	337	302	93	215	169
43	231	226	94	204	184
44	230	200	95	215	187
45	236	212	96	184	168
46	221	219	97	189	184
47	217	201	98	267	212
48	192	219	99	196	218
49	—	270	100	144	163
50	201	169	101	317	253
51	177	189			

the following:

For third graders:

$$\frac{44.85\sqrt{2(97)-2}}{\sqrt{2(97)-3}+1.96} < \sigma < \frac{44.85\sqrt{2(97)-2}}{\sqrt{2(97)-3}-1.96}$$

$$\frac{621.46}{15.78} < \sigma < \frac{621.46}{11.86}$$

$$39.38 < \sigma < 52.40$$

or $1550.78 < \sigma^2 < 2745.76$ where the sample standard deviation is 44.85.

For sixth graders:

$$\frac{38.41\sqrt{2(97)-2}}{\sqrt{2(97)-3}+1.96} < \sigma < \frac{38.41\sqrt{2(97)-2}}{\sqrt{2(97)-3}-1.96}$$

$$\frac{532.22}{15.78} < \sigma < \frac{532.22}{11.86}$$

$$33.73 < \sigma < 44.88$$

or $1137.71 < \sigma^2 < 2014.21$ where the sample standard deviation is 38.41.

Note that the N used in computing the confidence limits for both the third and sixth grade levels was 97. This is so because, as you can see from Table 13.2, there were third grade data missing from four schools and sixth grade data missing from four other schools.

Now we will examine these confidence intervals to see whether they lead to results that are consistent with those we got from the small sample estimates. Recall that the sample variances were 2011.73 for third graders and 1475.60 for sixth graders. Again, at first glance, it appears that the variance for third graders is larger than that for sixth graders; however, as was the case with the small sample confidence intervals, the large sample intervals overlap substantially. Again the results are inconclusive. Since the intervals do overlap it is entirely possible that both sample variances are estimates of a common population variance. Or, to put it differently, it is possible that there is no difference in the variability of third and sixth grade mean reading scores for the population of 184 elementary schools. The area of overlap in the two confidence intervals (1550.78 to 2014.21) provides a range of reasonable estimates of what a common population variance might be.

13.4.3 Comparisons of Sample Estimates and Parameters

Since the data on mean reading test scores are available for 184 elementary schools in Los Angeles, it is possible to compare the interval estimates based on the small and large samples with the actual parameters in order to see how good the estimates were. Table 13.3 presents a summary of the parameters and the point and interval estimates of variances for the third and sixth grade levels from the samples of 32 and 101 schools. Also included for the sake of comparison are population means and their point and interval estimates.

TABLE 13.3   PARAMETERS AND SAMPLE ESTIMATES OF MEAN READING TEST SCORES

Source of Data	Means	Variances	Interval Estimates of Means	Interval Estimates of Variances
Population (Parameters)				
Third Grade	226.89	2056.01		
Sixth Grade	216.44	1435.79		
Sample of 32 Schools				
Third Grade	224.10	3460.92	$202.13 < \bar{X} < 246.07$	$2159.15 < \sigma^2 < 6254.54$
Sixth Grade	218.86	2202.34	$201.01 < \bar{X} < 236.71$	$1386.96 < \sigma^2 < 4028.32$
Sample of 101 Schools				
Third Grade	226.34	2011.73	$217.30 < \bar{X} < 235.38$	$1550.78 < \sigma^2 < 2745.76$
Sixth Grade	217.68	1475.60	$209.94 < \bar{X} < 225.42$	$1137.71 < \sigma^2 < 2014.21$

Notice, first, that all the interval estimates of means actually include the parameters they are intended to estimate; that is, the population means, in every case, are in the intervals. Also, notice that the point estimates of the population means are very close to the parameters themselves. The same cannot be said for the point and interval estimates for variances. The third and sixth grade interval estimates based on the large sample procedure do, in fact, include the parameters (2056.01 for third grade and 1435.79 for sixth grade), and the point estimates of those parameters are reasonably close to the actual parameter values. However, when small sample procedures are used to do interval and point estimates, the results are not as convincing. The sixth grade interval estimate does include the parameter, but the third grade estimate does not. Furthermore, the point estimates of the parameters considerably overestimate their values.

Also, note that the population variances for third and sixth graders are different and that the variance for third graders is, in fact, larger than the one for sixth graders. These facts we were not able to confirm from the sample data because of the overlapping interval estimates.

Let us return now to the question of where our attempts to estimate the population variances fell short. How might the failure of the small sample procedure to include the population variance for third graders in the interval estimate be explained? Furthermore, why are the point estimates for both third and sixth graders considerably larger than the parameters? Could it be a coincidence that the small sample produced an unusual estimate of the population variance for third graders? To test this notion, two other small simple random samples were drawn and confidence intervals were computed. The first of these samples produced interval estimates that failed to include the population variances for either the third or the sixth graders. The second of these samples produced an interval estimate for the third grade level that included the population variance, but the interval estimate for sixth graders failed to do so. Thus, out of three independent interval estimates of the third grade population variance, two failed to include the parameter. The probability of this happening by chance with 95% confidence limits is

$$(0.05)(0.05)(0.95) = 0.0024$$

In other words, it is rather unlikely that this particular sequence of estimates may be attributed to chance.

It is interesting to note, in addition, that the first of the two additional samples produced point estimates of the parameters that were too small (846.02 for third grade and 698.03 for sixth grade) and the second produced point estimates that were too large (2972.13 for third grade and 2490.69 for sixth grade).

As a further exercise, two additional large simple random samples were drawn from the 184 schools and interval estimates of population variances for the third grade and sixth grade levels were computed from these. In every case (for third and sixth grade and for each of the two additional samples) the interval estimates included the population variances and the point estimates of the parameters were reasonably close to the parameters themselves.

Since it is unlikely that the failure of the small sample estimates to include the parameters (with the exception of one out of three of the third grade estimates and one out of three of the sixth grade estimates) was a coincidence, we sought another explanation. It was stressed in the last chapter (at the end of Section 12.5) that sampling distributions of variances are sensitive to the assumption that the population being sampled is normally or near normally distributed. To test whether this assumption had been violated by the reading data, we computed standard measures of skewness (as in Chapter 5) for the populations of mean reading scores for third and sixth graders. The skewness for third graders was $+1.64$ and for sixth graders, $+1.29$. A skewness score greater than $+1.00$ indicates relatively extreme skewness; therefore, the population distributions for both grade levels exhibited definite positive skewness. The skewness for the third grade level was more extreme than that for the sixth grade, however. These population distributions are graphed in Figures 13.2A and 13.2B. As you can see from comparison of these graphs, the right tail of the distribution of third grade mean reading scores is longer than the right tail of the distribution for the sixth grade. This accounts for the difference in the skewness scores.

The results of our exploration of alternate interval estimates of the parameters for these reading data illustrates well the importance of the assumption of normally distributed population data for interval estimates of variances. The small sample procedure for making interval estimates was particularly subject to misleading results, whereas the large sample estimates were consistent in their inclusion of the parameters, in spite of the violation of the assumption of normally distributed data. This difference between small sample and large sample estimates stems in large part from the magnitudes of the respective sampling errors. In accordance with the law of large numbers, sampling error decreases as sample size increases. Therefore, the relatively small sampling error associated with the larger samples offset the problem of skewed population distributions.

When a statistic provides reasonably accurate estimates of parameters in spite of the fact that the assumptions underlying its use are violated, that statistic is said to be **robust**. It should be obvious from the results reported in Table 13.3 that the mean is a robust statistic. Interval estimates of the population mean consistently included the parameter and all point estimates were reasonably close

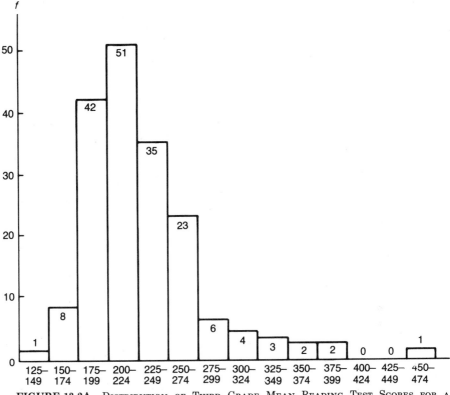

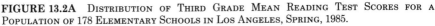

FIGURE 13.2A Distribution of Third Grade Mean Reading Test Scores for a Population of 178 Elementary Schools in Los Angeles, Spring, 1985.

to the parameters. Small sample interval and point estimates of variances were much more sensitive to violation of the normality assumption; large sample estimates, however, were fairly robust.

13.5 SUMMARY

In this chapter we have studied procedures for making interval estimates of proportions, means, and variances. Interval estimates may be computed for many other statistics as well. As a matter of fact, whenever a statistic has a known sampling distribution, it is possible to compute an interval estimate (see Noether, 1972). In the chapters that follow we will have occasion to consider a few other interval estimation procedures as they relate to the subject matter being discussed. We will consider, too, the use of interval estimates as an alternative to traditional hypothesis testing procedures.

Interval estimation has received, perhaps, less attention than it deserves as a tool for the analysis of sociological data. This chapter has demonstrated some of the possible uses of interval estimation procedures.

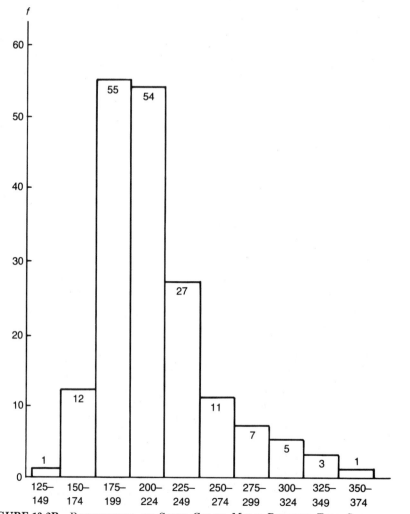

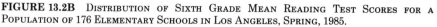

FIGURE 13.2B Distribution of Sixth Grade Mean Reading Test Scores for a Population of 176 Elementary Schools in Los Angeles, Spring, 1985.

CONCEPTS TO KNOW AND UNDERSTAND

point estimates   critical regions
interval estimates   the meaning of confidence
confidence interval   efficient estimate
confidence limits   robustness

QUESTIONS AND PROBLEMS

1. Explain the difference between a point estimate and an interval estimate of a parameter. Which is preferable? Why?

2. Under what circumstances may the normal distribution be used as an approximation to the binomial probability distribution?

3. What is the relationship between the standard error of the binomial probability distribution and the standard error of a proportion?

4. Explain what the probability figure (*e.g.*, 95%) associated with an interval estimate means.

5. What is the relationship between sample size and the magnitude of the standard error?

6. What are the steps that can be taken to decrease the width of a confidence interval?

7. How important is the normal distribution assumption for making interval estimates of variances? What are the consequences of violating the assumption?

8. If you have access to a computer and a data file of sample survey data, select three variables of interest to you. One variable should be dichotomous and another should be interval. The third variable should be one that would divide the data set into two to four interesting subgroups (*e.g.*, large, medium, small families; gender; age groups; geographic location; etc.). Now, find point and interval estimates of proportions, means, and variances (or standard deviations) for each of the subgroups defined by the third variable, using the two variables you selected for study. State what each point and interval estimate means in your own words. Then, compare and contrast the various measures across the subgroups. Show your work and indicate the formulas you used in your computations.

9. If you do not have access to a computerized data set, use data from Problem 14 in Chapter 12. The U.S. overall percentage of persons living in urban places of 2500 or larger in 1980 was 73.7. U.S. Bureau of the Census (1984:12, Table 12). Using careful random sampling procedures (*e.g.*, using a table of random numbers), draw a sample of size $N = 10$ from the numbered listing of 50 states plus the District of Columbia. Using the percentage urban as an interval score, compute the sample mean percentage urban and set up confidence limits around the mean. Do this for the sample variance of percentage urban. Finally, dichotomize the sample data, using the overall U.S. average 73.7 as the cutting point. Any state drawn in the sample with a higher figure call "highly urban" and any state in the sample with a figure at or below the U.S. average call "low urban." Then compute a sample percentage urban and set up confidence limits around this percentage (or treat it as a proportion). Then, state in your own words what these numbers mean. Compare them with the parameters.

10. Extend Problem 9 by computing the same statistics on a large sized sample as well. Use, say, a sample of size 35. What differences are there in procedure? Justify any similarity or difference in outcome.

11. Again, Problem 9 could be extended by drawing random samples from subgroups of states defined by region. Compare and contrast the confidence intervals and point estimates based upon samples drawn from the separate regions. Census regional classification of states is given below. You may wish

to combine these categories into larger categories for purposes of drawing a sample. Numbers refer to line numbers in problem 12, Chapter 12.

North East   (01 through 06)
Middle Atlantic   (07 through 09)
East North Central   (10 through 14)
West North Central   (15 through 21)
South Atlantic   (22 through 30)
East South Central   (31 through 34)
West South Central   (35 through 38)
Mountain   (39 through 46)
Pacific   (47 through 51)

GENERAL REFERENCES

Blalock, Hubert M., Jr. *Social Statistics*, rev. ed. (New York, McGraw-Hill), 1979.
    Chapter 12 of this book treats clearly the subject of point and interval estimates. Blalock argues convincingly that confidence intervals are implicit tests of a whole range of hypotheses.
Walker, Helen M., and Joseph Lev. *Statistical Inference* (New York, Holt, Rinehart and Winston), 1953.
    Chapter 8 deals with inferences concerning variances and standard deviations. This is a topic that is ignored in many other statistics books.

LITERATURE CITED

Labovitz, Sanford, "The Nonutility of Significance Tests: The Significance of Tests of Significance Reconsidered," *Pacific Sociological Review*, 13, 3 (Summer 1970), pp. 141–147.
Lerner, Monroe, and Odin W. Anderson, *Health Progress in the United States, 1900–1960* (Chicago, University of Chicago Press), 1963.
Morrison, Denton E., and Ramon E. Henkel, *The Significance Test Controversy* (Chicago, Aldine), 1970.
Morse, Nancy C., and Robert S. Weiss, "The Function and Meaning of Work and the Job," *American Sociological Review*, 20, 2 (April 1955), pp. 191–198.
Noether, Gottfried E., "Distribution-Free Confidence Intervals," *The American Statistician*, 26, 1 (February 1972), pp. 39–41.
Parsons, Talcott, *The Social System* (New York, The Free Press), 1951.
Thoits, Peggy A., "Multiple Identities: Examining Gender and Marital Status Differences in Distress," *American Sociological Review*, 51 (April 1986), pp. 259–272.
U.S. Bureau of the Census, *Statistical Abstract of the United States: 1985*, 105th ed. (Washington, D.C., U.S. Government Printing Office), 1984.
U.S. Public Health Service, National Center for Health Statistics, *Physician Visits. Volume and Interval Since Last Visit, United States—1975*, DHEW Pub. No. (PHS) 79-1556 (Washington, D.C., U.S. Government Printing Office), 1979.
————, National Center for Health Statistics, *Health: United States, 1985*, DHHS Pub. No. (PHS) 86-1232 (Washington, D.C., U.S. Government Printing Office), 1985.
U.S. Senate, *Developments in Aging 1959 to 1963: A Report of the Special Committee on Aging, United States Senate* (Washington, D.C., U.S. Government Printing Office), 1963.

CHAPTER 14

The Logic of
Hypothesis Testing

Suicide is an individual act with psychological overtones. Emile Durkheim, the famous French sociologist, maintained, however, that the regularity and predictability of suicide rates over time could not be explained by psychological variables. He was convinced that suicide rates were explainable, rather, in terms of "social facts." He said

> If, instead of seeing in them only separate occurrences, unrelated and to be separately studied, the suicides committed in a given society during a given period of time are taken as a whole, it appears that this total is not simply a sum of independent units, a collective total, but is itself a new fact *sui generis*, with its own unity, individuality, and consequently its own nature—a nature, furthermore, dominantly social. (Durkheim, tr. 1951:46.)

With the collective nature of the suicidal act in mind, Durkheim attributed the regularity of its rates for various populations to the social fact of group solidarity, or the lack thereof. He theorized that people lacking support from group solidarity are most vulnerable to suicide. His theory did not seek to explain the dynamics of individual suicides; it sought, rather, to explain suicide *rates* in terms of the differential vulnerability of various cohorts of people.

Although Durkheim's work was completed more than 90 years ago, it is considered a sociological classic because it illustrates the proper relationship between theory and data. In spite of soft spots in his methods, his general theoretical notions about suicide as a sociological phenomenon are still relatively sound.*

From his **theory**, Durkheim derived some specific hypotheses about the relationship between social solidarity and suicide rates. Even though he never defined social solidarity in a rigorous manner, he did indicate various indices of it that could be observed and measured. For instance, he felt that social solidarity varied

---

* Douglas (1967) critiques Durkheim's work and reviews further studies of suicide.

from one religious persuasion to another. He reasoned that social solidarity was highest among Jews, next among Catholics, and lowest among Protestants. He attributed these differences to differences in the degree to which the lives of individual members were dominated by their religions. Because Protestantism allowed for more free inquiry and placed more responsibility on the shoulders of the individual, it fostered less social solidarity than the other two. Accordingly, Durkheim **hypothesized** that suicide rates would be highest for Protestants and lower for Catholics and Jews. To test his hypothesis, he examined suicide statistics for various European countries. In general, he found that his hypothesis was supported by the data. For example, in the states of Germany suicide rates varied in direct proportion to the number of Protestants and in inverse proportion to the number of Catholics (Durkheim, tr. 1951:153). Moreover, those European countries that were predominantly Catholic (*e.g.*, Portugal, Spain, and Italy) had low suicide rates as compared with high rates for predominantly Protestant countries, and the rates for mixed Catholic–Protestant countries were intermediate (Durkheim, tr. 1951:152). When the rates of Protestant, Catholic, and Jewish groups were compared, Protestant rates were consistently higher than those of the other two, and Jewish rates were generally lower than those for Catholics (Durkheim, tr. 1951:155).*

Social solidarity $\longrightarrow$ suicide rates
measured by religion,
marital status, etc.

Another index of social solidarity examined was marital status. Durkheim reasoned that the married enjoyed more social solidarity than the unmarried or widowed; thus, suicide rates would be lower for them. Again, the data tended to support his hypothesis. Rather consistently, for each age category, married persons had lower rates than unmarried or widowed persons (Durkheim, tr. 1951: 176–177). Furthermore, suicide rates for married persons with children were consistently lower than those for married persons without children (Durkheim, tr. 1951:186 ff.).

Durkheim went on to examine several other variables that he took to be indices of social solidarity, in each case comparing suicide rates from several different sources. The data generally supported his theory.[†]

Durkheim's basic approach was to develop theory to account for the regularity and predictability of suicide rates in Europe in the latter part of the 19th century. His major explanatory concept was social solidarity. He *selected* a number of measurable variables to be indices of social solidarity, *deduced* a number of specific hypotheses relating these indices to suicide rates, *gathered* available

---

* It should be noted that Durkheim's statistics relating suicide rates and religion were for countries or areas of countries and not for individuals. If predominantly Protestant countries have high suicide rates, it does not necessarily follow that those who are committing suicide are Protestants (*see also* Robinson, 1950.)

[†] Although the statistical techniques known to us were not available to Durkheim, on a raw level he duplicated the reasoning underlying modern statistics. Sir Francis Galton invented correlation earlier (Galton, 1886), but the technique was not generally known nor understood at the time that Durkheim conducted his study.

data bearing on his hypotheses, and *examined* them to see whether they lent support to his predictions. He *concluded* that the data, on the whole, supported his hypotheses and theory. Therefore, he offered his theory as a fruitful explanation of suicide rates.

The point of this exercise in social science was to develop a theoretical explanation for a social phenomenon useful for predicting the same class of phenomena in the future. Durkheim did not *prove* his theory any more than any scientific theory is ever proven. He did demonstrate, however, that his findings were useful for predicting suicide rates. This is the way science uses theory. A theory is never proven; it is demonstrated to be useful or not useful, and is used or revised. If a competing theory is developed that predicts better, it is substituted for the previous theory and is used until it is replaced, in turn, by another theory that is an even better predictor.

The process of developing theoretical explanations of social behavior is what sociology is all about. Science is, after all, a continual interplay between theory and data. The examination of data in a systematic manner gives rise to theory; theory gives direction to the collection of new data; and new data either give additional support to the theory or contribute to its refutation. The game of science is concerned with the selection of the most useful theory from a number of competing ones. The *application* of scientific knowledge by practitioners and policymakers involves the practical use of the most fruitful theory currently in vogue.

When Durkheim developed his theory, he had in mind an explanation that would transcend particular populations. He did not, however, use sampling techniques nor did he worry whether his data were representative of either general or special populations. The data he used were descriptive of specific geographic areas at specific points in time. He did examine sets of data descriptive of a number of *different* special populations. In effect, he replicated his hypothesis tests. A very important principle of science is that hypothesis tests should be repeated independently a number of times to gauge their soundness. The impressive aspect of Durkheim's work was the consistency with which these separate sets of data upheld his hypotheses.

14.1   STATISTICS AND HYPOTHESIS TESTING

When concepts that appear in sociological theories are not defined in measurable form, the hypotheses linking the concepts are not directly testable. The usual procedure in such cases is to specify measurable indices of the concepts, frame hypotheses relating the indices, and test these **working hypotheses** against empirical data. Figure 14.1 illustrates the relationship between a general theoretical hypothesis and a working hypothesis using Durkheim's study of suicide as an example.

Robert E. Clark conducted a study designed to test the **general hypothesis** that incidence of mental disorders varies with occupational status (Clark, 1949). As indices of mental disorders Clark used diagnostic categories assigned to

Abstract concepts

GENERAL HYPOTHESIS: Social Solidarity ⟶ Level of Suicide
(Causal)

Measurable indices

WORKING HYPOTHESIS: Groups with ⟶ Suicide Rates
(Empirical)   different
levels of
solidarity
(*e.g.*, Jews,
Catholics,
Protestants)

FIGURE 14.1 RELATIONSHIP BETWEEN A GENERAL THEORETICAL HYPOTHESIS AND A WORKING HYPOTHESIS: DURKHEIM'S THEORY OF SUICIDE

patients in mental hospitals in the Chicago area. He looked separately at rates for alcoholic psychoses, senile psychoses, paresis, manic-depressive psychoses, and schizophrenia among patients whose occupations were known. He used two indices of occupational status: a measure of the prestige of the occupation on the North–Hatt scale,* and median income for the occupation in the Chicago area at the time of the study.

From one general hypothesis Clark framed several working hypotheses, relating indices of mental disorders to indices of occupational status. When he tested his working hypotheses against the data, he found that his general hypothesis had to be qualified. Alcoholic psychoses, senile psychoses, paresis, and schizophrenia were inversely related to occupational status (the higher the occupational status, the lower the incidence of the disorder), but manic-depressive psychoses were unrelated (Clark, 1949:440). Although Clark's substantive findings are interesting in themselves, we are concerned primarily with the procedures used to test the general hypothesis. Clark selected indices of his theoretical concepts and recast his general hypothesis in terms of these indices, thus deriving working hypotheses; then he subjected his working hypotheses to empirical tests and, on the basis of these tests, drew conclusions about the general hypothesis.

When population data are available, the indices serve as parameters, and decisions about hypotheses can be made simply by examining the parameters (no test of significance is needed). When Durkheim compared suicide rates in Catholic Bavaria with those in Protestant Prussia, he merely had to note that the Prussian rates were higher. Since these rates were parameters for his populations, the differences between them were actual differences (assuming that the data were error free). Therefore, descriptive statistics allow for direct tests of hypotheses without further complications.

Unfortunately, population data are not usually available. We are faced with the necessity of examining sample data and making generalizations about the

* For a description of the North–Hatt Scale, see Reiss (1961).

population. The procedure for testing hypotheses with sample data is as follows:

1. *Specify indices of the concepts included in the general hypothesis.*
2. *Derive working hypotheses that link the indices.*
3. *Draw a sample of data from the population to which the hypotheses apply.*
4. *Use statistical techniques to analyze the sample data.*
5. On the basis of that analysis, *decide whether the data support the working hypotheses.*
6. *Decide whether the general hypothesis fruitfully describes the population.*

When hypotheses are tested using sample data, the complication introduced is that **parameters must be estimated**, since they cannot be examined directly because they are not available to the researcher. For example, if Durkheim's comparisons of suicide rates in Catholic Bavaria and Protestant Prussia had been based on sample data rather than population data, he would have been faced with the necessity of estimating suicide rates from his sample data and then deciding whether the estimated parameters actually differed.

We found in Chapter 13 that estimating parameters involves the use of probability theory applied to sampling distributions. As you will see later, there is a close relationship between interval estimates of parameters and statistical tests of hypotheses. The difference between them is a difference in orientation rather than kind.

BOX 14.1   SAMPLING DISTRIBUTION

If the concept of the sampling distribution is not yet quite clear to you, perhaps you should go back and review Chapter 12, particularly the units under Section 12.2. An understanding of the concept of sampling distribution is essential to the discussion that follows.

14.1a   The Statistical Hypothesis

When we use sample data to test hypotheses, it is necessary to introduce a third type of hypothesis, the **statistical hypothesis**. *We start with a general hypothesis, which we translate into working hypotheses, and from our working hypotheses we derive statistical hypotheses.* Statistical hypotheses make statements about population parameters, but are tested by examination of statistics computed from sample data. As a result of the outcomes of tests of statistical hypotheses, we decide what conclusions about the working hypotheses are warranted and, in turn, these decisions help us to make decisions about the general hypothesis.

Again, we rely heavily on sampling distributions to help us make decisions. The questions we ask, however, are somewhat different from those raised in estimating parameters. We examine the following questions:

Is it reasonable to conclude that the statistic computed from the sample is an estimate of a specific, given parameter?

Are the statistics computed reasonably seen as separate sample estimates of a common parameter or do they estimate different, distinct parameters?

In each case, primary concern is with making decisions about hypotheses that refer to parameters or to relationships between parameters. In classical hypothesis testing we must choose between two competing hypotheses.

14.2   TESTING STATISTICAL HYPOTHESES

Thus far our discussion of hypothesis testing has been fairly abstract. It might be helpful at this point to take a concrete example, run through the process involved in testing a hypothesis, then analyze the procedure involved. In the process of presenting the example we will introduce the concepts that play an integral part in the testing of statistical hypotheses.

One characteristic that particularly distinguishes the developing countries of the world from the others is the rate at which infants and young children die. Infant and childhood diseases that have long since ceased to be serious killers in the industrialized countries still take a terrible toll of babies and small children in developing countries. According to World Health Organization estimates, 3,450,000 children die every year of diseases that are preventable through vaccination (United Nations Children's Fund, 1985). Measles alone is estimated to kill 2 million annually. Is the distinction between the developing and the developed countries merely that vaccinations are more common in the latter than in the former? According to McKeown (1976), malnutrition is an important factor in the whole equation. People who are malnourished are more vulnerable to infection, and thus fall victim to diseases that, in other circumstances, are much less virulent (1976:35).

One might assume that malnutrition would not be a common characteristic of countries that are largely rural because the inhabitants could subsist on foodstuffs they grew themselves. According to Bogue (1969:46), however, predominantly rural settlement patterns are characteristic of developing countries and it is in the developing countries that death rates are high.

To understand better this whole process whereby underdevelopment, malnutrition, infectious disease, and high infant death rates are linked, an informative preliminary step would be to establish the nature of the relationship between rural-urban settlement and the level of nutrition existing nationally. The Food and Agriculture Organization of the United Nations (U.N.) has collected international data on daily per capita calorie supply as a percentage of the requirement necessary for satisfactory nutrition (United Nations Children's Fund, 1985:134–35). These data may be used as a measure of the level of nutrition available for each country covered.

Consistent with what has been said thus far about underdevelopment, nutrition, disease, and death rates, **our working hypothesis will be that countries that are predominantly rural will be less likely to provide the required per capita daily calorie intake than will the countries of the world in general.** To test this hypothesis, we used the U.N. data to divide the countries of the world into two categories: those that meet or exceed the percentage requirement of daily calorie intake and those that do not. Using this distinction, we found that 65% of the countries of the world met or exceeded the daily percentage requirement.

Furthermore, we singled out the countries of the world in which more than 50% of their populations were rural and drew a simple random sample of 30 of those countries. When these countries were examined to determine whether their per capita daily calorie intake met or exceeded the percentage requirement, it was found that 9 of the countries did and 21 did not. The research question that we wish to answer is whether this distribution for the 30 predominantly rural countries provides evidence in support of our working hypothesis.

14.2.1   The Null Hypothesis

Framing a statistical hypothesis in a positive manner, we might come up with the following: **Countries that are more than 50% rural are significantly less likely to meet or exceed the daily per capita calorie intake requirement than are the countries of the world in general.** The kind of sample data we would consider as evidence of significance would have to be stated explicitly so that we could test the hypothesis.

Since statistical inference is based upon probability theory, **tests of statistical hypotheses are probabilistic rather than absolute**. It is not possible to prove or disprove statistical hypotheses in an absolute sense. The best that can be achieved is an estimate of their truth or falsity.

It so happens that the rejection of a statistical hypothesis is much more clear-cut than its acceptance. Therefore, the usual procedure is to frame a specific statistical hypothesis that is often contrary to that which we are hoping to prove. Such a hypothesis is known as a **null hypothesis**. If sample data warrant rejection of the null hypothesis, that is regarded as evidence for its alternatives— those hypotheses our theory predicted and those we proposed as explanations in the first place. The null hypothesis gets its name from the fact that it is the hypothesis to be nullified by statistical test.

The advantage of using the null hypothesis is that it serves as a basis for selecting a specific sampling distribution, that is, the sampling distribution that would be found if the null hypothesis were, in fact, true. This sampling distribution is then used to determine whether sample data warrant rejection of the null hypothesis in favor of some set of alternatives to it.

Since we wish to seek evidence in support of the contention that the rural countries of the world are less likely to meet daily calorie intake requirements than are countries in general, we wish to show that significantly fewer than 0.65 of the rural countries meet or exceed those requirements. We use the 0.65 because that is the proportion of *all* of the countries of the world that meet or exceed the requirements. It is, therefore, the parameter of interest to us in generating the sampling distribution necessary to test our null hypothesis. The null hypothesis would be formulated as follows: **The proportion of rural countries that meet or exceed the daily per capita calorie intake requirement will not differ significantly from 0.65.** If this null hypothesis were true, we would have a sampling distribution of proportions with an expected value of 0.65, which is the value for countries in which the daily per capita calorie intake met or exceeded the requirement.

The procedure we follow in testing the null hypothesis is (a) assume that it is true, (b) generate a sampling distribution from the null hypothesis, (c) draw a random sample, (d) collect the relevant sample data, (e) compute the relevant statistic, and (f) decide whether it is reasonable to assume that the statistic came from the given sampling distribution. If the probability that the statistic came from the given sampling distribution is as small as or smaller than some pre-determined level, we reject the null hypothesis in favor of its alternatives. If the probability is not as small as or smaller than the predetermined level, we fail to reject the null hypothesis.

Notice that we *fail to reject* the null hypothesis rather than accept it. If we accepted the null hypothesis, we would be saying, in effect, that it is true. How-ever, if we fail to find reason to reject the null hypothesis, it does not necessarily follow that it is true. We are saying, rather, that the data we collected did not provide us with sufficient basis for concluding that the null hypothesis is false. Perhaps, our data collection was merely inadequate.

By the same reasoning, if we *do* reject the null hypothesis, it does not follow that we accept its alternatives. All we imply by rejecting the null hypothesis is that some set of alternatives to it is more probable than the null hypothesis itself.

Note, also, that both the null hypothesis and its alternatives apply not to sample data but to population data—not to statistics but to parameters. We test the null hypothesis with sample data and generalize from statistics to parameters.

14.2.2 One-tailed and Two-tailed Tests of Hypotheses

If the null hypothesis claims that the proportion of countries meeting or exceeding the daily requirement will not differ significantly from 0.65, then the alternative hypothesis is that it will differ. Symbolically, the null hypothesis and the set of its alternatives can be stated as follows:

$$H_0 : P = 0.65$$

$$H_a : P \neq 0.65$$

where H_0 is the null hypothesis, H_a the set of alternatives, and P the proportion of rural countries in the world in which the daily per capita calorie intake is met or exceeded.

Actually, the alternatives to the null hypothesis consist of a whole range of possible hypotheses that **can be divided meaningfully into two distinct classes**. First, there is the set for which the proportion of rural countries meeting or exceeding the daily requirement is greater than 0.65. We will call this set **alternative 1** and represent it symbolically as

$$H_1 : P > 0.65$$

Second, there is a set for which the proportion is less than 0.65. We will call this **alternative 2** and represent it symbolically as

$$H_2 : P < 0.65$$

When a null hypothesis has two distinct sets of alternatives as this one does, *the test is called a two-sided or* **two-tailed test** *of the hypothesis.* This terminology is used because each set of alternatives to the null hypothesis is represented by a separate side or tail of the sampling distribution.

We are not interested in both sets of alternatives to the null hypothesis for the daily per capita calorie intake data. Rather, we are interested in testing whether rural countries are less likely than countries in general to meet or exceed the daily requirement. The set of alternatives we are interested in finding support for, therefore, is what we have labeled alternative 2 (H_2). Since we defined the null hypothesis earlier as that hypothesis that includes all except that which we are interested in demonstrating, the appropriate null hypothesis in our case would not be $H_0: P = 0.65$; rather it would be $H_0: P \geq 0.65$ (*i.e.*, P equal to or greater than 0.65). The null hypothesis and its set of alternatives would thus be represented as

$$H_0: P \geq 0.65$$

$$H_2: P < 0.65$$

In this case we have labeled the set of alternatives to the null hypothesis H_2 to be consistent with the labeling used above. We could just as well have labeled it H_1.

When there is only one set of alternatives to the null hypothesis as there is in this case, *the test of the null hypothesis is said to be a one-sided or* **one-tailed test**. Here we anticipate a significant deviation from chance in only one tail of the sampling distribution. The null hypothesis can only be rejected if the sample produces rural countries in which the proportion of countries meeting or exceeding the daily per capita calorie intake requirement is sufficiently smaller than the hypothesized 0.65 to meet some predetermined criterion for rejection. If it turned out that the proportion of rural countries meeting or exceeding the daily requirement were larger than 0.65, it would not be possible to reject the null hypothesis because that possibility is included as part of the null hypothesis.

Whenever research is done **to test theory**, the theory will generally suggest which set of alternatives to the null hypothesis the sample data are expected to support. The appropriate null hypothesis will be for a one-tailed test. Whenever **exploratory research** is undertaken, and there is no theory to suggest direction of relationship, the two-tailed test of the null hypothesis is ordinarily employed. As we shall see later, the one-tailed test enjoys certain advantages that make it the preferred test. The two-tailed test should be reserved for those cases where the one-tailed test is not feasible.

14.2.3   Significance Levels and Critical Regions

Once we have decided upon a one-tailed or a two-tailed test and have framed the null hypothesis and its alternatives, we must decide on criteria for rejecting the null hypothesis.

To establish criteria for rejecting the null hypothesis, we must specify the applicable sampling distribution. **To compute a sampling distribution, we need a specific value of P.** The null hypothesis for the one-tailed test, however, specifies a whole range of possible values (*e.g.*, those equal to or greater than 0.65). However, if we use the limiting value (in this case the lowest value, 0.65, included in the null hypothesis) and find that we could reject the null hypothesis for this limiting value, then we know that we could have rejected any other hypothesis included in the one-sided null hypothesis because the sample outcome would be shown to be even less likely for these more extreme hypotheses. On the other hand, if the sample outcome is not deemed unusually rare when testing the limiting value of the null hypothesis, then, at least for one value included under the one-sided null hypothesis, the hypothesis could not be rejected. The procedure, then, is to use the specific limiting value under the one-sided null hypothesis to construct the appropriate sampling distribution.

The appropriate sampling distribution for the calorie intake data is the **binomial probability distribution** with an expected value of 0.65. This expected value is taken from the stated null hypothesis. In this particular instance the value was set at 0.65 because that is the proportion of countries in the world that meet or exceed the daily calorie requirement (see Section 14.2). The specific shape of the sampling distribution will depend upon the size of the sample from which the statistic is computed. If we can specify the sample size before data are collected, we can determine exactly which outcomes will justify rejecting the null hypothesis. If sample size depends upon the outcome of data collection, then we must set up the criteria for rejecting the null hypothesis in a more general way. If the latter is the case, then we must make a decision such as the following: We will consider any sample *outcome* so infrequent as to occur by chance in the specified sampling distribution 5% of the time or less as warranting the rejection of the null hypothesis. This 5% figure is known as a **level of significance** (or **alpha**, symbolized by the lowercase Greek letter α), in this case, *the 5% level of significance*.

When we choose to use the 5% level of significance as a criterion for rejecting the null hypothesis, we are saying, in effect, if the outcome we observe is so unusual that it would be likely to occur in the sampling distribution less than 5 times out of 100 by chance, then we will decide to reject the null hypothesis.

There is nothing sacred about the 5% level of significance (Labovitz, 1968; Skipper *et al.*, 1967). It has merely been defined by convention to be a reasonably rare, chance, occurrence. The more stringent 1% level of significance is also used in social science research, although less frequently than the 5% level. It is reasonable to assume that the popularity of the 5% level relates to the relatively unsophisticated measurement levels of social science data. We will have more to say later about the choice of an appropriate level of significance.

Once we have decided upon the level of significance necessary to reject the null hypothesis we are ready to sample and collect data.

We will turn now to the data on calorie intake from our sample of 30 rural countries. The sampling distribution has been worked out using procedures for computing a binomial sampling distribution discussed in Chapter 12. In particular, Formula 12.2 was used to compute the probability of occurrence of each term of the distribution with $P = .65$, $Q = .35$, and $N = 30$. Table 14.1 is a tabular presentation of this sampling distribution.

Recall that we are interested in determining whether there are fewer rural countries meeting or exceeding the daily calorie requirement than is the case for all countries in the world; consequently, we are interested in a set of alternatives

TABLE 14.1 BINOMIAL PROBABILITY SAMPLING DISTRIBUTION OF NUMBER OF RURAL COUNTRIES MEETING OR EXCEEDING THE PER CAPITA DAILY CALORIE INTAKE REQUIREMENT ($N = 30$, $P = 0.65$)

No. of Countries Meeting or Exceeding		Probability[a]	
30		.0000024	
29		.0000395	
28		.0003077	
27		.0015467	
26		.0056214	
25		.0157401	
24	Not in the critical region	.0353143	
23		.0651957	
22		.1009279	
21		.1328452	
20		.1502173	
19		.1470659	
18		.1253831	
17		.0934809	
16		.0611222	
15		.0351060	
14		.0177218	
13		.0078585	
12		.0030561	
11		.0010394	
10		.0003078	
9		.0000789	
8	In the critical region	.0000174	.03 (near but not over $\alpha = .05$)
7		.0000032	
6		.0000005	
5		.0000001	
4		.0000000	
3		.0000000	
2		.0000000	
1		.0000000	
0		.0000000	
		Total = 1.0000000	

[a] These probabilities are approximate because of errors in rounding.

to the null hypothesis in only one tail of the sampling distribution. In this case we are interested in the tail in which the number of countries meeting or exceeding the daily requirement is less than 0.65. Since we have decided to use the 5% level of significance, we must examine the sampling distribution and determine which extreme outcomes are likely to occur by chance 5% of the time or less. It is obvious from Table 14.1 that the probability of 14 or fewer countries meeting or exceeding the daily per capita calorie intake requirement meets this criterion, since the probabilities from that term down to the term in which no countries meet or exceed the requirement sum to .03. If we included the probability of 15 countries meeting or exceeding the daily requirement in our sum, we would exceed the 5% level (including that term, the probability would be slightly over 6.5%). Therefore, any sample result in which there were 14 or fewer countries meeting or exceeding the daily requirement would warrant rejection of the null hypothesis. Since our sample of 30 rural countries resulted in only 9 of them meeting or exceeding the daily requirement, **rejection of the null hypothesis** would be warranted.

Looked at another way, the probability of occurrence of 9 out of 30 countries in which the daily calorie requirement is met or exceeded is 0.0000789 according to Table 14.1. However, since any event more extreme than 9 out of 30 (8 out of 30, for example) would also support our set of alternatives to the null hypothesis, the practice is to sum the probabilities of all events as extreme as or more extreme than the one observed in the sample distribution. Summing all the events from 9 out of 30 down to none out of 30 gives a probability of approximately 0.0001 (called the **associated probability** of a sample outcome as extreme or more extreme than the observed sample). This is the probability figure we would use to evaluate the test of our null hypothesis. As long as this sum of probabilities is equal to or less than 0.05, rejection of the null hypothesis would be warranted.

All outcomes that fall in the tail of the sampling distribution where rejection of the null hypothesis is warranted are said to be in the *critical region*. Thus, *the critical region is that portion of the tail (or tails) of a sampling distribution that meets the preestablished criteria for rejection of the null hypothesis.*

In a sampling distribution such as the normal distribution the tail in the critical region would be represented by an area because the normal distribution is a continuous distribution. Since the binomial sampling distribution is a discrete distribution, the critical region is made up of a finite set of points or outcomes rather than an area.

BOX 14.2 CHECKING BACK

Let us pause for a moment and take stock. Some rather abstract ideas have been presented to you thus far in Chapter 14. We have discussed the null hypothesis and its alternatives, one- and two-tailed tests of hypotheses, critical regions, and levels of statistical significance. If all of these ideas are still not clear to you, before you proceed further you had better go back and reread the chapter up to this point. In the next few pages we will be discussing some more very important abstruse concepts. It is advisable to read this entire chapter slowly and thoughtfully. If you can master the basic logic involved in testing statistical hypotheses, the rest of the volume will come easily to you.

When a one-sided test is called for, the critical region is located in one tail of the sampling distribution. When a two-sided test is in order, the critical region is divided evenly between both tails of the distribution. In effect, we are dividing up the area (or outcomes) in the sampling distribution into a portion where observed sample outcomes favor the null hypothesis and another portion (or portions) in which sample outcomes favor the alternatives to the null hypothesis.

If we wished to test the null hypothesis for our sample of 30 rural countries at the 1% level rather than the 5% level, we would require 12 or fewer countries for a sample of size 30 to meet or exceed the daily calorie requirement. The sum of the probabilities up to that point is approximately .0045. If we added the term in which 13 countries met or exceeded the daily requirement, the sum would be greater than 1% (or .01 in terms of proportions). It would be approximately .012. Note that the sample result we got would have also fallen in the critical region for the 1% level of significance and the null hypothesis would have been rejected had that been our decision rule.

14.2.4   The Normal Approximation to the Binomial Sampling Distribution

In Section 12.3.4, we pointed out that as the number of cases involved in a sample increases the normal distribution is, increasingly, a satisfactory approximation to the binomial sampling distribution. Since N's as large as 30 make computing values for the binomial sampling distribution tedious, it is recommended that the normal approximation of the binomial be used instead. As mentioned in Section 12.3.3, as long as both NP and NQ are equal to or greater than 5, the normal curve is a satisfactory approximation to the binomial distribution. In the present case this rule of thumb applies; consequently, we will use the normal distribution approximation to test our null hypothesis and carry on with that approximation in our discussion of hypothesis testing for the remainder of the chapter.

Recall that when the normal approximation is used instead of the binomial sampling distribution itself the mean (in frequency form) is computed with the formula $\mu_B = NP$ and the standard error with the formula $\sigma_B = \sqrt{NPQ}$ (see Sections 12.3.1 and 12.3.2). Substituting the information we have about daily per capita calorie intake into these formulas, we come up with the following:

$$\mu_B = NP = 30(0.65) = 19.5$$
$$\sigma_B = \sqrt{NPQ} = \sqrt{30(0.65)(0.35)} = 2.61$$

Thus, if the null hypothesis ($H_0: P \geq 0.65$) were true, we would have a normal sampling distribution with a mean of 19.5 and a standard error of 2.61.

Since we wish to do a one-tailed test at the 5% level of significance and we wish to put the critical region in the left (lower) tail of the normal distribution, we need to determine the standard (z) score that cuts 5% off the left tail of the curve. To determine this value we consult Appendix Table A looking in column C for the value closest to .0500. You will notice that there are two values equally close to .0500. The area in the tail for a standard score of 1.65 is .0495 and the area for 1.64 is .0505. If we interpolated between these two values, we would find that the standard score that cuts exactly .0500 off the tail is 1.645. Customarily, the value 1.64 is used as the appropriate one because rounding 1.645 to the nearest even number gives 1.64. The standard score 1.64, then, is the one we will use to bound the critical region and, since we are interested in the left tail of the curve the value will actually be -1.64.

Our job now is to convert our sample result (nine countries meeting or exceeding the daily calorie requirement) into a standard (z) score and compare that score with our critical z-score of -1.64. Any computed z-score of -1.64 or less will fall into the critical region and warrant rejection of the null hypothesis. The appropriate formula for computing the z-score from our data is Formula 12.5:

$$z = \frac{X - \mu_B}{\sigma_B}$$

where, in this case, X is the observed frequency of rural countries meeting or exceeding the daily calorie requirement.

As was explained in Section 12.3.4, the normal distribution is a continuous distribution but frequencies are discrete. Therefore, it is necessary to make a *correction for continuity* for the frequency of 9. Instead of using 9 for X, then, we will use 9.5 because that would be the upper limit of a continuous score of 9. We use the upper limit of the score in this case because the score is in the left tail of the sampling distribution. If it were in the right tail, we would use the lower limit. That procedure has the effect of decreasing the distance between the mean and the X, thus providing a conservative test of the null hypothesis.

BOX 14.3 CORRECTION FOR CONTINUITY

If you do not remember the *correction for continuity*, do not let it bother you. Go back and read the explanation of the correction in Section 12.3.4. We will have occasion to refer to the correction for continuity again in this volume. It is used generally in evaluating a discrete empirical distribution with a continuous sampling distribution.

Substituting our values for the mean, the standard error, and X into the z-score formula, we come up with the following result:

$$z = \frac{9.5 - 19.5}{2.61}$$

$$= -3.83$$

This z-score is definitely into the critical region and warrants rejection of the null hypothesis in favor of the alternative that the P for rural countries is lower than the P of 0.65 for all of the countries of the world. The null hypothesis would be rejected at the 5% level of significance since that is the level of significance we set as our criterion for rejection. The computed value of z (-3.83) is well beyond the z-score bounding the critical region (-1.64); consequently, the probability value associated with the chance occurrence of 9 or fewer out of 30 countries meeting or exceeding the daily calorie requirement is considerably less than 0.05. Examination of Appendix Table A for the area in the tail associated with a z-score of -3.83 provides an approximate value of 0.0001 (using $z = 3.80$ instead of 3.83 since the latter value is not tabled). If you look back at the result we got when we used the binomial sampling distribution to test our null hypothesis, you will discover that this is about the same value we arrived at there.

14.2.5   Decision Errors

The *decision-making process in scientific research is based upon the* **principle of uncertainty**. Whenever we make a decision about a null hypothesis, whether the decision is to reject or fail to reject, we make our decision in the face of uncertainty. We set up criteria for the rejection of the null hypothesis and reject if the criteria are met, or fail to reject if the criteria are not met. If we reject the null hypothesis, we are saying, in effect, that we believe it to be false. It is possible, however, to observe an outcome that warrants rejection of the null hypothesis according to our criteria even though the null hypothesis is true. Any outcome included in the sampling distribution can occur as a result of chance (that is what the sampling distribution is all about).

Some outcomes are, however, much less probable than others. It is these less probable outcomes that we include in our critical regions. When we get an outcome in the critical region and reject the null hypothesis, we are saying that we choose to believe that outcome was not a result of chance, but the null hypothesis was false. This could, perhaps, be due to the systematic operation of some variable that interests us. If we consistently reject the null hypothesis whenever we reach the critical region, over the long haul, we will be wrong 5% of the time in rejecting it at the 0.05 level of significance. We never know for sure whether a particular outcome was due to chance. That is where the uncertainty principle enters the picture. If the observed outcome falls in the critical region, and the critical region only includes 5% of the total possible occurrences in the sampling distribution, it is reasonable to suppose that we did not get such a result by chance (although it may, in fact, be a rare chance result). For example, we just found that the computed z-score value for our daily calorie intake data warranted rejection of the null hypothesis (the probability was approximately 0.0001 so the computed z-score was well within the critical region). Accordingly, we rejected the null hypothesis in favor of the alternative that the proportion of rural countries in the world meeting or exceeding the daily calorie intake requirement is less than the world figure (.65).

14.2.5a   Type I Error

Our conclusion might be *wrong*; the null hypothesis might be *true*. If this is the case and we have rejected the null hypothesis because our criteria for rejection were satisfied by the data, then we made what is known as a **type I error**. *A type I error is committed if the null hypothesis is rejected when it is, in fact, true.* Type I errors can be committed because it is possible to observe sample outcomes that diverge widely from the hypothesized parameter because of the vicissitudes of random sampling.

The potential probability of committing a type I error is no greater than the level of significance adopted. If the test is made at the 5% level of significance, then the potential probability of a type I error is 5%. This is so because we reject the null hypothesis whenever we observe a sample outcome that falls in the critical region; and, indeed, 5% of the time the sample outcomes can be expected to fall in the critical region even when the null hypothesis is true.

Five percent is the maximum possible type I error, but is not the error that will necessarily be risked each time the null hypothesis is tested. In the example above, the outcome had a probability of occurring by chance of approximately .0001; thus, for this particular test, that figure represents the probability of a type I error. The risk of a 5% probability of a type I error would result from a finding that actually was right on the boundary of the critical region and therefore had a probability of 5%.

14.2.5b   Type II Error

Whenever the sample outcome *does not* fall into the critical region, in accordance with our predetermined rules for decision making we will fail to reject the null hypothesis. If the sample outcome falls short of the critical region because the null hypothesis is true, then we have been correct in failing to reject it. *If, on the other hand, the null hypothesis is false and we fail to reject it, then we are guilty of a type II error.*

Notice that **type II error** arises whenever the sample outcome fails to reach the critical region, whereas type I error arises whenever the sample outcome falls in the critical region. It should be obvious that our decision rules do not subject us to risks of type I and type II errors simultaneously. Thus, when we reject the null hypothesis, we are subject to type I error; when we fail to reject the null hypothesis, we are subject to type II error. (Figure 14.2 diagrams the relationship between decisions and type I and type II errors.)

We found that the maximum possible type I error was equal to the area (or proportion of outcomes) in the critical region: 5% in the case of the 5% level of significance. Apparently, the maximum possible type II error is equal to the area in the sampling distribution that is not in the critical region: 95% in the case of the 5% level of significance. However, it is not quite as simple as all that. The maximum possible type II error depends upon the actual value of the parameter of interest and the sample size. This information, unfortunately, is not generally available to us.

Returning now to our data on daily per capita calorie intake, we will consider the type II error involved in the test of the null hypothesis. Since 65% of the countries of the world had daily per capita calorie intakes meeting or exceeding the requirement, we set up the sampling distribution with an expected value of .65, which translated into a mean value of 19.5 out of 30 countries sampled. Furthermore, since we expected the parameter for rural countries to be lower than .65, we used a one-tailed, 5% test of the null hypothesis with the critical region in the left tail of the normal sampling distribution bounded by a standard score of -1.64.

	Decision Made	
Null Hypothesis	Reject	Fail to Reject
True	Type I Error	Correct Decision
False	Correct Decision	Type II Error

FIGURE 14.2 Possible Outcomes of Statistical Decisions

We can translate the standard score of -1.64 into a value representing number of countries by multiplying it by the standard error (2.61) and adding that product to the mean. Thus:

$$\mu_B + z\sigma_B = 19.5 + (-1.64)(2.61)$$
$$= 19.5 + (-4.28) = 19.5 - 4.28 = 15.22$$

Therefore, any sample result equal to or less than 15.22 countries out of 30 would warrant rejection of the null hypothesis. In other words, 15.22 is the frequency value equivalent to the standard score of -1.64 that bounds the critical region in the left tail of the normal sampling distribution. Figure 14.3a is a graphic

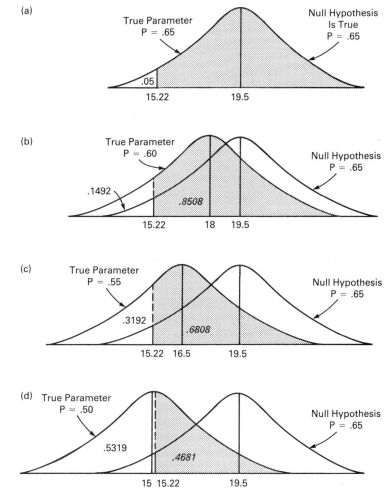

FIGURE 14.3 SAMPLING DISTRIBUTION OF THE FREQUENCY OF COUNTRIES WITH PER CAPITA DAILY CALORIE INTAKE MEETING OR EXCEEDING A MINIMUM REQUIREMENT. SAMPLES SIZE 30. NULL HYPOTHESIS IS P = .65 (AN EXPECTED 19.5 OUT OF 30 COUNTRIES).

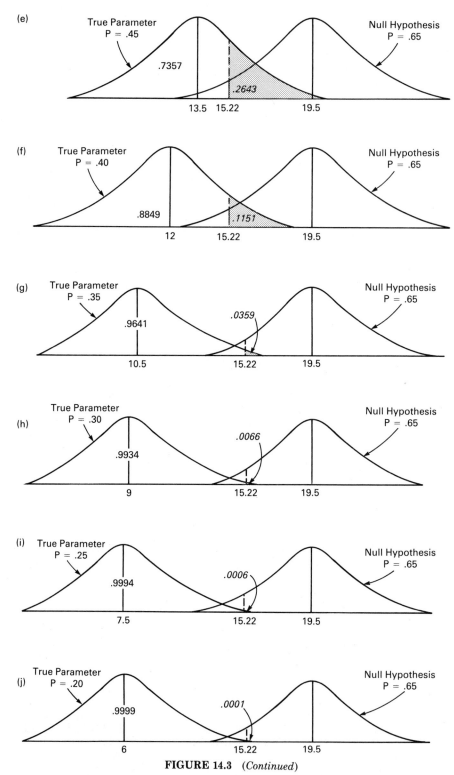

FIGURE 14.3 (*Continued*)

representation of this sampling distribution. The 0.05 in the left tail of this sampling distribution represents both the critical region for the test of the null hypothesis and the probability of making a type I error.

The probability of making a type II error, as was stated earlier, depends upon the actual value of the parameter for rural countries. In the sample of 30 rural countries, 9 of the 30 met or exceeded the daily per capita calorie intake. In terms of percentages, that represents 30% of the sampled countries (9/30 = 0.30, or 30%). For the sake of illustrating how the probability of type II error may be determined, we will assume that the parameter for all of the rural countries of the world is 30%. If the parameter were 30% instead of 65%, the sampling distribution would have a mean of 9 [30(0.30)] and a standard error of 2.51 [$\sqrt{30(0.30)(0.70)}$].

If we used the original sampling distribution with a mean of 19.5 and a standard error of 2.61 to test the null hypothesis and if the actual sampling distribution had a mean of 9 and a standard error of 2.51, then what would be the probability of making a type II error? First, note that if these were the actual circumstances, there would be no chance of making a type I error because the null hypothesis would be false. Now examine Figure 14.3H. The curve on the right represents the original sampling distribution and the one on the left represents the sampling distribution with a mean of 9. Notice that .9934 of the area in the curve on the left is to the right of the boundary of the critical region in the negative tail of the original sampling distribution. That is, 99.34% of the values in the sampling distribution on the left are in the critical region of the original sampling distribution. Only .0066, or .66%, of the values in the left curve fail to reach the critical region of the original distribution. Therefore, only .66% of the time would we fail to reject the null hypothesis when it was, in fact, false. This, of course, would be the probability of a type II error.

We were only able to come up with this probability for type II error by assuming that we knew the parameter for rural countries to be 30%. Of course, we do not know that this is the case, which is the crucial problem involved in attempting to compute the probability of a type II error. If we actually knew the parameter, we would not need to use inferential statistics to make statements about it. Since we really do not know the parameter, we are unable to get any clear idea of how large a type II error we are risking for a particular test of the null hypothesis.

Figure 14.3 illustrates an important principle about type II error. The probability of failing to reject (accepting) a false null hypothesis gets smaller as the distance between the parameter suggested by the null hypothesis and the real parameter increases. It is only when the null hypothesis is wrong by a small amount that the chance of making a type II error becomes large. As we shall see, researchers attempt to control type II error by opting for larger samples (so the standard error is smaller and a given difference between the parameter suggested by the null hypothesis and the true parameter is relatively large in z-score terms).

14.2.5c   Power of the Test

The complement of type II error is known as the **power of the test**. *Power of the test is defined, therefore, as 1 − Pr(type II error).* Since type II error stems

TABLE 14.2 DISTRIBUTIONS OF TYPE II ERRORS AND POWER FOR VARIOUS VALUES OF P FOR A ONE-TAILED, 5% LEVEL OF SIGNIFICANCE, WITH THE HYPOTHESIZED $P = 0.65$

P	μ_B	σ_B	z	Pr(Type II)	$Power$
0.65	19.5	2.61	−1.64	[Pr(Type I) $= 0.05$][a]	
0.60	18.0	2.68	−1.04	0.8508	0.1492
0.55	16.5	2.72	−0.47	0.6808	0.3192
0.50	15.0	2.74	0.08	0.4681	0.5319
0.45	13.5	2.72	0.63	0.2643	0.7357
0.40	12.0	2.68	1.20	0.1151	0.8849
0.35	10.5	2.61	1.80	0.0359	0.9641
0.30	9.0	2.51	2.48	0.0066	0.9934
0.25	7.5	2.37	3.26	0.0006	0.9994
0.20	6.0	2.19	4.21	0.0001	0.9999

[a] There are no entries in this row because the null hypothesis is true.

from *failing to reject* the null hypothesis when it is *false*, the *power of the test* refers to the probability of *rejecting* the null hypothesis when it is *false*. (Power of the test is represented in the lower-left cell of Figure 14.2.) In the specific example above, the probability of a type II error was .66%, so the power of the test was 99.34%. In that particular case the test of the null hypothesis would be very **efficient** since a false null hypothesis would be rejected almost all of the time.

The dilemma facing us in trying to make use of the concepts of type II error and power of the test is that in any one test of a null hypothesis we do not know the value of the parameter and, thus, are unable to compute type II error or power of the test.

However, even though we cannot compute the probability of type II error or power of the test for a particular test of a null hypothesis, we can assume various values for the parameter and compute probabilities of type II error and power for each. Table 14.2 gives the distributions of type II errors and power generated by assuming various actual values of the parameter for rural countries of the world. Figure 14.3 represents graphically the actual sampling distribution for each separate parameter as compared to the original sampling distribution for which P is assumed to be 0.65. Figure 14.4 plots the curves of the distributions of type II error and power given in Table 14.2.

To clarify how the values in Table 14.2 are arrived at, we will work through the case in which the actual parameter is .60. The first thing that we must do in order to get the desired values for type II error and power is to find the mean and standard error associated with a parameter of .60. Using the formulas for the mean and standard error as before, we come up with the following results:

$$\mu_B = NP = 30(.60) = 18$$

$$\sigma_B = \sqrt{NPQ} = \sqrt{30(.60)(.40)} = 2.68$$

Since 15.22 is the value that bounds the critical region in the sampling distribution suggested by the null hypothesis, we will substitute that value, along with the

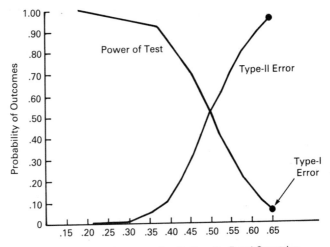

FIGURE 14.4 Distribution of Type II Error and Power for Various Values of P with $N = 30$ and Hypothesized $P = .65$

mean and standard error, into the standard score formula and solve for z as follows:

$$z = \frac{X - \mu_B}{\sigma_B} = \frac{15.22 - 18}{2.68}$$

$$= -1.04$$

Looking up the standard score of -1.04 in Appendix Table A, we find that the area in the tail of the curve (column C) is .1492. Since the standard score is negative, that area represents the area in the left tail of the sampling distribution with a parameter of .60 and $1 - .1492 = .8508$ represents the area in the remainder of the distribution. The area .1492 is the area in the sampling distribution with a parameter of .60 that is in the critical region of the original sampling distribution with a parameter of .65. Consequently, instead of rejecting the null hypothesis .05 of the time, we would reject it .1492 of the time if the actual parameter were .60. Therefore, the figure .1492 represents the power of the test for that particular parameter and its complement (.8508) represents the probability of a type II error. Each of the other values in Table 14.2 was computed in like manner, using each of the P's in column 1, in turn, as the actual parameter.

Note in Table 14.2 that when the actual value of P is close to the hypothesized value (.65) the probability of type II error is large and the power of the test is minimal. This happens because the sampling distribution around the actual parameter is very nearly superimposed upon the sampling distribution generated from the null hypothesis. For example, when the actual parameter is .60, the probability of type II error is .8508 and the probability for power of the test in only .1492. Figure 14.3B is a graphic representation of this particular case. Because the mean (18) associated with a parameter of .60 is relatively close to the hypothesized mean of 19.5, there is much overlap in the two sampling distributions and ample opportunity to confuse a sample value from one with a sample value from the other.

If the actual value of the parameter for rural countries were, in fact, .30 as our sample would lead us to expect, there would be little confusion between the actual sampling distribution and the original sampling distribution suggested by the null hypothesis. The distance between the mean of 9 for the actual sampling distribution and the mean of 19.5 is so large that there is little overlap in the two distributions (see Figure 14.3H); therefore, there would be little tendency to err in making a decision about the null hypothesis.

What is the point of generating theoretical distributions for type II error and power of the test such as those in Table 14.2 and Figure 14.4? These theoretical distributions are actually very useful to sociologists in planning their statistical analysis. For example, if there is a choice of alternative statistical techniques to analyze a body of data, and information about their respective power distributions is available, that information may be used as a basis for deciding which technique to use. Other things being equal, it would be reasonable to select the most powerful test (if one exists). Furthermore, it can generally be demonstrated that the power distribution for a one-tailed test will be more efficient than that for a two-tailed test. This useful bit of information comes from a comparison of theoretical power distributions such as the one we have been examining.

14.2.5d   Power Analysis and Determining Needed Sample Size

Investigators often face questions such as "What is the minimum sample size needed to discriminate between a null hypothesis and sample outcomes that differ by a certain amount?" or "How large a difference will this test discriminate at a given level of power?" Hypothesis-testing formulas can be examined to help answer these questions. To illustrate, we will use the z-score formula used in testing a one-sample test for proportions to estimate **needed sample size**:

$$z = \frac{p - P}{s_p}$$

or, substituting the formula for the standard error of proportions for s_p,

$$z = \frac{p - P}{\sqrt{PQ/N}}$$

where p is the sample proportion, P is the hypothesized proportion in the population from which the sample was presumably drawn, and N is the size of the simple random sample that was drawn. The z-score, then, indicates how deviant the sample is from the hypothesized value in terms of the standard error of the sampling distribution.

If we algebraically convert the formula above to solve for N, it becomes

$$N = \frac{PQz^2}{(p - P)^2}$$

where $(p - P)$ is the minimum expected or desired difference to be discriminated as statistically significant by the test, z is the value for the sampling distribution corresponding to the desired level of significance, and P and Q are hypothesized population proportions under the null hypothesis. We can use this formula to estimate the minimum sample size needed to reject a null hypothesis at some specified level of significance.

To illustrate, a 1989 survey by the National Center for Health Statistics reports that 13.9% of people in the United States are not covered by health insurance (Ries, 1991). We suspect that this varies by income and want to conduct a study to see if people in families with low annual incomes (say, less than $10,000 per year) are significantly more likely *not* to have health coverage. How large a sample would be required?

First, we need to decide on how large a difference we expect between a sample outcome (p) and the hypothesized population value (P, in this case, .139). Suppose we decide that the ($p - P$) difference we would expect is $+.10$. That is, people from low-income families will have a proportion without health coverage that is about .10 higher than people in general. We also need to decide on the level of significance we want to use; for example, we will use the 5% level of significance (.05). And since we are predicting direction, it will be a one-tailed test. Using the table of areas under a normal curve in Appendix A, the region of rejection under the null hypothesis will start at $+1.645$. Substituting in the formula above and solving for N, we find we need a sample size of at least 32.

$$N = \frac{(.139)(.861)1.645^2}{.10^2} = 32.4$$

This is a minimum estimate, of course, and would typically be increased to allow for nonresponse and to provide for the analysis of other variables and detailed contrasts that may also be included in an overall research project.

Similar use of formulas for other tests of significance can also help identify a minimum sample size in other situations. For example, sample size for a one-sample test of means would involve solving the following formula for N:

$$z = \frac{\bar{X} - \mu}{s/\sqrt{N}} \quad \text{so} \quad N = \frac{s^2 z^2}{(\bar{X} - \mu)^2}$$

where s is the estimated population standard deviation, z is the value for the sampling distribution that corresponds to the desired level of significance, and ($\bar{X} - \mu$) is the minimum expected or desired difference to be discriminated as statistically significant by the test. The final chapter reviews considerations in selecting sample size.

These formulas can also be used to conduct a **power analysis** to see, for example, how much difference between a hypothesized population value and a true population value could be discriminated at a given level of significance and power of the test (or how large a sample would be needed to discriminate a given difference between hypothesized and true parameters at a given level of power). Using the example of health care coverage, we want to test the null hypothesis that the population proportion of people without health care coverage, among those in low-income families, would be .139, the overall U.S. average. We will use a one-tailed test, since we expect that the proportion will be higher than .139. If this null hypothesis is false, could we reject it with a probability of .95 (*i.e.,* power = .95) where the true population proportion is at least 10 percentage points higher than our hypothesized value, or .239?

In this instance, assume we will use a simple random sample of size 100. This is large enough to permit us to use the normal approximation of the binomial sampling distribution (*i.e.*, both NP and NQ are equal to or larger than 5). If we use the .05 level of significance for a one-tailed test, the region of rejection of the null hypothesis that $P = .139$ is $+1.645$. The standard error of the sampling distribution of proportions is

$$s_p = \sqrt{\frac{PQ}{N}}$$

Here

$$\sqrt{\frac{(.139)(.861)}{100}} = .0346$$

Thus, the region of rejection of the null hypothesis would start at $.139 + 1.645(.0346) = .1959$. This is shown in the sampling distribution below:

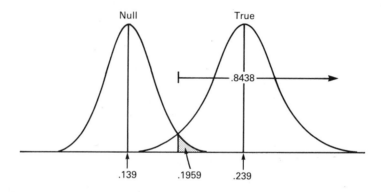

If the true population proportion were .239 instead of .139, the true standard error for samples of size 100 would be

$$\sigma_p = \sqrt{\frac{(.239)(.761)}{100}} = .0426$$

On the true sampling distribution, the previously determined region of rejection, which starts at .1959, would fall at $z = (.1959 - .239)/.0426 = -1.012$ below the true parameter. The power of that test, the probability of rejecting the false null hypothesis, would be all the area under the true sampling distribution that falls in the original region of rejection. In this case, using Appendix A for the areas under a normal sampling distribution, it is $.3438 + .5000 = .8438$. A sample of 100 would result in a test that would reject the false null hypothesis of $P = .139$ in favor of a true parameter that is .239 (.10 higher than the false value) 84% of the time (power equals .84). A test with this sample size is not as powerful as the .95 level we desired. Using the formulas above, we can determine that the sample size would have to be at least 265 for the test to have the desired .95 power.

The National Center for Health Statistics survey found that the percentage of people from families with low incomes (less than \$10,000 per year income) who do not have health coverage is about 27.4%. If this is true, a sample size of 265 from such a population would be more likely than .95 to be among those where the sample percentage without health coverage was high enough to reject the hypothesized value of 13.9% at the .05 level of significance.

Note that this approach to estimating sample size and the power of a test generally requires an investigator to make some population estimates, for example, the size of the population standard deviation or the difference between the actual proportion of successes and that which is hypothesized. These formulas also assume that sampling is simple random sampling. More advanced texts deal with procedures for other situations (*e.g.*, Hays, 1981; Kraemer and Thiemann, 1987).

14.2.6   Minimizing Decision Errors

In testing a null hypothesis it is desirable, if possible, to minimize the probabilities of committing both type I and type II errors. It is easy enough to control the maximum probability of a type I error by selecting the size of the critical region. If we decide to test the null hypothesis at the 5% level of significance and set the critical region accordingly, we are saying, in effect, that we are not willing to risk a type I error larger than 5%.

The size of the type I error we are willing to risk depends upon the seriousness of committing such an error as compared to a type II error. For example, if the null hypothesis states that couples who have had premarital counseling do not have lower divorce rates than the average for all couples, and we reject that hypothesis in favor of the alternative that their divorce rates are lower, the consequences of making a type I error will probably not be considered serious. If engaged couples were subjected to premarital counseling because of our findings and our conclusions were wrong, it would probably do them no harm. On the other hand, if we failed to reject the null hypothesis, and it was false (a type II error), failure to institute a premarital counseling program might lead to the dissolution of marriages that might have been saved. In this case, a type I error might be considered less serious than a type II error.

On the other hand, if our null hypothesis states that capital punishment does not deter people from committing crimes of violence, and rejection of the hypothesis will lead to capital punishment, a type I error would have serious

---

BOX 14.4   DECISION ERRORS AGAIN

Section 14.2.5 on decision errors is usually not an easy one to understand the first time around. If you are confused at this point, do not give up. Read the section over again carefully. Many students tend to glance at tables and figures without really reading them. The tables and graphs included in this section are an integral part of the explanation, so do not take them lightly. Study them carefully and they will help you to comprehend what is being said.

consequences whereas a type II error might not be considered as serious because of the alternative of life imprisonment. In a case such as this we would probably want to minimize the probability of type I error by reducing the critical region to 1% or less.

Unfortunately, reducing the critical region to 1% increases the proportion of the distribution outside the critical region to 99%. This has the potential of increasing type II error. What we wish to accomplish is the reduction of type I and type II error simultaneously.

There are two ways of reducing the probability of type I error without increasing the probability of type II error. The first is to **increase the sample size**. As sample size increases, sampling error decreases. The smaller the sampling error, the more sensitive a statistical test is to differences in parameters and the less likelihood of confusing actual differences with differences attributable to chance.

The second way to reduce the probability of a type I error is to **replicate** the study. If a study is conducted and the statistic in question just reaches the 5% level of significance, the null hypothesis can be rejected in favor of its alternative. The probability of a type I error is 0.05. There are 5 chances out of 100 that the statistic reached the critical region as a result of sampling error. If the same study is conducted a second time with an independent sample and with the same result as the first time, the probability of a type I error would again be 0.05. However, the probability that two type I errors were committed in succession in two independent tests of the same hypothesis would be $(0.05)^2 = 0.0025$. If the study were conducted a third time with the same result, the probability of a type I error would be $(0.05)^3 = 0.000125$. As you can see, the probability of type I error gets increasingly smaller with replication.*

There are, then, three ways of reducing the probability of a type I error: (a) reduce the size of the critical region, (b) increase the sample size, and (c) replicate the study. The first method is not very satisfactory since it has the effect of increasing the probability of a type II error. The second and third are preferable because they do not increase the probability of a type II error; as a matter of fact, they decrease the probability of a type II error. Increasing sample size reduces type II error because the reduced sampling error makes the statistical technique more sensitive to actual differences between parameters. Moreover, replication reduces type II error as well as type I error. If the probability of making a type II error was 0.42, then the probability of making two type II errors in a sequence of two independent studies would be $(0.42)^2 = 0.18$.

Of course, type II error can be reduced by increasing the area in the critical region. This, however, has the effect of increasing the probability of a type I error.

There is one other way of reducing the probability of a type II error without affecting the probability of a type I error. As was mentioned in Section 14.2.5, a one-tailed test of significance has more power than a two-tailed test. This implies that the type II error is smaller for a one-tailed than for a two-tailed test. Certainly, one must have found reasons for using a one-tailed rather than a two-tailed test. These reasons come from theories that are sufficiently well developed to permit

* For a discussion of the dangers of confusing type I errors with theoretically significant findings, see Sterling (1959).

one-tailed hypothesis tests. Otherwise, it is foolhardy to place the whole critical region in one tail of the sampling distribution.

It is possible, then, to reduce the probability of type II error four different ways: (a) increase the size of the critical region, (b) increase the sample size, (c) replicate the study, and (d) use a one-tailed test of significance. The last three steps can be taken without increasing the probability of a type I error and, incidentally, the second and third also reduce the probability of a type I error. More generally, one could look for more powerful statistical tests as well.

14.2.7 Steps in Testing Statistical Hypotheses

To recapitulate the process of testing statistical hypotheses, here is a list of the steps involved:

1. Select a hypothesis from a theory to be tested. If possible, select a crucial hypothesis; that is, one that is a keystone of the theory. This makes for an efficient and economical test of the theory.

2. Derive a working hypothesis or hypotheses from the general hypothesis. This is done by specifying measurable indices of the concepts of the general hypothesis and of the relationships between the indices.

3. State the null hypothesis or hypotheses. The null hypothesis is derived from the working hypothesis and is specific and often contradictory to the working hypothesis.

4. Decide which statistical technique will be used to test the null hypothesis.

5. Choose the appropriate sampling distribution. The sampling distribution that is appropriate is suggested by the null hypothesis and is determined by the nature of the data to be collected and by the statistical technique to be used.

6. Define the critical region. This includes deciding on the level of significance to be used and on whether a one-tailed or two-tailed test should be conducted. The two-tailed test of significance should be used primarily for exploratory research. The major motivation for doing research should be to test theory, and theory should lead to directional hypotheses. Therefore, one-tailed tests should be used more frequently than two-tailed tests.

7. Draw a random sample of observations. This includes the process of designing the sample and deciding on its size.

8. Compute a statistic from the data, and locate it on the sampling distribution.

9. If the statistic is in the critical region, reject the null hypothesis; if not, fail to reject.

10. Decide what bearing the results have on the working hypothesis and the general hypothesis.

11. Replicate the study. It is only through the repetition of a study that we develop a real feeling for how fruitful or unfruitful a theory is.

The fact that these steps are listed from 1 to 11 should not be construed to mean they are carried out in that order. They are listed in this way to suggest a flow of logic rather than a step-by-step guide for accomplishing the work. There is no simple, orderly sequence of steps involved in testing a statistical hypothesis. The order in which the steps are carried out may vary, and, frequently, several steps are carried out simultaneously.

It should also be noted that a rather comprehensive understanding of theory and research methods is a necessary prerequisite to carrying out these steps successfully. Statistical analysis, theory, and research methods are so integrally interrelated that they cannot really be separated.

14.2.8   The Importance of Replication

Scientists have long recognized the value of replicating studies. Some experiments in the physical and life sciences are repeated again and again until the researchers are convinced of the validity of their findings.

Unfortunately, replication receives all too little attention from sociologists. Sociologists seem to be reluctant to repeat studies that have already appeared in their professional literature. Most researchers seem to be determined to do something unique. The effect of not replicating studies is, of course, to perpetuate type I errors in the guise of scientific generalizations. For example, suppose that a sociologist does a study, tests a null hypothesis at a given significance level, rejects the null hypothesis, concludes that it is a valid finding, and publishes the results. Suppose further that rejection of the null hypothesis is based upon a type I decision error (the null hypothesis is rejected when it is in fact true). Once the study appears in print, other sociologists will be less likely to conduct similar studies because "the research has already been done." The spurious findings of the original study may be taken at face value and given the status of a scientific generalization (Sterling, 1959:30–34). Perhaps this explains why many so-called principles of sociology have little predictive value and little basis in fact.

Much more attention needs to be given to replication if sociologists are to develop theoretical models that are useful predictive and explanatory instruments. It is only when we replicate studies that we begin to realize how tenuous many of our *research findings* are.

14.3   CONFIDENCE INTERVALS AND TESTS OF HYPOTHESES

An alternative to the traditional method of testing null hypotheses described in this chapter is to use confidence intervals. Some authors have pointed out that the use of confidence intervals to test hypotheses enjoys some advantages over the traditional approach (Natrella, 1960; see also Blalock, 1979, General References). First we will consider how confidence intervals may be used to test hypotheses; then we will examine the advantages of using this approach.

We have just examined the test of a null hypothesis—that no fewer rural countries met or exceeded the daily per capita calorie intake requirement than did countries in the world in general. We rejected this null hypothesis because the computed z-score of -3.83 was well into the critical region. Instead of computing the z-score, we could have computed 99% confidence limits around the population mean of 19.5. Using the population mean and standard error (2.61), we can substitute into the following formula with $z = \pm 2.58$ (which bound the middle 99% of the normal sampling distribution; see Appendix Table A) and get the required limits:

$$cl = \mu_B \pm z(\sigma_B) = 19.5 \pm 2.58(2.61)$$

(14.1)
$$cl_1 = 19.5 - 6.73 = 12.77$$

$$cl_2 = 19.5 + 6.73 = 26.23$$

Therefore, the 99% confidence limits are 12.77 and 26.23, inclusive. Any number of rural countries in the sample of 30 falling between these limits is a reasonable expectation for a population with a mean of 19.5. Any number falling outside these limits may be considered unusual. The observed number of countries was 9. This result is outside the confidence interval and below the lower limit, so we would be justified in rejecting the null hypothesis.

The following advantages are offered for this approach to testing hypotheses over the traditional approach:

1. The approach divides all of the possible hypotheses into two sets, those that are tenable because they are in the confidence interval and those that are not because they are outside the interval. It is possible, therefore, to test simultaneously a whole set of hypotheses.

2. In rejecting the null hypothesis we not only discover what outcome is unlikely, but by placing confidence limits around the observed sample outcome (9) we have available to us, at the same time, a whole set of likely outcomes to offer as a parameter estimate.

3. The width of the confidence interval gives us some notion of how confident we can be about rejecting the null hypothesis. The narrower the confidence interval, the more assurance we can have in rejecting. Also important, of course, is the distance of the observed outcome from the confidence limits.

In the example above, the interval was not as narrow as we might have liked; but the observed outcome did fall more than 1 standard error below the lower limit of the interval.

The significance of the width of the confidence interval is that it gives one a notion of how powerful the test of the null hypothesis is. Narrow confidence intervals tend to be found in those cases where the power curve is efficient (Natrella, 1960:21).

14.4 SUMMARY

Chapter 14 has introduced you to the general process by which statistical hypotheses are tested. We have explained the use of the null hypothesis and pointed out how it can be used to specify the sampling distribution. A distinction was made between one-tailed and two-tailed tests, and applications of each were discussed. Significance level was defined, and its relationship to critical regions explained. Type I errors, type II errors, and power of the test were discussed; and techniques for reducing the probabilities of making decision errors were suggested. We listed the 11 steps in the logic of testing a statistical hypothesis and

504 INFERENTIAL STATISTICS: HYPOTHESIS TESTING

gave an example of how these steps are actually carried out. Finally, an alternative approach to hypothesis testing was discussed, using confidence intervals.

It is important to note that the inductive process in statistics is merely an aid to researchers in making substantive decisions. Ultimately, researchers must decide whether their data support or fail to support the theoretical notions they are investigating. Statistics is an aid for researchers to lean on in the process of making decisions. However, statistics alone are not capable of making substantive decisions.

Furthermore, tests of statistical significance are a crude way of determining whether there is anything that can be generalized from samples to populations. Tests of null hypotheses constitute a primitive but necessary step in the process of scientific investigation. The significant analysis, from the theoretical standpoint, is the descriptive analysis that is used. Such an analysis addresses itself to the meaning of data. Tests of null hypotheses help us determine whether such meanings should be generalized to populations.

Since tests of null hypotheses are an integral part of the generalization process, in the chapters that follow we will consider a number of statistical techniques that are used to test null hypotheses. In each case, the general hypothesis-testing process will be the one described here; but the detailed procedures will vary from one application to another.

CONCEPTS TO KNOW AND UNDERSTAND

general hypothesis
working hypothesis
statistical hypothesis
null hypothesis
one-tailed and two-tailed tests
level of statistical significance

critical region
type I error
type II error
power of the test
replication

QUESTIONS AND PROBLEMS

1. Why is Durkheim's study of suicide considered a classic example of sociological research? Is the generalization of Durkheim's findings justified? If yes, why? If not, why not?

2. What are the steps involved in using sample data to test statistical hypotheses?

3. Explain the logic involved in using the null hypothesis.

4. Explain the difference between a one-tailed and a two-tailed test of significance. Under which circumstances would you use a two-tailed test? Under which circumstances would you use a one-tailed test?

5. Explain the meaning of the concept of a *critical region*.

6. Essentially, what does *statistical significance* mean?

7. Distinguish between a type I and a type II error. How would you go about minimizing the probabilities of committing both types of errors?

8. What is the relationship between the type II error and the power of the test? How is the concept of *the power of the test* used in statistical analysis?

9. Why is it important to replicate a study? See if you can find examples of studies in sociology that have been replicated. Ask a physics or biology professor how often replication is used in his or her discipline.

10. Explain the relationship between finding a confidence interval and testing a statistical hypothesis. What are the advantages of using a confidence interval to test a statistical hypothesis?

11. Using the sample data on daily per capita calorie intake for rural countries, set up a two-tailed test of the null hypothesis (1% level of significance) and carry out the test with the normal approximation to the binomial sampling distribution (see Section 14.2.4). How do the results of this test compare with the test for one tail at the 5% level?

GENERAL REFERENCES

Blalock, Hubert M., *Social Statistics*, rev. ed. (New York, McGraw-Hill), 1979.
    In this excellent intermediate-level book, Blalock devotes Chapter 10 to a discussion of the steps involved in the test of a statistical hypothesis.
Edwards, Allen, *Statistical Methods for the Behavioral Sciences* (New York, Holt, Rinehart and Winston), 1954.
    This is another clearly written intermediate statistics book. The discussion of the concept of the power of the test is one of the best to be found anywhere.
Mohr, Lawrence B., *Understanding Significance Testing* (Beverly Hills, Calif., Sage), 1990.
Siegel, Sidney, *Nonparametric Statistics for the Behavioral Sciences* (New York, McGraw-Hill), 1956.
    The first three chapters of this book are especially recommended as an introduction to the subject of hypothesis testing.

LITERATURE CITED

Bogue, Donald J., *Principles of Demography* (New York, Wiley), 1969.
Clark, Robert E., "Psychoses, Income, and Occupational Prestige," *American Journal of Sociology*, 54 (1949), pp. 433–440.
Douglas, Jack D., *The Social Meaning of Suicide* (Princeton, N.J., Princeton University Press), 1967.
Durkheim, Emile, *Suicide: A Study in Sociology* (1897), translated by John A. Spaulding and George Simpson (Glencoe, Ill., The Free Press), 1951.
Galton, Francis, *Hereditary Stature*, Royal Society Proceedings, XL, 1886.
Hays, William L., *Statistics*, 3rd ed. (New York, Holt, Rinehard and Winston), 1981, sections 7.11 and 8.11.
Kraemer, Helena Chmura, and Sue Thiemann, *How Many Subjects? Statistical Power Analysis in Research* (Newbury Park, Calif., Sage), 1987.

Labovitz, Sanford, "Criteria for Selecting A Significance Level: A Note on the Sacredness of .05," *American Sociologist*, 3 (1968), pp. 220–222.

McKeown, Thomas, *The Modern Rise of Population* (London: Edward Arnold), 1976.

Natrella, Mary G., "The Relation Between Confidence Intervals and Tests of Significance," *The American Statistician*, 14, 1 (February 1960), pp. 20–22.

Reiss, Albert J., Jr., *Occupations and Social Status* (New York, The Free Press), 1961.

Ries, Peter. "Characteristics of Persons with and without Health Care Coverage: United States, 1989." Advance Data from Vital and Health Statistics, no. 201. (Hyattsville, M.: National Center for Health Statistics), 1991.

Robinson, William S. "Ecological Correlations and Behavior of Individuals," *American Sociological Review*, 15, 3 (June 1950), pp. 351–357.

Skipper, J. K., Jr., et al., "The Sacredness of .05; A Note Concerning the Uses of Significance in Social Sciences." *American Sociologist*, 2 (February 1967), pp. 16–19.

Sterling, T. D., "Publication Decisions and Their Possible Effects on Inferences Drawn from Tests of Significance—or Vice Versa," *Journal of the American Statistical Association*, 54 (1959), pp. 30–34.

United Nations Children's Fund, *The State of the World's Children 1986* (New York, Oxford University Press), 1985.

CHAPTER 15

Testing Hypotheses about Central Tendency, Dispersion, and Form

Sociology is concerned with the study of social systems such as groups and organizations. People are of interest not as individuals, but as members of social systems. At times sociologists focus their attention on social systems in toto; at other times they focus their attention on the members of systems. The members may be people, but they may also be subsystems of larger systems (*e.g.*, departments of an organization).

Whether they focus their attention on the study of total social systems, subsystems, or people as members of systems, sociologists characteristically collect the kinds of data that lend themselves to analysis of central tendency and variability.

Analysis that utilizes measures of central tendency may focus on questions such as the following: Do graduates of private colleges have annual incomes that are higher on the average than those of graduates of American colleges in general? Is the average age of members of the Democratic party lower than that for members of the Republican party? Do lower income families spend larger proportions of their incomes for food than higher income families? Do industrialized countries have higher population density than nonindustrialized countries? Do companies with diversified product lines tend to have higher average volumes of sales than companies with less diversified product lines?

Popular as studies that focus on central tendency are in sociology, the sociologist can hardly afford to ignore questions about variability. The relative homogeneity or heterogeneity of social systems may be crucial to certain theoretical questions. In the past, sociologists have given less attention to questions of variability than they have to questions of central tendency. Accordingly, there are many significant questions involving variability begging answers. For example, sociologists might investigate such questions as the following: Are income differentials greater in larger cities than they are in small ones? Is the age structure of the population of central cities more or less variable than that of the suburbs? Does the per capita expenditure on education vary more among states with growing populations than among states with stable or declining populations? Does size of membership vary more among locals of craft unions than the locals of industrial unions?

These questions suggest just a few of the kinds of studies that sociologists undertake in which they may be interested in testing hypotheses about central tendency or variability. This chapter will be devoted to a consideration of the techniques available for testing hypotheses of these kinds. We will specifically discuss **one-sample tests** for means, proportions, and variances; **tests of two independent samples** for means, proportions, and variances; and *k*-sample **tests** for means.

Sociologists have turned increasingly to mathematical models to predict social behavior. Consequently, in addition to central tendency and variability, consideration of the form of a distribution has become more important. In recognition of this fact, we have devoted part of this chapter to the chi-square goodness-of-fit test as a technique for evaluating the forms of distributions. That discussion will not only provide you with a technique for testing hypotheses about form; it will also serve an introduction to hierarchical analysis, which is covered in Chapter 16. Finally, a few nonparametric tests and alternative approaches are considered.

15.1 ONE-SAMPLE TEST FOR MEANS

The Bureau of Labor Statistics of the U.S. Department of Labor carries out area wage surveys annually in selected metropolitan areas to provide information on earnings for standard work weeks in occupations common to a variety of industries. The data are used to generalize to all metropolitan areas in the country. A standard of pay is computed on the basis of the average weekly salary (or hourly rate) for the occupations in all standard metropolitan areas combined; then the rates for particular metropolitan areas are expressed as percentages of the standard rate. For example, the standard rate of pay for clerical office workers in all industries is set at 100. The rate for the Chicago metropolitan area in the year from January to December 1980 was 104, whereas for New Orleans it was 93 (U.S. Department of Labor, 1983:295–296).

These data can be used to seek answers to questions relevant to an understanding of various labor markets. For purposes of illustration, attention will be directed to clerical office workers, although the data are also available for workers in electronic data processing, skilled maintenance, and unskilled plant labor.

One question we might explore is this: What are the consequences of size of metropolitan area for the pay scales of clerical office workers? Do the pay scales of these workers in the smaller metropolitan areas of the country (areas with populations of less than 1 million) compare favorably or unfavorably with the nationwide standard of 100?

We might argue that the results should come out in either of two ways. First, we might suspect that the pay scales would be below standard in smaller metropolitan areas because of scarcity of jobs and a relatively abundant labor supply. Then too, the cost of living might be lower in those areas, thus keeping down the pay scales. On the other hand, there may be a shortage of qualified workers in the smaller metropolitan areas, leading to higher pay scales.

To see which of these alternatives is supported by the data, we might compare the average pay scale for a sample of 43 smaller metropolitan areas with the standard of 100 for all metropolitan areas. In this case we do not have a specific theory from which to work. Accordingly, we will conduct an exploratory analysis of the data to see which alternative is supported. The following null hypothesis will be tested:

$$H_0 : \mu_s = 100$$

where μ_s is the mean pay rate for clerical office workers in metropolitan areas with populations less than 1 million.

This hypothesis will be tested by the **single sample mean test** in which the sample mean is compared to the expected value of 100 to see whether it is significantly different. Since the study is exploratory, the null hypothesis is set up for a two-tailed test. The two sets of alternatives to the null hypothesis are the following:

$$H_1 : \mu_s > 100$$

$$H_2 : \mu_s < 100$$

It should be noted that, in order to justify use of the one-sample test for means to test the hypothesis, we must have reason to believe that we can satisfy, sufficiently, those assumptions of the technique discussed below in Section 15.1.2.

Table 15.1 presents the pay scale indices for the 43 smaller metropolitan areas in the Bureau of Labor Statistics sample. Because the null hypothesis will be tested with a sample of only 43 cases, the appropriate sampling distribution will be Student's t-distribution (see Chapter 12 on selecting sampling distributions). The

TABLE 15.1   PAY RATES (JANUARY–DECEMBER, 1980) FOR CLERICAL OFFICE WORKERS IN ALL INDUSTRIES FOR A SAMPLE OF METROPOLITAN AREAS WITH LESS THAN 1 MILLION POPULATION[a]

101	94	86	93
87	127	93	100
98	85	123	95
100	85	92	91
97	87	90	98
94	92	93	83
85	94	94	93
90	104	86	89
99	118	95	84
110	96	108	96
106	95	91	

Source: U.S. Department of Labor (1983:295–296, Table 103).
[a] These rates are in terms of percentages of the standard rate of 100.

formula for degrees of freedom for the one-sample case is $df = N - 1$.* If the null hypothesis is tested at the 5% level of significance using a two-tailed test, the t's that bound the critical regions in the tails of the distribution are -2.021 and $+2.021$. These t scores were found by entering Appendix Table B with 42 degrees of freedom and finding entries for the 5% level of significance, two-tailed test.[†]

Using the survey data in Table 15.1, we compute the sample mean and standard error of the mean and solve for t with the following formula:

(15.1)
$$t = \frac{\bar{X} - \mu}{s_{\bar{X}}}$$

where $\bar{X}$ is the sample mean, μ is the population mean, and $s_{\bar{X}}$ is the standard error of the mean based on sample data (see Formula 12.20).

Table 15.2 gives a step-by-step account of testing the null hypothesis and presents the results.

We will focus our attention on the test of the null hypothesis summarized in Table 15.2. This hypothesis states that the pay scales for clerical office workers in the smaller metropolitan areas will not differ, on the average, from the standard scale of 100 (the expected mean).

As was stated above, there are two sets of alternatives to the null hypothesis. The first set is that μ_s is actually greater than 100 ($\mu_s > 100$) and the second is that μ_s is actually less than 100 ($\mu_s < 100$). According to the previously established criteria for testing the null hypothesis, we will reject the null hypothesis in favor of the first set of alternatives if $t \geq +2.021$, or we will reject the null hypothesis in favor of the second set if $t \leq -2.021$. If t does not reach either of these two critical regions, we will fail to reject the null hypothesis.

Using the data in Table 15.1, we compute the sum of scores (ΣX) and the sum of squared scores (ΣX^2). These sums are used to compute the sample mean ($\bar{X}$) and the sample standard deviation (s). Because the sample standard deviation is a biased estimator, we use Table 12.4 to correct it. Then, we compute the sample estimate of the standard error of the mean ($s_{\bar{X}}$).

Substituting into Formula 15.1, we compute t. As you can see from Table 15.2, $t = -2.82$, which is well into the critical region in the left tail of the sampling distribution. Therefore, we reject the null hypothesis in favor of the set of alternatives that the mean pay scale of clerical office workers in the smaller metropolitan areas is lower than the overall population mean of 100.

Let us analyze the process used to test the null hypothesis. We determined that the proper sampling distribution to use is Student's t-distribution, with 42 degrees of freedom. This is so because when N is as small as 43 sample means tend to distribute themselves in a t-distribution rather than a normal one. We decided to do a two-tailed test of the null hypothesis at the 5% level of significance, so we split the critical region evenly between the two tails of the sampling distribution, putting 2.5% in each tail. We found that the critical regions in the tails were bounded by t scores of $+2.021$ on the right and -2.021 on the left. The two-tailed

---

* See Section 12.5.7 for a general explanation of the concept of degrees of freedom.
[†] Since Table B does not have entries for 42 df, we used the entries for 40 df instead, thus erring on the side of conservatism.

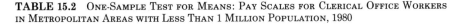

TABLE 15.2   ONE-SAMPLE TEST FOR MEANS: PAY SCALES FOR CLERICAL OFFICE WORKERS IN METROPOLITAN AREAS WITH LESS THAN 1 MILLION POPULATION, 1980

Null Hypothesis

$H_0 : \mu_s = 100$

Alternatives to H_o

$H_1 : \mu_s > 100$ designate this set of alternatives
if $t \geq +2.021$.

$H_2 : \mu_s < 100$ designate this set of alternatives
if $t \leq -2.021$.

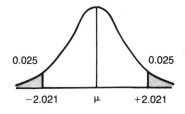

Computations

$\Sigma X = 4117$ and $\Sigma X^2 = 398225$

$$\bar{X} = \frac{\Sigma X}{N} = \frac{4117}{43} = 95.74$$

$$s = \sqrt{\frac{\Sigma X^2 - (\Sigma X)^2/N}{N-1}} = \sqrt{\frac{398225 - (4117)^2/43}{42}} = \sqrt{96.3378} = 9.82$$

(Correcting the standard deviation using Table 12.4 of Chapter 12 we get: $(s)(a_2) = (9.82)(1.00597) = 9.88$.)

$$s_x = \frac{s}{\sqrt{N}} = \frac{9.88}{\sqrt{43}} = 1.51$$

$$t = \frac{\bar{X} - \mu}{s_{\bar{x}}} = \frac{95.74 - 100}{1.51} = \frac{-4.26}{1.51} = -2.82$$

Conclusion

Therefore, rejection of null hypothesis in favor of the set of alternatives $H_2 : \mu_s < 100$ is warranted.

test was selected because this exploratory study did not provide a specific directional expectation.

Given that the area under the curve in the sampling distribution consists of sample means drawn from random samples of the population for which $\mu = 100$ is the parameter, we then agreed to call any sample mean that fell into one of the critical regions a "rare" or unusual sample mean.

The standard error of the mean measures the variability of sample means around the population mean in a sampling distribution. The formula used to compute t is a ratio between sampling error (as measured by the standard error) and the observed difference between the sample mean and the expected value of the population mean. If the observed difference is at least 2.021 times as large as the standard error, then the resulting t score will be in one or the other of the critical regions, and the null hypothesis will be rejected. The probability that the t score will fall into one of the critical regions by chance is 5%. We are thus risking a maximum probability of type I error of 5% by rejecting the null hypothesis whenever the computed t is in one of the critical regions. Since this is an exploratory analysis, a type I error of 5% is considered an acceptable risk. Of course, it is only through replication of the study that the researcher can develop confidence that the finding is not due to chance.

Given that the results of this exploratory analysis hold up under replication, they could serve as the basis for design of further studies to determine the factors that explain why pay scales are lower in the smaller metropolitan areas.

15.1.1 Use of the Normal Approximation to the *t*-distribution

In the example just given, Student's *t*-distribution was used as the sampling distribution to evaluate the null hypothesis. This was necessary because N was small and s was used as an estimate of the unknown σ in the formula for the standard error of the mean.* Note that the use of Student's *t*-distribution is based upon the assumption that the population being sampled is normally distributed.

Given that the normal population assumption is a reasonable one, and the additional condition that the sample N is large, the normal sampling distribution may be used to approximate the *t*-distribution.† In this case, *large* will be defined as 100 or more cases. If the normal approximation is used, the process of analyzing the data is the same, but the standard scores bounding the critical regions are taken from the normal curve instead of one of the *t*-distributions. Formula 15.1 is still used to compute the standard score, but the standard score is designated z instead of t to indicate that the normal curve is the sampling distribution being used.

15.1.2 Assumptions of the One-sample Test for Means

To justify using the one-sample test for means, it is necessary to make the following assumptions: (a) there is interval measurement of the variable, (b) there are independent observations, (c) the population being sampled is normally distributed, and (d) the data being tested are from a simple random sample of the population. If these assumptions are met, the test is a powerful one. If the assumptions are violated, one must be very careful about the kinds of interpretations of data that are made. The test, however, is fairly robust. That is, small violations of assumptions have little impact on conclusions.

15.2 TWO-SAMPLE TEST FOR MEANS

Although the one-sample test for means is used occasionally by sociologists (*e.g.*, Verwaller, 1970), the two-sample test is found much more frequently in the literature.

---

* See Section 12.5.6 for an explanation of the effects of substituting s for σ in computing the standard error.
† An examination of the last entry in the table of the *t*-distribution under the .05 column indicates that the normal curve is the limiting case of Student's *t*-distribution. However, when degrees of freedom reach 60, the *t* score, which cuts 2.5% of the distribution out of each tail of the curve, is already down to 2.00; therefore, it is fast approaching the limit of 1.96.

Basically, the two-sample test involves drawing *two random samples* that differ on some dichotomous variable and then comparing their central tendencies on a second variable to see whether any difference appearing is large enough to be called unusual or rare. For example, one might wish to compare the mean incomes of a sample of college graduates versus the mean incomes of a sample of high school graduates, hoping to see whether there is a systematic difference, and whether such a difference can be generalized to the populations from which the samples were drawn. The solution to the problem is again a comparison of observed differences with expected differences and the key to the comparison is a sampling distribution and a standard score with the appropriate standard error term in its denominator.

15.2.1   The Sampling Distribution of Differences

To arrive at a standard error for comparing two sample means, we must consider the sampling distribution that is generated by repeatedly drawing a pair of simple random samples from a common population and finding the difference between their means. Just for the sake of discussion, assume that we draw pairs of simple random samples of size N from a normally distributed population, continuing until we exhaust all of the unique samples of size N in the population. Furthermore, assume that we compute the differences between all those pairs of means, add up all the difference scores, and divide by the number of difference scores to find the mean difference. Because all the simple random samples are drawn from the same population, and all the sample means are estimates of the same population mean, the differences found between pairs of sample means are merely random differences; the sum of the differences will be 0, and the mean difference will be 0. Furthermore, the differences will distribute themselves normally around the mean difference of 0. Thus, we will have a normal sampling distribution of differences between pairs of sample means. *Of course, we do not actually go through the work of generating the sampling distribution of differences.* Knowing what would happen if we did generate such a distribution gives us the necessary information to use it without having to generate it ourselves.

15.2.1a   Standard Error of the Difference

The measure of variability of the differences between pairs of sample means around their mean difference of 0* is known as the **standard error of the difference**. Its interpretation is the same as that of the standard deviation or the standard error of the mean. That is, approximately 68% of the differences between pairs of sample means will fall within 1 standard error of the expected mean difference, assuming a normal distribution. Let the expression, $\sigma_{\bar{X}_1 - \bar{X}_2}$, be the symbolic name for the standard error of the difference between pairs of sample

---

* The expected value of the mean difference is 0 only if $\mu_1 = \mu_2$. It is possible to test a hypothesis where the mean difference is specified at some other value than 0.

> **BOX 15.1** SOME BASIC CONCEPTS
>
> It will help you considerably to understand this section if you feel com-
> fortable with the following concepts: standard deviation, variance, and
> standard error of the mean. You should know how they differ and how
> they are related. The material covered in Sections 12.5.2 through 12.5.5 in
> Chapter 12 discusses these three concepts in some detail. If you feel the
> need to refresh your memory, go back and review those sections and then
> proceed with this section.

means; then the formula for the standard error of the difference (for independent
samples) is as follows.*

(15.2)
$$\sigma_{\bar{X}_1 - \bar{X}_2} = \sqrt{\frac{\sigma^2}{N_1} + \frac{\sigma^2}{N_2}}$$

or

$$= \sqrt{\sigma^2 \left(\frac{1}{N_1} + \frac{1}{N_2} \right)}$$

where σ^2 is the population variance, N_1 is the number of cases in the first sample,
and N_2 is the number of cases in the second sample.

The problem with this formula is that it requires that the population variance be known. It is very seldom true that the sociologist working with sample
data will know the population variance. Given that two samples have been drawn,
there will be two sample variances, s_1^2 and s_2^2, available, however. Presumably,
the two sample variances are independent, unbiased estimates of the unknown
population variance.

An assumption of the test of significance of difference between means is that
the populations from which the two samples are drawn have the same variance.
This assumption is especially pertinent given our hypothesis that the two samples
were drawn from the same underlying population. Yet, typically, we have two
sample estimates, s_1^2 and s_2^2, of that common population variance σ^2, and the
sample estimates are not identical to each other, presumably differing by chance.

Instead of having two sample variances as separate estimates of the same
parameter, it is efficient to combine them into a more reliable single estimate of the
parameter. The single estimate will be more reliable because it will be based on N_1
+ N_2 cases, whereas the individual estimates are based on N_1 and N_2 cases,
respectively. To combine the two separate sample estimates, s_1^2 and s_2^2, into a single
pooled estimate that we will call simply s^2, we may use the following formula:

(15.3)
$$s^2 = \frac{\Sigma x_1^2 + \Sigma x_2^2}{N_1 + N_2 - 2}$$

---

* The general formula for the standard error of the difference includes a correlation term, but
in the case of independent samples the correlation is 0 and the term disappears.

where

$$\Sigma x_1^2 = (N_1 - 1)s_1^2 = \Sigma(X - \bar{X}_1)^2$$

and

$$\Sigma x_2^2 = (N_2 - 1)s_2^2 = \Sigma(X - \bar{X}_2)^2$$

Once s^2 is found, it can be substituted for the population variance in the formula for the standard error of the difference (Formula 15.2), giving us the following sample estimate of the standard error:

(15.4)
$$s_{\bar{X}_1 - \bar{X}_2} = \sqrt{s^2\left(\frac{1}{N_1} + \frac{1}{N_2}\right)}$$

where $s_{\bar{X}_1 - \bar{X}_2}$ is the sample estimate of the standard error of the difference. It is also known as the *standard error of the difference*.

To pool the sample variances using Formula 15.3, we must have some indication that the variances are homogeneous (*i.e.*, similar enough to be considered to have come from the same population). As we shall see later in this chapter (Section 15.7), it is possible to test for the homogeneity of sample variances. If the test for homogeneity of variances leads us to conclude that they are heterogeneous, they cannot be pooled, and we must use the following formula for the standard error of the difference instead:

(15.5)
$$s_{\bar{X}_1 - \bar{X}_2} = \sqrt{\frac{s_1^2}{N_1} + \frac{s_2^2}{N_2}}$$

Since we substituted s^2 for σ^2 in Formula 15.4 for the standard error of the difference, we are faced with the same situation we encountered with the sampling distribution for the one-sample test. That is, instead of having a constant, σ^2, in the numerator of each fraction in the formula and a variable (N_1 or N_2) in the denominator, we have variables in both the numerators and the denominators.* Thus, we find that the sampling distribution is the symmetrical but platykurtic Student's t-distribution rather than the normal distribution. The appropriate formula for degrees of freedom in the two-sample case is $df = N_1 + N_2 - 2$ if the population standard deviations are equal.† However, when $N_1 + N_2 \geq 100$, the normal sampling distribution is a good approximation to Student's t-distribution.

With Formula 15.4, given the data from two random samples, we can get an estimate of the sampling error for the sampling distribution of differences between pairs of sample means. Armed with this estimate, we can test hypotheses about empirically determined differences between sample means using the following test statistic:

(15.6)
$$t = \frac{(\bar{X}_1 - \bar{X}_2) - (\mu_1 - \mu_2)}{s_{\bar{X}_1 - \bar{X}_2}}$$

---

* Note that $s^2(1/N_1 + 1/N_2) = s^2/N_1 + s^2/N_2$.
† If $\sigma_1 \neq \sigma_2$ (tested by a test for homogeneity), the test is only approximately a t-distribution with df a weighted average of $N_1 - 1$ and $N_2 - 1$.

where $\bar{X}_1$ and $\bar{X}_2$ are the two sample means, μ_1 and μ_2 reflect any expected difference between population means (in this case the expected difference is 0 so that term can be ignored), and $s_{\bar{X}_1 - \bar{X}_2}$ is the standard error of the difference between means.

BOX 15.2   DEGREES OF FREEDOM AGAIN

Notice that the appropriate degrees of freedom formula for this two-sample case is $df = N_1 + N_2 - 2$. The formula for the standard error of the difference essentially combines the standard error of the mean for the first sample with the standard error for the second sample. This can be seen most clearly in Formula 15.5. Actually, the standard errors are squared [thus, $(s_1/\sqrt{N_1})^2 = s_1^2/N_1$], summed, and the square root of their sum is taken. The combination is handled this way because standard errors are not additive, but their squares are (an example of the geometric mean).

Since the standard errors of the two samples are combined in the formula for the standard error of the difference, the degrees of freedom associated with each sample standard error are also combined. That is, the degrees of freedom for the first sample, $N_1 - 1$, are added to the degrees of freedom for the second sample, $N_2 - 1$, to arrive at the proper degrees of freedom for the standard error of the difference; i.e., $(N_1 - 1) + (N_2 - 1) = N_1 + N_2 - 2$.

Perhaps this would be a good point at which to go back to Section 12.5.7 and review the general discussion of the concept of degrees of freedom.

15.2.2   An Example from the Literature

Spaeth (1985) studied job power of a sample of working people in Illinois in 1982. He was interested in the extent to which their job power, as measured by their control over monetary resources and personnel, acted as a determinant of their earnings. He found that control over monetary resources was the strongest determinant of the earnings of men, but monetary control and control over personnel were joint determinants of the earnings of women.

Among the quantitative variables of his study were scores for men and women on the Duncan Socioeconomic Index (a social class measure based on occupation). Spaeth pointed out that previous studies were equivocal on whether men or women have higher scores on this variable (1985:610).

To illustrate how to do a test of differences between means from two independent samples, we will focus on the Socioeconomic Index (SEI) data Spaeth collected to see whether his results shed light on the relative standings of men and women on this variable. The data, based upon a telephone sample, included SEI scores for 311 men and 245 women. Since there is no indication from theory or past studies whether we should expect higher scores for men or women, we will set up a two-tailed test of the null hypothesis as follows:

$$H_0 : \mu_1 = \mu_2$$

where μ_1 is the mean socioeconomic index for men and μ_2 is the mean for women. The hypotheses for the two sets of alternatives to the null hypothesis, accordingly, are the following:

$$H_1 : \mu_1 > \mu_2$$

$$H_2 : \mu_1 < \mu_2$$

Selecting the 5% level of significance for the test, we divide the critical region equally between the tails of the normal sampling distribution. The normal distribution is the appropriate sampling distribution in this case because of the relatively large sizes of the two samples (311 and 245). In fact, any combination of two sample N's (N_1 and N_2) that sums to 100 or more justifies use of the normal sampling distribution.

Examination of Appendix Table A indicates that standard scores of $+1.96$ and -1.96 are the appropriate ones to bound the critical regions for a two-tailed 5% level test of significance using the normal sampling distribution. The question to be answered is whether Spaeth's data warrant rejection of the null hypothesis.

The mean and standard deviation for the SEI scores of the men in the sample were 40.34 and 20.71, respectively; the corresponding mean and standard deviation for women were 43.35 and 17.48. In Table 15.3 we have presented the essential steps in the test of the null hypothesis using these data. The standard score formula for evaluating the data against the normal sampling distribution is as follows (assuming no expected difference between μ_1 and μ_2):

(15.7)
$$z = \frac{\bar{X}_1 - \bar{X}_2}{s_{\bar{X}_1 - \bar{X}_2}}$$

where $\bar{X}_1$ is the sample mean for men and $\bar{X}_2$ is the sample mean for women.

This is another variation of the familiar standard score formula we have seen. If one of the sample means is larger than the other so that it is at least 1.96 times as large as the standard error of the difference, we will reject the null hypothesis in favor of one or the other of its sets of alternatives.*

In using this technique to test the null hypothesis, we must assume that the conditions are such that the statistical model underlying the technique (this model is discussed in Section 15.2.3) is reasonably satisfied.

As you can see from Table 15.3, the computed z-score is -1.82, which is not into the critical region in the left tail of the sampling distribution. Therefore, the null hypothesis is not rejected. Spaeth's data do not support the notion that the SEI scores of men and women differ.

---

* Note that hypotheses are stated in terms of parameters but are tested in terms of statistics.

TABLE 15.3   TWO-SAMPLE TEST FOR MEANS: SOCIOECONOMIC INDICES OF MEN AND WOMEN

Null Hypothesis

$$H_0 : \mu_1 = \mu_2$$

Alternatives to H_o

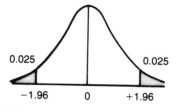

$H_1 : \mu_1 > \mu_2$ designate this set of alternatives
   if $z \geq +1.96$.

$H_2 : \mu_1 < \mu_2$ designate this set of alternatives
   if $z \leq -1.96$.

Computations

Sample 1: Men	Sample 2: Women
$\bar{X}_1 = 40.34$	$\bar{X}_2 = 43.35$
$s_1^2 = 428.90$	$s_2^2 = 305.55$
$\Sigma x_1^2 = 132959$	$\Sigma x_2^2 = 74554.2$
$N_1 = 311$	$N_2 = 245$

Pooled Variance

$$s^2 = \frac{\Sigma x_1^2 + \Sigma x_2^2}{N_1 + N_2 - 2} = \frac{132959 + 74554.2}{311 + 245 - 2} = 374.57$$

Standard Error of Difference

$$s_{\bar{X}_1 - \bar{X}_2} = \sqrt{s^2 \left(\frac{1}{N_1} + \frac{1}{N_2} \right)} = \sqrt{374.57 \left(\frac{1}{311} + \frac{1}{245} \right)}$$

$$= \sqrt{374.57(.007297)}$$

$$= \sqrt{2.733} = 1.65$$

Standard Score

$$z = \frac{\bar{X}_1 - \bar{X}_2}{S_{\bar{X}_1 - \bar{X}_2}} = \frac{40.34 - 43.35}{1.65} = -1.82$$

Conclusion

   Therefore, fail to reject the null hypothesis.

15.2.3   Assumptions of the Test for Differences between Means

The test for differences between means is a powerfull statistical tool for testing null hypotheses. Its use is limited, however, by the following set of assumptions: (a) the level of measurement of the variable being studied is at least interval scale, (b) the observations are independent, (c) the samples are independent, (d) the samples are simple random samples from a normally distributed population, and (e) the sample variances are homogeneous. This last assumption is particularly important because the technique is sensitive to differences in variability as well as central tendency. If variances are not equal or nearly so, it is difficult to establish that statistically significant differences are due to differences in central tendency alone.

15.3   *k*-SAMPLE TEST FOR MEANS OF INDEPENDENT SAMPLES

Thus far we have examined techniques for testing statistically significant differences between a sample mean and a population mean, and between means from two independent samples. It is conceivable that a researcher may be interested in testing for differences between means from more than two independent samples (that is, k samples where k is any number greater than 2). For instance, it may be desirable to compare the means from three samples to see if they differ.

It might seem that the way to handle such a situation would be to take the means in pairs and test the various combinations to see if they differ significantly. Thus, mean 1 would be tested against mean 2, mean 2 against mean 3, and mean 1 against mean 3, using the test described in the previous section. The problem with this approach is that the various tests of pairs of sample means would not all be independent tests, and the resulting probabilities would be overlapping. For example, if mean 1 is tested against mean 2 and mean 2 is tested against mean 3, the test of mean 1 against mean 3 is not independent of the other two tests. In such a case the usual sampling distributions are not applicable.

15.3a   Analysis of Variance

What must be done instead is a test that makes possible a simultaneous comparison of more than two sample means and has a known sampling distribution. The **analysis of variance** is such a test. Analysis of variance (sometimes called ANOVA) is, in effect, an extension of the test for differences between means to cover the case where more than two samples (k samples) are involved. It has a known sampling distribution, the **F-distribution** (see Section 15.3.2). Although the analysis of variance is a test of differences in central tendencies, it is based upon a comparison of variances, thus, the name.

The basis for the analysis is a comparison of the amount of heterogeneity within samples with the amount between samples. It stands to reason that if subjects are exposed to the same conditions as others in their own group, but to conditions that differ from those to which subjects in other groups are exposed (and those conditions make a difference), then subjects within groups will be more alike than subjects from different groups.

This is the kind of reasoning behind the concept of social class. Those who defend the concept assume there is homogeneity within classes and heterogeneity between classes. For example, they claim that upper-class persons think and act similarly, but differently from persons who belong to other classes (Warner *et al.*, 1949). On the other hand, critics of the class concept argue that the concept is unfruitful because there is as much heterogeneity within classes as there is between classes (see Faris, 1954; Svalastoga, 1964).

Perhaps the best way to make the analysis of variance understandable and to explain the procedures used in doing such analysis is to work through an example from the literature.

15.3.1   An Example from the Literature

Coverman (1983) studied the effects of domestic and wage labor on men's and women's wages. She hypothesized that domestic labor (*e.g.*, housework and child care) negatively affects income and is an explanatory factor in the relative inequality in compensation of men and women. Her data came from the *Quality of Employment Survey, 1977: Cross Section* (Quinn and Staines, 1979). From a national probability sample of 1515 respondents, she singled out white, currently employed, currently married subjects for study, 698 men and 240 women. To assess possible class differences in the hypothesized relationship, the subjects were divided into working class and nonworking class men and women. Table 15.4 summarizes that part of Coverman's data we will focus on to illustrate the analysis of variance. Domestic labor and spouse's child care are both expressed in terms of hours per week and wages per week is expressed as weekly earnings in dollars. Because the distributions of wages were skewed, Coverman transformed the raw dollar data into natural logs for purposes of analysis, thus normalizing the distributions. It is these logged wage values we will use for our illustration.

Before we continue with our analysis, however, it should be pointed out that the data in Table 15.4 on domestic labor and spouse's child care are consistent with the hypothesis of the study. Women in the sample spent much more time than men both in domestic labor and in child care and earned less per week in dollars. As a matter of fact, through the use of regression analysis (see Chapter 7), Coverman came to the conclusion that the data supported her hypothesis (1983:634).

Returning now to our analysis, we will examine the logged wage rates for working class and nonworking class men and women to determine whether the observed sample differences for these four categories justify rejecting the null hypothesis and generalizing to the respective populations the samples represent.

In this case we wish to compare four means for differences among them. Accordingly, the null hypothesis we wish to test is that there are no differences among the four means; that is, they are all equal to the mean of the total population. Symbolically, the null hypothesis may be stated as follows:

$$H_0: \mu_1 = \mu_2 = \mu_3 = \mu_4 = \mu$$

where μ_1 is the mean for working class women, μ_2 is the mean for nonworking class women, μ_3 is the mean for working class men, and μ_4 is the mean for nonworking class men. The final mean without the subscript represents the hypothesized mean for the total population.

If the analysis of variance leads us to believe that there is a difference between any, some, or all of the means for the four categories of subjects, we will

TABLE 15.4 MEANS AND STANDARD DEVIATIONS OF SELECTED VARIABLES FOR WHITE, CURRENTLY MARRIED, EMPLOYED WOMEN AND MEN, BY CLASS

	Women		Men	
	Working Class (N = 100)	*Non-Work. Class (N = 109)*	*Working Class (N = 196)*	*Non-Work. Class (N = 397)*
Domestic labor				
Mean	47.42	47.60	27.19	24.06
s.d.	29.72	26.34	19.18	17.16
Spouse's child care				
Mean	11.33	12.41	32.84	33.69
s.d.	13.77	15.80	35.20	35.67
Wages/week				
Mean	161.92	177.20	334.59	425.33
s.d.	102.76	111.65	375.44	260.95
Logarithms of wages/week				
Mean	4.91	4.95	5.62	5.90
s.d.	0.59	0.73	0.59	0.56

Source: Adapted from Table 3 of the original article (Coverman, 1983:631).

reject the null hypothesis in favor of the set of alternatives that two or more of the means are significantly different.

Using the information provided in the Coverman article, we have computed the summary statistics necessary to carry out the analysis of variance, (Table 15.5). In addition to the statistics provided by Coverman, we have computed variances and sums of squares for each of the four categories and have computed the corresponding mean, standard deviation, variance, and sum of squares for all four categories combined. The sums of squares were found using the following formula:

(15.8)
$$\Sigma x^2 = (n - 1)s^2$$

The standard deviations are only included in Table 15.5 because they were

TABLE 15.5 SUMMARY STATISTICS FOR LOGGED WAGES OF WORKING CLASS AND NONWORKING CLASS MEN AND WOMEN

Respondent Category	$\bar{X}$	s	s^2	Σx^2	n
Working class women	4.91	.59	.35	34.65	100
Non-working class women	4.95	.73	.53	57.24	109
Working class men	5.62	.59	.35	68.25	196
Non-working class men	5.90	.56	.31	122.76	397
Total respondents	5.58	.72	.51	412.02	802

included in the original statistics. In doing analysis of variance, it is the *variance* rather than the standard deviation in which we are interested.

The first four variances in Table 15.5 measure the variability of the logged wages within each of the four respondent categories around their respective means. The fifth variance (for total respondents) measures the variability of logged wages of all 802 respondents around the overall mean of 5.58.

15.3.1a   Total Sum of Squares

The *sum of squares* (short for the sum of the squared deviations of scores around their mean) for all respondents (412.02) around the overall mean, interestingly enough, can be partitioned into two separate sources of variation, the variation among logged wages within each of the four respondent categories (called the sum of squares within groups), and the variation of the mean logged wages for each of the four respondent categories around the mean logged wages for total respondents (called the sum of squares between groups). Let Σx_t^2 be the total sum of squares, Σx_w^2 the within-groups sum of squares, and Σx_b^2 the between-groups sum of squares. We can say then that

(15.9)
$$\Sigma x_t^2 = \Sigma x_w^2 + \Sigma x_b^2$$

The *total sum of squares* may be computed by taking the deviation of each logged wage for each of the 802 respondents from the mean logged wage of all those respondents, squaring the deviations, and summing them. That is,

(15.10)
$$\Sigma x_t^2 = \sum_{i=1}^{802} (X_i - \bar{X}_t)^2$$

where X_i is the logged wage of the ith respondent, $\bar{X}_t$ is the mean logged wage for all 802 respondents, and where $\Sigma_{i=1}^{802}$ means the squared deviations are summed for all 802 respondents.

15.3.1b   Within-groups Sum of Squares

The *sum of squares within groups* may be computed by finding the sum of squared deviations of logged wages within each of the respondent categories around their sample means and then summing these sums of squares for all k of the samples. That is, in general,

(15.11)
$$\Sigma x_w^2 = \Sigma x_1^2 + \Sigma x_2^2 + \Sigma x_3^2 + \cdots + \Sigma x_k^2$$

where Σx_1^2 is the sum of squares of the first sample, Σx_2^2 is the sum of squares of the second sample, etc., and k is the number of samples.

For the data in Table 15.5 we find the within-groups sum of squares by substituting into the formula and summing as follows:

$$\Sigma x_w^2 = 34.65 + 57.24 + 68.25 + 122.76 = 282.90$$

Since we could easily find the variances for these data by squaring the standard deviations (which had already been computed), we computed the sums of squares for the four respondent categories using Formula 15.8. However, if we were working with raw data, we would compute the sum of squares with the following computational formula:*

(15.12)
$$\Sigma x^2 = \Sigma X^2 - \frac{(\Sigma X)^2}{N}$$

15.3.1c   Between-groups Sum of Squares

The *sum of squares between groups* is computed by taking the deviation of each sample mean from the total mean, squaring the deviation, multiplying the squared deviation by the sample n, and summing these products for all k samples. That is,

(15.13)
$$\Sigma x_b^2 = \sum_{i=1}^{k} n_i (\bar{X}_i - \bar{X}_t)^2$$

where n_i is the number of cases in the ith sample, $\bar{X}_i$ is the mean of the ith sample, $\bar{X}_t$ is the total mean, and k is the number of samples.

For the data at hand, the sum of squares between groups may be computed as follows:

$$\Sigma x_b^2 = 100(4.91 - 5.58)^2 + 109(4.95 - 5.58)^2 + 196(5.62 - 5.58)^2 + 397(5.90 - 5.58)^2$$

$$= 129.12$$

Since it is true that $\Sigma x_t^2 = \Sigma x_w^2 + \Sigma x_b^2$, we can check the results by adding the sum of squares within groups to the sum of squares between groups to see whether their sum is equal to the total sum of squares:

$$282.9 + 129.12 = 412.02$$

15.3.1d   Within-groups and Between-groups Variances

To test the null hypothesis, it is necessary to compute the variance between groups and the variance within groups. The between-groups variance is then divided by the within-groups variance to get the computed F. The between-groups variance is computed by dividing the between-groups sum of squares by $k - 1$ degrees of freedom where k is the number of samples:

(15.14)
$$s_b^2 = \frac{\Sigma x_b^2}{k - 1}$$

The within-groups variance is computed by dividing the sum of squares within groups by $N - k$ degrees of freedom, where N is the number of scores in the total sample (*i.e.*, $N = n_1 + n_2 + n_3 + \cdots + n_k$):

(15.15)
$$s_w^2 = \frac{\Sigma x_w^2}{N - k}$$

* This formula should be familiar to you; it appeared as Formula 12.14.

Substituting the between-groups sum of squares we computed above (129.12) and $k - 1$ (3) into Formula 15.4 to solve for the between-groups variance, we get:

$$s_b^2 = \frac{129.12}{3} = 43.04$$

Then, substituting the corresponding within-groups sum of squares (282.9) and $N - k$ (798) into Formula 15.5 to solve for the within-groups variance, produces

$$s_w^2 = \frac{282.9}{798} = 0.3545$$

BOX 15.3 WATCH THE SUMS OF SQUARES

The reasoning involved here is not at all complicated. However, if you do not follow the explanation carefully, you may not understand the total sum of squares, the between-groups sum of squares, and the within-groups sum of squares. As you read this material, pay close attention to Table 15.5 and relate the numbers there to the formulas for sums of squares. It may be very helpful to you to take paper and pencil and actually substitute the numbers from Table 15.6 for the symbols in the formulas for the within-groups and the between-groups sums of squares.

15.3.2 The *F*-Distribution

If the null hypothesis that two sample variances are independent estimates of a common population variance ($s_1^2 = s_2^2 = \sigma^2$) is true, then the ratio of the two sample variances has an F sampling distribution. The F-distribution, like the chi-square (χ^2) distribution and the t-distribution, is actually a family of related sampling distributions rather than a single distribution. Consequently, the F table (see Appendix Table D) is set up with degrees of freedom to represent individual distributions from the family. In the table of F, however, there are entries for degrees of freedom across the top of the columns and down the edge of the rows. The df_1 across the top of the table refers to the number of degrees of freedom for the variance in the *numerator* of the ratio, and the df_2 down the first column of the table refers to the number of degrees of freedom for the variance in the *denominator* of the ratio.

The first section of the F table gives values of F that cut 5% of the area out of the right tail of the distribution, and the second section gives values of F that cut 1% of the area out of the right tail. The F-distributions are such that the critical region usually appears only in the right tail. If the two variances that form the ratio are equal, the F will equal 1. The F values in the critical region will always be greater than 1. The larger the F values, the farther they are into the right tail of the distribution. In general, the distributions in the family of F-distributions are skewed to the right. The shape of a particular F-distribution varies with different values of df_1 and df_2.

15.3.3   Test of the Null Hypothesis

To test the null hypothesis for the Coverman data, we must compute F from the following formula:

(15.16)
$$F = \frac{s_b^2}{s_w^2}$$

where s_b^2 is the between-groups variance and s_w^2 is the within-groups variance.

The between-groups variance always appears in the numerator of the F ratio, and the within-groups variance in the denominator. The F ratio essentially compares the observed differences between the sample means of the groups being studied (as measured by the between-groups variance) with an "error" term consisting of measures of variability of scores within the groups being studied (the within-groups variance). Since we are interested in rejecting the null hypothesis, and the null hypothesis can only be rejected when differences appear between the sample means for the various groups studied, we always divide the error term into the between-groups variance. Only when the between-groups variance is *larger* than the within-groups variance will there be a possibility of rejecting the null hypothesis. The F ratio is a value larger than 1 where the groups are quite different (represented by differences in their means), compared to differences between scores within groups.

The computed F ratio is evaluated against the tabled F value that bounds the critical region of the sampling distribution. Suppose that we decide to test the null hypothesis for the Coverman data at the 1% level of significance. To find the particular F value for the 1% level of significance, we enter the second section of Appendix Table D with $df_1 = k - 1$, or 3, and $df_2 = N - k$, or 798. The $k - 1$ degrees of freedom are those for the between-groups variance, and the $N - k$ degrees of freedom are those for the within-groups variance. Since the F table does not have an entry for 798 degrees of freedom, we will use the value for the next smaller entry (120 degrees of freedom), thus erring on the side of conservatism, because the tabled F will be slightly larger than the one for 798 degrees of freedom would have been. Looking in the table with 3 and 120 degrees of freedom, we find that the F value that bounds the 1% critical region is 3.95. Thus, if the computed F value is as large as or larger than 3.95, we will reject the null hypothesis that all the sample means are estimates of a common population mean and will conclude that at least two of them estimate different parameters.

Table 15.6 is a summary table of the analysis of variance for the logged wage data.* Figure 15.1 is a graph of the F-distribution of interest. Note that the computed F value is 121.41, which is quite far into the critical region. Therefore, rejection of the null hypothesis is warranted. We are now willing to say, with some confidence because of the size of the computed F, that at least two of the four means estimate separate parameters.

---

* Note that the layout of this table is traditional and useful because it provides relevant data on F.

TABLE 15.6   SUMMARY TABLE OF ANALYSIS OF VARIANCE FOR LOGGED WAGE DATA

Source of Variation	Σx^2	df	s^2	F
Between groups	129.12	3	43.04	121.41*
Within groups	282.90	798	0.3545	
Total	412.02			

* $P < 0.01$.

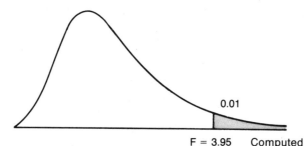

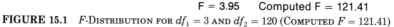

F = 3.95   Computed F = 121.41

FIGURE 15.1 F-DISTRIBUTION FOR $df_1 = 3$ AND $df_2 = 120$ (COMPUTED $F = 121.41$)

15.3.4   Assumptions of One-way Analysis of Variance

The analysis of variance discussed here is the simplest case, called the **one-way analysis of variance**. The technique may be elaborated in a number of ways that we will not discuss (see Edwards, 1960, General References). These variations of analysis of variance naturally make more restrictive assumptions. For example, it is generally necessary to be able to assign subjects randomly to different experimental conditions. This random assignment is most easily handled in a laboratory situation. Because sociologists usually engage in survey research and field studies in which it is not possible to assign subjects randomly to conditions, the one-way analysis of variance is most applicable.

The one-way analysis of variance makes basically the same assumptions as the test for differences between means. It is, as a matter of fact, an extension of that technique to the case in which there are more than two means to compare. Interval measurements are assumed. It is assumed that the observations are independent. The samples should be independent simple random samples from a normally distributed population. And the population variances are assumed to be equal (an assumption that can be tested).

If the sample N's are large, deviation from the assumption of a normally distributed population will not result in serious errors of interpretation. The **assumption of homogeneous variances is crucial**, however, if we are interested in establishing that differences between samples are due to differences in central tendency. The analysis of variance is sensitive to differences in variability as well as differences in central tendency. Therefore, if variances are not homogeneous, it will be difficult to sort out differences in variability from differences in central tendency.

15.3.5   Comparing Specific Means

It should be reiterated that the analysis of variance does not isolate specific differences between means. It tests a whole set of sample means simultaneously, and rejection of the null hypothesis merely indicates that there is a statistically significant difference between at least two means in the set. However, there could be differences between more than two means or even between all the means in the set.

A number of techniques are available that test for significance of difference of specific contrasts between two or more sample means, without having to worry about overlapping probabilities arising from non-independent comparisons [see Edwards, 1960 (General References); Harris, 1985; Norusis, 1986; SPSS Inc., 1986]. Remember that at a given level of significance we would expect a certain number of significant differences by chance. These techniques are based upon joint sampling distributions appropriate for such multiple comparisons.

One of these techniques that is particularly useful because it makes possible tests of all possible combinations of means in a set is **Scheffe's Test for Contrasts**. Furthermore, it can be used to make either *a priori* or *a posteriori* comparisons. We will use this test to examine the specific differences between the four means for logged wages.

There are a number of logical comparisons we might wish to make between the means for working class and nonworking class men and women. In particular, the following comparisons would be informative:

1. Working class women versus nonworking class women ($\bar{X}_1$ vs. $\bar{X}_2$).
2. Working class women versus working class men ($\bar{X}_1$ vs. $\bar{X}_3$).
3. Nonworking class women versus nonworking class men ($\bar{X}_2$ vs. $\bar{X}_4$).
4. Working class men versus nonworking class men ($\bar{X}_3$ vs. $\bar{X}_4$).
5. Working and nonworking class women versus working and nonworking class men ($\bar{X}_1$ and $\bar{X}_2$ vs. $\bar{X}_3$ and $\bar{X}_4$).
6. Working class women and men versus nonworking class women and men ($\bar{X}_1$ and $\bar{X}_3$ vs. $\bar{X}_2$ and $\bar{X}_4$).

To test such comparisons as these, we must come up with a set of null hypotheses that allow us to make linear, independent comparisons between the means of interest. In general, these null hypotheses take the following form:

$$H_0 : c_1\mu_1 + c_2\mu_2 + \cdots + c_k\mu_k = 0$$

where the c's represent weighting coefficients such that **their sum is 0**. For example, in comparing the mean for working class women with that for nonworking class women, the null hypothesis would be

$$H_0 : 1\mu_1 + (-1\mu_2) = 0$$

where $+1$ and -1 are the values of the coefficients used to compare two means. In comparing one mean with the combination of three others the appropriate coefficients would be 1, $-1/3$, $-1/3$, $-1/3$. Thus, the null hypothesis for such a comparison would be

$$H_0:1\mu_1 + (-1/3\mu_2) + (-1/3\mu_3) + (-1/3\mu_4) = 0$$

Notice that the sum of these four coefficients $(1, -1/3, -1/3,$ and $-1/3)$ is equal to 0. When the coefficients involved in such a comparison equal 0 the comparison is said to be orthogonal, or independent. This requirement overcomes the problem of overlapping probabilities mentioned earlier.

To make the six specific comparisons that we decided would be informative earlier, we will test the following six null hypotheses:

1. $H_0:1\mu_1 - 1\mu_2 = 0$
2. $H_0:1\mu_1 - 1\mu_3 = 0$
3. $H_0:1\mu_2 - 1\mu_4 = 0$
4. $H_0:1\mu_3 - 1\mu_4 = 0$
5. $H_0:1/2\mu_1 + 1/2\mu_2 - 1/2\mu_3 - 1/2\mu_4 = 0$
6. $H_0:1/2\mu_1 + 1/2\mu_3 - 1/2\mu_2 - 1/2\mu_4 = 0$

These hypotheses refer to the population means for the four response categories, but we test them using the sample means because they are the ones available.

15.3.5a Scheffe's Test

Scheffe's Test involves computing what amounts to a between-groups variance (which we shall refer to as a contrast variance), dividing it by the within-groups variance, and evaluating it against the F-distribution. Accordingly, we must first compute the required contrast (between-groups) variances for the comparisons we wish to make, then solve for the F's. The following formula is the one used to compute each contrast variance we need:

(15.17)
$$s^2_{contr} = \frac{(\Sigma c_j \bar{X}_j)^2}{\Sigma(c_j^2/n_j)}$$

where s^2_{contr} represents the variance for the desired contrast between means to be examined, c_j represents the weighting coefficient of the jth group, $\bar{X}_j$ represents the mean of the jth group, and n_j represents the number of cases in the jth group.

To compute the contrast variance to compare the mean for working class women $(\bar{X}_1)$ with the mean for nonworking class women $(\bar{X}_2)$, we substitute into Formula 15.17 as follows:

$$S^2_{contr} = (1\bar{X}_1 - 1\bar{X}_2)^2 \bigg/ \left(\frac{1^2}{n_1} + \frac{1^2}{n_2}\right)$$

$$= [(1)4.91 - (1)4.95]^2 \bigg/ \left(\frac{1}{100} + \frac{1}{109}\right)$$

$$= \frac{.0016}{.0192} = .0833$$

In comparing the combination of means for working class and nonworking class women with those for working class and nonworking class men (hypothesis 5 above), we substitute into the formula as follows:

$$S^2_{contr} = (\tfrac{1}{2}\bar{X}_1 + \tfrac{1}{2}\bar{X}_2 - \tfrac{1}{2}\bar{X}_3 - \tfrac{1}{2}\bar{X}_4)^2 \Bigg/ \left(\frac{(\tfrac{1}{2})^2}{n_1} + \frac{(\tfrac{1}{2})^2}{n_2} + \frac{(\tfrac{1}{2})^2}{n_3} + \frac{(\tfrac{1}{2})^2}{n_4} \right)$$

$$= [(\tfrac{1}{2})4.91 + (\tfrac{1}{2})4.95 - (\tfrac{1}{2})5.62 - (\tfrac{1}{2})5.90]^2 \Bigg/ \left(\frac{.25}{100} + \frac{.25}{109} + \frac{.25}{196} + \frac{.25}{397} \right)$$

$$= \frac{.6889}{.0067} = 102.82$$

The within-group variance we use to compute the F's is the same one we used earlier in our analysis of variance (0.3545).

15.3.5b   A Priori and A Posteriori Tests

The critical F value used to test the null hypotheses depends upon whether the comparisons you wish to make between means are planned before (*a priori*) or after (*a posteriori*) the data are collected and the sample means are computed. If the comparisons are planned in advance and the null hypotheses are formulated before the data are collected, then the appropriate F to bound the critical region of the sampling distribution is the one with 1 and $N - k$ degrees of freedom. For example, for the data we are analyzing the critical F for the 5% level of significance would be 3.92 (where we would use $df_2 = 120$ instead of 798 because the latter entry is not included in Appendix Table D).

If, on the other hand, we decided which comparisons to make after doing the analysis of variance and examining the sample means (*a posteriori*), the appropriate F to bound the critical region would be the one equivalent to $k - 1$ times the tabled F for $k - 1$ and $N - k$ degrees of freedom. In this case, that F would be $3(2.68) = 8.04$ for the 5% level of significance.

The test of the null hypothesis is much more conservative when the comparison is *a posteriori* than when it is *a priori*. It is reasoned that the *a posteriori* test should be more conservative because the researcher can examine the sample means before choosing which comparisons to make. If Scheffe's Test is used *a priori*, it is not really necessary to do analysis of variance first. The results of the test for contrasts are equally legitimate with or without a preliminary test of the null hypothesis using analysis of variance and they initially provide more information.

To test the contrasts between means for the logged wage data, we will assume the comparison to be *a posteriori* and use 8.04 [3(2.68)] for the critical F value. Table 15.7 summarizes the results of the analysis using Scheffe's Test for Contrasts. Notice that the difference between mean 1 and mean 2 was the only one that was not significant. In the other five, each null hypothesis was rejected (hypotheses 2, 3, 4, and 5 at the 1% and 6 at the 5% level). Table 15.7 reports the associated probability of each F-test.

TABLE 15.7 SUMMARY TABLE OF SCHEFFE'S TEST FOR CONTRASTS OF LOGGED WAGES

Means Compared	s^2_{contr}	F	Pr
$\bar{X}_1$ vs. $\bar{X}_2$	0.08	0.23	—
$\bar{X}_1$ vs. $\bar{X}_3$	33.38	94.16	<0.01
$\bar{X}_2$ vs. $\bar{X}_4$	77.14	217.60	<0.01
$\bar{X}_3$ vs. $\bar{X}_4$	10.29	29.03	<0.01
$\bar{X}_1$ & $\bar{X}_2$ vs. $\bar{X}_3$ & $\bar{X}_4$	102.82	290.04	<0.01
$\bar{X}_1$ & $\bar{X}_3$ vs. $\bar{X}_2$ & $\bar{X}_4$	3.82	10.78	<0.05
Within-group (error)	0.3545		

The results reported in Table 15.7 appear to provide evidence in support of the following conclusions:

1. The logged wages of working class women do not differ from those of nonworking class women.

2. The logged wages of working class men are higher than those of working class women.

3. The logged wages of nonworking class men are higher than those of nonworking class women.

4. The logged wages of nonworking class men are higher than those of working class men.

5. The logged wages of working class and nonworking class men, combined, are higher than those of working class and nonworking class women, combined.

6. The logged wages of nonworking class men and women combined are higher than those of working class men and women combined.

Diagram 15.1 (p. 564) summarizes the essential features of tests of hypotheses about means for the one-sample, two-sample, and k-sample cases. Study the diagram and refer back to our discussions of these techniques for more detailed information.

15.4 ONE-SAMPLE TEST FOR PROPORTIONS

Much of the data with which sociologists work lend themselves most readily to presentation as proportions or percentages.* The occasion arises, therefore, to test hypotheses about proportions. In this section we will examine procedures used to test a sample proportion against a known or theoretical population proportion to determine whether the sample proportion is a likely estimate of it. In the following

---

* A percentage is easily converted to a proportion by dividing it by 100. Since proportions are more convenient to use for computations, it is common to do tests of hypotheses for proportions rather than percentages.

section (15.5) we will examine procedures used to test for differences between proportions from two independent samples.

Hogan and Kitagawa (1985) conducted a study of the fertility of black female adolescents in the city of Chicago. They examined various ethnographic explanations (such as less defined and enforced norms of sexual behavior in ghetto areas) of the sexual behavior of a sample of over 1,000 black females between the ages of 13 and 19.

One of the hypotheses they investigated was that "girls who see their sisters become teenage parents are more likely to accept single parenthood as a way to achieve adult status" (Hogan and Kitagawa, 1985:825). Although Hogan and Kitagawa used a sophisticated, multivariate analysis to test the hypotheses of their study, we will use their data to illustrate how the hypothesis above could be tested with a simple one-sample test for proportions.

Of the 1,071 girls in their Chicago sample, 16.9% had experienced at least one pregnancy. We will compare this overall proportion with the proportion of girls with a sister who was a teenage mother and who had, themselves, been pregnant. Since the authors suggest a tendency for these girls to be pregnant more often than the girls in general, the null hypothesis will be as follows:

$$H_0 : P_s \leq P_u$$

where P_s is the population proportion for girls with a sister who was a teenage mother and who had, themselves, been pregnant and P_u is the proportion of all girls in the sample who had been pregnant.

The set of alternatives to the null hypothesis, then, is the following:

$$H_1 : P_s > P_u$$

Since 16.9% of all of the girls in the sample had been pregnant, P_u will be .169 for this test of the null hypothesis. There were 300 girls in the sample who had one or more sisters who were teenage mothers. Since $(.169)(300) = 50.7$, the rule of thumb applies,* and we can use the normal sampling distribution as an approximation of the binomial sampling distribution of the sample proportion p. If we do a one-tailed test of the null hypothesis at the 0.05 level of significance, the z-score that bounds the critical region will be $+1.64$. Consequently, if the standard score computed from the sample data is equal to or greater than $+1.64$, we will reject the null hypothesis in favor of its set of alternatives.

As was pointed out in Section 13.2.2, the standard error of a proportion with P_u known (as it is in this case) is given by

(15.18)
$$\sigma_p = \sqrt{\frac{P_u Q_u}{N}}$$

Given the standard error, a standard score can be computed from

(15.19)
$$z = \frac{p - P_u}{\sigma_p}$$

---

* Both NP and NQ are equal to or greater than 5 (see Section 12.3.3).

TABLE 15.8   ONE-SAMPLE TEST FOR PROPORTIONS: PREGNANT BLACK ADOLESCENTS WITH SISTERS WHO WERE TEENAGE MOTHERS

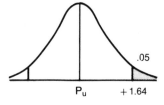

Null Hypothesis

$$H_0 : P_s \leqq P_u$$

Set of Alternatives to H_o

$H_1 : P_s > P_u$ designate this set of alternatives
if $z \geqq +1.64$.

Computations

$$P_u = 0.169$$
$$Q_u = 0.831$$
$$p = 0.27$$
$$N = 300$$

$$\sigma_p = \sqrt{\frac{P_u Q_u}{N}} = \sqrt{\frac{(.169)(.831)}{300}} = \sqrt{.0004681} = .0216$$

$$z = \frac{p - P_u}{\sigma_p} = \frac{.27 - .169}{.0216} = \frac{.101}{.0216} = +4.68$$

Conclusion

Therefore, reject the null hypothesis in favor of the set of alternatives $H_1 : P_s > P_u$.

Assuming that all the necessary conditions of the statistical model can be met (see the next section), we can use Formula 15.19 to test the null hypothesis. All we need do is substitute $P_u = 0.169$, $Q_u = 1 - P_u = 0.831$, and $N = 300$ into Formula 15.18 to solve for σ_p. Then p (the sample proportion), P_u, and σ_p can be substituted into Formula 15.19 to compute z. Hogan and Kitagawa found that 27% of the 300 girls in the sample with sisters who were teenage mothers had, themselves, been pregnant; therefore, the sample proportion (p) is .27 in this case. Table 15.8 traces the process of testing the null hypothesis step by step.

As can be seen from the table, $\sigma_p = .0216$. The computed standard score comes to +4.68, which is well into the critical region beyond +1.64, so rejection of the null hypothesis is warranted. As a matter of fact, the probability of getting z as large as 4.68 by chance (given the assumptions of the test) is less than .0000034 (see the normal curve table, Appendix Table C). It can be concluded with some degree of confidence, therefore, that the rate of pregnancy for adolescent black girls in Chicago with sisters who were teenage mothers was higher at the time of study than it was for all adolescent black girls in the city.

15.4a   Assumptions of the One-sample Test for Proportions

The one-sample test for proportions is really just another version of the test for frequencies that makes use of the binomial probability distribution. Therefore, the assumptions are basically the same. It is assumed that the observations are independent and that the data constitute a simple random sample from the population. If it is possible to meet the rule of thumb that NP and NQ are both equal to or greater than 5, then the binomial sampling distribution can be approximated by the normal distribution.

It is not necessary to assume that the population is normally distributed. As a matter of fact, the shape of the population distribution need not be specified. Although most appropriate for nominal data, the test can also be applied to ordinal or interval data. At the ordinal and interval levels the test wastes information since it does not take account of order or interval size.

15.5   TWO-SAMPLE TEST FOR PROPORTIONS

Sociologists have been concerned for some time with assessing the consequences of matriarchal family structures in black families for children. It is sometimes argued that the dominance of black mothers leads to poor academic achievement by black boys and increased emphasis upon the education of black girls. Kandel conducted a study of a sample of adolescents (black and white) from a large working-class urban high school to test some of these notions (Kandel, 1971). She examined interaction within comparable black and white families, both intact and mother-headed, and examined the consequences for the educational aspirations of adolescents and parents.

Among other things, she compared mothers' educational aspirations for their boys and girls by race and type of family. She reports proportions (percentages) of black mothers and white mothers from intact and mother-headed families who aspire to college educations for their boys and girls. To determine whether the differences between levels of aspirations of black and white mothers are significant, we will test her data for significance of the difference between proportions from independent samples. To justify the use of this test, the assumptions related to it must be met reasonably well (see Section 15.5.3). In making such a test, we must first determine the appropriate sampling distribution and the measure of sampling error.

15.5.1   The Sampling Distribution of Differences between Proportions

If we were actually to generate a sampling distribution of differences between proportions, we would do it by drawing pairs of simple random samples from a population, computing a proportion for each sample, and taking the difference between each pair of proportions. If we were to do so until all pairs of unique samples of size N were exhausted, and then sum all the differences, the differences would sum to 0 and the mean difference would be 0. This is so because any differences between pairs of proportions, if from simple random samples of the same population, are nothing more than random differences. If we were to plot all of the differences on a curve, they would distribute themselves approximately normally about the mean difference of 0, with a standard error equal to the following:

(15.20)
$$\sigma_{p_1 - p_2} = \sqrt{PQ\left(\frac{1}{N_1} + \frac{1}{N_2}\right)}$$

where P is the population proportion, $Q = 1 - P$, N_1 is the number of cases in the first sample, and N_2 is the number of cases in the second sample.

This standard error is known as the **standard error of the difference between proportions**.

Unfortunately, when we only have data from two samples, P and Q are unknown parameters. The null hypothesis states that $P_1 = P_2 = P$. If the sample proportions, p_1 and p_2, are estimates of P_1 and P_2, then by the null hypothesis they are estimates of the common parameter, P. Thus, we have two independent estimates of P, one based on N_1 cases and the other on N_2 cases. We can get a more reliable common estimate of P by pooling p_1 and p_2 using the following formula:

(15.21)
$$p = \frac{N_1 p_1 + N_2 p_2}{N_1 + N_2}$$

We justify this pooling procedure by assuming, for the time being, that the null hypothesis is true.

The pooled estimate of Q can be arrived at by subtraction, since $q = 1 - p$. These pooled estimates can be substituted into the formula for the standard error of the difference in place of the unknown parameters, P and Q. The resulting standard error will be an estimate of $\sigma_{p_1 - p_2}$. Its formula is as follows:

(15.22)
$$s_{p_1 - p_2} = \sqrt{pq\left(\frac{1}{N_1} + \frac{1}{N_2}\right)}$$

This standard error has the same interpretation as the other standard errors we have talked about so far. That is, in a *normal* sampling distribution, approximately 68% of the differences between pairs of sample proportions will be within 1 standard error of the mean difference of 0.

With the use of Formula 15.22 and the knowledge that the sampling distribution of differences between proportions will be approximately normal, we are ready to test the null hypothesis.

15.5.2   Testing the Null Hypothesis

We will return to the Kandel study now and use her data to illustrate how to test for significance of difference between proportions. We will compare the educational aspirations that black and white mothers of intact families had for their boys and girls.

According to Kandel, empirical evidence on this question is sparse, and past definitions of relevant concepts have been ambiguous. In a sense, then, her study was an exploration of this and related questions. Since Kandel does not explicitly set up one-tailed tests of the null hypotheses, we have set up two-tailed tests at the .05 level of significance. Table 15.9 presents the essential steps in the test of the null hypothesis for boys, and Table 15.10 does the same for girls.

TABLE 15.9 TWO-SAMPLE TEST FOR PROPORTIONS: EDUCATIONAL ASPIRATIONS THAT MOTHERS OF INTACT FAMILIES HELD FOR THEIR SONS

Null Hypothesis

$$H_0: P_1 = P_2 = P$$

Sets of Alternatives to H_o

$H_1: P_1 > P_2$ designate this set of alternatives
     if $z \geq +1.96$.

$H_2: P_1 < P_2$ designate this set of alternatives
     if $z \leq -1.96$.

.025   .025
-1.96 0 $+1.96$

Computations

	Sample 1 White Mothers	Sample 2 Black Mothers
	$p_1 = .78$	$p_2 = .90$
	$q_1 = .22$	$q_2 = .10$
	$N_1 = 276$	$N_2 = 49$

Pooled Estimates of P and Q

$$p = \frac{N_1 p_1 + N_2 p_2}{N_1 + N_2} = \frac{(276)(.78) + (49)(.90)}{276 + 49} = \frac{259}{325} = .80$$

$$q = 1 - p = 1 - .80 = .20$$

Standard Error of the Difference

$$s_{p_1 - p_2} = \sqrt{pq\left(\frac{1}{N_1} + \frac{1}{N_2}\right)} = \sqrt{(.80)(.20)\left(\frac{1}{276} + \frac{1}{49}\right)} = \sqrt{(.16)(.024)} = \sqrt{.0038} = .06$$

Standard Score

$$z = \frac{p_1 - p_2}{s_{p_1 - p_2}} = \frac{.78 - .90}{.06} = \frac{-.12}{.06} = -2.00$$

Conclusion

Therefore, reject the null hypothesis in favor of the set of alternatives $H_2: P_1 < P_2$.

Source: Data from Kandel (1971).

To compute the standard score from the sample data we use the following formula:

(15.23)
$$z = \frac{p_1 - p_2 - (P_1 - P_2)}{s_{p_1 - p_2}}$$

According to the null hypothesis $(P_1 - P_2) = 0$; therefore, the expression drops out of the formula and it simplifies to

(15.24)
$$z = \frac{p_1 - p_2}{s_{p_1 - p_2}}$$

TABLE 15.10   TWO-SAMPLE TEST FOR PROPORTIONS: EDUCATIONAL ASPIRATIONS THAT MOTHERS OF INTACT FAMILIES HELD FOR THEIR DAUGHTERS

Null Hypothesis

$$H_0 : P_1 = P_2 = P$$

Sets of Alternatives to H_o

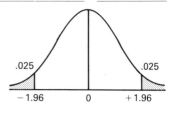

$H_1 : P_1 > P_2$ designate this set of alternatives
if $z \geq +1.96$.

$H_2 : P_1 < P_2$ designate this set of alternatives
if $z \leq -1.96$.

Computations

	Sample 1 White Mothers	Sample 2 Black Mothers
	$p_1 = .50$	$p_2 = .86$
	$q_1 = .50$	$q_2 = .14$
	$N_1 = 290$	$N_2 = 52$

Pooled Estimates of P and Q

$$p = \frac{N_1 p_1 + N_2 p_2}{N_1 + N_2} = \frac{(290)(.50) + (52)(.86)}{290 + 52} = \frac{190}{342} = .56$$

$$q = 1 - p = 1 - .56 = .44$$

Standard Error of the Difference

$$s_{p_1 - p_2} = \sqrt{pq\left(\frac{1}{N_1} + \frac{1}{N_2}\right)} = \sqrt{(.56)(.44)\left(\frac{1}{290} + \frac{1}{52}\right)} = \sqrt{(.246)(.023)} = \sqrt{.0057} = .075$$

Standard Score

$$z = \frac{p_1 - p_2}{s_{p_1 - p_2}} = \frac{.50 - .86}{.075} = \frac{-.36}{.075} = -4.80$$

Conclusion

Therefore, reject the null hypothesis in favor of the set of alternatives $H_2 : P_1 < P_2$.

Source: Data from Kandel (1971).

As can be seen from an examination of Tables 15.9 and 15.10, the null hypotheses with respect to educational aspirations for both boys and girls were rejected in favor of their sets of alternatives.* Black mothers had higher aspirations for their children than white mothers. Furthermore, the difference

* When the computed standard score falls near the boundary of the critical region, but within the region, it may be advisable to correct for continuity. This is suggested because the normal distribution is a continuous distribution, whereas proportions are discrete. The correction is accomplished by recomputing the proportions used in the numerator of the z-score formula.

(continues)

between white and black mothers was much more pronounced for girls than for boys. Whereas 78% of white mothers and 90% of black mothers aspired to college educations for their sons, only 50% of white mothers as compared to 86% of black mothers aspired to college educations for their daughters.

15.5.3 Assumptions of the Test for Differences between Proportions

To justify use of the test for differences between proportions, we must assume that the observations are independent and that the samples are independent. It is also assumed that the samples are simple random samples of the population. The assumption that the null hypothesis is true serves as a rationale for getting a pooled estimate of the population proportion and using it to compute the standard error.

It is not necessary that the population being sampled be normally distributed; in fact, it is not necessary to specify its form at all. The technique is most appropriately used for nominal data, but it can also be used for ordinal or interval data, although it is wasteful of information for such data.

Diagram 15.2 (p. 569) summarizes the essential features of the one-sample and two-sample tests for proportions.

15.6 ONE-SAMPLE TEST FOR VARIANCES

It is sometimes of interest to test a sample variance against a known or hypothesized population variance to see whether it is an estimate of the given population variance. To illustrate how this kind of problem is handled, we will use data on accidental deaths during 1981 from a random sample of 25 of the United States.[†] We will compare the variability of these sample data with the variability of deaths from all causes to see whether deaths from accidents are more or less variable from state to state. Table 15.11 lists the 25 sample states and their accidental death rates for 1981. The variance for these death rates is 191.11. The question is whether this sample variance can be seen as an estimate of the

---

(Footnote continues from p. 536)
Since the proportions are arrived at by f/N, the proportions are converted back to these fractions. Then the f for the smaller of the two proportions is increased by .5 and the f for the larger of the two is decreased by .5. The proportions are recomputed with these corrections. The effect will be to reduce the numerator of the z-score formula slightly. The corrections treat the frequencies as though they were continuous rather than discrete scores.

[†] The sample data were d: . from U.S. Bureau of the Census (1984:72, Table 107; and 76, Table 112). We could have used the data for all of the states, but we drew a random sample of 25 states in order to illustrate the process of testing a sample variance against a known or theoretical population variance. Note that in this case we drew a random sample without replacement, so we need to recognize the fact that we do not have a simple random sample when we interpret the results. The decision to sample without replacement in this instance was based upon substantive rather than statistical grounds. It was considered desirable for the problem being investigated to get the widest possible representation of states without resorting to a total enumeration of the population.

TABLE 15.11 ACCIDENTAL DEATH RATES FOR RANDOM SAMPLE OF 25 STATES, 1981[a]

State	Death Rate
South Carolina	53.6
Maine	40.8
Utah	43.4
Alaska	90.1
Nebraska	49.3
New Mexico	56.7
Nevada	58.3
Ohio	34.9
Alabama	55.0
Connecticut	34.3
Iowa	44.1
Missouri	47.4
Wyoming	82.7
Oregon	49.3
California	43.0
Georgia	52.4
Wisconsin	37.2
Michigan	36.8
Idaho	60.2
Texas	53.2
Arkansas	47.2
North Carolina	50.5
Hawaii	32.4
New York	32.8
West Virginia	48.9

Source: Based on data from U.S. Bureau of Census (1984:76, Table 112).
[a] Rates per 100,000 estimated midyear population in each area.

population variance for the overall death rates. To answer this question, we will set up a test of the following null hypothesis:

$$H_0 : \sigma_A^2 = \sigma_D^2 = 59.66$$

where σ_A^2 is the population variance for accidental death rates and σ_D^2 is the population variance for overall death rates.* There are two sets of alternatives to this null hypothesis:

$$H_1 : \sigma_A^2 > \sigma_D^2$$

and

$$H_2 : \sigma_A^2 < \sigma_D^2$$

* Since the overall death rates were not comparable to the rates for accidental deaths in units of measurement, it was necessary to transform the overall rates into comparable units and compute a transformed σ^2. The conversion was accomplished by using the ratio of the sample mean for accidental deaths to the population mean for all causes of deaths ($\bar{X}_A / \mu_D = .0589$) as a conversion factor. Therefore, the σ^2 for the null hypothesis represents what the population variance for overall death rates would be, if those rates were in units comparable to those for accidental deaths (*i.e.*, number per 100,000 population).

or, more simply,

$$H_1 : \sigma_A^2 \neq \sigma_D^2$$

Assuming the necessary conditions have been met (see Section 15.6a), the null hypothesis may be tested using the following formula:

(15.25)
$$\chi^2 = \frac{(N-1)s^2}{\sigma^2}$$

The value in Formula 15.25 has a chi-square sampling distribution with $df = N - 1$ for random samples of size N drawn from a normal population with variance σ^2. If we use the .05 level of significance for the test, the χ^2 values that bound the critical regions (with $df = 24$) will be 39.364 for the set of alternatives H_1 (on the right tail of the χ^2 distribution) and 12.401 for the set of alternatives H_2 (on the left tail). See Table C in the Appendix.

If accidental death rates were comparable in variability to the overall death rates, the population variance for accidental death rates would be 59.66 (the transformed rate for overall deaths). We will use this figure for the hypothesized population variance and see whether the observed sample variance is significantly different. Table 15.12 summarizes the test of the null hypothesis.

TABLE 15.12 ONE-SAMPLE TEST FOR VARIANCES: ACCIDENTAL DEATH RATES FROM A SAMPLE OF 25 STATES[a]

Null Hypothesis

$H_0 : \sigma_A^2 = \sigma_D^2$

Sets of Alternatives to H_o

$H_1 : \sigma_A^2 > \sigma_D^2$ designate this set of alternatives if $\chi^2 \geq 39.364$.

$H_2 : \sigma_A^2 < \sigma_D^2$ designate this set of alternatives if $\chi^2 \leq 12.401$.

.025   .025

12.401   df $-$ 2 = 22   39.364

Computations

$N = 25$

$df = N - 1 = 24$

$\sigma_D^2 = 59.66$, the hypothesized population variance based on a transformation of the population variance for overall death rates.

$s^2 = 191.11$, the sample variance of accidental death rates for the 25 randomly drawn states.

$$\chi^2 = \frac{(N-1)s^2}{\sigma^2} = \frac{(24)191.11}{59.66} = 76.88$$

Conclusion

Therefore, reject the null hypothesis at the .05 level of significance in favor of the set of alternatives $H_1 : \sigma_A^2 > \sigma_D^2$.

---

[a] Note that in the pictured curve of the chi-square distribution, the point at the mode of the distribution is marked $df - 2 = 22$. As was mentioned in Chapter 12 (Section 12.6a), with the exception of the curve for $df = 1$, the mode of a chi-square distribution will be at $df - 2$.

As was mentioned previously, the sample variance for the 25 states is 191.11. Substituting this and the population variance (59.66) into the χ^2 formula, we get

$$\chi^2 = \frac{(N-1)s^2}{\sigma^2} = \frac{(24)(191.11)}{59.66} = 76.88$$

Since this χ^2 is into the critical region in the right tail of the distribution, we reject the null hypothesis in favor of the set of alternatives H_1, which implies that the sample variance estimates a population variance larger than that for the overall death rates. This result suggests the need for further study to determine why accidental death rates are more variable.

15.6a   Assumptions of the One-sample Test for Variances

The one sample test for variances is based on the following set of assumptions: (a) the level of measurement is at least interval scale, (b) the population is normally distributed, (c) the data are from a simple random sample of that population, and, thus, (d) the observations are independent.

Especially important is the assumption that the population is normally distributed. Variances are sensitive to deviations from normality and serious errors of interpretation can result. The assumption is less crucial, however, as N gets larger. Therefore, if there is some doubt about the normality of a population distribution, it is wise to take as large a sample as is feasible.

15.7   TWO-SAMPLE TEST FOR VARIANCES

With growing concern over the population explosion, it is of special interest to attempt to isolate those factors that influence trends in birth rates. Some of these have already been identified, though they are not necessarily fully understood. For example, birth rates usually tend to decline as countries become industrialized and urbanized. This finding may be attributed, in part, to such factors as changing attitudes about the desirability of having a large family as opposed to being able to afford a higher standard of living. Children, who were assets in the rural setting as cheap sources of labor, have become financial liabilities in the urban setting.

Also related to level of the birth rate is the level of education of a country's population. Among other things, it appears that more education leads to more knowledge about and more use of birth control techniques. If literacy rate is taken as an index of education, it stands to reason that countries whose populations are more literate should have lower birth rates, on the average, than countries with less literate populations.

Table 15.13 presents data on the birth rates (live births per 1,000 population) of a random sample of 20 countries in which a majority of the population is literate

TABLE 15.13 LIVE BIRTHS (PER 1,000 POPULATION) FOR RANDOM SAMPLES OF HIGH-LITERACY AND LOW-LITERACY COUNTRIES

High-Literacy Countries		Low-Literacy Countries	
Country	Birth Rate	Country	Birth Rate
Tanzania	50.9	Algeria	47.0
Belize	38.8	Botswana	50.5
Dominican Republic	34.6	Cape Verde	35.5
Trinidad & Tobago	25.7	Congo	44.7
Equador	41.6	Ethiopia	49.3
Surinam	28.0	Ghana	47.1
Uruguay	18.3	Liberia	49.8
Venezuela	36.9	Madagascar	46.0
Cyprus	19.6	Mali	43.2
Hong Kong	15.4	Mauritania	50.0
Korea	25.3	Togo	45.5
Macao	15.0	Upper Volta	48.1
Philippines	33.9	Guatemala	41.2
Singapore	17.3	Bahrain	34.4
Turkey	39.6	Bangladesh	47.2
Italy	10.9	Iran	42.5
Poland	19.7	Sabah	34.8
Hungary	11.9	Pakistan	42.8
San Marino	10.8	Nepal	44.6
New Caledonia	25.2	Morocco	45.4
$\bar{X}_1 = 25.97$		$\bar{X}_2 = 44.48$	
$N = 20$		$N = 20$	

and 20 countries in which a majority is illiterate.* For convenience, we will refer to these two sets of countries as "high-literacy" and "low-literacy" countries. As you can see from the means of the two samples, birth rates are considerably higher in low-literacy countries than in high-literacy countries. If put to a formal test (two-sample test for means from independent samples, Section 15.2), it will be found that the difference between the sample means is statistically significant. An assumption of this test is that the variances are equal. This assumption will be tested below.

Another significant question that might be asked about high-literacy versus low-literacy countries is whether there are differences in between-country birth-rate variability. If we were to find that the variability in birth rates was either larger or smaller for high-literacy countries than for low-literacy countries,

* These samples were drawn from all of those countries for which data were available; therefore, the population of countries is limited to those that keep records. Sources for these rates were United Nations *1982 Statistical Yearbook* (New York, 1985). Here again the random samples were drawn without replacement. The reason for using this procedure is basically the same as it was for the previous example. Because of the nature of the problem and the availability of data, it was considered desirable to get as broad a representation of countries as possible in the samples.

that would tell us that there is something about the social dynamics of changes in birth rates that requires further explanation. For example, if it so happens that the variance for birth rates is larger for low-literacy countries than it is for high-literacy countries, that would seem to indicate that there are several diverse combinations of variables that influence birth rates in such countries. On the other hand, if the variance turns out to be larger for high-literacy countries, that may indicate that the diverse combination of influences is present in high-literacy countries, whereas conditions influencing birth rates are more uniform in low-literacy countries. Either finding would lead to fruitful channels of further research. A statistical reason for this test is to check the assumption of the test of means that the variances of populations from which the two samples were drawn are equal.

To determine whether the two groups of countries *do* differ in variability of birth rates and also to demonstrate the test of significance of difference between the two variances, we will test the following null hypothesis:

$$H_0: \sigma_1^2 = \sigma_2^2$$

where σ_1^2 is the variance for birth rates of countries in which a majority of the population is literate and σ_2^2 is the variance for birth rates of countries in which a majority of the population is illiterate. The test will be made by comparing the two sample variances, s_1^2 and s_2^2, to see whether they are likely to be estimates of a common population variance σ^2 (the variance for birth rates of all countries) or estimates of two separate parameters σ_1^2 and σ_2^2. The two sets of alternatives to the null hypothesis that are consistent with the notion that there are two separate population variances are as follows:

$$H_1: \sigma_1^2 > \sigma_2^2$$

and

$$H_2: \sigma_1^2 < \sigma_2^2$$

or

$$H_1: \sigma_1^2 \neq \sigma_2^2$$

We have chosen to use a two-tailed test of the null hypothesis because our investigation of the variability of birth rates is an exploratory one, and we do not have a theory that suggests which variance should be larger. We also choose to test the null hypothesis at the .01 level of significance.

As we found earlier in this chapter (Section 15.3.2), the F-distribution is the appropriate sampling distribution to test a null hypothesis about two variances. To find the F that bounds the critical region, we enter the F table with $df_1 = N_1 - 1$ and $df_2 = N_2 - 1$, where N_1 is the sample size of the sample with the larger variance, and N_2 is the sample size of the sample with the smaller variance. This is so because in a two-tailed test of this kind, the F ratio is always computed by dividing the smaller variance into the larger one.* For a one-tailed test of sig-

---

* You should recall that in the analysis of variance, the between-group variance was always put in the numerator and the within-group or error variance in the denominator.

nificance, the variance that the theory predicts will be larger is put in the numerator of the F ratio. In either case, the critical region for the F-distribution is put in the right tail of the distribution, and the computed F approaches the critical region as it gets larger than 1. Therefore, the computed F will only be significant if the variance in the numerator of the F ratio is sufficiently larger than the variance in the denominator so that the ratio will reach the critical region.

Using the data from Table 15.13 to compute sample variances for the birth rates of high-literacy and low-literacy countries, we get

(15.26)
$$s^2 = \frac{\Sigma X^2 - [(\Sigma X)^2/N]}{N-1}$$

For high-literacy countries,

$$s_1^2 = \frac{16068.22 - [(519.4)^2/20]}{19}$$

$$= 135.76$$

For low-literacy countries,

$$s_2^2 = \frac{40020.72 - [(889.6)^2/20]}{19}$$

$$= 23.75$$

Assuming that we can meet the conditions necessary for performing an *F test* for two-sample variances (see Section 15.7a), we may now substitute the two-sample variances into the F ratio and solve for F as follows:

$$F = \frac{s_1^2}{s_2^2} = \frac{135.76}{23.75}$$

$$= 5.72$$

To find the F that bounds the 1% critical region, we enter Table D (in the Appendix) with $df_1 = N_1 - 1 = 20 - 1 = 19$ and $df_2 = N_2 - 1 = 20 - 1 = 19$. However, since the table does not have an entry for $df_1 = 19$, we will use the entry for $df_1 = 15$, thereby erring on the side of conservatism. The value for $df = 15$ is slightly larger than it would be for $df = 19$; thus the computed value would have to be slightly larger to reach the critical region. We find from the table that the F bounding the critical region is 3.15; therefore, our computed F of 5.12 is well into the critical region and rejection of the null hypothesis is warranted in favor of the set of alternatives $H_1 : \sigma_1^2 > \sigma_2^2$. The steps in the test of this null hypothesis are summarized in Table 15.14.

We may conclude from these results that the birth rates of the high-literacy countries are more variable than those of the low-literacy countries. This difference in variances deserves more study. Preliminary indications are that those factors related to high birth rates operate more uniformly than the factors related to lower birth rates. Since it was our intention only to demonstrate the use of a statistical technique, we will not pursue this investigation further. It appears, however, to be a subject worth pursuing.

TABLE 15.14   Two-sample Test for Variances: Data on Birth Rates of High-literacy and Low-literacy Countries

Null Hypothesis

   $H_0 : \sigma_1^2 = \sigma_2^2$

Sets of Alternatives to H_o

   $H_1 : \sigma_1^2 > \sigma_2^2$ designate this set of alternatives
         if $F \geq 3.15$ when $F = s_1^2 / s_2^2$.

   $H_2 : \sigma_1^2 < \sigma_2^2$ designate this set of alternatives
         if $F \geq 3.15$ when $F = s_2^2 / s_1^2$.

0.01

F = 3.15

Computations

Sample Group	Variance	N	df
High-literacy countries	135.76	20	19
Low-literacy countries	23.75	20	19

$$F = s_1^2 / s_2^2 = 135.76 / 23.75 = 5.72$$

Conclusion

   Therefore, reject the null hypothesis in favor of the set of alternatives $H_1 : \sigma_1^2 > \sigma_2^2$.

15.7a   Assumptions of the Two-sample Test for Variances

The assumptions underlying the two-sample test for variances are very much like those for the one-sample test: (a) an interval level of measurement, (b) independent observations, and (c) simple random samples from normally distributed populations.

   Again, the assumption of a normal population distribution is crucial. Violation of this assumption can lead to serious problems of interpreting results. Large samples will help somewhat in reducing errors of interpretation.

   Diagram 15.3 (p. 571) summarizes the step-by-step procedures for testing hypotheses about variances.

15.8   CHI-SQUARE (χ^2) GOODNESS-OF-FIT TEST

The **chi-square (χ^2) goodness-of-fit test** is by no means a new technique, nor is its use in sociology novel. With the increasing interest in mathematical models of prediction, however, the technique becomes more important as *a means of testing the fit of such theoretical models to empirical data*. That, in a nutshell, is the major function of the goodness-of-fit test. It is designed to tell how well a theoretical model such as a normal curve, or a Poisson distribution, or an S-curve predicts the form of some body of empirical data.

15.8.1   Test of a Theoretical S-Shaped Distribution

Spilerman has concerned himself with explanation of the location of racial disorders that occurred during the 1960s and has attempted to develop a structural equation predictive of such disorders (Spilerman, 1971). His structural equation was derived from an analysis of the disorders occurring between 1961 and 1967. He then used his equation to predict the location of racial disturbances occurring in 1968. Spilerman found the two variables predictive of disorders for a city were size of the black population and whether the city was in the South. He speculated that other characteristics of the cities studied were not predictive because the blacks involved were responding to frustrations generated outside their own communities (Spilerman, 1971:428).

The prediction curve used was a general S-shaped function with the following formula:

$$S = \alpha - \frac{\beta}{N^2}(e^{-\gamma N}) - \frac{\delta}{N}(1 - e^{-\gamma N})$$

where N is the black population size, e is the number of disorders in a city, and α, β, γ, and δ are empirically derived parameters.

With 1961–1967 disturbance data, Spilerman computed separate parameters for nonsouthern and southern cities and made separate predictions about the 1968 disturbances. In Table 15.15 the nonsouthern cities have been divided into six separate categories according to percentage of black population. The f_e's are expected frequencies computed from the prediction formula for S using the parameters for the nonsouthern data for 1961–1967. The f_o's are observed frequencies of disturbances in the cities studied during 1968. With these expected and observed frequencies, it is possible to compute a χ^2 using the formula

(15.27)

$$\chi^2 = \sum \frac{(f_o - f_e)^2}{f_e}$$

TABLE 15.15   Chi-square (χ^2) Test of Predicted and Actual Number of Disorders for 1968 by Nonsouthern Black Population

Population Size*	f_o	f_e	$f_o - f_e$	$(f_o - f_e)^2$	$(f_o - f_e)^2/f_o$
1	29	32.6	−3.6	12.96	0.40
2	15	13.0	2.0	4.0	0.31
3	14	13.5	0.5	0.25	0.02
4	15	13.5	1.5	2.25	0.17
5	11	13.3	−2.3	5.29	0.40
6	15	13.2	1.8	3.24	0.25
Totals	99	99.1			$\chi^2 = 1.55$

Source: Based on data from Spilerman (1971).
* Category 1 has the largest black population and category 6 the smallest.

Table 15.15 includes the computations necessary to find χ^2. We are interested in how well the predicted frequencies fit the observed frequencies of disturbances. If there was perfect prediction, χ^2 would be 0. The larger the χ^2 the greater the deviation of the observed frequencies from those predicted. Rather than set up a specific critical region to reject the null hypothesis that the observed frequencies are equal to the expected frequencies, it is reasonable to find the approximate probability for the computed χ^2 and use it to evaluate the goodness of fit of the observed to the expected frequencies.

The computed χ^2 for the nonsouthern cities was 1.55. For the goodness-of-fit test, the proper number of degrees of freedom is found by starting with the number of categories in which the data are grouped and subtracting 1 degree of freedom for each parameter estimated and an additional degree of freedom for the requirement that the sum of the expected frequencies must equal the sum of the observed frequencies.* Since the data in Table 15.15 are grouped into six categories, four parameters were estimated, and 1 degree of freedom was lost because of the requirement that $\Sigma f_o = \Sigma f_e$, we have $6 - 5 = 1$ degree of freedom for these data. Entering the χ^2 table (Appendix Table C) with the computed value of 1.55 and 1 degree of freedom we find that the probability associated with that χ^2 is between 0.20 and 0.30. Since the differences found between observed and expected frequencies could have occurred by chance as often as 20 or 30% of the time, it may be concluded that the fit, although not perfect by any means, is reasonably good. In fact, by current social science standards, it is quite good. The S-shaped curve was relatively efficient for predicting disturbances in the nonsouthern cities.

Table 15.16 presents the predictions, the observed disturbances, and the χ^2 compuations for the southern cities. In this case the computed χ^2 was 20.93 with 1 degree of freedom. The probability of the chance occurrence of deviations as great as those observed is less than 0.001. In this case the S-shaped curve did

TABLE 15.16 CHI-SQUARE (χ^2) TEST OF PREDICTED AND ACTUAL NUMBER OF DISORDERS FOR 1968 BY SOUTHERN BLACK POPULATION

Population Size	f_o	f_e	$f_o - f_e$	$(f_o - f_e)^2$	$(f_o - f_e)^2/f_e$
1	3	12.2	−9.2	84.64	6.94
2	5	6.6	−1.6	2.56	0.39
3	5	6.1	−1.1	1.21	0.20
4	6	6.0	0.0	0.0	0.00
5	11	5.7	5.3	28.09	4.93
6	12	5.3	6.7	44.89	8.47
Totals	42	41.9			$\chi^2 = 20.93$

Source: Based on data from Spilerman (1971).

* The sum of the expected frequencies in Table 15.15 is 99.1, whereas the sum of the observed frequencies is 99. This difference is due to errors in rounding. Note that there are also slight differences in Tables 15.16 and 15.17.

TABLE 15.17 Predicted and Actual Disorders for South in 1968 from Non-southern Curve by Black Population

Population Size	f_o	f_e	$f_o - f_e$
1	3	4.4	−1.4
2	5	2.7	2.3
3	5	3.5	1.5
4	6	5.2	0.8
5	11	10.2	0.8
6	12	15.9	−3.9
Totals	42	41.9	

Source: Based on data from Spilerman (1971).

not efficiently predict disorders. The communities with the largest black populations (category 1 cities) had less than 25% as many disorders as were predicted. On the other hand, those communities with a relatively small proportion of blacks (category 5 and 6 cities) had roughly twice as many disorders as were predicted.

Because the prediction formula for the southern data did not work very well, Spilerman decided to try the nonsouthern parameters as predictors for the southern cities. Table 15.17 presents the frequencies observed and the frequencies predicted, this time using the prediction formula for nonsouthern cities on the southern cities. The χ^2 was not computed for these data because some of the expected frequencies were too small to meet the assumptions of the χ^2 test. It is possible by a visual inspection of Tables 15.16 and 15.17, however, to see that the nonsouthern parameters predicted better for southern cities than did the southern parameters. Spilerman (1971:440) suggested that a change may be taking place in the South that is leading the southern blacks to react to frustration increasingly more as do the blacks outside the South.

What Spilerman has done, and what is done generally in using the χ^2 goodness-of-fit test, is to develop a theoretical model to predict some sort of social behavior and to test it against data to see how well it predicts. The expected frequencies represent predictions derived from the theoretical model and the observed frequencies come from the data. If the model predicts well, the differences between observed and expected frequencies will be small. The smaller the differences, the better the fit. In this case, then, sociologists hope to find the computed χ^2 to be minimal so they can conclude that their theoretical model is an efficient predictor.

15.8.2   A Test for Normality

The goodness-of-fit test can just as readily be used to test the normality of a distribution as it can to test an S-shaped distribution. There are occasions when sociologists may use the normal curve as a predictive model and other occasions

TABLE 15.18   FREQUENCY DISTRIBUTION OF RING-TO-CITY COMMUTER RATES, UNITED STATES, 1960

Ring-to-City Commuter Rates	SMSAs f
0-9.99	1
10-19.99	5
20-29.99	17
30-39.99	36
40-49.99	23
50-59.99	11
60-59.99	1
70-79.99	1
	Total = 95

Source: Yu (1972:78, Table 2).

when they are interested in testing the normality of an observed distribution to determine whether certain kinds of statistical analyses are appropriate (*e.g.*, whether the standard deviation is an appropriate measure of variability).

We will illustrate the goodness-of-fit test as a test for normality with data used by Yu in a study of commuter rates between the central cities of "standard metropolitan statistical areas" and their suburban rings (Yu, 1972). Analyzing census data for 95 standard metropolitan statistical areas, Yu attempted to relate differences in commuter rates to variations in characteristics of the metropolitan areas studied. One of the measures he used, called the "ring-to-city commuter rate," was defined as the number of ring inhabitants working in the central city expressed as a percentage of the total number of workers living in the ring (Yu, 1972:76). Table 15.18 is a frequency distribution of the ring-to-city commuter rates for the 95 standard metropolitan statistical areas studied (the data are from the 1960 census).

We might legitimately ask whether the sample of 95 standard metropolitan statistical areas represents a population of SMSAs that has a normal distribution of ring-to-city commuter rates. To answer this question, it is necessary (a) to generate a set of expected frequencies for Table 15.18 that would be found if the rates were normally distributed and (b) to compare those expected frequencies with the observed frequencies to determine goodness of fit.

15.8.2a   Generating Normal Expected Frequencies

The data summarized in the frequency distribution of Table 15.18 have a mean of 37.32 and a standard deviation of 11.89 (these were computed from the 95 sample scores). The frequencies in Table 15.18 are the observed numbers of SMSAs in the sample that fall into each of the classes of ring-to-city rates. To test the goodness of fit of these frequencies to the normal distribution, it is necessary to generate a set of expected frequencies for a normal distribution having the same

TABLE 15.19 GENERATION OF NORMAL EXPECTED FREQUENCIES FOR RING-TO-CITY COM-
MUTER RATE DATA FROM TABLE 15.18

Lower Limits of Classes	Standard Scores	Areas from Mean	Areas Between	f_e
		0.5000		
			0.0107	1.02
10	-2.30	0.4893		
			0.0614	5.83
20	-1.46	0.4279		
			0.1955	18.57
30	-0.62	0.2324		
			0.3234	30.72
40	$+0.23$	0.0910		
			0.2667	25.34
50	$+1.07$	0.3577		
			0.1142	10.85
60	$+1.91$	0.4719		
			0.0251	2.38
70	$+2.75$	0.4970		
			0.0030	0.29
		0.5000		
				Total = 95.00

Source: Data from Yu (1972).

mean, standard deviation, and N as the empirical distribution. In Table 15.19, we have generated these expected frequencies through the following procedures:

1. We took the lower limit of each class and solved for z using the formula

(15.28)
$$z = \frac{X - \bar{X}}{s}$$

where X successively takes the **value of each lower limit**, beginning with the second one, 10, and ending with 70. (The reason for ignoring the first lower limit, 0, will be explained later.) To find the z-score for the lower limit, 10, for example, we substituted into the formula as follows:

$$z = \frac{10 - 37.32}{11.89}$$

$$= -2.30$$

2. After we solved the z-scores for each of the lower limits, we turned to the table of the normal curve (Appendix Table A) and found the areas from the mean to each of these respective z-scores.

3. We took the differences between those adjoining areas, when the signs of the z-scores were the same, and the sum in the single case, where the two adjoining z-scores had different signs (*i.e.*, $z = -0.62$ and $z = +0.23$). At the lower

end of the distribution the area corresponding to the z-score of -2.30 (*i.e.*, 0.4893) was subtracted from 0.5000 because we wished to find the area to the end of the distribution; half of the normal curve includes an area of 0.5000. At the upper end of the distribution we did the same thing: finding the area to a z of $+2.75$ and subtracting from 0.5000 (thus, $0.5000 - 0.4970 = 0.0030$). Once we found all these areas we treated them as proportions of cases that would fall into each class if the distribution were normal.

4. We converted each of these proportions to expected frequencies by multiplying by $N = 95$.

15.8.2b Evaluating Goodness-of-Fit

In Table 15.20, χ^2 has been computed comparing the normally distributed expected frequencies with the observed frequencies. Notice that some of the classes of the original frequency distribution have been combined. The classes 0–9.99 and 10–19.99 have been combined to form the class 0–19.99, and the classes 50–59.99, 60–69.99, and 70–79.99 have been combined to form the class 50–79.99. This was done because the *rule of thumb for computing χ^2 says that no expected frequency should be less than 5.* To avoid small expected frequencies at the ends of the frequency distribution, it was necessary to combine classes.

To determine the appropriate number of degrees of freedom for evaluating χ^2, we begin with the number of classes into which the data have been categorized after any necessary combining of classes has been carried out (5); subtract 1 degree of freedom for each parameter estimated (μ and σ were estimated by $\bar{X}$ and s), and 1 degree of freedom for the requirement that the sum of the expected frequencies should equal the sum of the observed frequencies. Thus, there are $5 - 3 = 2$ degrees of freedom. The computed χ^2 is 1.54. If we enter the χ^2 table (Appendix Table C) with 2 degrees of freedom, we find that the computed χ^2 can be expected to occur by chance between 30 and 50% of the time.

It may be concluded from this analysis that a normal distribution gives a good fit to the actual distribution of ring-to-city commutation rates.

TABLE 15.20 CHI-SQUARE (χ^2) GOODNESS-OF-FIT TEST OF RING-TO-CITY COMMUTER RATE DATA

Commutation Rate	f_o	f_e	$f_o - f_e$	$(f_o - f_e)^2$	$(f_o - f_e)^2/f_e$
0–19.99	6	6.85	−0.85	0.7225	0.11
20–29.99	17	18.57	−1.57	2.4649	0.13
30–39.99	36	30.72	5.28	27.8784	0.91
40–49.99	23	25.34	−2.34	5.4756	0.22
50–79.99	12	13.52	−1.52	2.3104	0.17
					$\chi^2 = 1.54$

Source: Data from Yu (1972).

15.8.3   Assumptions of the χ^2 Goodness-of-Fit Test

The χ^2 goodness-of-fit test can be used to evaluate many different kinds of theoretical models as long as the data meet the following assumptions: (a) the observations are independent, (b) the data constitute a simple random sample of the population, and (c) the expected frequencies are no smaller than the minimum allowable size. The rule of thumb is that *no expected frequency should be less than 5*. Although we have not applied the principle to the data analysis reviewed in this chapter, some statisticians use another rule that the smallest expected frequency should be no less than 10 when there is only 1 degree of freedom. It was because Spilerman's data in Table 15.17 did not meet the first rule of thumb (no expected frequency should be less than 5) that chi-square was not computed.

Diagram 15.4 (p. 573) summarizes the procedures that are used to generate normal expected frequencies from empirical data in determining whether the data are normally distributed.

15.9   NONPARAMETRIC STATISTICS AND ALTERNATIVE APPROACHES TO TESTS

A number of tests of significance that we have discussed imply some knowledge or assumptions about the population from which the samples were drawn. For example, the single-sample means test requires that the variable whose mean is of concern be normally distributed in the population. The test of significance of difference between means for two-samples requires normal populations, equal population variances, and that the two samples be independent of each other. Some assumptions, such as the normality assumption, can be relaxed where large samples are used. What test can be used when these **parametric** assumptions are not reasonable? They might not be reasonable because we do not have information upon which to base a realistic assumption, or we might have some evidence that the assumptions cannot be met. Simply listing the assumptions does not make them true or the test appropriate.

A number of statistical tests have been developed to meet some of these problematic situations. Some of these tests simply avoid the need to make assumptions about population parameters and thus are called **nonparametric** tests. In some cases, a problem can be avoided by rethinking how to approach the problem. We will briefly introduce four tests that illustrate some of the inventiveness of statistical reasoning for otherwise difficult problems. Computer programs are available that specialize in computing these tests (*e.g.*, StatXact).

15.9.1   Randomization Test

The randomization test for independent samples is a test of difference between means of two (usually small) samples (Siegel, 1956). It helps to solve the problem of a test of difference between means where one cannot legitimately assume that

---

TABLE 15.21   LIVE BIRTHS PER 1,000 POPULATION (RANDOM SAMPLE FROM TABLE 15.13 DATA)

High-literacy Countries		Low-literacy Countries	
Poland	19.7	Guatemala	41.2
Belize	38.8	Congo	44.7
Venezuela	36.9	Liberia	49.8
Singapore	17.3	Cape Verde	35.5
Philippines	33.9	Morocco	45.4

---

the samples come from normal-shaped populations, and the sample sizes are too small to relax the normality assumption. We will use a small random sample of five cases selected from each of the two groups of countries in Table 15.13. The two random samples are listed in Table 15.21.

The rationale of the randomization test is that the high–low literacy labels could be assigned randomly to countries so that those in one group are merely randomly different from those in the other group. On the other hand, if the groups differ systematically, then most of the high scores would be in one group and most of the low scores would be in the other. Thus we can think of a long list of possible ways of ordering the scores, between these extremes. For this study, the list would extend from a situation where the high-literacy group included the five lowest scores and the low-literacy group included the five highest scores, all the way to the reversed extreme, where the high-literacy group included the five highest scores out of the ten scores we have here. The total number of different possible combinations of scores, divided into n_1 scores in the first group and n_2 scores in the second group, can be computed using the formula for combinations (see Section 10.4.2), as follows:

$$\text{Number of combinations:} \quad \frac{N!}{(N - n_1)!\, n_1!}$$

where $N = n_1 + n_2$, the total of the two sample sizes. Here the number of combinations is

$$\frac{10!}{(10 - 5)!\, 5!} = 252$$

There are 252 different ways in which these ten scores could be assigned to two groups with five in each group. We will think of these as ordered from combinations with high scores in one group at one extreme to combinations with high scores in the opposite group at the other extreme.

The null hypothesis is that the two groups of scores come from populations that do not differ in terms of mean number of children per 1,000 population. Our alternative hypothesis is that in countries with *higher* literacy there will be a *lower* average number of children, calling for a one-tailed test. In this case, the region of rejection of the null hypothesis constitutes the combinations at the extreme where low scores are assigned primarily to the high-literacy group.

First, we need to identify the region of rejection, and then we need to see whether our particular sample is in the region of rejection. If we choose to use the 5% level of significance, then 5% of the 252 possible combinations that are in this extreme of the list of possible combinations will constitute the region of rejection.

$$(.05)(252) = 12.6$$

Thus, 12 (using a slightly conservative figure) of the extreme combinations (where low scores are mostly in the high-literacy group) is the desired region of rejection.

Now we need to identify these five combinations. First, we assign the ten scores, five per group, starting with the most extreme combination. Then we list the next most extreme assignment of the ten scores, and so on, until we have listed the 12 or 13 most extreme combinations constituting the region of rejection. It is convenient to use the sum of scores in one group as an indicator of the ordering of the extreme combinations:

	High-literacy Group	High Group Sum	Low-literacy Group
1.	17.3 19.7 33.9 35.5 36.9	143.3	38.8 41.2 44.7 45.4 49.8
2.	17.3 19.7 33.9 35.5 38.8	145.2	36.9 41.2 44.7 45.4 49.8
3.*	**17.3 19.7 33.9 38.8 36.9**	**146.6**	**35.5 41.2 44.7 45.4 49.8**
4.	17.3 19.7 33.9 35.5 41.2	147.6	36.9 38.8 44.7 45.4 49.8
5.	17.3 19.7 33.9 36.9 41.2	149.0	35.5 38.8 44.7 45.4 49.8
6.	17.3 19.7 33.9 38.8 41.2	150.9	35.5 36.9 44.7 45.4 49.8
7.	17.3 19.7 36.9 38.8 41.2	153.9	33.9 35.5 44.7 45.4 49.8
8.	17.3 19.7 33.9 35.5 49.8	156.2	36.9 38.8 41.2 44.7 45.4
9.	17.3 19.7 35.5 41.2 44.7	158.4	33.9 36.9 38.8 45.4 49.8
10.	17.3 19.7 35.5 41.2 45.4	159.1	33.9 36.9 38.8 44.7 49.8
11.	17.3 19.7 35.5 44.7 45.4	162.8	33.9 36.9 38.8 41.2 49.8
12.	17.3 19.7 35.5 41.2 49.8	163.5	33.9 36.9 38.8 44.7 45.4
13.	17.3 19.7 35.5 45.4 49.8	167.7	33.9 36.9 38.8 41.2 44.7

This is the region of rejection for our one-tailed test. If we had used a two-tailed test, then half of the 12 cases would be in this tail, and we would have to list the other extreme, where the *highest* scores were in the *high*-literacy group, and half of the region of rejection would be in that tail of the sampling distribution list.

It is clear that the sample we drew (marked by the asterisk and in bold type) is among those in the one-tailed, 5% region of rejection. Thus we would reject the null hypothesis that the two groups came from populations with equal mean scores and conclude that the two small samples must have come from populations where the high-literacy countries had a lower number of live births per 1,000 population than the low-literacy countries.

This nonparametric test does not require assumptions about the form of the population distributions. Comparison of nonparametric and parametric tests are made by computing the ratio of the number of cases needed by a parametric test to the number of cases needed by a nonparametric test in order for it to be equally powerful, a measure called *power efficiency*. Often, nonparametric tests require more cases and thus are less efficient in this sense (although they solve other assumption problems). In this case, the randomization test uses all the available information so that, even if the assumptions of the usual test of significance of difference of means could be met, it would be as efficient as that test (a power efficiency of 100%).

15.9.2 Direct Difference Test

Some research involves comparison of two groups where the observations in one group are related in some way to the observations in the other group. Thus, the assumption of two independent samples is not met. This can happen in a number of ways. For example, an investigator could match cases in one group with cases in another group in order to control as much as possible for unwanted differences. If one group (*e.g.* an experimental group to be given some special treatment) had a white, middle-aged, married female, then the investigator might select a matching case with these same characteristics for the comparison group (*e.g.*, the control group). Another way in which cases in two groups might be related is if the cases were related in some patterned way, such as husbands and their wives, parents and their children, or employer and employee. Special consideration needs to be given to statistical tests involving related samples.

Table 15.22 contrasts years of schooling completed by a random sample of

TABLE 15.22 YEARS OF SCHOOL COMPLETED BY A SIMPLE RANDOM SAMPLE OF MOTHERS AND A 30+-YEAR-OLD DAUGHTER (FICTITIOUS DATA)

	School Years Completed		
	Mothers	*Daughters*	*Difference*
	14	16	+2
	12	16	+4
	11	13	+2
	11	10	−1
	17	20	+3
	8	12	+4
	12	12	0
	12	16	+4
	16	17	+1
	20	12	−8
	16	16	0
	12	14	+2
	13	12	−1
	16	20	+4
	12	16	+4
	16	16	0
	12	13	+1
	14	12	−2
	12	10	−2
Means	13.5	14.4	+.89

mothers and a sample consisting of one of their daughters aged 30 or older. Rather than two independent samples, one might suspect that there is a systematic relationship between mothers and daughters. To test the significance of difference between these two related samples, we will, instead, think of them as a single sample of pairs of scores. Instead of two samples of 19 cases each, we will think of these data as one sample of 19 pairs of scores.

The direct difference test uses the algebraic difference between the scores for each pair to test the null hypothesis that the population mean difference is zero. The alternative hypotheses could be one or two tailed. Here, we will test the hypothesis that daughters have a higher average number of years of education than mothers; that is, there is a positive difference.

If the assumption of a normal population of differences can be assumed, then we would use the usual one-sample t-test of means with degrees of freedom equal to $N - 1$, where N is the number of pairs ($df = 19 - 1 = 18$). Using the corrected standard deviation (3.04), the standard error of differences is

$$s_d = \frac{s}{\sqrt{N}} \quad \text{or} \quad s_d = \frac{3.04}{\sqrt{19}} = .697$$

which is substituted into the formula for the t-test:

$$t = \frac{(\bar{X}_d - \mu_d)}{s_d} = \frac{(+.89 - .0)}{.697} = +1.28$$

where $\bar{X}_d$ is the mean difference found in the sample, μ_d is the hypothesized population difference (here, zero), and s_d is the standard error of difference. At the .05 level of significance for a one-tailed test, the critical value of Student's t with $df = 18$ is $+1.734$. Since the computed value is smaller than the critical value, we cannot reject the null hypothesis that the population mean difference is zero. Thus, based on this small sample, we would conclude that in the population of mother–daughter pairs, daughters have the same mean number of years of schooling as their mothers.

The direct difference test illustrates how a statistical problem can be reformulated into a problem that can be readily handled.*

15.9.3   Sign Test

The sign test also illustrates how a statistical problem can be handled by rethinking how the problem is approached. In the preceding example of mothers' and daughters' education, the sign test would be useful if the data were ordinal, rather than interval. It is also a useful test if the normality assumption cannot be made.

Using the data in Table 15.22, instead of using algebraic differences we would focus on only the sign of the difference between the 19 pairs of mothers and daughters in the matched sample. Our null hypothesis would be that the population from which these data were drawn had 50% plus differences and 50% minus

* In the case of the two-sample test of means, the standard error formula given in Section 15.2.1a can be revised to include a correlation term that would also address the problem of matched samples.

differences. That is, $P_+ = .50$. Eliminating the tied pairs, we have 16 signed pairs ($N = 16$) and the proportion of pluses in the sample is $p_+ = .69$.

We will use the single sample test of proportions to test the null hypothesis. Again, since we are predicting daughters to be higher on education than mothers, we will use a one-tailed test. The standard error and z-score results are as follows:

$$s_p = \sqrt{\frac{(.50)(.50)}{16}} = .125$$

$$z = \frac{(.69 - .50)}{.125} = +1.52$$

If we use the .05 level of significance, the critical value of z would be $+1.645$ and we could *not* reject the null hypothesis that in the population the proportion of positive differences in education between mothers and daughters is .50. Notice that in this test we are not using (or cannot use) the detailed interval-level score data about the magnitude of difference. We are using only the sign of the difference.

15.9.4 Runs Test

The final example of a test for special situations to be discussed here is called the runs test. It is used where there is an ordered sequence of dichotomous data, such as scores over time or trend data. The following scores will serve as an example. As ten students enter the classroom door, their gender is recorded:

M F F F M M F M M

Is this a random sequence or not? We will test the idea of order by counting the number of runs in the sequence. A run is a consecutive series of the same type of score. In this example, there are five runs, counted as follows:

M F F F M M F M M

Now, the question is whether this is the number of runs one would expect of a random sequence. If all males arrived at the same time and all females arrived together, then we would have two runs, certainly an unexpectedly small number. If they alternated M F M F, etc., that would be an unusually high number of runs. The null hypothesis is that it is a random sequence and the test can be a one- or two-tailed test. Since we have no expectation about how students enter classrooms by gender, we will use a two-tailed test.

Tables of critical values for the runs test are available (see Siegel, 1956) and selected values are shown in Table 15.23. In this table, n_1 is the number of one type of score and n_2 is the number of the other type of score. The values are given only for the .05 level of significance for a two-tailed test (or the .025 level for a one-tailed test).

In this instance, the five observed runs for $n_1 = 4$ and $n_2 = 5$ are not sufficiently extreme in either direction and we cannot reject the null hypothesis of a random sequence.

TABLE 15.23 SELECTED CRITICAL VALUES FOR THE RUNS TEST[a]

n_1	n_2	Lower Limit	Upper Limit
4	5	2	9
5	5	2	10
6	5	3	10
8	10	5	15
10	10	6	16
20	20	14	28

[a] If the number of runs is equal to or more extreme than these limits (equal to or smaller than the lower limit or equal to or bigger than the upper limit), then reject the null hypothesis of a random sequence at the .05 level of significance for a two-tailed test. If the number of runs equals or is more extreme in the predicted direction, the null hypothesis can be rejected at the .025 level of significance for the one-tailed test.

15.10 SUMMARY

In this chapter we have examined techniques for testing hypotheses about central tendency, variability, and form. We discussed procedures for testing hypotheses about means in the one-sample case, the two-sample case, and the k-sample case. For the k-sample case, we discussed the simple one-way analysis of variance. We pointed out that there are many types of analysis of variance procedures, but that most of those are limited in their use to laboratory experimentation. Certain fields of sociology, such as small group research, are amenable to laboratory experimentation, but others are not. Therefore, space was not devoted to a consideration of the more elaborate analysis of variance designs.

Tests of hypotheses of proportions for the one- and two-sample cases were described and discussed. It was explained that these techniques have more general applicability because the assumptions underlying their use are much less stringent than the assumptions underlying tests about means.

Next, we discussed tests of hypotheses about variances. We explained procedures for making tests for the one- and two-sample case. It is important to note that the assumption of a normal population distribution is a crucial one, because variances are sensitive to departures from normality. As a matter of fact, variances are much more sensitive to such departures than are means.*

In the latter part of the chapter we explained how to test the fit of a theoretical distribution to an empirical distribution. The chi-square goodness-of-fit test was used first to test the fit of an S-curve to an empirical distribution; then a normal distribution was generated and χ^2 used to test its fit to an empirical distribution.

Finally, we introduced selected tests that illustrate how otherwise difficult statistical problems can be reformulated into testable form. Some of these tests

* There is a test for this assumption of equal variances, called Bartlett's Test (see Dixon and Massey, 1957).

are nonparametric in that they do not require certain assumptions about population parameters (*e.g.*, the normality assumption in the means test). Some tests involve reformulating a statistical problem so that a simpler test will apply. The runs test is a test for random sequences of a dichotomous variable, and it serves to illustrate the broader range of hypotheses for which there are statistical tests.

Much attention was devoted earlier (Parts III and IV) to measures of association. In the next chapter the discussion will focus on techniques for generalizing statements about association from sample data to population data.

CONCEPTS TO KNOW AND UNDERSTAND

sampling distribution of differences
between means
standard error of the difference
between means
analysis of variance (ANOVA)
test for contrasts
total sum of squares
within-groups sum of squares
chi-square goodness-of-fit test
between-groups sum of squares

within-groups variance
between-groups variance
F distribution
F ratio
one-way analysis of variance
sampling distribution of differences
between proportions
standard error of the difference
between proportions
nonparametric statistics

QUESTIONS AND PROBLEMS

1. If the formula $t = (\bar{X} - \mu)/s_{\bar{x}}$ is a ratio, explain the nature of the ratio. That is, what does the numerator represent, what does the denominator represent, and what is the significance of the ratio between the numerator and denominator?

2. Explain the part played by the null hypothesis in generating the sampling distribution of differences between pairs of sample means.

3. Explain what is measured by the standard error of the difference between means.

4. In a test for significance of difference between sample means, why is the assumption that the variances are homogeneous an important one?

5. What is the rationale behind the analysis of variance? That is, what kind of comparison does the test imply?

6. If an analysis of variance test is conducted to test the null hypothesis $\mu_1 = \mu_2 = \mu_3 = \mu_4 = \mu$, and the null hypothesis is rejected, what can we say about differences between the means?

7. What justifies the pooling of two sample proportions (p_1 and p_2) to get a common estimate of the population proportion (P)?

8. Which assumption is crucial for testing hypotheses about variances? Why?

9. At this point, one of the most effective ways to firm up your knowledge of the various hypothesis testing techniques is to try them out on real problems of interest to you. It is important to understand thoroughly not just the calculations but the logic and usefulness of the procedures in drawing conclusions about worthwhile ideas or problems of relevance to your subject matter interest. This can be achieved in any number of ways such as:

a. Design and conduct (perhaps with others in the class) a manageable survey or observational study of some phenomenon of interest to you. It would be helpful to include in your study some reading or review of at least the main recent work in your topic area. Consult your course instructors for ideas or use work started in a past term paper. You will want to pay particular attention to the interesting variables and how they are measured plus subgroupings of subjects that you will use in making contrasts and comparisons. What do you expect the data to show? How would you explain that? How would you explain results that were exactly opposite? Then, carefully state your problem, define the relevant population and sampling frame, specify how it is you will draw the sample, set up measures, and conduct the data gathering. Try to draw a sample sufficiently large (say 30 or even 100) so that large sample statistical assumptions can be met. You can treat a random small sample from these data as a data set for small sample procedures to contrast with large sample procedures. Be sure that some measures suit themselves to percentage or proportion use and some suit themselves to interval level treatment. Your instructor can show you how to use computer facilities to process your data. Then, test a variety of hypotheses you have using the range of techniques discussed in this chapter. You may be able to use the same data to test hypotheses using techniques in the next chapter. Be sure to state clearly your procedures and conclusions and the reasons behind these.

b. If you do not have resources or time to conduct a study, you could use available data from published articles on topics of interest to you, governmental data (a wealth exists in the government documents section of your library), or data sets already stored on the computer (such as data from an instructor's research or data from national studies). In some instances where, for example, information is given for each of the states in the United States, you can draw a random sample from these data for your use. In that case, compare your results with the population parameters if they are available to you. Cities, counties, states, organizations, or countries are interesting units of analysis for sociological studies.

10. Some statistical reports, such as the one selectively reproduced below, include information on sampling error. The following table provides numbers and rates of deaths for the United States by gender, color, and age for 2 years plus the information about sampling error.

a. Using this information, state confidence intervals for the rate and for the frequency of deaths for the United States and for males and females. Pick a confidence interval of your choice for which information is given.

b. Test the hypothesis that male death rates are not higher than the overall rate.

c. Test the hypothesis that White and Other background are equal in terms
   of frequency of deaths, in terms of death rate.
d. Test the hypothesis that, overall, the death rate has not declined between
   the two years.

In each instance, organize your work in terms of the steps of the procedure
you use, show your work, and state conclusions in your own words. There are
many other hypotheses you may want to test on these data in terms of change
from year to year, and comparisons of gender, color, and age groups with each
other or over time. In general, what conclusions can one draw from the frequencies and rates shown in the table?

DEATHS IN THE SAMPLE AND ESTIMATED DEATH RATES, BY AGE, COLOR, AND SEX: UNITED
STATES, 12 MONTHS ENDING WITH JANUARY 1977 AND 1978
(Rates on an annual basis per 1,000 estimated population residing in area for specific group)

| | 12 Months Ending with January | | | |
| | 1978 | | 1977 | |
Age, Color, and Sex	Number	Rate	Number	Rate
Total	190,541	8.8	191,102	8.9
Male	104,976	10.0	105,366	10.1
Female	85,565	7.7	85,735	7.8
White	167,513	8.9	167,812	9.0
Male	91,867	10.0	92,009	10.1
Female	75,646	7.9	75,803	8.0
All other	23,028	8.0	23,290	8.2
Male	13,109	9.5	13,357	9.8
Female	9,919	6.5	9,933	6.7
Under 1 year	4,589	14.5	4,744	15.7
1–14 years	2,153	0.4	2,136	0.4
15–24 years	4,801	1.2	4,544	1.1
25–34 years	4,496	1.4	4,193	1.3
35–44 years	5,776	2.5	5,811	2.5
45–54 years	14,328	6.1	15,051	6.4
55–64 years	29,313	14.4	29,538	14.7
65–74 years	44,659	30.6	44,578	31.4
75–84 years	49,373	72.4	49,548	73.2
85 years and over	30,995	149.0	30,923	156.3
Not stated	58	...	36	...

Since the estimates of deaths and death rates presented in this report are based
on a sample of the death certificates, they are subject to sampling variability.
The percent errors shown in the following table are measures of sampling
variability. They show the percent difference between figures based on a sample
and those based on a complete count that may result from sampling variation.
Values are shown for two different sized samples. The first column refers to the

monthly sample size, the second to the annual; cumulative monthly totals shown during the year will be between the two. The percent differences apply to both frequencies and rates. Chances are about 2 out of 3 that the percent difference between a sample estimate and a complete count is less than the percent shown. Chances are about 19 out of 20 that the percent difference due to sampling variation is less than twice the percent shown. In this report a change in an estimated monthly rate from one year to the next is regarded as significant only when the percent change is at least 3 times the percent error of the most recent rate.

	Percent Error of Estimate	
Number of Deaths in Sample for a Group	15,000 Total Deaths in Sample Each Month	180,000 Total Deaths in Sample Each Year
5	42.4	42.4
10	30.0	30.0
20	21.2	21.2
50	13.4	13.4
100	9.5	9.5
200	6.7	6.7
500	4.2	4.2
1,000	2.9	3.0
2,000	2.0	2.1
5,000	1.1	1.3
10,000	0.5	0.9
20,000	...	0.6
50,000	...	0.4
100,000	...	0.2

Source: U.S. Public Health Service (1978).

11. As part of a study, "The Causes and Cost of Racial Exclusion from Job Authority," James R. Kluegel (1978) developed a measure of job authority from answers by individuals in his sample to a series of questions about kinds of supervisory behavior they exercise. The higher the index the higher the amount of job authority. For our purposes, we will consider the data he gathered from two samples of employed men in urban Wisconsin as a simple random sample. Job authority (and other statistics) are presented below for five occupational categories in his sample. You may want to consult his article to see how he tested his main conclusions.

From these data, test the following hypotheses:

a. That the mean job authority for self-employed managers (or another category you select) is equal to the overall mean job authority of these groups, 1.591.

b. That salaried managers are not higher than clerical workers.

c. That clerical and sales workers are equal.

d. That there is no difference in mean job authority for these five groups.

In each case, show the steps in your work, interpret the results in your own words, and discuss the conclusions that can be drawn.

Occupational Category	$\bar{X}$ Job Authority	s	n	Σx^2
Self-employed managers	4.254	1.293	64	105.326
Salaried managers	2.459	1.164	111	149.039
Sales	1.433	1.249	67	102.960
Clerical	.834	1.124	103	128.864
Laborer	.462	.778	158	95.030
Overall (for these groups)	1.591	1.657	503	1380.799

12. Explain how the chi-square goodness-of-fit test is used to test the predictability of a mathematical model.

13. Find a frequency distribution in a sociology journal such as the *American Sociological Review* or *The American Journal of Sociology*, generate expected frequencies for a normal distribution from the data in the table, and use the chi-square goodness-of-fit test to determine whether the normal (expected) frequencies fit the observed frequencies.

GENERAL REFERENCES

Blalock, Hubert M., Jr., *Social Statistics*, rev. ed. (New York, McGraw-Hill), 1979.
    If you read the sections in Blalock dealing with tests of hypotheses about central tendencies, it will reinforce what you have learned in this chapter.

Edwards, Allen, *Experimental Design in Psychological Research,* 2nd ed. (New York, Holt, Rinehart and Winston), 1960.
    This is an excellent source to read on the analysis of variance. It discusses those more complex analysis of variance applications for laboratory experiments that were not discussed in this chapter.

Freeman, Linton, C., *Elementary Applied Statistics,* (New York, Wiley), 1965.

Freund, John E., *Modern Elementary Statistics*, 3rd ed. (Englewood Cliffs, N.J., Prentice Hall), 1967.
    In Chapter 10 of this book there is a discussion of tests of hypotheses about variability. Reading the material in Freund may help to reinforce what you have learned in this chapter.

Pierce, Albert, *Fundamentals of Nonparametric Statistics* (Belmont, Calif., Dickenson Publishing Co.) 1970.

Siegel, Sidney, *Nonparametric Statistics for the Behavioral Sciences* (New York, McGraw-Hill), 1956.
    A basic reference containing many nonparametric tests.

StatXact, a computer program for nonparametric statistics by Cytel Software Corporation, 137 Erie Street, Cambridge, MA 02139.

LITERATURE CITED

Coverman, Shelley, "Gender, Domestic Labor Time, and Wage Inequality," *American Sociological Review*, 48 (October 1983), pp. 623–637.

Dixon, Wilfrid J., and Frank J. Massey, Jr., *Introduction to Statistical Analysis* (New York, McGraw-Hill), 1957, pp. 179–180.

Faris, Robert E. L., "The Alleged Class System in the United States," *Research Studies, State College of Washington,* 22 (1954) pp. 77–83.

Harris, Richard J., *A Primer of Multivariate Statistics,* 2nd ed. (Orlando, Fl., Academic Press, 1985), pp. 150–156.

Hogan, Dennis P., and Evelyn M. Kitagawa, "The Impact of Social Status, Family Structure, and Neighborhood on the Fertility of Black Adolescents," *American Journal of Sociology,* 90 (January 1985), pp. 825–855.

Kandel, Denise B., "Race, Maternal Authority, and Adolescent Aspiration," *American Journal of Sociology,* 76, 6 (May 1971), pp. 999–1018.

Kluegel, James R., "The Causes and Cost of Racial Exclusion from Job Authority," *American Sociological Review,* 43 (June 1978), pp. 285–301.

Norusis, Marija J., *The SPSS Guide to Data Analysis* (Chicago, SPSS Inc.), 1986.

Quinn, Robert, and Graham Staines, *Quality of Employment Survey, 1977: Cross Section* (Ann Arbor, Michigan: Inter-University Consortium for Political and Social Research), 1979.

Spaeth, Joe L., "Job Power and Earnings," *American Sociological Review,* 50 (October 1985), pp. 603–617.

Spilerman, Seymour, "The Causes of Racial Disturbances: Tests of an Explanation," *American Sociological Review,* 36, 3 (June 1971), pp. 427–442.

SPSS Inc., *SPSS User's Guide,* 2nd ed. (Chicago, SPSS Inc.), 1986.

Svalastoga, Kaare, "Social Differentiation," in Robert E. L. Faris, editor, *Handbook of Modern Sociology* (Chicago: Rand McNally), 1964, pp. 547–550.

U.S. Bureau of the Census, *Statistical Abstract of the United States: 1985* (Washington, D.C., U.S. Government Printing Office), 1984.

U.S. Department of Labor, Bureau of Labor Statistics, *Handbook of Labor Statistics,* Bulletin 2175 (Washington, D.C., U.S. Government Printing Office), 1983.

U.S. Public Health Service, "Births, Marriages, Divorces, and Deaths for February, 1978," *Monthly Vital Statistics Report,* Vol. 27, No. 2, (May 18, 1978), p. 11, Table 5, DHEW Publication No. (PH5) 78–1120 (Washington, D.C., National Center for Health Statistics), 1978.

Verwaller, Darrel J., "Social Mobility and Membership in Voluntary Associations," *American Journal of Sociology,* 75, 4 (January 1970), pp. 481–495.

Warner, W. Lloyd, Marchia Meecker, and Kenneth Eells, *Social Class in America* (Chicago, Science Research Associates), 1949.

Yu, Eui-Young, "Correlates of Commutation between Central Cities and Rings of SMSA's," *Social Forces,* 51 (September 1972), pp. 74–86.

DIAGRAM 15.1   Tests of Hypotheses about *Means*

THE ONE-SAMPLE TEST

Problem

    To determine whether a given sample mean is an estimate of a known or theoretical population mean.

Assumptions

1. Interval level of measurement
2. Independent observations
3. Normally distributed population
4. A simple random sample of that population

Null and Alternate Hypotheses

$$H_0 : \mu = \text{(some specified value)}$$

$$H_1 : \mu \neq \text{(specified value)}$$

or

$$\mu < \text{(this specified value) or } \mu > \text{(this specified value)}$$

Standard Error of the Mean

$$\sigma_{\bar{X}} = \frac{\sigma}{\sqrt{N}} \qquad \text{if } \sigma \text{ is known.}$$

$$s_{\bar{X}} = \frac{s}{\sqrt{N}} \qquad \text{if } \sigma \text{ is unknown.}$$

Standard Score Formulas

$$z = \frac{\bar{X} - \mu}{\sigma_{\bar{X}}} \qquad \text{if } \sigma_{\bar{X}} \text{ is known.}$$

$$z = \frac{\bar{X} - \mu}{s_{\bar{X}}} \qquad \text{if } \sigma_{\bar{X}} \text{ is unknown but } N \text{ is large.}$$

$$t = \frac{\bar{X} - \mu}{s_{\bar{X}}} \qquad \text{if } \sigma_{\bar{X}} \text{ is unknown and } N \text{ is small.}$$

Sampling Distribution

1. The normal curve if σ is known or N is large (100 or more).
2. Student's t-distribution with $N - 1$ degrees of freedom if σ is unknown and N is small.

Step 1: List assumptions.

Step 2: Set up the null hypothesis. Specify whether the test is one-tailed or two-tailed. Specify the level of significance. Determine critical region.

Step 3: Draw a simple random sample from the population and collect data.

DIAGRAM 15.1 *(Continued)*

Step 4: From the sample data, compute the mean, standard deviation, and standard error of the mean.

Step 5: Substitute into the appropriate standard score formula and compute the standard score. If the standard score is in the critical region, reject the null hypothesis; if not, fail to reject it.

THE TWO-SAMPLE TEST

Problem

To determine whether the means of two independent samples estimate a common parameter.

Assumptions

1. Interval level of measurement
2. Independent observations
3. Normally distributed population
4. Two independent simple random samples
5. Homogeneity of population variances

Null and Alternate Hypotheses

$$H_0: \mu_1 = \mu_2$$

$$H_1: \mu_1 \neq \mu_2$$

or
$$\mu_1 < \mu_2 \quad \text{or} \quad \mu_1 > \mu_2$$

Standard Error of the Difference

$$\sigma_{\bar{X}_1 - \bar{X}_2} = \sqrt{\sigma^2 \left(\frac{1}{N_1} + \frac{1}{N_2} \right)} \qquad \text{if } \sigma \text{ is known.}$$

$$s_{\bar{X}_1 - \bar{X}_2} = \sqrt{s^2 \left(\frac{1}{N_1} + \frac{1}{N_2} \right)} \qquad \text{if } \sigma \text{ is unknown and the sample variances are homogeneous.}$$

Formula for Finding Pooled Variance

$$s^2 = \frac{\sum x_1^2 + \sum x_2^2}{N_1 + N_2 - 2}$$

Standard Score Formulas

$$z = \frac{\bar{X}_1 - \bar{X}_2}{\sigma_{\bar{X}_1 - \bar{X}_2}} \qquad \text{if } \sigma \text{ is known.}$$

$$z = \frac{\bar{X}_1 - \bar{X}_2}{s_{\bar{X}_1 - \bar{X}_2}} \qquad \text{if } \sigma \text{ is unknown but } N_1 + N_2 \geqq 100.$$

$$t = \frac{\bar{X}_1 - \bar{X}_2}{s_{\bar{X}_1 - \bar{X}_2}} \qquad \text{if } \sigma \text{ is unknown and } N_1 + N_2 < 100.$$

DIAGRAM 15.1 *(Continued)*

Sampling Distribution

1. The normal curve if σ^2 is known or if $N_1 + N_2 \geq 100$.
2. Student's t-distribution with $N_1 + N_2 - 2$ degrees of freedom if σ is unknown and $N_1 + N_2 < 100$.

Step 1: List assumptions.

Step 2: Set up the null hypothesis. Specify whether the test is one-tailed or two-tailed. Specify the level of significance. Determine critical region.

Step 3: Draw two simple random samples and collect data.

Step 4: From the sample data compute means, variances, the pooled variance, and the standard error of the difference.

Step 5: Substitute into the appropriate standard score formula and solve. If the standard score is in the critical region, rejection of the null hypothesis is warranted. If not, rejection is not warranted.

THE k-SAMPLE TEST

Problem

To determine whether the means of k independent samples (where k is any number greater than 2) estimate a common parameter.

Assumptions

1. Interval level of measurement
2. Independent observations
3. Normally distributed population
4. Independent simple random samples
5. Homogeneous population variances

Null and Alternate Hypotheses

$$H_0 : \mu_1 = \mu_2 = \mu_3 \ldots = \mu_k$$

$$H_1 : \mu_1 \neq \mu_2 \neq \mu_3 \ldots \neq \mu_k$$

The Sums of Squares

Total sum of squares:

$$\Sigma x_t^2 = \sum_{i=1}^{N} (X_i - \bar{X}_t)^2$$

Between-groups sum of squares:

$$\Sigma x_b^2 = \sum_{i=1}^{k} n_i (\bar{X}_i - \bar{X}_t)^2$$

DIAGRAM 15.1 *(Continued)*

Within-groups sum of squares:

$$\Sigma x_w^2 = \Sigma x_1^2 + \Sigma x_2^2 + \cdots + \Sigma x_k^2$$

where

$$\Sigma x_1^2 = \sum_{i=1}^{n_i} (X_i - \bar{X}_1)^2, \quad \text{etc.}$$

As an alternative:

$$\Sigma x_w^2 = \Sigma x_t^2 - \Sigma x_b^2$$

Variances

Between groups:

$$s_b^2 = \frac{\Sigma x_b^2}{k - 1}$$

Within groups:

$$s_w^2 = \frac{\Sigma x_w^2}{N - k}$$

F-Ratio

$$F = \frac{s_b^2}{s_w^2}$$

Sampling Distribution

The F-distribution with $df_1 = k - 1$ and $df_2 = N - k$.

Step 1: List assumptions.

Step 2: Set up the null hypothesis. Specify the level of significance. Determine critical region.

Step 3: Draw simple random samples and collect data.

Step 4: From the sample data compute total sum of of squares, sum of squares between groups, and sum of squares within groups. Compute the within-groups variance and the between-groups variance.

Step 5: Substitute into the F ratio and solve. Enter the F table with $k - 1$ and $N - k$ degrees of freedom. If the computed F is as large as or larger than the tabled F, reject the null hypothesis. If not, fail to reject.

SCHEFFE'S TEST FOR CONTRASTS

Null Hypothesis for Contrasts

To compare two means:

$$H_0 : 1\mu_1 - 1\mu_2 = 0$$

DIAGRAM 15.1 *(Continued)*

To compare one mean with two:

$$H_0 : 1\mu_1 - 1/2\mu_2 - 1/2\mu_3 = 0$$

To compare one mean with three:

$$H_0 : 1\mu_1 - 1/3\mu_2 - 1/3\mu_3 - 1/3\mu_4 = 0$$

To compare two means with two other means:

$$H_0 : 1/2\mu_1 + 1/2\mu_2 - 1/2\mu_3 - 1/2\mu_4 = 0$$

General form of null hypothesis:

$$H_0 : c_1\mu_1 + c_2\mu_2 + c_3\mu_3 + \cdots + c_k\mu_k = 0$$

Contrast Variance

$$s^2_{\text{contr}} = (\Sigma\, c_j\bar{X}_j)^2 / \Sigma\, (c_j^2/n_j)$$

F-Ratio

$$F = s^2_{\text{contr}} / s^2_w$$

Sampling Distribution

The *F*-distribution:

1. For *a priori* contrasts: F with $df_1 = 1$ and $df_2 = N - k$

2. For *a posteriori* contrasts: $F = (k - 1)F$, with $df_1 = k - 1$ and $df_2 = N - k$

For *a priori* use:

Step 1: Set up null hypothesis. Specify level of significance. Determine critical region.

Step 2: Draw random sample and collect data.

Step 3: From sample data, compute total sum of squares, within-groups sum of squares, contrast variances, and within-groups variance.

Step 4: Substitute into F ratio and solve. Enter F table with 1 and $N - k$ degrees of freedom. For each computed F as large as or larger than the tabled F, reject the null hypothesis. For each not as large, fail to reject.

For *a posteriori* use:

Step 1: If analysis of variance warrants rejection of null hypothesis, generate null hypotheses for desired comparisons. Specify level of significance. Determine critical region.

Step 2: Compute contrast variances.

Step 3: Compute F ratios. Enter F table with $k - 1$ and $N - k$ degrees of freedom. Multiply tabled F by $k - 1$ to find critical value of F. If computed F's are as large or larger than $(k - 1)$ times tabled F, reject null hypothesis. If not, fail to reject.

DIAGRAM 15.2 Tests of Hypotheses about Proportions

The One-sample Test

Problem

     To determine whether a given sample proportion is an estimate of a known or theoretical population proportion.

Assumptions

1. Independent observations
2. Simple random sample of the population
3. NP and NQ are both equal to or greater than 5

Null and Alternative Hypotheses

$$H_0 : P = \text{(some specified value)}$$

$$H_1 : P \neq \text{(that specified value)}$$

or

$$P > \text{(that value)} \quad \text{or} \quad P < \text{(that value)}$$

Standard Error

$$\sigma_p = \sqrt{\frac{PQ}{N}}$$

Standard Score Formula

$$z = \frac{p - P}{\sigma_p}$$

Sampling Distribution

     Normal curve.

Step 1: List assumptions.

Step 2: Set up the null hypothesis. Specify whether the test is one-tailed or two-tailed. Specify the level of significance.

Step 3: Draw a simple random sample from the population and collect data.

Step 4: From the sample data compute the sample proportion. Compute the standard error of the proportion.

Step 5: Substitute into the standard score formula and solve for z. If the computed z is in the critical region, reject the null hypothesis. If not, fail to reject.

The Two-sample Test

Problem

     To determine whether two sample proportions from two independent random samples estimate a common population proportion.

DIAGRAM 15.2 *(Continued)*

Assumptions

1. Independent observations
2. Independent simple random samples

Null and Alternative Hypotheses

$$H_0 : P_1 = P_2$$

$$H_1 : P_1 \neq P_2$$

or

$$P_1 > P_2 \quad \text{or} \quad P_1 < P_2$$

Standard Error of the Difference

Pool sample estimates p_1 and p_2 to get p; find q by subtraction ($q = 1 - p$), and substitute into following standard error formula:

$$s_{p_1 - p_2} = \sqrt{pq\left(\frac{1}{N_1} + \frac{1}{N_2}\right)}$$

Standard Score Formula

$$z = \frac{p_1 - p_2}{s_{p_1 - p_2}}$$

Sampling Distribution

The normal curve with a mean difference between proportions of 0.

Step 1: List assumptions.

Step 2: Set up the null hypothesis. Specify whether the test is one tailed or two tailed. Specify the level of significance.

Step 3: Draw two simple random samples and collect data.

Step 4: From the samples compute sample proportions. Compute the standard error of the difference between proportions.

Step 5: Substitute into the standard score formula and compute z. If the computed z is in the critical region, reject the null hypothesis. If not, fail to reject.

DIAGRAM 15.3   Tests of Hypotheses about Variances

THE ONE-SAMPLE TEST

Problem

   To determine whether a sample variance is an estimate of a known or theoretical population variance.

Assumptions

1. Interval level of measurement
2. Independent observations
3. Normally distributed population
4. Simple random sample of that population

Test Statistic

$$\chi^2 = \frac{(N-1)s^2}{\sigma^2}$$

Sampling Distribution

   χ^2-distribution with $N-1$ degrees of freedom.

Step 1: List assumptions.

Step 2: Set up the null hypothesis. Specify whether the test is one-tailed or two-tailed. Specify the level of significance.

Step 3: Draw a simple random sample from the population and collect data.

Step 4: From the sample data, compute the sample variance.

Step 5: Substitute sample variance and population variance into χ^2 formula and solve. If χ^2 is in the critical region, reject the null hypothesis. If not, fail to reject.

THE TWO-SAMPLE TEST

Problem

   To determine whether two sample variances estimate a common population variance.

Assumptions

1. Interval level of measurement
2. Independent observations
3. Normally distributed populations
4. Simple random samples from those populations

Test Statistic

$$F = s_1^2/s_2^2$$

DIAGRAM 15.3 *(Continued)*

If the null hypothesis is for a one-tailed test, the sample variance, which the alternative hypothesis predicts will be larger, is put in the numerator of the F ratio.

If the null hypothesis is for a two-tailed test, the larger sample variance will always be the numerator.

Sampling Distribution

F-distribution with $N_1 - 1$ and $N_2 - 1$ degrees of freedom. Degrees of freedom for variance in numerator of F ratio will be df_1 and in denominator df_2.

Step 1: List assumptions.

Step 2: Set up the null hypothesis. Specify whether the test is one tailed or two tailed. Specify the level of significance.

Step 3: Draw two simple random samples and collect data.

Step 4: Fom the sample data compute the two-sample variances.

Step 5: Substitute the sample variances into the F formula. If the computed F is in the critical region, reject the null hypothesis. If not, fail to reject.

DIAGRAM 15.4   Generating Normal Expected Frequencies from Data

Problem

To determine what the frequencies would be for a frequency distribution of sample data if the data were representative of a normally distributed population.

Assumptions

1. Independent observations
2. Interval level of measurement
3. Data constitute a simple random sample of the population
4. Sample mean and standard deviation are estimates of population mean and standard deviation
5. Sum of the normal expected frequencies is equal to N

Step 1:   Draw a simple random sample from the population and collect data.

Step 2:   Compute the mean and the standard deviation. Set up a frequency distribution.

Step 3:   Find the lower limit of each class. Find the standard score corresponding to each lower limit (except for the lower limit of the first class at the low end of the distribution). Use the following z-score formula:

$$z = \frac{X - \bar{X}}{s}$$

where X is a lower limit of a class, $\bar{X}$ is the sample mean, and s is the sample standard deviation.

Step 4:   Using the table of the normal curve (Appendix Table A), find the areas from the mean to each standard score. Add an additional area to each end of the distribution equal to .5000. These two areas are the areas from the mean to the end of the distribution in each tail of the normal curve.

Step 5:   Wherever adjoining z-scores have the same sign, take the difference between their areas. In the middle of the distribution where one z-score is negative and the adjoining one positive, add the two areas (this is the area around the mean).

Step 6:   Multiply each difference by N to convert the proportions to frequencies. The resulting frequencies are the frequencies we would expect to find if the distribution was, in fact, normal. (Study Table 15.19 in conjunction with these instructions to learn how the expected frequencies were computed there.)

Step 7:   Use the expected frequencies and the observed frequencies to compute χ^2.

CHAPTER 16

Hypothesis Tests of Association

Lethal aggressive behavior, resulting in suicides and homicides, has long been of interest to sociologists. Durkheim's classical work on suicide was an early study in this tradition. Whitt, Gordon, and Hofley conducted a cross-national study of the relationship of suicide and homicide rates to level of industrialization and religious tradition (Whitt, Gordon, and Hofley, 1972). They postulated that suicide and homicide are alternative aggressive responses to frustration. Frustration is thought to arise in a population as an effect of incongruencies between religious and economic values. The researchers formed a "lethal aggression rate" by adding together suicide and homicide rates for countries, and they hypothesized that the lethal aggression rate would be negatively correlated with industrialization in those countries where religious traditions are congruent with industrialization and positively correlated in those where religious traditions are incongruent. The proportion of economically active males not engaged in agriculture was used as an index of industrialization.

The researchers argued that Protestantism is congruent with industrialization, and, therefore, in a sample of Protestant countries the correlation between lethal aggression rate and industrialization should be negative. When they analyzed their data for some Protestant countries they found the coefficient of correlation between lethal aggression rate and level of industrialization to be $-.53$, a moderate negative correlation. The authors concluded that this result supported their theoretical expectations.

If Whitt, Gordon, and Hofley wished only to reach conclusions about countries for which they had data, merely computing the correlation coefficient would have been sufficient. If they desired, on the other hand, *to generalize their finding* beyond the countries for which they had data, they would be obligated to turn to **inferential hypothesis testing procedures**. They were, in fact, interested in generalizing and did proceed to test inferential statistical hypotheses about the particular coefficient of -0.53 and others computed in the course of their study.

Since statistics based on sample data are subject to sampling error, it is necessary to determine whether the association found for the sample can be generalized to the population or whether it is attributable to sampling error. The usual procedure used is to test the null hypothesis of zero association, using our knowledge about the sampling distribution of the statistic in question. For example, we might wish to test the null hypothesis that the Whitt, Gordon, and Hofley coefficient of $-.53$ is a sample estimate of a population coefficient of correlation, which is actually 0.

$$H_0 : \rho = 0$$

where rho (ρ) is the population correlation coefficient.

If the data were such that rejection of this null hypothesis was warranted, we might choose to conclude that the correlation found for the sample data would be generalizable to the population from which the sample came. If the null hypothesis is not rejected, we are saying, in effect, that the correlation found may be attributable to sampling error and is, therefore, not generalizable. The correlation of $-.53$ does, by the way, warrant rejection of the null hypothesis. A technique to test the null hypothesis regarding a correlation coefficient will be described in Section 16.4.

In this chapter we will examine procedures for testing null hypotheses for various measures of association. First, we will look at a technique appropriate to test the null hypothesis about various measures of nominal association. Then we will discuss tests of the null hypothesis for measures of ordinal association. Finally, we will explain how to test the null hypothesis for interval level correlation and regression.

16.1   CHI-SQUARE (χ^2) TEST FOR INDEPENDENCE

Sociologists are frequently interested in measuring association between two nominal scale variables. Many of the variables with which they work are measured by crude instruments, and these measurements lend themselves best to nominal scale interpretation. For example, in studying gender as a variable sociologists commonly make a gross distinction between male and female. They seldom attempt to measure degree of maleness or femaleness.

A number of measures of nominal association, which are based upon the χ^2 statistic (delta based), are important subjects for study in the field of descriptive statistics (see Chapter 6). The χ^2 statistic can be used to test the null hypothesis of no association, when one of these measures of association is used. The χ^2 technique that is appropriate for this kind of application is known as the **chi-square (χ^2) test for independence**. The null hypothesis that is tested is, in fact, that the two variables in question are independent rather than interrelated.

It should be noted at the outset that chi-square (χ^2) is both the term for a statistical technique used to test hypotheses and the term for a family of sampling distributions. Chi-square sampling distributions were discussed at some length in Section 12.6a, and were used in Chapter 13 to compute a confidence interval for the variance (Section 13.4), in Chapter 15 in the one-sample test for the variance (Section 15.6), and again in Chapter 15 to test goodness-of-fit (Section 15.8).

To illustrate the chi-square test for independence, we will turn to a study of language retention and labor-force participation conducted by Robinson

---

BOX 16.1 INDEPENDENCE

The discussion that follows assumes that you understand the concept of independence and the special multiplication rule of probability related to it. If this material seems hazy to you, it might be wise to go back and review Section 10.3.5.

(1985). Robinson investigated the relationship between retention of an Indian mother tongue and various aspects of labor-force activity using the Canadian Public Use Sample from the 1971 census. Rather than concentrate on the total sample of the Canadian Indian ethnic group, she focused only on those who had first spoken and still understood their mother tongue. There were 390 males who fell into this category. As part of a more elaborate analysis (which included females), Robinson carried out a chi-square test of independence between language retention and labor-force participation for her male subjects. The null hypothesis she tested was, in effect, that there was no difference in tendency to participate in the labor-force among those who retained their mother tongue and those who did not. It is this analysis that we will examine to illustrate the use of the chi-square technique. Table 16.1 presents Robinson's data for the 390 Indian males.

16.1.1 Computation of Expected Frequencies*

If the 390 males for whom data were available represent a simple random sample of all Indian males in the population who first spoke and still understood their Indian mother tongue, then we can use the information in Table 16.1 to estimate certain parameters. The four marginal totals of Table 16.1 (108, 282, 168, and 222) give the distributions of the 390 men in the sample on retention or nonretention of their mother tongue and participation or nonparticipation in the labor force. Thus, 282 of the men retained their mother tongue and 108 did not; 222 participated in the labor force and 168 did not. These marginal totals can be converted to proportions by dividing each by $N = 390$. Then, if the sample is representative of the

---

TABLE 16.1 LABOR FORCE PARTICIPATION BY LANGUAGE RETENTION AMONG CANADIAN MEN WITH INDIAN MOTHER TONGUE

	Nonretainers	Retainers	Totals
Not in labor force	32	136	168
In labor force	76	146	222
Totals	108	282	390

Source: Adapted from Robinson (1985:521, Table 1).

---

* Also discussed in Section 6.4.1a.

population, these proportions will be estimates of the corresponding parameters. That is, the proportion $282/390 = .72$ will be an estimate of the proportion of men in the population who retained their mother tongue; and $108/390 = .28$ will be an estimate of the proportion who did not. Likewise, the proportion $222/390 = 0.57$ will estimate the proportion of men in the population who participated in the labor force and $168/390 = 0.43$, the proportion who did not.

For the random sample of 390 men, the four entries in Table 16.1 represent the observed number of men who did not retain their mother tongue and were not in the labor force (32), of men who did retain their mother tongue and were not in the labor force (136), of men who did not retain their mother tongue and were in the labor force (76), and of men who did retain their mother tongue and were in the labor force (146). The question to which the chi-square test of independence addresses itself is whether these *observed frequencies* resemble those that would be obtained if language retention and labor force participation were actually independent of each other.

Earlier, in Section 10.3.5, we found that *the probability of the joint occurrence of independent events is equal to the product of their separate probabilities.* Assuming the null hypothesis to be true, we can use this rule of probability to determine what the entries in Table 16.1 would look like if the two variables in question *were* actually independent. This second set of entries we will call *expected frequencies*. The four proportions that were computed from the *marginal totals* of Table 16.1 (.72, .28, .57, and .43) can be treated as probabilities of occurrence of events in the population. Thus, .72 represents the probability of men in the population retaining their mother tongue, .28 represents the probability of their not retaining it, .57 represents the probability of their participating in the labor force, and .43 represents the probability of their not participating.

If the null hypothesis were true and language retention were, in fact, independent of labor-force participation, then by the probability rule of the joint occurrence of independent events the probability of language nonretainers not participating in the labor force would be $(.28)(.43) = .1204$, the probability of language retainers not participating would be $(.72)(.43) = .3096$, the probability of language nonretainers participating in the labor force would be $(.28)(.57) = .1596$, and, finally, the probability of language retainers participating would be $(.72)(.57) = .4104$. These latter probabilities (.1204, .3096, .1596, and .4104) can now be used to compute the expected frequencies for the four entries in Table 16.1. For convenience in discussing the entries and marginal totals for a 2 by 2 table, let them be labeled as they are below:

a	b	$a + b$
c	d	$c + d$
$a + c$	$b + d$	N

Thus, .1204 is the expected proportion for cell a, .3096 for cell b, .1596 for cell c, and .4104 for cell d. These expected proportions may be converted to expected frequencies by multiplying each by $N = 390$. The resulting expected frequencies are those that would occur if the two variables in question were independent.

To summarize how we computed the expected frequency for the "language nonretainers who were not in the labor force": First we assumed that the 390 cases represented a random sample of the population and that the proportions computed from the marginal totals could be used to estimate the corresponding population proportions. Next, we assumed that the null hypothesis of independence was true, allowing us to apply the special multiplication rule of probability for independent events. This justified multiplying 108/390 by 168/390 to get the expected proportion of cases; then we multiplied this expected proportion by N to get the expected frequency. That is,

$$\left(\frac{108}{390}\right)\left(\frac{168}{390}\right)(390)$$

Note that one of the N's (390) in the denominators of the fractions can be canceled with the final N to give the following result:

$$\frac{(108)(168)}{390}$$

In general, the expected frequencies for each cell can be computed by multiplying the marginal totals for the cell and dividing by N.

To find the expected frequency for cell a in the 2 by 2 table above, the following formula can be used:

$$f_e(a) = \frac{(a + c)(a + b)}{N}$$

For cell b the formula is

$$f_e(b) = \frac{(b + d)(a + b)}{N}$$

For cell c the formula is

$$f_e(c) = \frac{(a + c)(c + d)}{N}$$

Finally, for cell d the formula is

$$f_e(d) = \frac{(b + d)(c + d)}{N}$$

Using the formula for cell a with the data from Table 16.1 gives

$$\frac{(108)(168)}{390} = 46.52$$

In a 2 by 2 table such as Table 16.1, it is not necessary to compute all four expected frequencies in this way. Because the sums of the various pairs of expected frequencies must equal the marginal totals, once we find the expected frequency for one cell we can find the other three by subtracting the known number from the marginal totals. For example, we can get the expected frequency for cell c of Table 16.1 by subtracting the expected frequency for cell a (46.52) from the mar-

TABLE 16.2   EXPECTED FREQUENCIES FOR DATA OF TABLE 16.1

	Nonretainers	Retainers	Totals
Not in labor force	46.52	121.48	168
In labor force	61.48	160.52	222
Totals	108	282	390

ginal total $a + c$ (108). Then the expected frequencies for cells b and d can be found, in turn, by subtracting from marginal totals $a + b$ and $c + d$. Table 16.2 presents the expected frequencies (f_e's) we calculated from Table 16.1.

16.1.2   Test of the Null Hypothesis

Once all of the expected frequencies have been computed, they can be compared with the observed frequencies to see whether they are similar. Essentially, chi-square measures the extent to which the observed frequencies in a contingency table deviate from those frequencies that would be expected if the null hypothesis were true. The null hypothesis could be stated as follows: H_0:*no association between variables in the population* or, in this case, the population probability of labor-force participation given one has retained one's mother tongue, is simply equal to the probability of retaining one' mother tongue (*i.e.*, they are independent). The alternative to the null hypothesis would be simply H_a:*there is some association between variables in the population.*

To test the null hypothesis, the following formula is used to compute χ^2:

(16.1)
$$\chi^2 = \sum \frac{(f_o - f_e)^2}{f_e}$$

where f_o is the frequency observed for a particular cell and f_e is the frequency expected for the same cell. The large summation sign, Σ (capital sigma), tells us to compute the fractions for each cell and then sum over all cells to get χ^2. In Table 16.3 we have set up the computations necessary to get χ^2 for the Robinson data.

A careful examination of Table 16.3 will reveal that what was done to compute χ^2 was to (a) find the difference between each observed frequency and the

TABLE 16.3   CHI-SQUARE (χ^2) COMPUTATIONS FOR DATA FROM TABLE 16.1

Cell	f_o	f_e	$f_o - f_e$	$(f_o - f_e)^2$	$(f_o - f_e)^2/f_e$
a	32	46.52	-14.52	210.83	4.53
b	136	121.48	14.52	210.83	1.74
c	76	61.48	14.52	210.83	3.43
d	146	160.52	-14.52	210.83	1.31
				$\chi^2 =$	11.01

corresponding expected frequency for each cell in the table, (b) square each difference, (c) divide each squared difference by its respective expected cell frequency, and (d) add the resulting quotients for all cells in the table. The sum of these quotients is the computed χ^2.

Once χ^2 has been obtained, it must be compared to the χ^2 value in the appropriate chi-square sampling distribution that bounds the desired critical region. If the computed χ^2 is in the critical region, it is appropriate to reject the null hypothesis. The computed χ^2 for Table 16.3 was 11.01.* To determine whether that value is in the critical region of the sampling distribution, we must turn to the table of chi-square (Appendix Table C).

As we pointed out in Section 12.6a, the chi-square sampling distribution is actually a family of related distributions, and the table must be entered after the degrees of freedom and level of significance have been determined. The first column of the table lists degrees of freedom and the headings list various areas in the right tail of the distribution. The values in the body of the table represent the χ^2 scores that bound the critical region with the areas noted in the column headings.

Assuming that we have decided to test the null hypothesis at the 5% level of significance, the column headed .05 is the one we must look at. We also, however, must determine the number of degrees of freedom for the data we have analyzed so that we will know which of the many chi-square distributions in the table to use.

In the chi-square test for independence the number of categories in which the data have been classified (on both variables) is the determinant of the number of degrees of freedom. The applicable formula is as follows:

(16.2)
$$df = (r - 1)(c - 1)$$

where r equals the number of categories into which the horizontally listed (row) variable is classified, and c equals the number of categories into which the vertically listed (column) variable is classified. In the example, labor-force participation is the row variable, and it is divided into two categories: not in labor force and in labor force; thus $r = 2$. The column variable represents retention of mother tongue, and it is divided into two categories as well: nonretention and retention; thus $c = 2$. Therefore, the number of degrees of freedom for this case is

$$df = (2 - 1)(2 - 1) = 1$$

A restriction placed on the computation of the expected frequencies for the test for independence was that the sums of the expected frequencies equal the marginal totals. Therefore, when the first expected frequency was found for the 2 by 2 table, the other three expected frequencies could be determined simply by subtraction. The number of degrees of freedom for a 2 by 2 table is 1. This corresponds to the number of entries in the table that are not determined by the need for the sums of the expected frequencies to agree with the marginal totals. It can be said, in general, that the number of degrees of freedom in a chi-square test of independence will be equal to the number of cells in the contingency table not determined by the marginal totals of the table.

---

* Robinson reported a corrected chi-square of 10.3 in her article (1985:521). She used the correction for continuity, which is explained in Section 16.1.5.

We will return now to the chi-square table with a .05 critical region and 1 degree of freedom to see what the tabular chi-square value is that bounds the critical region. The table shows a value of 3.841. Therefore, any computed χ^2 as large as or larger than 3.841 will be in the critical region and warrant the rejection of the null hypothesis. The χ^2 computed from Robinson's data was 11.01; thus rejection of the null hypothesis was warranted. Consequently, Robinson could conclude with some confidence that language retention and labor-force participation were not independent in the population to which she wished to generalize.

16.1.3 χ^2 for Larger Tables

The χ^2 test for independence is not limited to 2 by 2 tables. The χ^2 can be computed for tables of any size as long as the assumptions we will discuss below are met. The amount of computing involved increases as the size of the table increases. However, in any case, the computations should be done by computer or calculator.

To illustrate the computation of χ^2 from a larger table, we will look at a study conducted by Crane (1965). She studied a sample of scientists (biologists, political scientists, and psychologists) at major and minor universities to explore factors related to productivity and recognition. Note that Crane's data do not constitute a simple random sample of a specified population. She apparently assumed that they were a random sample from some population, and thereby justified her use of a test of statistical significance. For purposes of illustrating the technique in question, we will also make this assumption.

Among other things, Crane was interested in finding whether there was an association between scholarly productivity and the current academic affiliation of the scientists in the population. Current academic affiliation was assigned to one of three categories: major university, high-minor university, and low-minor university, based on the work of Berelson (1960). These categories were cross-classified against high and low productivity and the null hypothesis of independence was tested at the .01 level of significance. Crane's data and our test of her null hypothesis are presented in Table 16.4.

In Table 16.4, as in Table 16.1, the expected frequencies were computed by multiplying the appropriate marginal totals and dividing by N. Thus, the expected frequency for the high-productivity–major-university entry was found as follows:

$$\frac{(72)(54)}{150} = 25.92$$

Likewise, the high-productivity–high-minor university entry expected frequency was found by

$$\frac{(36)(54)}{150} = 12.96$$

Since Table 16.4 is a 2 by 3 table, the number of degrees of freedom is $df = (r-1)(c-1) = (2-1)(3-1) = 2$. With 2 degrees of freedom, two of the expected frequencies are free to vary and must be computed by multiplying marginal totals and dividing by N. The other expected frequencies may be found by subtracting from the marginal totals. Therefore, once the expected frequencies

TABLE 16.4   CHI-SQUARE (χ^2) TEST OF PRODUCTIVITY BY CURRENT ACADEMIC AFFILIATION USING CRANE DATA

Productivity	Current Academic Affiliation			Totals
	Major University	High-Minor University	Low-Minor University	
High	33 (25.92)	14 (12.96)	7 (15.12)	54
Low	39 (46.08)	22 (23.04)	35 (26.88)	96
Totals	72	36	42	150

Computations for χ^2

Cell	f_o	f_e	$f_o - f_e$	$(f_o - f_e)^2$	$(f_o - f_e)^2/f_e$
a	33	25.92	7.08	50.1264	1.93
b	14	12.96	1.04	1.0816	0.08
c	7	15.12	−8.12	65.9344	4.36
d	39	46.08	−7.08	50.1264	1.09
e	22	23.04	−1.04	1.0816	0.05
f	35	26.88	8.12	65.9344	2.45
					$\chi^2 = 9.96$

Source: Adapted from Crane (1965:704, Table 2).

25.92 and 12.96 (for example) are found as above, all the others can be found by subtraction.

Once all the expected frequencies were found for Table 16.4, the differences between observed and expected frequencies were computed and squared. Each squared difference was divided by its expected frequency, and the quotients were summed to obtain χ^2.

You may have noticed that all the differences in the 2 by 2 Table 16.1 of the previous example were the same (although two were negative and two were positive). In the 2 by 3 Table 16.4, the differences were not all the same. Differences for high and low productivity were the same for pairs in the same column because the productivity variable was dichotomized, but differences across rows for current academic affiliation were not the same. Only when variables are dichotomized will the differences between observed and expected frequencies be the same for different cells in the table. This is because rows and columns must add up to their marginal totals.

The computed χ^2 in Table 16.4 was 9.96. Entering the χ^2 table (Appendix Table C) with 2 degrees of freedom and a .01 level of significance, we find that the χ^2 value bounding the critical region is 9.210. The computed χ^2 is in the critical region, so rejection of the null hypothesis was warranted. Crane reached the conclusion that current academic affiliation was associated with scholarly productivity.

To determine the nature of the relationship, it is necessary to examine the differences between observed and expected frequencies for the individual cells of the contingency table. The differences reveal that scientists affiliated with major universities produced more scholarly work than chance would lead one to expect and those affiliated with low-minor universities produced less. The expected and observed frequencies for high and low productivity were very much alike for those scientists affiliated with high-minor universities so those cells contributed little to the χ^2. The statistically significant result is attributable to differences between scientists in major universities and those in low-minor universities.

Crane also collected data on universities at which her subjects did their graduate work and found that variable was related to productivity as well. To explore further the relationship between productivity and current academic affiliation, it would be possible to use scientist's graduate school as a control variable. Thus, a **set of conditional tables** could be examined in which prestige level of the scientist's graduate school was controlled at the three levels—major university, high-minor university, and low-minor university—and current academic affiliation was compared with level of productivity. Separate χ^2's could be computed for the three conditional tables to test them for independence. Furthermore, the three resulting χ^2's could be **summed to give an overall test** of the independence of current academic affiliation and the level of productivity, holding graduate school constant. That is, adding the three χ^2's for the conditional tables tests the independence of the two main variables while sorting out the effects of the control variable. To evaluate this **pooled χ^2**, the degrees of freedom for each of the conditional tables are also summed. Thus, if we were to pool the χ^2's from three 2 by 3 conditional tables, the resulting χ^2 would be evaluated against a chi-square sampling distribution with 6 degrees of freedom, again, assuming independent random samples.

Note that the value of the pooled χ^2 would not necessarily be the same as the overall χ^2 of 9.96, which was computed from the data of Table 16.4. That χ^2 ignored data on the scientist's graduate school; the pooled χ^2 would reflect the influence of that variable.

BOX 16.2   CONDITIONAL TABLES

If you don't quite understand this discussion about the additive nature of χ^2, perhaps it is because you are unclear about the conditional table concept. If this is the problem, you must review material on conditional tables (see early portions of Chapter 8).

16.1.4   Interpretation of a Significant χ^2

The χ^2 computed from Crane's data fell in the critical region but fairly close to the boundary of that region. It is possible for a computed χ^2 to extend quite far into the right tail of the sampling distribution. The larger the χ^2 (depending upon the number of degrees of freedom involved), the farther it will be into the right tail of the sampling distribution.

What does it mean when the computed χ^2 is large? It does *not* mean that the relationship between the variables being investigated is strong. The χ^2 *does not measure strength of relationship*. It merely measures whether there is a relationship that is not likely to be due to chance. When the value of χ^2 is large, it means that we can be more confident about rejecting the null hypothesis and concluding that the variables are related.

The level of statistical significance of the χ^2 test is, in fact, independent of the strength of the relationship that exists between the variables studied. It is possible to get a large and very significant χ^2 when the association between the variables of interest is very weak. It is also possible to get a nonsignificant χ^2 when the association between the variables of interest is very high. This latter situation would mean, in effect, that although there is a strong association between the variables for the data at hand, generalization beyond the data would not be warranted.

If the uncertainty coefficient U_{yx},* is computed for Crane's data in Table 16.4 the resulting coefficient is .05. This is a modest association, and yet χ^2 was found to be statistically significant at the .01 level.

If the sizes of the deviations between the observed and expected frequencies are held constant and size of sample is increased, the size of the resulting χ^2 will be decreased. For example, if the differences between observed and expected frequencies in a 2 by 2 table were 10 and $N = 100$, holding the differences constant at 10 and increasing N to 200 would have the effect of decreasing χ^2. *Deviations of 10 are less likely in smaller samples (e.g., 100 cases) than in larger ones.*

On the other hand, if both the deviations between observed and expected frequencies and sample size are doubled, the resulting χ^2 will also be doubled, thereby moving it farther into the critical region of the sampling distribution. The principle to consider here is that *large deviations found in a large sample provide more evidence for rejecting the null hypothesis than proportionately the same deviations from a smaller sample.*

16.1.5 Assumptions of the χ^2 Test for Independence

There are a number of assumptions that are implied by the χ^2 test for independence. *First*, it is assumed that the data being analyzed are a simple random sample of the population. This assumption makes it possible to use the marginal totals to compute expected frequencies and test the null hypothesis of no association.

Second, it is assumed that the observations are independent, a property that simple random sampling assures. This rules out the use of χ^2 for comparing observations of the same subjects at two different points in time or for comparing observations of matched subjects.

Third, it is assumed that no expected frequency in the contingency table being analyzed will be less than 5. When expected frequencies are too small, the

---

* The uncertainty coefficient, U_{yx} is a measure of association for nominal data mentioned in Section 7.2.2.

chi-square sampling distribution does not represent adequately the distribution of the test statistic:

$$\sum \frac{(f_o - f_e)^2}{f_e}$$

It is assumed that the underlying distribution of the computed chi-square statistic is continuous because the chi-square sampling distribution is a continuous distribution. In fact, the distribution of the chi-square statistic is discrete; however, when degrees of freedom are more than 1, the assumption of continuity is reasonable.

There is some disagreement about how reasonable the assumption is for the 2 by 2 table with 1 degree of freedom (Grizzle, 1967; Mantel and Greenhouse, 1968). It is suggested that when the computed χ^2 is in the critical region but near the boundary of that critical region a correction for continuity should be made. The correction is effected by reducing the absolute differences between observed and expected frequencies by .5. The formula for the corrected χ^2 is as follows:

(16.3)
$$\chi_c^2 = \sum \frac{(|f_0 - f_e| - .5)^2}{f_e}$$

The correction has the effect of reducing the magnitude of the computed χ^2. Therefore, a χ^2 that is very near the boundary of the critical region may be made nonsignificant by the correction. If the χ^2 is well within the critical region, the correction will make no difference in the results of the analysis.

In a table that is larger than a 2 by 2 table, it is often possible to combine categories, reduce the size of the table, and increase the size of the expected frequencies. Any such collapsing of categories will reduce the amount of information to be obtained from the analysis and should not be attempted unless the combination of categories has some logical justification (grouping error may also result, see Section 6.6). For example, if the expected frequencies had not been large enough in the 2 by 3 Table 16.4 we just completed analyzing, it would be logically defensible to combine the high-minor universities and the low-minor universities into a single minor university category if that would serve the purpose of increasing the sizes of the expected frequencies sufficiently. It would not be logically defensible, however, to combine the major universities and the low-minor universities and compare them to the high-minor universities.*

Fourth, the sum of all of the expected frequencies must be equal to the sum of all the observed frequencies. If this is not so, then it would be impossible to get a χ^2 of 0. In testing for independence a χ^2 of 0 (representing no difference between observed and expected frequencies) must be a distinct possibility. This is assured by the way expected cell frequencies are computed under H_0.

Two conditions that can be relaxed in using the χ^2 test are the measurement requirement, and the requirement that the shape of the population distribution must be specified. The χ^2 lends itself to the analysis of nominal data. As a matter of fact, it is insensitive to order; and when it is used to analyze ordinal or interval data, information on ordering of categories is ignored.

---

* When the expected frequencies are too small in a 2 by 2 table, Fisher's Exact Test can be used instead of χ^2 (see, for example, Blalock, 1979.)

Neither is χ^2 limited to use with a normally distributed population variable. It can be used no matter what the shape of the population distribution. It is a distribution-free or nonparametric technique.

These two latter points make the χ^2 technique an extremely popular one among sociologists. It should be pointed out, however, that χ^2 should not be used in lieu of another more powerful technique just because it is rather widely applicable. The sociologist should always use the most appropriate and powerful technique called for by the research problem and warranted by the data. Furthermore, χ^2 should be used in conjunction with one of the measures of association appropriate to nominal data, such as the uncertainty coefficient discussed in Chapter 7.

If a measure of association is computed for a body of sample data and that measure does not reveal a level of association strong enough to be theoretically or practically worthwhile, there seems to be little point in computing χ^2. This is because the only function that χ^2 serves is to help the researcher decide whether sample findings can be generalized to the population from which the sample was drawn.

Unfortunately, there seems to be widespread misunderstanding of what a statistically significant χ^2 means. That fact, plus the fact that χ^2 is easy to compute and has a set of assumptions that are not too demanding, has led to a greater use of the technique in the literature than seems warranted.

Diagram 16.1 (p. 608) summarizes step by step the procedures used in the χ^2 test of independence.

16.2   TESTING HIERARCHICAL MODELS

In Section 8.1.4b we had occasion to discuss the use of odds ratios in hierarchical analysis. As we pointed out, the concept is very useful for analyzing cross-classifications of three or more nominal or ordinal level variables. We demonstrated how the odds ratios may be used to sort out relationships between and among such variables for data at hand.

We mentioned, in the process of discussing odds ratios, that Goodman (1972a, 1972b) has provided techniques to test models that specify the particular nature of the relationships between and among categorical variables. It is appropriate to examine these model-testing techniques at this point because (a) they are used inferentially, (b) they involve the process of hypothesis testing, (c) the hypotheses posit associations between variables, and (d) the chi-square technique is used to test the models.

BOX 16.3   ODDS RATIOS AND HIERARCHICAL MODELS

Because odds ratio analysis is so integral to testing hierarchical models, it is recommended that you go back and review Section 8.1.4b before studying Section 16.2. After reviewing Section 8.1.4b and studying Section 16.2, it will become clear to you that odds ratio analysis is the descriptive counterpart of the inferential tests of hierarchical models covered in this chapter.

Just as odds ratios deal with hierarchies of relationships from univariate to multivariate, so too inferential tests of the models can be hierarchical in nature. At the top of the hierarchy are what Goodman refers to as **saturated models**, which include every conditional relationship possible for a given set of data, and at the bottom is an overall model with no variables controlled and no conditions specified. The point of the whole model-testing exercise is to identify the theoretical model somewhere between the top and the bottom that best fits the empirical data.

16.2.1   Effects

We have been using the term *relationship* in discussing these models; however, that term is too narrow to describe all that can be going on in a theoretical model. The more general term used by Goodman in discussing the variables in a model is the term **effect**. Effect is used in a theoretical sense to explain why the frequencies in a contingency table vary from one cell to another. There are, according to Goodman, single-variable, two-variable, and multiple-variable effects. A single-variable effect on a cell frequency is attributable to the marginal distribution of the variable in question. For example, in a table comparing the death rates of blacks and whites, differences in cell frequencies occur, in part, because whites greatly outnumber blacks in the U.S. population. The marginal totals for number of whites and blacks, in other words, will reflect the distribution of those two racial groups in the population. A two-variable effect is a difference in cell frequencies attributable to association or relationship between two variables. For example, when we did the chi-square test for independence on Robinson's data dealing with language retention and labor-force participation (Section 16.1), we found the two variables were associated; consequently, there were fewer retainers of the mother tongue in the labor force than expected and more nonretainers than expected.

Multiple-variable effects are due to interactions among three or more variables that influence the distribution of frequencies in the cells of a table. For example, in Chapter 8 when we examined Messer's (1967) data on morale, interaction, and age environment, we found that the level of interaction affected the relationship between age environment and morale (Table 8.2E). In a low-interaction situation, morale was lower in a mixed-age environment than in an age-concentrated environment; however, in a high interaction situation, age environment and morale were unrelated.

Odds ratios were, in fact, used in Chapter 8 to analyze these different levels of effects: first-order odds ratios that deviate from 1.00 indicate single-variable effects, second-order odds ratios that deviate from 1.00 indicate two-variable effects (associations), and third- and higher order odds ratios that deviate from 1.00 indicate multiple-variable effects (interactions). Consequently, an odds ratio equal to 1.00 indicates that there is no effect.

16.2.2   Hierarchical Models

As was mentioned previously, models in the Goodman system are arranged hierarchically, with the saturated model at the top and what we may now describe as the **no-effects** model at the bottom. The saturated model includes all effects possible for the particular variables included in the model. Between the saturated model and the no-effects model are a number of other models that become increasingly more simple (include fewer effects) as they move down the hierarchy. For example, if we were to develop models for two variables, A and B, the saturated model would include the two single-variable effects $\{A\}$ and $\{B\}$ and the two-variable effect $\{AB\}$.* The no-effects model would, naturally, not include any of the variables and between these two extremes would be three other possible models: the one including both single-variable effects, $\{A\}\{B\}$; the one including a single-variable effect for A alone, $\{A\}$; and the one including a single-variable effect for B alone, $\{B\}$. Arranging these models in order would give us the following hierarchy:

Model	Effects Included
$\{AB\}$	$\{AB\}\{A\}\{B\}$
$\{A\}\{B\}$	$\{A\}\{B\}$
$\{A\}$	$\{A\}$
$\{B\}$	$\{B\}$
No effects	none

Notice that the saturated model is identified as $\{AB\}$. Since these models are hierarchical, the saturated model includes those below it in the hierarchy; therefore, the inclusion of $\{A\}$ and $\{B\}$ is implied by $\{AB\}$. As a matter of fact, the models in the list above are *not strictly* hierarchical. Model $\{A\}$ and model $\{B\}$ are partial models and the order in which they are listed is arbitrary. Thus, model $\{A\}$ does not imply model $\{B\}$, nor vice versa.

16.2.3 The Basic Formula for a Model

Although there is some similarity between the process involved in regression analysis and the testing of these hierarchical models, they are not the same. In regression analysis we attempt to account for the variability in a set of scores on a dependent variable in terms of the influence of one or more independent variables. In these hierarchical models we are trying to explain the influence of a number of variables on the frequencies that occur in a contingency table. The frequencies are not variables, but are considered to be the result of the effects of a number of variables, singly or in some combination. The basic formula for a hierarchical model, accordingly, seeks to reconstruct an expected or theoretical frequency, that is, the frequency that would be found in a particular cell if the theoretical model were true. The basic formula for the saturated two-variable model is as follows:

(16.4)
$$F_{ij} = \eta \mathrm{T}(A)_i \mathrm{T}(B)_j \mathrm{T}(AB)_{ij}$$

---

* We will adopt the use of braces to symbolize the different levels of effects. A three-variable effect involving variables A, B, and C, accordingly, would be symbolized as $\{ABC\}$.

where F_{ij} is the expected frequency for a cell of the contingency table (in this case a 2 by 2 table), η (eta) is the geometric mean of the expected cell frequencies, $T(A)_i$ and $T(B)_j$ (taus) are the single-variable effects of variables A and B, and $T(AB)_{ij}$ is the two-variable or association effect of A and B.

BOX 16.4   NOTATION FOR HIERARCHICAL MODEL TESTING

An explanation of the notation system used by Goodman in discussing the testing of hierarchical models is in order here. Take note of the fact that the symbols used in this discussion differ in some respects from symbols used in earlier parts of the book, and symbols that are the same as some of those used earlier do not mean the same here as they did before. You will find that this is a continuing problem with notation systems in statistics. Unfortunately, there is no standard system of notation.

For the discussion of testing hierarchical models:

1. Capital letters such as A, B, and C stand for variables. Note that in the earlier discussion of odds ratios (Section 8.1.4b), X stood for the independent variable, Y for the dependent variable, and T for the test or control variable. Because the variables in hierarchical models may take different roles, depending upon which particular model is being tested, the more flexible A, B, C notation is used.

2. Lowercase letters (*e.g.*, a, b, c) stand for the number of categories into which the variables (*e.g.*, A, B, C) are divided.

3. The lower case Greek letter eta (η) stands for the geometric mean of the expected cell frequencies (see computation formula 16.5).

4. The capital Greek letter tau (T) stands for the effect of a variable or combination of variables.

5. F_{ij} stands for the expected frequency for a cell of a contingency table.

6. f_{ij} stands for the observed frequency for a cell of a contingency table.

7. N stands for the number of cases in a sample.

8. The capital Greek letter pi (Π) stands for the operation of multiplication.

9. The subscripts i and j represent rows and columns, respectively.

Take note of the fact that we have changed our notation system from that we used in discussing the chi-square test for independence. In that instance we used f_o to represent an observed frequency and f_e to represent an expected frequency. Because of the system of subscripts used in discussing hierarchical models, we will use f with the appropriate subscripts to represent an observed frequency and F (again with the appropriate subscripts) to represent an expected frequency.

Formula 16.4 is a multiplicative formula that accounts for the expected frequencies in terms of a constant (η), the effect of variable A [$T(A)_i$], the effect of variable B [$T(B)_j$], and the effect of the association of variables A and B [$T(AB)_{ij}$]. The formula is multiplicative because it involves the multiplication rule of probabilities (Chapter 10).

The constant η, in actuality, defines the no-effects model in which the expected frequencies are the same for every cell of the model. The formula for η is as follows:

(16.5)

$$\eta = \prod_{ij}^{ab} \left(\frac{\sum_{i}^{a} \sum_{j}^{b} f_{ij}}{ab} \right)^{1/ab}$$

where the symbol $\prod_{ij}^{ab}$ indicates that the expression $\left(\sum_{i}^{a} \sum_{j}^{b} f_{ij}/ab \right)$ should be multiplied over all cells. The expression $\sum_{i}^{a} \sum_{j}^{b} f_{ij}$ is merely the sum of all the cell frequencies, which is equal to N (the number of cases in the sample). The a and b refer to the numbers of categories in each of the A and B variables. Formula 16.5 can be simplified as follows:

$$\eta = \left[\left(\frac{N}{ab} \right)^{ab} \right]^{1/ab}$$

$$= \frac{N}{ab}$$

Accordingly, for a 2 by 2 table the formula reduces to the quotient of N divided by 4. If we had a 2 by 2 table with 100 cases, for example, $N/ab = 100/4 = 25$. Therefore, the expected frequency would be 25 for each of the four cells. As we will see this equal distribution of expected frequencies is characteristic of the no-effects model, no matter how many cells there are in the table.

Returning to our hierarchy of possible models for two variables, Formula 16.4 as was indicated above would be the basic formula for the saturated model $\{AB\}$. The formula for the model with two single-variable effects $\{A\}\{B\}$ would be

$$F_{ij} = \eta T(A)_i T(B)_j$$

The formulas for the single-variable effect $\{A\}$ and the single-variable effect $\{B\}$ would be

$$F_{ij} = \eta T(A)_i$$

$$F_{ij} = \eta T(B)_j$$

Finally, the formula for the no-effects model would be

$$F_{ij} = \eta$$

Take note of the fact that the no-effects model only includes the constant (η). As a result, all the expected frequencies in the model are equal. This is what is meant by *no effects*.

Formula 16.4 can be generalized to three or more variables. For example, in the case of three variables A, B, and C the formula for the saturated model would be

$$F_{ijk} = \eta T(A)_i T(B)_j T(C)_k T(AB)_{ij} T(AC)_{ik} T(BC)_{jk} T(ABC)_{ijk}$$

Introducing a third variable increases the number of possible models in the hierarchy to 16 and, as the number of variables is increased, the possibilities increase rapidly. **Rather than test all the possible models, the researcher should use theory as a basis for choosing those models that should be subjected to test.** In general, there will be particular models among the possibilities that will not be theoretically interesting and that, therefore, can be eliminated from consideration. Although the computer makes it relatively easy to test all possible models, this course of action is not recommended. Theory should provide direction to the analysis that is conducted!

16.2.4   Characteristics of the Models

Davis (1974; see General References) distinguishes between the no-effects model, the single-variable effects model, the two-variable effects model, and higher-order models in terms of the following characteristics:

1. The no-effects model is an equal probability model in which all of the cells have the same expected frequency and all odds ratios in the table are 1.00.

2. The single-variable effects model is one in which the expected frequencies are not necessarily identical and the marginal frequencies are not necessarily identical, but the expected frequencies in the cells are exactly proportional to the marginal frequencies. The first-order odds ratios may depart from 1.00, but all higher-order odds ratios are equal to 1.00. In other words, there are no associations between or interactions among the variables in the table.

3. In the two-variable model, single-variable effects and two-variable associations may occur, but no interactions or higher-level variations in associations may occur. First- and second-order odds ratios may depart from 1.00, but all higher-order odds ratios must equal 1.00.

4. In higher-order variable models, single-variable effects, two-variable associations, interactions among three or more variables, and higher-level variations in associations may all occur. First-, second-, and higher-order odds ratios may depart from 1.00, but the number of variables in the model limits the order of the odds ratios that may depart from 1.00. Thus, in a three-variable model, for example, no fourth- or higher-order odds ratio may deviate from 1.00.

16.2.5   Fitting the Models

In point of fact, the process of carrying out tests of hierarchical models is never done by hand. Once you go beyond the simplest cases, the work involved in testing the models is impractical without a computer. All the major statistical programs available on mainframe or minicomputers (*e.g.*, SPSSX, SAS, and BMDP) provide for testing such models. Furthermore, the procedures appear in several statistical packages for microcomputers (including SPSS/PC, SAS, BMDP DC90, and SYSTAT). To illustrate the procedures involved in testing hierarchical models, however, we will take Messer's (1967) data referred to earlier (Sections 8.1.1 and 16.2.1) and use them to demonstrate the steps involved in setting up some

TABLE 16.5 FREQUENCY DISTRIBUTION OF INTERACTION BY AGE ENVIRONMENT BY MORALE

			Morale-(Y)			
(X) Inter-action	*Low Age Environment (T)*		*Medium Age Environment (T)*		*High Age Environment (T)*	
	Concentr.	*Mixed*	*Concentr.*	*Mixed*	*Concentr.*	*Mixed*
High	14	13	29	33	15	20
Low	4	29	18	51	8	9
Total	18	42	47	84	23	29

Source: Data from Table 8.1 (Messer, 1967).

simple models and testing them for goodness of fit. For convenience we have re-produced Table 8.2F here as Table 16.5. This table presents the simultaneous distribution of Messer's 243 subjects on the three variables of his study: morale, age environment, and interaction.

Using these data, we will develop three levels of models: the no-effects model, the single-variable model, and the two-variable model. In general, these and higher-order models can be derived from the data without directly using Formula 16.4. The version of Formula 16.4 for the no-effects model is straightforward, however, and can be used conveniently in its simplified form ($F_{ij} = N/ab$) to compute the expected frequencies for the no-effects model. Recall (from Section 16.2.4) that the no-effects model is an equal probability model; consequently, the simplified version of Formula 16.4 extended to three variables will give us the needed expected frequencies for all of the cells of the table. Since three variables are involved here instead of two, the simplified version of Formula 16.4 will be as follows:

$$F_{ijk} = \frac{N}{abc}$$

where N is the number of cases in the sample and a, b, and c are the number of categories of data for variables A, B, and C. Here we will equate variable A with interaction, which is designated X in Table 16.5; B with morale, (designated Y); and C with age environment (designated T). Since there are two categories of interaction, three of morale, and two of age environment, $abc = (2)(3)(2) = 12$, which agrees with the number of cells in Table 16.5. Given this figure and the N of 243, we find that $F_{ijk} = 243/12 = 20.25$. Therefore, the expected frequency in each of the 12 cells will be 20.25. These frequencies are represented in Table 16.6A as the no-effects model.

Now that we have computed the no-effects model, we must compare it with the observed distribution in Table 16.5 to see whether it fits. To do this, we use the chi-square formula (Formula 16.1) that was discussed earlier in the chapter. That formula is reproduced below with the slight shift in notation mentioned earlier (*i.e.*, with f_{ijk} representing an observed frequency and F_{ijk} representing an expected frequency).

(16.6)

$$\chi^2 = \sum \frac{(f_{ijk} - F_{ijk})^2}{F_{ijk}}$$

Applying Formula 16.6 to the data in the original Table 16.5 and the expected frequencies in Table 16.6A, we can compute χ^2 to see whether the no-effects model fits the raw data. The resulting computed value of χ^2 is 95.12. The immediate question is what that means. To answer the question, we find the χ^2 value that bounds the critical region with the 5% level of significance and compare the computed χ^2 with it. If the computed value is in the critical region, we conclude that the model (shown by the expected frequencies) does not fit the raw data. On the other hand, if the computed value is not in the critical region, we tentatively accept that as evidence that the model does fit—tentatively, because we may later find another model that fits better.

TABLE 16.6   MODELS FOR DISTRIBUTION OF INTERACTION BY AGE ENVIRONMENT BY MORALE—EXPECTED FREQUENCIES

A. No-effects Model

(A) Inter-action	Morale-(B)					
	Low Age Environment (C)		Medium Age Environment (C)		High Age Environment (C)	
	Concentr.	Mixed	Concentr.	Mixed	Concentr.	Mixed
High	20.25	20.25	20.25	20.25	20.25	20.25
Low	20.25	20.25	20.25	20.25	20.25	20.25

B. Single-variable Effects Model

(A) Inter-action	Morale-(B)					
	Low Age Environment (C)		Medium Age Environment (C)		High Age Environment (C)	
	Concentr.	Mixed	Concentr.	Mixed	Concentr.	Mixed
High	11.09	19.53	24.21	42.64	9.61	16.93
Low	10.64	18.74	23.23	40.92	9.22	16.24

C. Two-Variable Effects Model

(A) Inter-action	Morale-(B)					
	Low Age Environment (C)		Medium Age Environment (C)		High Age Environment (C)	
	Concentr.	Mixed	Concentr.	Mixed	Concentr.	Mixed
High	10.93	16.07	28.84	33.15	17.83	17.17
Low	7.19	25.81	18.12	50.88	5.07	11.93

To find the χ^2 value that bounds the 5% critical region, we need to determine the appropriate number of degrees of freedom. In general, degrees of freedom for the no-effects model are found by subtracting 1 from the total number of cells in the table (*i.e., abc* − 1), which in the present case is (2)(3)(2) − 1 = 11. In Appendix Table C, the tabled value for 11 degrees of freedom in the .05 column is 19.675. Therefore, any computed χ^2 as large as or larger than 19.675 will warrant rejecting the no-effects model as a reasonable representation of the raw data in Table 16.5. Since the computed value for these data is 95.12, our conclusion is that the no-effects model does not fit.

Now let us turn to the single-variable effects model to see how it is derived and tested against the raw data. Recall (from Section 16.2.4) that the single-variable effects model has expected cell frequencies that are exactly proportional to the marginal frequencies. This is basically the same condition as for the chi-square test for independence (Section 16.1). In this situation, however, we are working with three variables rather than two. Nevertheless, the procedure for finding expected frequencies is analogous.

If you look at the upper left cell in Table 16.5 (with a frequency of 14), you will notice that it represents subjects who are low on morale, high on interaction, and in a concentrated-age environment. If we find the marginal totals for these particular categories of the three variables, convert them to proportions, and multiply them by N, we will get the expected frequency for that particular cell of the table. Checking back to Tables 8.2B and 8.2C, we find that, out of the $N = 243$ subjects in the study 60 had low morale, 124 had high interaction, and 88 were in an age concentrated environment. Therefore, the expected frequency for the upper-left cell of the table would be

$$\left(\frac{60}{243}\right)\left(\frac{124}{243}\right)\left(\frac{88}{243}\right)243 = 11.09$$

Following the same line of reasoning, we can fill in the expected frequencies for each of the other cells in the table. Unfortunately, we cannot solve for some of the expected frequencies and find the others by subtraction as simply as we did in the test for independence because Table 16.5 is a three-dimensional table.

The procedure we just used to find the expected frequency in the upper-left cell of the raw-data table can be simplified somewhat by canceling N's and generalized as follows:

(16.7)
$$F_{ijk} = \frac{\Sigma A_i \times \Sigma B_j \times \Sigma C_k}{N^{v-1}}$$

where ΣA_i is the marginal total for the appropriate category of variable A, ΣB_j is the marginal total for the appropriate category of variable B, ΣC_k is the marginal total for the appropriate category of variable C, N is the number of cases in the sample, and v is the number of variables. Here N is raised to the 3 − 1, or second, power. Of course, this formula can be extended to cover any number of variables.

When Formula 16.7 is applied to the data in Table 16.5, the single-variable effects model results in the expected frequencies in Table 16.6B. As before, we use Formula 16.6 to test this model for goodness of fit to the raw data. The resulting χ^2 in this case is 26.46. Again we must determine the appropriate number of degrees of freedom to test the model. The procedure for finding degrees of freedom for the single-variable effects model is to start with the total number of cells in the table [$abc = (2)(3)(2) = 12$], subtract 1, and then subtract $k - 1$ for each variable, where k is the number of categories into which the variable is divided.

In the present instance we have two dichotomous variables and one variable with three categories. Therefore, degrees of freedom will be equal to

$$df = (2)(3)(2) - 1 - (2 - 1) - (3 - 1) - (2 - 1) = 7$$

The χ^2 value that bounds the critical region with 7 degrees of freedom at the .05 level is 14.067 (see Appendix Table C). The computed value of χ^2 is 26.46; therefore, we conclude that the single-variable effects model does not fit the raw data well.

We still have the two-variable effects model to derive and test for goodness of fit. At this point it is no longer easy to determine the appropriate expected frequencies to test the model. This is really a job for the computer! The iterative algorithm used by the computer to generate these expected frequencies can be done by hand for a three-variable model and Davis (1974) illustrates the technique. The algorithm involves a number of steps by which successive approximations to the expected frequencies are made until (a) they sum to the marginal totals for the variables and (b) all higher-order odds ratios equal 1.00. The entries in Table 16.6C were derived on the fifth iteration through the use of this algorithm.

We can test these two conditions of the model. From Table 8.2B we found that there were 60 respondents with low morale. If we sum the four expected frequencies in the low-morale category in Table 16.6C, they also sum to 60 (10.93 + 16.07 + 7.19 + 25.81 = 60). If you try other combinations of expected frequencies in Table 16.6C, you will find that they, too, meet this condition. The second condition is that all higher-order odds ratios must equal 1.00. The expected frequencies were used to compute the odds ratios in Table 16.7. Notice that the third-order odds ratios in the right column are all 1.00 or very close to it (within rounding error). These and all higher-order odds ratios will be 1.00, thus meeting the second condition.

Applying Formula 16.6 to the raw data and the expected frequencies in Table 16.6C, we get a computed χ^2 of 6.59. To get the appropriate number of degrees of freedom for the two-variable effects model, we do the following:

1. Determine the total number of cells in the table [$abc = (2)(3)(2) = 12$] and subtract 1 ($12 - 1 = 11$).
2. Subtract $k - 1$ for each variable in the model [$(a - 1)$, $(b - 1)$, and $(c - 1)$] ($11 - (2 - 1) - (3 - 1) - (2 - 1) = 7$).
3. Subtract $(r - 1)(c - 1)$ for each pair of variables, where r is the number of categories of one variable and c is the number of categories of the other. Thus:

For morale and interaction, $(r - 1)(c - 1) = (3 - 1)(2 - 1) = 2$
For morale and age environment, $(r - 1)(c - 1) = (3 - 1)(2 - 1) = 2$
For interaction and age environment, $(r - 1)(c - 1) = (2 - 1)(2 - 1) = 1$

Therefore, $7 - 2 - 2 - 1 = 2$.

If we look in Appendix Table C with 2 degrees of freedom in the .05 column, we find that the χ^2 value that bounds the critical region is 5.991. Therefore, the two-variable effects model also fails to fit the raw data well. If we were to go on

TABLE 16.7 ODDS RATIOS FOR MORALE, INTERACTION, AND AGE ENVIRONMENT
COMPUTED FROM EXPECTED FREQUENCIES OF TWO-VARIABLE EFFECTS MODEL (TABLE 16.6)

Interaction and Morale, Controlling Age Environment

Condition	Second-Order Odds Ratio	Third-Order Odds Ratio	Fourth-Order Odds Ratio
(H H Concentr.) I M|E = (L M Mixed)	2.210 2.209	1.000	
(H M Concentr.) I M|E = (L L Mixed)	1.047 1.046	1.001	
(H H Concentr.) I M|E = (L L Mixed)	2.313 2.312	1.000	

Interaction and Environment, Controlling Morale

Condition	Second-Order Odds Ratio	Third-Order Odds Ratio	Fourth-Order Odds Ratio
(H C High) I E|M = Moderate) (L M Low)	2.444 2.443 2.442	1.000 1.000	1.000

Environment and Morale, Controlling Interaction

Condition	Second-Order Odds Ratio	Third-Order Odds Ratio	Fourth-Order Odds Ratio
(C H High) E M|I = (M M Low)	1.194 1.193	1.001	
(C M High) E M|I = (M L Low)	1.279 1.278	1.001	
(C H High) E M|I = (M L Low)	1.527 1.526	1.001	

to the next higher-order model, that model would be the saturated model with cell entries the same as those for the observed data. Thus it would fit the data perfectly. Three-variable interaction is what the data show. Remember, however, that there are 16 different models possible for three variables. Using one of the computer programs for this type of hierarchical analysis, we could examine a number of these alternative models to see which does fit the data best.

We will not pursue that line of analysis here, however, because our purpose was to give you an introduction to and illustrate the mechanics of this type of model testing.

16.2.6   Log-linear Models

The kind of hierarchical model testing we have been discussing is usually referred to as **log-linear analysis** because, instead of the frequencies themselves, their natural logarithms are analyzed.* One advantage of using the natural log transformation is that the basic formula for computing the expected frequencies (Formula 16.4) in its logarithmic form is additive and linear, which is a much more familiar and arithmetically simple form than the multiplicative form of Formula 16.4. Furthermore, the analogy between such other familiar techniques of analysis as analysis of variance and regression analysis becomes much more apparent. Also, in the log-linear version of the analysis, a different expression of the formula for chi-square is used in testing the models. In general, however, both approaches give the same results. Computer programs use the log-linear approach, but the process described here is basically the same as that used in the log-linear approach.

16.3   TEST FOR GAMMA (γ)

A frequently used measure of association for ordinal data is gamma (γ) (introduced in Chapter 7). The *parameter* γ (symbolized by the Greek letter) may be estimated for the total population by using the *statistic* G, as computed from sample data. When G is used as an estimate of the corresponding γ parameter, it is appropriate to test the null hypothesis, $\gamma = 0$, to evaluate the possibility that the computed G is merely due to sampling error.

Armer and Youtz investigated the extent to which a western-style education has modernizing effects on African youth (Armer and Youtz, 1971). Interviews were conducted with a random sample of 591 young men in Kano City, Nigeria, to determine the extent of their education and their commitment to "modern value orientations." The six value orientations taken as symptomatic of modernity were independence from family, ethnic equality, empiricism, mastery over nature, futurism, and receptivity to change. A composite measure of modernity based on these value orientations was dichotomized into high- and low-modernity categories and these were cross-tabulated against three educational levels: no education, some primary education, and some secondary education. Table 16.8 summarizes the data for these cross-tabulations and the computation of G. The sample G formula is as follows:

(16.8)
$$G = \frac{N_s - N_d}{N_s + N_d}$$

where N_s = the number of pairs of cases ranked in the same way on both variables, and N_d = the number of pairs ranked in the opposite way on the two variables (see Section 7.3).

* See Kennedy (1983; see General References), Knoke and Burke (1980), Gilbert (1981), and Fienberg (1980) for this approach.

TABLE 16.8 LEVEL OF WESTERN EDUCATION AND INDIVIDUAL MODERNITY

Western Education Level	Individual Modernity		Totals
	Low	High	
No Education	194	118	312
Some Primary Education	94	117	211
Some Secondary Education	11	57	68
			$N = 591$

Computation of G

$$G = \frac{N_s - N_d}{N_s + N_d}$$

$$N_s = 194(117 + 57) + 94(57) = 39{,}114$$

$$N_d = 11(118 + 117) + 94(118) = 13{,}677$$

$$G = \frac{39{,}114 - 13{,}677}{39{,}114 + 13{,}677}$$

$$= +0.48$$

Source: Adapted from Armer and Youtz (1971:611, Table 2).

As you can see from the computations carried out in Table 16.8, the G for these data was $+.48$. Since the motive for sampling the young men of Kano City was to generalize about all young men of Kano City (and perhaps beyond), it is appropriate to determine whether the association of .48 is likely to have come from a population in which the association between education and modernity is 0. Therefore, one may test the null hypothesis

$$H_0 : \gamma \leq 0$$

If this null hypothesis can be rejected, there will be reason to believe the computed sample G reflects a population association, which differs from 0. Notice that the null hypothesis, as stated, is for a one-tailed test. This is because we are interested in the thesis that γ is greater than 0. Generally, tests of the hypothesis of zero correlation or association will be one-tailed tests because the theoretically expected direction of association is specified before the tests of significance are conducted.

Goodman and Kruskal have worked out a normal approximation of the sampling distribution of G that makes tests of the null hypothesis possible (Goodman and Kruskal, 1963; see General References).* They give the following formula for converting G to a standard score:

(16.9)
$$z = (G - \gamma)\sqrt{\frac{N_s + N_d}{2N(1 - G^2)}}$$

* It can be used with an N as small as 50.

Assuming that the null hypothesis is true, the γ in the formula will be 0. We can then substitute the necessary sample data into the formula and solve for z. Making these substitutions, we get the following results:

$$z = (.48 - 0)\sqrt{\frac{39,114 + 13,677}{2(591)(1 - .2304)}}$$

$$= .48\sqrt{\frac{52,791}{909.67}} = (.48)(7.62) = +3.66$$

If we chose to test the null hypothesis at the .01 level with a one-tailed test, the z-score that bounds the critical region of the sampling distribution would be $+2.33$. Since the computed z-score is $+3.66$, it extends well into the right tail of the sampling distribution, so the null hypothesis can be rejected in favor of the set of alternatives that γ is greater than 0. The interpretation would be that there is a positive relationship between educational level and individual modernity.

Formula 16.9 gives a conservative estimate of z. Goodman and Kruskal present another formula that gives a more accurate estimate, but it is very cumbersome to work with without a computer program (Goodman and Kruskal, 1963:325). It involves the use of computing matrices to find the necessary values for the formula. Formula 16.9 is so much more convenient to handle that it may be used for all but those cases in which rejection of the null hypothesis is in doubt. Formula 16.9 generally underestimates z; thus a test of the null hypothesis that uses that formula and that falls short of reaching the critical region could presumably reach the critical region if the more accurate formula was used.

The procedure just described for testing the significance of γ (with Formula 16.9) applies as well to conditional tables for which the G's have been computed. Each conditional table is merely treated as a separate case and each G is used to test the null hypothesis: $H_0 : \gamma = 0$. It is entirely possible that an overall γ that is statistically significant may yield a set of conditional γ's, some of which are statistically significant and others of which are not. This information would be used in making interpretations discussed in Chapter 8.

16.3a Assumptions of the Test for Significance of γ

In using the sample G to test the null hypothesis that $\gamma = 0$, it is necessary to make the following assumptions: (a) the measures upon which the sample G is based are independent, (b) the measures of both variables are ordinal in nature, (c) the sample is a simple random sample from the population, and (d) the sample is large enough to justify using a normal approximation to the sampling distribution. It is not necessary to assume that the population is normally distributed, however. The γ is another distribution-free or nonparametric technique.

Diagram 16.2 (p. 609) lists the essential steps in the test of significance for γ.*

---

* Tests of significance for Somers's d_{yx} and Tau-b are based on the numerator ($S = N_S - N_D$). These tests are performed automatically when Somers's d_{yx} or Tau-b are computed by a statistical package such as the Statistical Package for the Social Sciences (SPSS).

BOX 16.5 PARAMETRIC AND NONPARAMETRIC TECHNIQUES

It is important, at this point, that you understand the difference between a parametric and a nonparametric technique. The more traditional statistical techniques (*e.g.*, mean, standard deviation, Pearson's correlation coefficient) rely on the assumption that the populations sampled for the data are normally distributed. Consequently, these techniques are known as distribution-bound or parametric statistics.

There is another class of techniques, most of which have been developed more recently, that make no assumptions about the shape of the population distribution sampled. These techniques are known as distribution-free or nonparametric statistics. Representative of these nonparametric techniques are the chi-square tests, lambda, gamma, Somers's d_{yx}, and Spearman's rho. Section 15.9 introduced other nonparametric tests: the randomization test, sign test, and runs test.

The parametric statistics, in addition to being distribution-bound, are generally applicable to more sophisticated levels of measurement such as interval or ratio scales, whereas the nonparametric techniques generally apply to less sophisticated levels of measurement such as nominal or ordinal scales.

16.4 TEST FOR PRODUCT–MOMENT COEFFICIENT OF CORRELATION AND THE REGRESSION COEFFICIENT

Featherman studied the socioeconomic achievement of a sample of the white male metropolitan population in the United States (Featherman, 1971). His sample included 715 men from the Princeton Fertility Study for whom longitudinal data were available over a 10-year period. He sought to develop a model to represent the process of socioeconomic achievement through the middle years of the work career. Among the variables he considered were education and income during the 35–44 age period. He found that the **product moment coefficient of correlation** for these two variables was .478 (Featherman, 1971:297, Table 1).

Since his data represented a random sample of a larger population, we may legitimately ask whether a correlation of .478 is likely to indicate the presence of a relationship between education and income for that age period in the population as a whole. The accepted procedure is to test the null hypothesis that the population correlation (symbolized by the lower case Greek letter rho, ρ) is 0 or less ($H_0: \rho \leq 0$). If we can reject the null hypothesis on the basis of the sample data and accept the set of alternatives that $\rho > 0$, we will feel justified in concluding that the computed value of .478 is not just a quirk of sampling error. A two-tailed test may be appropriate in certain situations where no directional hypothesis is available ($H_0: \rho = 0$).

The sample coefficient of correlation (r) will only have a symmetrical sampling distribution when ρ is actually 0. This is so because r has absolute limits of -1.00 and $+1.00$. When ρ is 0 the sample r's can deviate from the parameter an equal distance above and below. However, as ρ deviates from 0 the sampling distribution becomes progressively skewed. For example, if ρ were actually $+.5$, the sample r's could go as low as -1.00 (a distance of 1.5) on the left side of the sampling distribution, but they could go as high as $+1.00$ (a distance of .5) on the right side. The sampling distribution, therefore, would be negatively skewed.

An approach that avoids the problem of the skewed sampling distribution is one that involves transforming r into a statistic that has a normal sampling distribution. The formula for making this transformation from r to Z' is as follows:*

(16.10)
$$Z' = 1.151 \log \frac{1 + r}{1 - r}$$

The sampling distribution of Z' will be approximately normal even though the sampling distribution of r is not. Furthermore, the standard error of Z' depends only upon N and can be found with the following simple formula:

(16.11)
$$\sigma_{Z'} = \frac{1}{\sqrt{N - 3}}$$

To test the null hypothesis that $\rho \leq 0$, we will convert r to Z' and find 95% confidence limits around Z'; then we can convert those limits to r's.† If the confidence interval does not include 0, the null hypothesis can be rejected at the .05 level of significance. If the interval does include 0, we will not reject the null hypothesis. It is not necessary to use Formula 16.10 to convert r to Z' since the Z's and their corresponding r's are tabulated in Appendix Table E.

Since Featherman's sample consisted of an N of 715, that figure can be substituted into Formula (16.11) to solve for the standard error of Z':

$$\sigma_{Z'} = \frac{1}{\sqrt{N - 3}} = \frac{1}{\sqrt{715 - 3}} = \frac{1}{26.683} = 0.0375$$

According to Appendix Table E, the Z' that corresponds to an r of .478 is .5204.

Since we are interested in a one-tailed test (of the hypothesis that the population correlation coefficient is not greater—in the positive direction—than 0), we would reject this hypothesis when the *lower* confidence limit around our estimate of $+r$ does not fall below an r of 0. To find the one-sided, 95% confidence limit, we would look for the standard score that cuts off 5% of the area under a normal curve in one tail of that distribution (or, in effect, use the standard score for 90%, two-sided confidence intervals). That value is -1.645 (see the normal curve table, Appendix Table A). Equipped with this information, we can substitute into the following formula for confidence limits and compute the one-sided, 95% confidence limit around Z':

(16.12)
$$cl = Z' - 1.645(\sigma_Z)$$

---

* Note that this Z' is not the standard z-score that we have used frequently throughout this volume.
† Here confidence limits are used to test the null hypothesis since there is no table of critical values of Z' itself.

> **BOX 16.6** CONFIDENCE INTERVALS
>
> Notice that Formula (16.12) for finding confidence limits is very similar to the formulas we used for making interval estimates in Chapter 13 (*e.g.*, Formulas 13.4 and 13.5). In Chapter 14 we pointed out that confidence intervals may be used to test hypotheses (Section 14.3). Formula 16.12 is an example of such an application.

Thus

$$cl_1 = .5204 - 1.645(.0375) = .5204 - .0617 = .4587$$

Now that the lower confidence limit has been computed for Z', it can be converted back to r by returning to Appendix Table E and reading the r that corresponds to Z of 0.4587. This r is .429. Since this r is above 0, the null hypothesis can be rejected in favor of the set of alternatives that $\rho > 0$. If the lower limit had been below 0, rejection of the null hypothesis would not have been warranted.

If we had computed the upper confidence limit ($cl = .5204 + 1.645(.0375) = .5821$), you would note that the confidence limits in terms of Z' (.4587 and .5821) are symmetric about the Z' of .5204. When these limits are reconverted to r's, the limits become .429 and .524. These limits are not symmetrical about the original r of .478. The sampling distribution of r's is slightly negatively skewed. Figure 16.1 represents graphically the sampling distributions of Z' and r for this particular case.

Formula 16.12 includes a standard (z) score of $+1.645$ for the one-sided 95% confidence limit. Other z-scores may be used in the formula to find other confidence intervals. For example, if a one-sided 99% confidence limit was desired, the z-score used would be $+2.33$.* Two-sided standard scores would be ±1.96 and ±2.58, respectively.

16.4.1 Test of Significance for the Regression Coefficient (for One Independent Variable)

Since the regression coefficient $\beta = 0$, when the correlation coefficient $\rho = 0$, the null hypothesis can be expressed as follows:

$$H_0 : \beta = \rho = 0$$

Therefore, the procedure of converting r to Z' and finding confidence limits is also a test of significance for the regression coefficient. We need only to test $\rho = 0$, and, if that null hypothesis can be rejected, we know that the regression coefficient also differs from 0.

---

* Whether the sign should be $+$ or $-$ depends on the one-tailed hypothesis being considered.

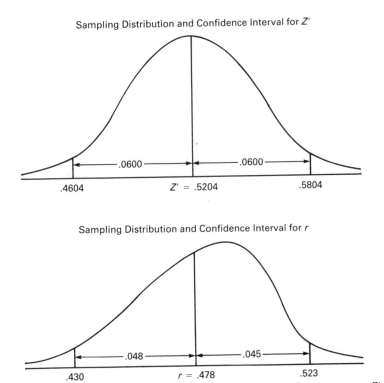

FIGURE 16.1 SAMPLING DISTRIBUTIONS WITH 90% CONFIDENCE INTERVALS FOR Z' AND r
WHEN $r = .478$, $N = 715$

16.4.2   Assumptions of the Test of Significance Using Z'

In order to test the null hypothesis about ρ and the regression coefficient, it is necessary to make the following assumptions: (a) the observations are independent, (b) both variables are measured at the interval level, (c) both variables are normally distributed (bivariate normal distribution), (d) the variables are linearly related, (e) N is large, and (f) the data constitute a simple random sample from some population. If we are not able to satisfy these assumptions, we must proceed with caution and consider carefully whether we are still willing to make generalizations beyond the sample data.

   Computer statistical packages that perform correlation and regression analyses (*e.g.*, BMDP, SAS, and SPSS) do not generally use the Z' transformation and the normal sampling distribution to test for significance. Either the F-distribution, Student's t-distribution, or both are used. Since Student's t-distribution is a symmetrical sampling distribution, it is most applicable when the population parameter, rho (ρ), is zero (see Hayes, 1963:520–521). When the actual parameter is low (near zero), the sampling distribution of r is not seriously skewed. However, the higher the correlation is, the more skewed its sampling distribution and the less appropriate the t-distribution is for testing significance.

The F-distribution and the t-distribution are directly related. As a matter of fact, the value of F is the square of the t-value; consequently, the F and t tests used for correlation are essentially identical. Results of both the F and t tests must be interpreted with caution, therefore, whenever there is reason to believe that ρ is high.

16.4.3   Tests of Significance for Partial and Multiple Correlation Coefficients

The Z'-transformation can also be used to compute confidence intervals for the **partial** and the **multiple correlation coefficients**. The only difference in procedure is that the standard error of Z' has a slightly different formula. The appropriate standard error formula for both partial r and multiple R is as follows:

(16.13)
$$\sigma_{Z'} = \frac{1}{\sqrt{N - m - 1}}$$

where N is the number of cases in the sample, and m is the total number of variables involved. Thus, for a first-order partial r (e.g., $r_{12 \cdot 3}$), m would equal 3; and for a multiple R with three independent variables (e.g., $R_{1 \cdot 234}$), m would equal 4.

Once the standard error of Z' is found, the partial or multiple correlation coefficient can be transformed to Z' using Appendix Table E, and Formula 16.12 can be applied to compute confidence limits.

The assumptions behind the test of significance are the same as for the product moment coefficient of correlation, but they are extended to as many variables as are used in the analysis.

Diagrams 16.3A (p. 610) and 16.3B (p. 611) summarize the essential steps in the tests of null hypotheses concerning the product moment coefficients of correlation, regression coefficients, partial coefficients of correlation, and multiple coefficients of correlation.

16.5   SUMMARY

A study of descriptive statistics must devote considerable time to a discussion of measures of association and their use for finding the meanings of data (as in Chapters 6 and 7). The present chapter was devoted to a discussion of procedures for generalizing these measures beyond the sample data from which they are computed.

First, we considered the *chi-square test for independence*, which serves as a test of significance for a number of the delta-based measures of association. We found that if the null hypothesis of independence could be rejected through chi-square (χ^2) analysis, a measure of association computed could be generalized to the population sampled.

Second, we examined the procedures used to derive and test hierarchical models for cross-classifications of nominal variables (or ordinal and interval variables treated as nominal variables). Here we found that the *chi-square goodness-of-fit test* was the appropriate one to test the various models.

Next, we considered procedures for testing the significance of *gamma* (G), a measure of association of ordinal variables.

Finally, we considered procedures for testing the significance of the *product moment coefficient of correlation* and its associated regression coefficient. We examined the skewed sampling distribution usually associated with r and the reasons for its skewness. We circumvented the problem of skewness by converting r to Z', which has a normal sampling distribution, and by finding a confidence interval. It was pointed out that this test is a simultaneous test for the coefficient of correlation and the regression coefficient. Furthermore, the same test with a slightly altered standard error formula can be used to test the significance of a partial or a multiple coefficient of correlation.

In the next and final chapter of this volume we will attempt to put statistics into context as an integral part of the sociological endeavor.

CONCEPTS TO KNOW AND UNDERSTAND

chi-square (χ^2) test for independence
observed frequencies
expected frequencies
marginal totals
G
product moment r
partial r

multiple R
effect
hierarchical model
Z'-transformation
standard error of Z'
goodness of fit

QUESTIONS AND PROBLEMS

1. Explain when and why you would test the null hypothesis that a measure of association or correlation is 0.

2. Why is it important to have a simple random sample of data when you are doing a χ^2 test for independence?

3. If you make a χ^2 test for independence at the 5% level of significance with 1 degree of freedom, and the computed χ^2 is 48.6, what does the magnitude of the χ^2 mean?

4. Why does the test of significance for the product moment coefficient of correlation simultaneously test the significance of the regression coefficient?

5. When and why is the sampling distribution of product moment r likely to be skewed rather than symmetrical?

6. The sociological journals contain a wealth of studies in which various measures of association are used and their significance tested. Your task is to examine

some of these studies in your library's journal collection and, showing your steps and reasoning, compute the following tests of significance.

a. A chi-square test on a table cross-classifying nominal variables. You may have to multiply column n's by percentages in order to get cell frequencies.

b. A test of an ordinal measure of association.

c. A test of a Pearsonian correlation coefficient.

d. A test of a multiple correlation coefficient.

e. A test of a regression coefficient.

Alternatively, you could compute tests of significance on tables selected from earlier examples in this text (see chapters on descriptive statistics).

GENERAL REFERENCES

Davis, James A., "Hierarchical Models for Significance Tests in Multivariate Contingency Tables: An Exegesis of Goodman's Recent Papers," in Herbert L. Costner, editor, *Sociological Methodology 1973–1974* (San Francisco, Jossey-Bass), 1974.
The best single source of explanation of Goodman's system for testing hierarchical models for multivariate contingency tables.

Goodman, Leo A., and William H. Kruskal. "Measures of Association for Cross Classifications III: Approximate Sampling Theory." *Journal of the American Statistical Association*, 58, 302 (June 1963), pp. 310–364.
This technical article includes a discussion of the sampling distribution of *G*. The relevant part of the article (pp. 322–330) is readable even for the statistically unsophisticated.

Kennedy, John J., *Analyzing Qualitative Data: Introductory Log Linear Analysis for Behavioral Research* (New York, Praeger), 1983.
A good source for a nonmathematical introduction to procedures for testing hierarchical models. It is also a good source of information on applications of chi-square.

Lieberman, Bernhardt, ed., *Contemporary Problems in Statistics* (New York, Oxford University Press), 1971.
This book contains many articles of interest for the student of statistics. Section 5, "The Use and Misuse of Chi-Square," is particularly relevant to this chapter.

LITERATURE CITED

Armer, Michael, and Robert Youtz, "Formal Education and Individual Modernity in an African Society," *American Journal of Sociology*, 76, 4 (January 1971), pp. 604–626.

Berelson, Bernard, *Graduate Education in the United States* (New York, McGraw-Hill), 1960.

Blalock, Jr., Hubert M., *Social Statistics*, rev. ed. (New York, McGraw-Hill), 1979.

Crane, Diana, "Scientists at Major and Minor Universities: A Study of Productivity and Recognition," *American Sociological Review*, 30, 5 (October, 1965), pp. 699–714.

Featherman, David L., "A Research Note: A Social Structural Model for the Socioeconomic Career," *American Journal of Sociology*, 77, 2 (September 1971), pp. 293–304.

Fienberg, Stephen E., *The Analysis of Cross-Classified Categorical Data*, 2nd ed. (Cambridge, Mass., MIT Press), 1980.

Gilbert, G. Nigel, *Modelling Society: An Introduction to Loglinear Analysis for Social Researchers* (London, Allen & Unwin), 1981.

Goodman, Leo A., "A General Model for the Analysis of Surveys," *American Journal of Sociology*, 77 (May 1972), pp. 1035–1086. (a)

————, "A Modified Multiple Regression Approach to the Analysis of Dichotomous Variables," *American Sociological Review*, 37 (February 1972), pp. 28–46. (b)

Grizzle, James E., "Continuity Correction in the χ^2 Test for 2×2 Tables," *American Statistician*, 21, 4 (October 1967), pp. 28–32.

Hays, William L., *Statistics for Psychologists* (New York, Holt, Rinehart and Winston), 1963.

Knoke, David, and Peter J. Burke, *Log-Linear Models* (Beverly Hills, Calif., Sage), 1980.

Mantel, Nathan, and Samuel W. Greenhouse, "What Is the Continuity Correction?" *American Statistician*, 22, 5 (December 1968), pp. 27–30.

Messer, Mark, "The Possibility of an Age-concentrated Environment Becoming a Normative System," *Gerontologist*, 7 (December 1967), pp. 247–251.

Robinson, Patricia, "Language Retention among Canadian Indians: A Simultaneous Equations Model with Dichotomous Endogenous Variables," *American Sociological Review*, 50 (August 1985), pp. 515–529.

Whitt, Hugh P., Charles C. Gordon, and John R. Hofley, "Religion, Economic Development, and Lethal Aggression," *The American Sociological Review*, 37, 2 (April 1972), pp. 193–201.

DIAGRAM 16.1   Chi-Square (χ^2) Test for Independence

PROBLEM

To determine whether two nominal variables are related.

ASSUMPTIONS

1. Independent observations
2. Nominal measurements
3. Simple random sample
4. Underlying continuous distribution of the χ^2 statistic, no expected frequency is less than 5.

FORMULA

$$\chi^2 = \sum \frac{(f_o - f_e)^2}{f_e}$$

If the significance of the computed χ^2 is marginal, and $df = 1$, you may use the following corrected formula:

$$\chi_c^2 = \sum \frac{(|f_o - f_e| - 0.5)^2}{f_e}$$

SAMPLING DISTRIBUTION

The χ^2-distribution with degrees of freedom equal to $(r - 1)(c - 1)$.

Step 1: Set up the null hypothesis. Specify the level of significance.

Step 2: Draw a simple random sample from the population and collect the data.

Step 3: Put the sample data in a contingency table and compute the expected frequencies from the marginal totals and N.

Step 4: Compute χ^2. If the computed χ^2 is in the critical region, reject the null hypothesis. If not, fail to reject it.

Step 5: If the null hypothesis is rejected, examine the differences between observed and expected frequencies to determine the nature of the relationship.

DIAGRAM 16.2 Test of Significance for Gamma (γ)

PROBLEM

To determine whether the population association of two ordinal variables is different from 0.

ASSUMPTIONS

1. Independent observations
2. Ordinal measurement on both variables
3. Simple random sample
4. N is large enough to justify using a normal approximation to the sampling distribution of G.

G FORMULA

$$G = \frac{N_s - N_d}{N_s + N_d}$$

STANDARD SCORE FORMULA

$$z = (G - \gamma)\sqrt{\frac{N_s + N_d}{2N(1 - G^2)}}$$

SAMPLING DISTRIBUTION

The normal curve as an approximation of the sampling distribution of G.

Step 1: Set up the null hypothesis. Specify the level of significance.

Step 2: Draw a random sample from the population and collect data.

Step 3: Compute G from the sample data.

Step 4: Compute the z-score using the sample data. If computed z-score is in the critical region, reject the null hypothesis. If not, fail to reject it.

Step 5: If the computed z-score does not reach the critical region but is close, use Goodman and Kruskal's alternate formula (see Goodman and Kruskal, 1963; General References).

DIAGRAM 16.3A Tests for Product Moment Coefficient of Correlation and the Regression Coefficient (for One Independent Variable)

PROBLEM

1. To determine whether the association between two interval variables as measured by r is significantly different from 0.
2. To determine whether the corresponding regression coefficient is significantly different from 0.

ASSUMPTIONS

1. Independent observations
2. Both variables measured at interval level
3. Both variables normally distributed (bivariate normal distribution)
4. Variables are linearly related
5. Simple random sample of the population

Z'-TRANSFORMATION

$$Z' = 1.151 \log \frac{1 + r}{1 - r}$$

STANDARD ERROR OF Z'

$$\sigma_{Z'} = \frac{1}{\sqrt{N - 3}}$$

FORMULA FOR CONFIDENCE LIMITS OF Z'

$$cl = Z' \pm z(\sigma_{Z'})$$

SAMPLING DISTRIBUTION

Normal sampling distribution when r is transformed to Z'.

Step 1: Set up the null hypothesis. Specify the level of significance. If 0.05, use the 95% confidence interval. If 0.01, use the 99% confidence interval, etc.

Step 2: Draw a simple random sample from the population and collect data.

Step 3: Compute r, b (the sample estimate of the regression coefficient), etc.

Step 4: Convert r to Z' using Appendix Table E. Find the standard error of Z'. Find confidence limits for Z'.

Step 5: Convert the confidence limits for Z' to r's. If 0 is included within the limits, fail to reject the null hypothesis. If not, reject the null hypothesis.

DIAGRAM 16.3B Tests for Partial and Multiple Coefficients of Correlation

PROBLEM

1. To determine whether the association between two interval variables as measured by partial r is significantly different from 0 after additional variables have been partialed out.
2. To determine whether the association between one dependent interval variable and two or more independent interval variables as measured by multiple R is significantly different from 0.

ASSUMPTIONS

1. Independent observations
2. All variables measured at interval level
3. All variables normally distributed
4. All variables linearly related
5. A large sample N
6. A simple random sample of the population

Z' TRANSFORMATION

$$Z' = 1.151 \log \frac{1 + \text{partial } r}{1 - \text{partial } r} \quad \text{or} \quad Z' = 1.151 \log \frac{1 + \text{multiple } R}{1 - \text{multiple } R}$$

STANDARD ERROR OF Z':

$$\sigma_{Z'} = \frac{1}{\sqrt{N - m - 1}}$$

FORMULA FOR CONFIDENCE LIMITS OF Z'

$$cl = Z' \pm z(\sigma_{Z'})$$

SAMPLING DISTRIBUTION

Normal sampling distribution when partial r or multiple R is transformed to Z':

Step 1: Set up the null hypothesis. Specify the level of significance. If .05, use the 95% confidence interval. If .01, use the 99% confidence interval, etc.

Step 2: Draw a simple random sample from the population and collect data.

Step 3: Compute partial r or multiple R.

Step 4: Convert partial r or multiple R to Z' using Appendix Table E. Find the standard error of Z'. Find the confidence limits for Z'.

Step 5: Convert the confidence limits for Z' to partial r's or multiple R's. If 0 is included within the limits, fail to reject the null hypothesis. If not, reject it.

CHAPTER 17

Statistical Analysis in Sociology

The field of descriptive statistics is concerned with ways of making sense out of the data at hand. Tables, graphs, and statistics (or parameters) are used analytically to investigate the distributions of single variables and the relationships between two or more variables. In Parts II, III, and IV we considered various ways in which data are analyzed and described. Part II focused on techniques for analyzing one variable, Part III considered techniques for analyzing two variables, and Part IV dealt with techniques for handling more than two variables.

In Parts V and VI we attempted to show how inferential leaps are made from descriptive analysis of the data at hand to larger bodies of data. We proceeded to explain the basic logic of statistical inference, building upon the three fundamental topics of (a) probability, (b) sampling, and (c) the sampling distribution. Also, we focused attention upon the two major inferential strategies of estimating parameters and testing statistical hypotheses and considered a number of techniques useful for following these strategies.

It is our intent in this chapter to stand back and look at all that has gone before in terms of how it relates to sociology and to the work of sociologists. It is our contention that, when put into proper perspective, statistics is and will continue to be a valuable tool for sociologists in their quest for understanding and predicting social behavior.

17.1 THEMES IN DESCRIPTIVE STATISTICS

We have covered most of the primary techniques sociologists use to describe their data statistically. Although sociologists are most often interested in multivariate description simply because their ideas usually involve more than two variables, we have seen that the differences among techniques unfold from a few basic themes.

Starting with simple differentiation of a common collection of cases, we moved into techniques for analyzing two or more groups on the basis of some characteristic that interests us.

17.1.1   Quality of Data

The first basic theme is that *the results of analysis depend heavily on the quality of data.* "Garbage in, garbage out" is a snide remark sometimes made about those who unthinkingly "computerize" data. As with any analytic process, no matter how impressive or compact the result, if the original data are biased or measurements invalid or unreliable, no amount of statistical summary will remove these defects. Investigators should force themselves to confront their original data and ask: Are these data worth *any* analysis? The answer should be a firm yes or no, and investigators should firmly accept the result of that decision.

17.1.2   The Defined Meaning of Scores

A related theme here has been that *the scores an investigator examines have a specific meaning* that has been defined from the start. Scores are always measurements from certain kinds of cases collected at specific points in time and place. Scores thus measure variables whose particular meanings have already been defined and limited in advance. Certain information about the state of a case, on the scale of a particular variable, is the meaning a score conveys. In the measurement process some essential elements may not be gathered, and thus some information may be lost. Exactly which components of the meaning of a score are most important to preserve depends upon the research project. Often it seems that measurement problems loom so large that other elements of the meaning of a score shrink beside them. We have emphasized level of measurement (nominal, ordinal, interval, and ratio), scale continuity (continuous, discrete), and role in research (independent, control, intervening, dependent) as useful features of the meaning of scores. We have also devoted considerable space to discussing the different levels of units that may be examined—individual and group.

17.1.3   Identifying Relevant Data

The third basic theme is that *it is positively useful to ignore or drop some data.* In fact, we tended to divide data into two parts: the information the investigators want and need for a certain problem, and the other details—perhaps equally sound in every other sense—that they do not want. The general objective of descriptive statistics is to provide tools that permit investigators to distinguish basic research information from annoying or irrelevant details. What is basic information, of course, stems directly from the research questions asked, and what is good information to one investigator may be irrelevant or annoying detail to another. Worse, the details, if they are not carefully controlled and separated from interesting information, may turn out to be basic contamination that makes valid comparisons impossible. Many of the statistical techniques are of the sort that permit unwanted sources of variation to be controlled. It is probably painfully clear by now that no complete description of a set of scores is either possible or wise. There are just too many different descriptions that could be made for this to be a useful research strategy.

17.1.4 The Importance of Comparison

A fourth theme here is *the central importance of comparison and contrast* to research, in general, and to the statistical techniques we have covered. Contrasts are made between an individual score and the group from which the score comes. The z-score or standard score was a statistical description that permitted us to make some of the valid comparisons here. Frequently, contrasts are made between groups—either groups selected separately for comparison or subgroups within one general sample. Means and standard deviations, percentages and rates, histograms and scattergrams, correlation coefficients, odds ratios, and path coefficients—all were statistical means for making these contrasts. Finally and importantly, contrasts were made between the actual data on hand and models of the way the data might be or should be. Measures of association, to take one example, were often designed to contrast real data with some abstract model, such as a model of no association or a model of perfect association. In multivariate analysis, investigators often graphically lay out relationships between variables that they expect on the basis of their theory. Elaboration, standardization, partial correlation, and path analysis are techniques that help describe the fit between actual data and model. This theme is so essential that one could summarize it by saying that *where there is no contrast to be made, there is no study to be done.* Carried further, where there is no variation between scores, no statistical analysis is meaningful.

17.1.5 The Usefulness of Ratios

A fifth theme has been *the importance of the simple ratio* as a means for handling and describing contrasts of interest. Virtually all the statistical descriptions presented in this book are, at root, only ratios. The trick, of course, is to pick the numerator and denominator so that the measure is useful. Often the numerator is a measured subpart of a whole represented in the denominator. *PRE* measures of association, for example, were ratios of predictive error. How much error might be made in predicting the dependent variable with only its distribution as information, and how much of this potential error can be reduced or eliminated if certain added information on an independent variable is available as a basis of prediction? Most of the association measures we discussed were simply contrasts between different methods for making predictions.

17.1.6   Working with Variation

A sixth theme, mentioned earlier, is that *statistics is designed to handle variation in scores*—to describe in the presence of variation. Variation, however, comes from many sources, not all of which may be interesting or helpful to consider. Clearly, if there is no variation in the dependent variable, there is nothing to explain. Thus there are relatively few studies of the number of heads humans are born with. Given some variation, we can ask why, and introduce independent variables to help explain. There is also no progress to be made in explanation unless independent variables have some variation. If there were no variation in independent variables, then only one prediction would always be made for the dependent variable, which would hardly help account for differences in scores on the dependent variable. Investigators jealously guard and try to increase variation in the scores of dependent and explanatory variables. Other sources of variation are not desired and a segment of multivariate description dealt with ways to remove statistically unwanted sources of variation in data. For example, if an investigator wants to examine the relationship between ethnicity and residential mobility, then variation in home ownership and rentership has to be removed by standardization or partial correlation coefficients or, perhaps, by elaboration techniques.

17.1.7   The Active Role of the Investigator

Finally, the most important theme is that *the investigator's ideas take command* of the statistical description from start to finish. Without a research question there is no study, without a knowledge of what is information and what is contaminating detail the proper measures cannot be selected, without knowing what is to be measured appropriate descriptions cannot be made, and without some sound question or theory the contrasts most useful to answering the question cannot be determined. The role of an investigator's ideas and expectations is very clearly seen at key points in the research: (a) in the choice of what is relevant as a control variable or independent variable; (b) in the decision to use standardization or partial correlation techniques rather than various forms of elaboration; (c) in setting up a path diagram and the resulting computations to evaluate it; (d) in the choice of measures of association that contrast observed data with specific models of independence or perfect association. Thus, for example, the concern with multivariate descriptive statistics in sociology is a direct result of the way sociologists develop theory.

From start to finish, the organizing core of statistical analysis in sociology is the sociologist's ideas. This, in fact, is the primary reason for a statistics course in sociology or any discipline. The tools of inquiry must be closely woven into the fabric of the inquiry itself. Of course, this is not to deny the value of the expert in tools or the tool designer as a separate specialist, any more than it is to disparage the role of computer operator or clerical mathematician or computer designer or professional interviewer. Within a field like sociology, however, the problem is organized around substantive inquiry and not around the formal completeness of systems of logical tautology. Naturally, the more refinements we develop in appropriate kinds of computers and math and formal statistics and data-gathering mechanics, the more efficiently sociologists can get on with the task of using these analytic tools to perform substantive analysis. It cannot be emphasized too strongly, particularly with respect to the use of computer statistical software, that the researcher must be knowledgeable enough about this technology to avoid its misuse or abuse.

17.2   THE PLACE OF STATISTICS IN THE RESEARCH PROCESS

It is frustrating to try to discuss statistics in isolation from theory and research methods. The three are really inseparable. Statistics is only meaningful when it relates to theory and research. Interaction among these three subjects is so important that even leaving one out for the sake of pedagogical simplification creates a gap in the flow of logic that is the very essence of science.

For sociologists to identify themselves as a theorist or a researcher and deny the need for a grasp of theory, statistics, and research methods is both unrealistic and unproductive. The sociologist must be informed about and, indeed, must use the full range of conceptual and analytical tools available to advance the discipline. Durkheim is commonly pointed to as a prototype for sociologists because, in spite of the undeveloped nature of sociology in the nineteenth century, he was able to illustrate the necessary interaction between conceptualization and analysis in his work. Following the manner of the scientific method, he generated theory, collected data bearing on his theory, analyzed his data, and reached conclusions about the implications of his data for his theory.

In broad outline, sociological investigations should progress through these stages:

1. Generate a theory to explain some sociological phenomenon.
2. Derive from the theory some hypotheses to be tested against empirical observations.
3. Define relevant concepts to make them observable and measurable.
4. Specify the population to which the theory and hypotheses apply. The population may be either finite or infinite.
5. Design the sample from which the observations will be taken.
6. Specify strategies to be employed in collecting empirical observations that are relevant to testing the hypotheses.
7. Specify or construct instruments to be used for collecting the data.
8. Collect the data.
9. Describe analytically the sample data.
10. Interpret the findings and evaluate their implications for the hypotheses and theory. This is where the decision to make the inferential leap from sample to population is made.
11. Reconsider the hypotheses and theory in light of the findings. This is where any necessary alterations in the theory are made.
12. Replicate the process starting with step 2.

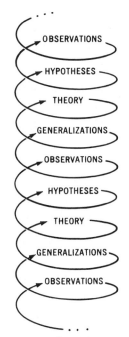

FIGURE 17.1   THE SOCIOLOGICAL INVESTIGATION AS A CYCLICAL PROCESS

It should be understood that listing these steps is an oversimplification of the process as it exists in reality. The process really has no beginning or end. Rather, it is a cyclical process that keeps returning to steps already completed, but, it is hoped, each time at a higher level representing a progression toward a more adequate explanatory system. Figure 17.1 attempts to portray this cyclical process graphically.

Moreover, it should not be assumed that the steps listed above follow in sequence as neatly as they have been presented here. Sometimes the steps interchange, but, usually, they overlap, or even occur simultaneously. And not every investigation necessarily includes all the steps enumerated. This presentation is in fact an idealized version of what can actually take place when sociologists are at work.

When researchers get to the stage of data analysis, statistical tools begin to play a major part. But statistical considerations also enter in at various other points in the process. As a matter of fact, if researchers do not concern themselves with statistics until they reach the data analysis stage, they are likely to have difficulty finding statistical techniques appropriate for the data. The use of statistics must be planned early in the research process if data analysis is to proceed smoothly.

When hypotheses are derived and concepts are defined, it is necessary to begin thinking ahead about the kind of statistical analysis that will come later so that the hypotheses are stated appropriately and the concepts defined in measurable ways. At this point it is appropriate to set up the statistical (null) hypotheses to be tested, to decide whether the tests will be one tailed or two tailed, and to select the level of significance (see Chapter 14). Furthermore, it is useful to set up dummy tables (tables without data) and dummy analyses to help anticipate what data might look like once they are collected. This constitutes a check system to determine whether the kinds of data necessary for the statistical analysis will be obtained.

BOX 17.1   LOOK BACK

This discussion assumes that you have a good grasp of such concepts as sampling distribution and sampling error. If you feel the need to refresh your memory, turn back and review Sections 12.2 and 12.2.2.

When the population is specified and the sample is designed it is necessary to decide how large the sample should be. Again statistics is relevant because a determinant of sample size is the amount of sampling error that is tolerable. Section 14.2.5d discussed ways to use standard error formulas to estimate sample size needed to discriminate a given difference between a null hypothesis and the true parameter, at a specified level of significance and probability of rejecting a false null hypothesis (*i.e.*, power of the test).

Of course, solving for the appropriate sample size assumes that we have also considered steps 6 and 7 in the research process because we would also have to settle on our strategies for collecting data and have in mind the kinds of data collection instruments we will use. The nature of the data collection instruments is relevant to the amount of sampling error tolerable. For example, if a measurement scale with a 10-point range were to be used, we would need to keep the sampling error small enough so that an observed difference of 10 points or less would result in a computed statistic with a possibility of reaching the critical region. If the sampling error is too large, the statistical test will be insensitive to even the largest difference that can be expected from the data.

Another important consideration in deciding how large a sample needs to be is the type of multivariate analysis of the data planned. The greater the number of control variables introduced into an analysis, the more cases required. For example, if a dependent variable was divided into three response categories, the independent variable into three, and a control variable into two, the sample would need to be large enough to distribute data over 18 cells of contingency tables. Adding a second control variable with two response categories would increase the number of cells in the tables to 36.

The dummy tables and trial analysis suggested above can be particularly useful in deciding minimal sample size.

When decisions are made about data collection strategies and data collection instruments, the level of measurement of the variables is crucial. The data

TABLE 17.1 Steps in the Research Process and Parallel Statistical Considerations

Steps in the Research Process	Statistical Considerations
1. Generation of theory	
2. Derivation of hypotheses	Construct statistical hypotheses
3. Definitions of relevant concepts	
4. Specification of population	Select appropriate sample size
5. Design of sample	
6. Specification of data collection strategies	Determine maximum sampling error tolerable; determine measurement levels; select appropriate statistical tests
7. Construction of data collection instruments	
8. Collection of data	
9. Analysis of data	Perform necessary computations
10. Interpretation of findings	Evaluate statistical results in terms of amount of information generated
11. Reconsideration of hypotheses and theory	
12. Replication of the whole process	

collection instruments must produce scores compatible with the measurement requirements of the statistical techniques employed to analyze the data.*

Even after the statistical analysis of the data has been completed, a knowledge of statistics plays a direct part in interpreting the findings. We must be thoroughly informed about the statistical tools used to analyze data and the kinds of information that they make available to us. Otherwise, we will not be in a position to express what it all means.

Table 17.1 summarizes the points we have tried to make here to illustrate the interaction between theory, research methods, and statistics. In the left column of the table are listed the 12 steps enumerated earlier and in the right column, the statistical considerations that most nearly parallel the steps. The attempt to portray these parallel statistical considerations in Table 17.1 demonstrates again the point stressed earlier in this chapter that the steps in research are not mutually exclusive and sequential; rather they overlap and sometimes occur simultaneously.

In a carefully planned program of research, most of the important decisions about statistical analysis are made before the data are collected. Decisions made after the data have been collected are usually stopgap measures that may lead to less than desirable solutions to problems arising from poor planning, or no planning at all. Of course, one of the most universally applicable generalizations of science, Murphy's law, can be expected to hold true even for the most carefully designed research. Murphy's law states, "If anything can go wrong, it will." Nevertheless, a well-planned project including a carefully thought out statistical analysis is likely to encounter fewer problems than would otherwise be the case.

---

* Throughout this volume the measurement requirements of each technique discussed have been made explicit and have been included among the lists of assumptions.

17.3 PLACING DESCRIPTIVE AND INFERENTIAL STATISTICS IN PERSPECTIVE

Sociologists and other behavioral scientists have questioned the necessity and desirability of inferential statistical techniques in sociological and behavioral science research. They have claimed that sociological data are not appropriate for inferential analysis or that inferential analysis gives no more than trivial results (Labovitz, 1970, 1971; Morrison and Henkel, 1970; Rozeboom, 1960; Selvin, 1957). If these criticisms are correct, then it is proper to ask what the justification is for discussing statistical inference.

Our position on this matter is as follows. As sociologists, we are interested in developing generalizations about social behavior. We commonly work with sample data but are not inherently concerned with samples. Rather, we are interested in the populations that they represent, whether those populations are finite or infinite. The logic of inferential statistics gives us a basis for making inferential leaps from samples to populations. When we analyze sample data and perform a test of statistical significance or estimate a parameter, we are seeking to apply the results of the sample analysis to the population the sample represents. A test of statistical significance merely supplies us with a probability figure we can weigh in making the decision whether or how to generalize beyond the sample. The test does not make the decision for us. It is merely a crutch that we may use to help us make the decision. As researchers we are responsible for our decisions, and we cannot blame wrong decisions on statistics. Likewise, we cannot legitimately pass on to others the responsibility of making the decisions for us. It is not proper, in other words, to conduct a test of statistical significance, publish the results of the test, and expect the reader to decide whether the inferential leap is justified. We should report the results of statistical tests to inform the reader of the information used to make decisions, but as researchers we must take responsibility for the decisions reached.

Some critics of tests of significance have complained that statistical significance will almost always be found if the sample size is large enough. It is true that larger samples have smaller sampling errors and are, thus, sensitive to smaller observed differences. **It is not necessary, however, to decide in favor of the inferential leap merely because a statistically significant result is obtained.** The decision to make an inferential leap should depend heavily upon *the substantive nature of the finding*. Statistical significance is only one criterion that should be considered in the decision process and "significant" does not mean "important."

In the past there have been a number of studies concerned with the subject of selective migration (Gist and Clark, 1938; Gist *et al.*, n.d.; Pihlblad and Gregory, 1954). These studies investigated the question whether migration from farm to city is selective with respect to intelligence—whether there is a tendency for the more able persons to migrate to cities and the less able to remain in rural areas. One of the best known of these studies was conducted by Gist and Clark (1938). They secured the I.Q. scores of a large number of students who had been enrolled in rural Kansas high schools in 1922–1923. In 1935, the year in which they actually conducted the study, they were able to locate 2,544 of these students by place of residence. They found 964 of them had moved to urban areas and 1,580 were still living in rural areas. The researchers computed the mean I.Q. for those who had moved to urban areas and compared it to that for those who remained in rural areas. They found the mean I.Q. of urban migrants to be 98.26 with a standard deviation of 11.52 and of rural nonmigrants to be 94.78 with a standard deviation of 11.36.

Because of their large samples the standard errors of the means of the two groups were small (.385 for the urban migrants and .286 for the rural nonmigrants) and the standard error of the difference was .48.* When the standard score is computed using the mean difference of 3.48 in the numerator and the standard error of the difference in the denominator, we get the following results:

$$z = \frac{\bar{X}_u - \bar{X}_r}{s_{\bar{X}_1 - \bar{X}_2}} = \frac{98.26 - 94.78}{.48} = \frac{3.48}{.48} = +7.25$$

The probability of occurrence of a z-score as large as $+7.25$ by chance is very small (see Appendix Table C). Therefore, Gist and Clark conclude, "The chances are overwhelming that the observed difference in the means of these samples is a true difference and not to be explained as arising from the accidental conditions of sampling" (1938:45–46).

Gist and Clark should not be criticized for rejecting the null hypothesis of no difference in the means because their data warranted its rejection. On the other hand, they might well be reprehended for conducting a test of significance at all, in the light of the small difference of 3.48 I.Q. points they observed. Taking into consideration the modest reliability of I.Q. scores, a difference of only 3.48 in mean I.Q. is substantively unimportant and theoretically uninteresting. Such a difference could certainly not justify the conclusion that the rural nonmigrants are inferior to the urban migrants. If, on the other hand, the researchers had found that the rural nonmigrants had a mean I.Q. of 90 and the urban migrants a mean I.Q. of 120, for example, then the difference might be of substantive importance and theoretical interest.

Although the researchers may be correct in concluding that, "the observed difference in the means of these samples is ... not to be explained as arising from the accidental conditions of sampling," the small difference observed may well have been attributed to the vicissitudes of individual performance on the I.Q. test itself. Furthermore, since it is reasonable to suppose that individuals' scores on an I.Q. test will vary a few points from one administration of the test to another, a mean difference between groups of 3.48 points is an unimportant one. Even if the observed difference is a real difference, in this case it is so small that it really does not make any difference as far as potential achievement of the two groups involved is concerned.

We believe that the critics of inferential statistics should be leveling their criticism at the users of statistics and at the way in which they use statistics rather than at the techniques themselves. Statistical techniques are tools that help

---

* When Gist and Clark conducted their study, they used "probable errors"—a measure of sampling error that is no longer used. We computed standard errors instead, and computed the standard score with the standard error of the difference between means as the measure of sampling error (see Section 15.2.1a).

researchers make decisions about their data. Statistical tools, like any other tools, can be used or misused. If used properly, they can be a valuable aid to the researcher. If misused, they can lead to serious errors of interpretation.

The critics of statistical inference are correct when they claim that tests of the null hypothesis represent a very primitive level of analysis. Rejection of the null hypothesis merely allows one to assume that whatever is happening at the sample level is probably paralleled by what is happening at the population level (Section 14.2.1). The crucial analysis of data is that which takes place at the sample level to explain the nature of observed relationships. This type of analysis is the subject of descriptive statistics. The techniques of descriptive statistics are those that are designed to make sense of what is happening in the data. It is upon these techniques that the emphasis must be placed if we are to advance sociology as a science. What inferential statistics adds to the process is a logic and a set of techniques for determining the probability that what we have learned about a sample also applies to the population that the sample represents.

17.4 USES AND ABUSES OF STATISTICS

Wisely handled, statistics can be a powerful tool of analysis for sociologists. They have a storehouse of techniques available to them to attack a huge variety of research problems. In this book we have done no more than scratch the surface. Serious students of sociology will go on to take more advanced methods and statistics courses and will consult the literature suggested in the bibliography in order to acquaint themselves with the many analytical tools available.

It should be kept in mind that most of the statistical techniques in the storehouse were not developed by sociologists with sociological applications in mind. For example, the analysis of variance was applied first to agricultural research. It was used to analyze the relative effectiveness of different types of fertilizers. The sociologist users of statistics often must adapt the techniques to their needs. If the data do not quite fit the assumptions of the technique they elect to use, then they should use the technique with caution, keeping in mind those assumptions they failed to satisfy, the importance of satisfying those assumptions (*i.e.*, robustness), and interpret the results accordingly.

If you examine the sociological literature for use of the techniques we have discussed, you will usually find that the data analyzed do not quite fit the assumptions of the techniques used. For example, you will frequently find tests of significance run on data from samples that are not quite random samples or means and standard deviations computed on data that are not clearly interval scale data. When statistics are used in such circumstances and the researcher who has done the analysis really understands the techniques used and the implications of their use for the analysis of the data, we must assume that these factors have been taken into consideration in interpreting the findings. The general principle is to be conservative in drawing conclusions.

Unfortunately, however, statistical techniques are frequently used by those who are ignorant of the function of the techniques and of the information that can be gleaned from their use. A little knowledge, particularly in statistics, can be a dangerous thing. The person who knows a little about χ^2, for example, may go around blithely computing χ^2 on whatever data are available, completely oblivious to whatever abuses the technique may be perpetrating (Section 16.1.4), and more "sophisticated" researchers may apply the latest statistical fad to any data they can find! Beware of the researcher who thinks that a test of significance is a magic technique for grinding out findings and that the .05 level of significance is sacred (Section 14.2.3). Such a person is no less a hindrance to the advancement of a sound, scientific sociology than the person who, through complete ignorance of statistics, says that sociological data are inherently unquantifiable and statistical techniques are completely useless in sociology.

17.5   IS STATISTICS NECESSARY FOR SOCIOLOGY?

When we put the matter into its proper perspective, it is rather meaningless to debate whether statistics is necessary for sociology. Statistics is basically a body of logic and techniques useful for organizing, analyzing, and making sense of information. It is an information-processing tool. Sociology, like all scientific disciplines, is an information-processing endeavor. We, as sociologists, are interested in making sense out of human social behavior, in understanding that behavior, and in predicting it. The basic postulate of sociology is that human social behavior is understandable and predictable. If we accept the postulate, then we should consider using *any* method available to us that will help us to understand and predict human social behavior.

The use of statistics in sociology does not imply that all data have to be reduced to numbers. Those aspects of data that are quantifiable should be quantified, but qualitative data need not be discarded. Quantitative and qualitative data can supplement and complement each other. It is often useful to analyze quantitative data, come up with findings, and return to qualitative data—such as observation narratives—to try to make further sense of the quantitative findings.

Sociologists should focus on theory and attack significant sociological problems in the best way possible, using any tools at their disposal to bring light to those problems. They should be problem oriented, not technique oriented. They should choose to investigate not only problems they can handle with the techniques they know and favor but, rather, significant problems to which they should apply any techniques available that will be helpful to them in their investigation.

17.6   DEVELOPMENTS AND PROSPECTS

The widespread availability of computers has shifted the emphasis in sociology from the study of univariate and bivariate problems to the study of multivariate problems. To be efficient predictors, sociological theories need to be multivariate to reflect the general complexity of the social world. Before computers, multi-variate statistical techniques were so tedious they were not commonly used. The computer has now made these techniques accessible and practical. In response to this breakthrough, sociological theories have increasingly become multivariate in form and additional multivariate statistical techniques have been developed. Furthermore, comprehensive packages of statistical procedures are now available for microcomputers as well as mainframes and minicomputers. Sociologists no longer need access to major computer centers to avail themselves of sophisticated statistical software. There is always the danger, however, of misusing these tools because of the ease with which they can be called upon. A solid, basic knowledge of the principles of statistics is still a prerequisite to intelligent use of statistical packages.

There has been a continuation of the trend toward the integration of theory, research methods, and statistics in recent work within the field of sociology. This integration was a long time in coming but seems to be firmly established now. Computers have facilitated this development, and the discipline, accordingly, has advanced at an accelerated pace. The impact of this integrated approach is increasingly being realized in more fruitful theory, more meaningful research, and more important results.

Another development that promises to have a continuing significant impact on sociology and on the academic preparation of future sociologists is the shift from a structural to a process approach to theorizing and researching. In the recent past social behavior has been viewed in static terms, and the bulk of the research done focused on single points in time, like stopping a movie to study a single frame. Much of current research still reflects this perspective. Sociologists have come to realize more and more, however, that what is orderly about social behavior may be the way in which it changes rather than the way in which it resists change. This perspective focuses attention on time series and longitudinal analysis. Accordingly, techniques for these kinds of analysis (including stochastic processes and trend analysis) are receiving increasing use. These developments, plus the formalization that is taking place in the area of theory construction, make a solid background in mathematics of change an important requirement in the academic training of future sociologists.

An interesting, related development that has grown out of sociologists' increasing appreciation of the importance of process has been a tendency to turn to the past. Because trend analysis is most meaningful when the data points are numerous, sociologists have come to appreciate the fact that they must go backward as well as forward in time to collect sufficient data points. This has caused many professionals in the field to move into areas of study previously reserved almost exclusively to historians. As a result, there is increasing interaction between sociologists and social historians and many of the techniques that previously were used to study current data are being applied to historical data. This promises to lead to significant methodological changes in the study of social history that will undoubtedly affect both the fields of history and sociology.

It is becoming increasingly apparent that the most pressing national and international problems are essentially sociological in nature. Sociologists have the potential to play central roles in the solution of many of those problems and will, increasingly, be given the opportunity to do so. To deliver on their potential, they will need to be well versed in theory, research methods, and statistics. What you have covered in this volume is only an introduction to what is available and to what lies ahead for serious students of sociology.

CONCEPTS TO KNOW AND UNDERSTAND

major themes in descriptive statistics

the interaction of theory, research methods, and statistics

the stages of sociological investigation

dummy tables and dummy analyses

estimation of sample size

responsibility for research decisions

the substantive nature of a finding

abuses of statistics

impact of computers on statistical analysis

integration of theory and research process analysis

QUESTIONS AND PROBLEMS

1. Describe the process through which a sociological investigation should progress. Read a sociological article or monograph and determine to what extent this process was followed.

2. Why is it important to consider statistical analysis at an early stage in planning a study?

3. What is meant by the statement that statistical analysis is merely a crutch upon which the sociologist may lean?

4. Make a clear-cut distinction between statistical significance and substantive importance. Which is more important to the sociologist?

5. What is meant by the statement that the sociologist should be problem oriented rather than technique oriented?

6. Describe the interaction that takes place between the researcher and the research.

GENERAL REFERENCES

Blalock, Hubert M., Jr., *Theory Construction: From Verbal to Mathematical Formulations* (Englewood Cliffs, N.J., Prentice Hall), 1969.
> The marriage between theory and research that is currently taking place in sociology has led to the kinds of theory construction activities described in this book.

Coleman, James S., *Longitudinal Data Analysis* (New York, Basic Books), 1981.
> For a discussion of techniques for the analysis of longitudinal data, consult this volume by Coleman.

Cozby, Paul C., *Using Computers in the Behavioral Sciences* (Palo Alto, Calif., Mayfield), 1984.

This short book will give you an overview of the uses of computers in sociology and the other behavioral sciences.

Jaffe, A.J., and Herbert F. Spirer, *Misused Statistics* (New York, Marcel Dekker), 1987.
This is an excellent case book of misuses and abuses of statistics across a broad spectrum of areas of application.

Kerlinger, Fred N., *Foundations of Behavioral Research, 3rd ed* (New York, Holt, Rinehart and Winston), 1986.
This excellent text attempts to integrate theory, research methods, and data analysis techniques. It addresses itself to many of the methodological issues that we have raised in this book.

Land, Kenneth C., "Formal Theory," in Herbert L. Costner, ed., *Sociological Methodology 1971* (San Francisco, Jossey-Bass), 1971, pp. 175–220.
This is an outstanding paper on the formalization of theory in sociology. It shows the relationship between theory, mathematical models, and data. Process models of social behavior are discussed.

Lieberman, Bernhardt, *Contemporary Problems in Statistics* (New York, Oxford University Press), 1971.
This book is a perceptive treatment of some important methodological issues in applied statistics.

Schrodt, Philip A., *Microcomputer Methods for Social Scientists* (Beverly Hills, Calif., Sage), 1984.
This book is written for the novice to computing. It focuses on the microcomputer and its applications to social science.

LITERATURE CITED

Gist, Noel P., and Carroll D. Clark, "Intelligence as a Selective Factor in Rural–Urban Migrations," *American Journal of Sociology*, 44 (July 1938), pp. 36–58.

Gist, Noel P., C.T. Pihlblad, and C.L. Gregory, *Selective Factors in Migration and Occupation*, University of Missouri Studies, XVIII, no. 2, n.d.

Labovitz, Sanford, "The Nonutility of Significance Tests: The Significance of Tests of Significance Reconsidered," *Pacific Sociological Review*, 13 (Summer 1970), pp. 141–148.

———, "The Zone of Rejection: Negative Thoughts on Statistical Inference," *Pacific Sociological Review*, 14, 4 (October 1971), pp. 373–381.

Morrison, Denton E., and Ramon E. Henkel, *The Significance Test Controversy* (Chicago, Aldine), 1970.

Pihlblad, C.T., and C.L. Gregory, "Selective Aspects of Migration among Missouri High School Graduates," *American Sociological Review*, 19, 3 (June 1954), pp. 314–324.

Rozeboom, William W., "The Fallacy of the Null-Hypothesis Significance Test," *Psychological Bulletin*, 57, 5 (1960), pp. 416–428.

Selvin, Hanan, "A Critique of Tests of Significance in Survey Research," *American Sociological Review*, 22, 5 (October 1957), pp. 519–527.

Appendix: Computer Program and Statistical Tables

COMPUTER PROGRAM TO GENERATE PSEUDORANDOM NUMBERS

Computers generate random digits by an arithmetic process. So, given the same starting conditions, the same digits will be produced; thus most computer random number programs start with some arbitrary number such as the time (see line 60 below, where seconds are used) or date or ask the user to provide an arbitrary number (see line 25 below). These are called *pseudo-random* numbers because the process creates a series of numbers that behave like random numbers. That is, tests show that each digit (0–9) is equally likely to appear over a large number of trials, and it is equally likely to appear in any position in a pseudo-random number. Furthermore, there are no systematic cycles or series of numbers. Tests of pseudo-random numbers help determine which random number computer program is the best. Most computer languages include a random number procedure such as the RND function in the BASIC program below (see line 85).

Most home and larger computers use a language called BASIC, which has in it a pseudo-random number generator. A program such as the following will provide random digits, one at a time, between specified limits. The same string of random digits will be given unless a different starter digit is given. Check with your computer's BASIC manual on how to enter and use the following program.

```
10 PRINT "******** RANDOM NUMBERS ********"
15 PRINT
20 PRINT "GIVE ANY STARTER NUMBER";
25 INPUT N2
30 PRINT "HOW MANY RANDOM NUMBERS";
35 INPUT N1
40 PRINT "RANGE = (LOWEST)";
45 INPUT N3
50 PRINT "          (HIGHEST)";
55 INPUT N4
60 SEC = VAL(RIGHT$(TIME$, 2)) + N2
65 FOR I = 1 TO SEC
70 DUMMY = RND(1)
75 NEXT I
80 FOR I = 1 TO N1
85 NO = RND(1)*(N4 − N3)
90 NO = INT(NO) + N3
100 PRINT "PRESS ENTER";
105 INPUT X
110 NEXT I
115 PRINT "MORE NUMBERS";
120 N$ = LEFT$(R$, 1)
125 IF N$ = "Y" THEN 10
130 END
```

Note: Omit lines 100 and 105 if you do not want to press enter to get each random number.

THE USE OF TABLE A (AREAS UNDER A NORMAL CURVE)

The use of Table A requires that the raw score be transformed into a *z*-score and that the variable be normally distributed.

The values in Table A represent the proportion of area in the standard normal curve which has a mean of 0, a standard deviation of 1.00, and a total area also equal to 1.00.

Since the normal curve is symmetrical, it is sufficient to indicate only the areas corresponding to positive *z*-values. Negative *z*-values will have precisely the same proportions of area as their positive counterparts.

Column *B* represents the proportion of area between the mean and a given *z*.

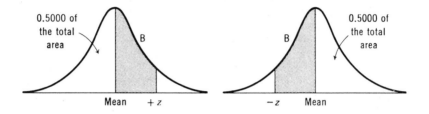

Column *C* represents the proportion of area beyond a given *z*.

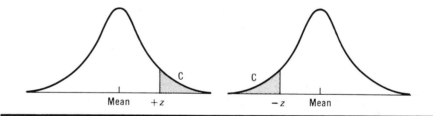

Source: Table A, Richard P. Runyon and Audrey Haber, *Fundamentals of Behavioral Statistics*, Second Edition, Addison-Wesley Publishing Company, Reading, Massachusetts, 1971. Reprinted by permission.

TABLE A   Areas under a Normal Curve

(A) z	(B) area between mean and z	(C) area beyond z	(A) z	(B) area between mean and z	(C) area beyond z	(A) z	(B) area between mean and z	(C) area beyond z
0.00	.0000	.5000	0.55	.2088	.2912	1.10	.3643	.1357
0.01	.0040	.4960	0.56	.2123	.2877	1.11	.3665	.1335
0.02	.0080	.4920	0.57	.2157	.2843	1.12	.3686	.1314
0.03	.0120	.4880	0.58	.2190	.2810	1.13	.3708	.1292
0.04	.0160	.4840	0.59	.2224	.2776	1.14	.3729	.1271
0.05	.0199	.4801	0.60	.2257	.2743	1.15	.3749	.1251
0.06	.0239	.4761	0.61	.2291	.2709	1.16	.3770	.1230
0.07	.0279	.4721	0.62	.2324	.2676	1.17	.3790	.1210
0.08	.0319	.4681	0.63	.2357	.2643	1.18	.3810	.1190
0.09	.0359	.4641	0.64	.2389	.2611	1.19	.3830	.1170
0.10	.0398	.4602	0.65	.2422	.2578	1.20	.3849	.1151
0.11	.0438	.4562	0.66	.2454	.2546	1.21	.3869	.1131
0.12	.0478	.4522	0.67	.2486	.2514	1.22	.3888	.1112
0.13	.0517	.4483	0.68	.2517	.2483	1.23	.3907	.1093
0.14	.0557	.4443	0.69	.2549	.2451	1.24	.3925	.1075
0.15	.0596	.4404	0.70	.2580	.2420	1.25	.3944	.1056
0.16	.0636	.4364	0.71	.2611	.2389	1.26	.3962	.1038
0.17	.0675	.4325	0.72	.2642	.2358	1.27	.3980	.1020
0.18	.0714	.4286	0.73	.2673	.2327	1.28	.3997	.1003
0.19	.0753	.4247	0.74	.2704	.2296	1.29	.4015	.0985
0.20	.0793	.4207	0.75	.2734	.2266	1.30	.4032	.0968
0.21	.0832	.4168	0.76	.2764	.2236	1.31	.4049	.0951
0.22	.0871	.4129	0.77	.2794	.2206	1.32	.4066	.0934
0.23	.0910	.4090	0.78	.2823	.2177	1.33	.4082	.0918
0.24	.0948	.4052	0.79	.2852	.2148	1.34	.4099	.0901
0.25	.0987	.4013	0.80	.2881	.2119	1.35	.4115	.0885
0.26	.1026	.3974	0.81	.2910	.2090	1.36	.4131	.0869
0.27	.1064	.3936	0.82	.2939	.2061	1.37	.4147	.0853
0.28	.1103	.3897	0.83	.2967	.2033	1.38	.4162	.0838
0.29	.1141	.3859	0.84	.2995	.2005	1.39	.4177	.0823
0.30	.1179	.3821	0.85	.3023	.1977	1.40	.4192	.0808
0.31	.1217	.3783	0.86	.3051	.1949	1.41	.4207	.0793
0.32	.1255	.3745	0.87	.3078	.1922	1.42	.4222	.0778
0.33	.1293	.3707	0.88	.3106	.1894	1.43	.4236	.0764
0.34	.1331	.3669	0.89	.3133	.1867	1.44	.4251	.0749
0.35	.1368	.3632	0.90	.3159	.1841	1.45	.4265	.0735
0.36	.1406	.3594	0.91	.3186	.1814	1.46	.4279	.0721
0.37	.1443	.3557	0.92	.3212	.1788	1.47	.4292	.0708
0.38	.1480	.3520	0.93	.3238	.1762	1.48	.4306	.0694
0.39	.1517	.3483	0.94	.3264	.1736	1.49	.4319	.0681
0.40	.1554	.3446	0.95	.3289	.1711	1.50	.4332	.0668
0.41	.1591	.3409	0.96	.3315	.1685	1.51	.4345	.0655
0.42	.1628	.3372	0.97	.3340	.1660	1.52	.4357	.0643
0.43	.1664	.3336	0.98	.3365	.1635	1.53	.4370	.0630
0.44	.1700	.3300	0.99	.3389	.1611	1.54	.4382	.0618
0.45	.1736	.3264	1.00	.3413	.1587	1.55	.4394	.0606
0.46	.1772	.3228	1.01	.3438	.1562	1.56	.4406	.0594
0.47	.1808	.3192	1.02	.3461	.1539	1.57	.4418	.0582
0.48	.1844	.3156	1.03	.3485	.1515	1.58	.4429	.0571
0.49	.1879	.3121	1.04	.3508	.1492	1.59	.4441	.0559
0.50	.1915	.3085	1.05	.3531	.1469	1.60	.4452	.0548
0.51	.1950	.3050	1.06	.3554	.1446	1.61	.4463	.0537
0.52	.1985	.3015	1.07	.3577	.1423	1.62	.4474	.0526
0.53	.2019	.2981	1.08	.3599	.1401	1.63	.4484	.0516
0.54	.2054	.2946	1.09	.3621	.1379	1.64	.4495	.0505

TABLE A AREAS UNDER A NORMAL CURVE *(Continued)*

(A)	*(B)* area between mean and	*(C)* area beyond	*(A)*	*(B)* area between mean and	*(C)* area beyond	*(A)*	*(B)* area between mean and	*(C)* area beyond
z	z	z	z	z	z	z	z	z
1.65	.4505	.0495	2.22	.4868	.0132	2.79	.4974	.0026
1.66	.4515	.0485	2.23	.4871	.0129	2.80	.4974	.0026
1.67	.4525	.0475	2.24	.4875	.0125	2.81	.4975	.0025
1.68	.4535	.0465	2.25	.4878	.0122	2.82	.4976	.0024
1.69	.4545	.0455	2.26	.4881	.0119	2.83	.4977	.0023
1.70	.4554	.0446	2.27	.4884	.0116	2.84	.4977	.0023
1.71	.4564	.0436	2.28	.4887	.0113	2.85	.4978	.0022
1.72	.4573	.0427	2.29	.4890	.0110	2.86	.4979	.0021
1.73	.4582	.0418	2.30	.4893	.0107	2.87	.4979	.0021
1.74	.4591	.0409	2.31	.4896	.0104	2.88	.4980	.0020
1.75	.4599	.0401	2.32	.4898	.0102	2.89	.4981	.0019
1.76	.4608	.0392	2.33	.4901	.0099	2.90	.4981	.0019
1.77	.4616	.0384	2.34	.4904	.0096	2.91	.4982	.0018
1.78	.4625	.0375	2.35	.4906	.0094	2.92	.4982	.0018
1.79	.4633	.0367	2.36	.4909	.0091	2.93	.4983	.0017
1.80	.4641	.0359	2.37	.4911	.0089	2.94	.4984	.0016
1.81	.4649	.0351	2.38	.4913	.0087	2.95	.4984	.0016
1.82	.4656	.0344	2.39	.4916	.0084	2.96	.4985	.0015
1.83	.4664	.0336	2.40	.4918	.0082	2.97	.4985	.0015
1.84	.4671	.0329	2.41	.4920	.0080	2.98	.4986	.0014
1.85	.4678	.0322	2.42	.4922	.0078	2.99	.4986	.0014
1.86	.4686	.0314	2.43	.4925	.0075	3.00	.4987	.0013
1.87	.4693	.0307	2.44	.4927	.0073	3.01	.4987	.0013
1.88	.4699	.0301	2.45	.4929	.0071	3.02	.4987	.0013
1.89	.4706	.0294	2.46	.4931	.0069	3.03	.4988	.0012
1.90	.4713	.0287	2.47	.4932	.0068	3.04	.4988	.0012
1.91	.4719	.0281	2.48	.4934	.0066	3.05	.4989	.0011
1.92	.4726	.0274	2.49	.4936	.0064	3.06	.4989	.0011
1.93	.4732	.0268	2.50	.4938	.0062	3.07	.4989	.0011
1.94	.4738	.0262	2.51	.4940	.0060	3.08	.4990	.0010
1.95	.4744	.0256	2.52	.4941	.0059	3.09	.4990	.0010
1.96	.4750	.0250	2.53	.4943	.0057	3.10	.4990	.0010
1.97	.4756	.0244	2.54	.4945	.0055	3.11	.4991	.0009
1.98	.4761	.0239	2.55	.4946	.0054	3.12	.4991	.0009
1.99	.4767	.0233	2.56	.4948	.0052	3.13	.4991	.0009
2.00	.4772	.0228	2.57	.4949	.0051	3.14	.4992	.0008
2.01	.4778	.0222	2.58	.4951	.0049	3.15	.4992	.0008
2.02	.4783	.0217	2.59	.4952	.0048	3.16	.4992	.0008
2.03	.4788	.0212	2.60	.4953	.0047	3.17	.4992	.0008
2.04	.4793	.0207	2.61	.4955	.0045	3.18	.4993	.0007
2.05	.4798	.0202	2.62	.4956	.0044	3.19	.4993	.0007
2.06	.4803	.0197	2.63	.4957	.0043	3.20	.4993	.0007
2.07	.4808	.0192	2.64	.4959	.0041	3.21	.4993	.0007
2.08	.4812	.0188	2.65	.4960	.0040	3.22	.4994	.0006
2.09	.4817	.0183	2.66	.4961	.0039	3.23	.4994	.0006
2.10	.4821	.0179	2.67	.4962	.0038	3.24	.4994	.0006
2.11	.4826	.0174	2.68	.4963	.0037	3.25	.4994	.0006
2.12	.4830	.0170	2.69	.4964	.0036	3.30	.4995	.0005
2.13	.4834	.0166	2.70	.4965	.0035	3.35	.4996	.0004
2.14	.4838	.0162	2.71	.4966	.0034	3.40	.4997	.0003
2.15	.4842	.0158	2.72	.4967	.0033	3.45	.4997	.0003
2.16	.4846	.0154	2.73	.4968	.0032	3.50	.4998	.0002
2.17	.4850	.0150	2.74	.4969	.0031	3.60	.4998	.0002
2.18	.4854	.0146	2.75	.4970	.0030	3.70	.4999	.0001
2.19	.4857	.0143	2.76	.4971	.0029	3.80	.4999	.0001
2.20	.4861	.0139	2.77	.4972	.0028	3.90	.49995	.00005
2.21	.4864	.0136	2.78	.4973	.0027	4.00	.49997	.00003
						4.50	.4999966	.0000034
						5.00	.4999997	.0000003
						5.50	.4999999	.0000001

TABLE B Student's t-Distribution*

	Level of Significance for one-tailed test					
	.10	.05	.025	.01	.005	.0005
	Level of Significance for two-tailed test					
df	.20	.10	.05	.02	.01	.001
1	3.078	6.314	12.706	31.821	63.657	636.619
2	1.886	2.920	4.303	6.965	9.925	31.598
3	1.638	2.353	3.182	4.541	5.841	12.941
4	1.533	2.132	2.776	3.747	4.604	8.610
5	1.476	2.015	2.571	3.365	4.032	6.859
6	1.440	1.943	2.447	3.143	3.707	5.959
7	1.415	1.895	2.365	2.998	3.499	5.405
8	1.397	1.860	2.306	2.896	3.355	5.041
9	1.383	1.833	2.262	2.821	3.250	4.781
10	1.372	1.812	2.228	2.764	3.169	4.587
11	1.363	1.796	2.201	2.718	3.106	4.437
12	1.356	1.782	2.179	2.681	3.055	4.318
13	1.350	1.771	2.160	2.650	3.012	4.221
14	1.345	1.761	2.145	2.624	2.977	4.140
15	1.341	1.753	2.131	2.602	2.947	4.073
16	1.337	1.746	2.120	2.583	2.921	4.015
17	1.333	1.740	2.110	2.567	2.898	3.965
18	1.330	1.734	2.101	2.552	2.878	3.922
19	1.328	1.729	2.093	2.539	2.861	3.883
20	1.325	1.725	2.086	2.528	2.845	3.850
21	1.323	1.721	2.080	2.518	2.831	3.819
22	1.321	1.717	2.074	2.508	2.819	3.792
23	1.319	1.714	2.069	2.500	2.807	3.767
24	1.318	1.711	2.064	2.492	2.797	3.745
25	1.316	1.708	2.060	2.485	2.787	3.725
26	1.315	1.706	2.056	2.479	2.779	3.707
27	1.314	1.703	2.052	2.473	2.771	3.690
28	1.313	1.701	2.048	2.467	2.763	3.674
29	1.311	1.699	2.045	2.462	2.756	3.659
30	1.310	1.697	2.042	2.457	2.750	3.646
40	1.303	1.684	2.021	2.423	2.704	3.551
60	1.296	1.671	2.000	2.390	2.660	3.460
120	1.289	1.658	1.980	2.358	2.617	3.373
∞	1.282	1.645	1.960	2.326	2.576	3.291

*Adapted from Table III of R. A. Fisher and F. Yates, *Statistical Tables for Biological, Agricultural and Medical Research*, 1948 Edition (Edinburgh and London: Oliver & Boyd Limited) by permission of the authors and publishers.

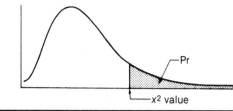

TABLE C VALUES OF CHI-SQUARE*

				Probability			
df†	.99	.975	.95	.90	.80	.70	.50
1	.000157	.000982	.00393	.0158	.0642	.148	.455
2	.0201	.0506	.103	.211	.446	.713	1.386
3	.115	.216	.352	.584	1.005	1.424	2.366
4	.297	.484	.711	1.064	1.649	2.195	3.357
5	.554	.831	1.145	1.610	2.343	3.000	4.351
6	.872	1.237	1.635	2.204	3.070	3.828	5.348
7	1.239	1.690	2.167	2.833	3.822	4.671	6.346
8	1.646	2.180	2.733	3.490	4.594	5.527	7.344
9	2.088	2.700	3.325	4.168	5.380	6.393	8.343
10	2.558	3.247	3.940	4.865	6.179	7.267	9.342
11	3.053	3.816	4.575	5.578	6.989	8.148	10.341
12	3.571	4.404	5.226	6.304	7.807	9.034	11.340
13	4.107	5.009	5.892	7.042	8.634	9.926	12.340
14	4.660	5.629	6.571	7.790	9.467	10.821	13.339
15	5.229	6.262	7.261	8.547	10.307	11.721	14.339
16	5.812	6.908	7.962	9.312	11.152	12.624	15.338
17	6.408	7.564	8.672	10.085	12.002	13.531	16.338
18	7.015	8.231	9.390	10.865	12.857	14.440	17.338
19	7.633	8.907	10.117	11.651	13.716	15.352	18.338
20	8.260	9.591	10.851	12.443	14.578	16.266	19.337
21	8.897	10.283	11.591	13.240	15.445	17.182	20.337
22	9.542	10.982	12.338	14.041	16.314	18.101	21.337
23	10.196	11.689	13.091	14.848	17.187	19.021	22.337
24	10.865	12.401	13.848	15.659	18.062	19.943	23.337
25	11.524	13.120	14.611	16.473	18.940	20.867	24.337
26	12.198	13.844	15.379	17.292	19.820	21.792	25.336
27	12.879	14.573	16.151	18.114	20.703	22.719	26.336
28	13.565	15.308	16.928	18.939	21.588	23.647	27.336
29	14.256	16.047	17.708	19.768	22.475	24.577	28.336
30	14.953	16.791	18.493	20.599	23.364	25.508	29.336

*Adapted from Table IV of R. A. Fisher and F. Yates, *Statistical Tables for Biological, Agricultural and Medical Research*, 1948 edition (Edinburgh and London: Oliver & Boyd Limited), by permission of the authors and publishers.
†For larger values of df, the expression $\sqrt{2x^2} - \sqrt{df^2 - 1}$ may be used as a normal deviate with unit variance.

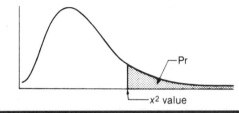

TABLE C VALUES OF CHI-SQUARE *(Continued)*

				Probability			
df	*.30*	*.20*	*.10*	*.05*	*.025*	*.01*	*.001*
1	1.074	1.642	2.706	3.841	5.024	6.635	10.827
2	2.408	3.219	4.605	5.991	7.378	9.210	13.815
3	3.665	4.624	6.251	7.815	9.348	11.345	16.268
4	4.878	5.989	7.779	9.488	11.143	13.277	18.465
5	6.064	7.289	9.236	11.070	12.832	15.086	20.517
6	7.231	8.558	10.645	12.592	14.449	16.812	22.457
7	8.383	9.803	12.017	14.067	16.013	18.475	24.322
8	9.524	11.030	13.362	15.507	17.535	20.090	26.125
9	10.656	12.242	14.684	16.919	19.023	21.666	27.877
10	11.781	13.442	15.987	18.307	20.483	23.209	29.588
11	12.899	14.631	17.275	19.675	21.920	24.725	31.264
12	14.011	15.812	18.549	21.026	23.337	26.217	32.909
13	15.119	16.985	19.812	22.362	24.736	27.688	34.528
14	16.222	18.151	21.064	23.685	26.119	29.141	36.123
15	17.322	19.311	22.307	24.996	27.488	30.578	37.697
16	18.418	20.465	23.542	26.296	28.845	32.000	39.252
17	19.511	21.615	24.769	27.587	30.191	33.409	40.790
18	20.601	22.760	25.989	28.869	31.526	34.805	42.312
19	21.689	23.900	27.204	30.144	32.852	36.191	43.820
20	22.775	25.038	28.412	31.410	34.170	37.566	45.315
21	23.858	26.171	29.615	32.671	35.479	38.932	46.797
22	24.939	27.301	30.813	33.924	36.781	40.289	48.268
23	26.018	28.429	32.007	35.172	38.076	41.638	49.728
24	27.096	29.553	33.196	36.415	39.364	42.980	51.179
25	28.172	30.675	34.382	37.652	40.646	44.314	52.620
26	29.246	31.795	35.563	38.885	41.923	45.642	54.052
27	30.319	32.912	36.741	40.113	43.194	46.963	55.476
28	31.391	34.027	37.916	41.337	44.461	48.278	56.893
29	32.461	35.139	39.087	42.557	45.722	49.588	58.302
30	33.530	36.250	40.256	43.773	46.979	50.892	59.703

TABLE D   *F*-TABLE FOR THE .05 LEVEL OF SIGNIFICANCE

$\frac{df_1}{df_2}$	1	2	3	4	5	6	8	10
				Numerator Degrees of Freedom				
1	161.4	199.5	215.7	224.6	230.2	234.0	238.9	241.9
2	18.51	19.00	19.16	19.25	19.30	19.33	19.37	19.40
3	10.13	9.55	9.28	9.12	9.01	8.94	8.85	8.79
4	7.71	6.94	6.59	6.39	6.26	6.16	6.04	5.96
5	6.61	5.79	5.41	5.19	5.05	4.95	4.82	4.74
6	5.99	5.14	4.76	4.53	4.39	4.28	4.15	4.06
7	5.59	4.74	4.35	4.12	3.97	3.87	3.73	3.64
8	5.32	4.46	4.07	3.84	3.69	3.58	3.44	3.35
9	5.12	4.26	3.86	3.63	3.48	3.37	3.23	3.14
10	4.96	4.10	3.71	3.48	3.33	3.22	3.07	2.98
11	4.84	3.98	3.59	3.36	3.20	3.09	2.95	2.85
12	4.75	3.89	3.49	3.26	3.11	3.00	2.85	2.75
13	4.67	3.81	3.41	3.18	3.03	2.92	2.77	2.67
14	4.60	3.74	3.34	3.11	2.96	2.85	2.70	2.60
15	4.54	3.68	3.29	3.06	2.90	2.79	2.64	2.54
16	4.49	3.63	3.24	3.01	2.85	2.74	2.59	2.49
17	4.45	3.59	3.20	2.96	2.81	2.70	2.55	2.45
18	4.41	3.55	3.16	2.93	2.77	2.66	2.51	2.41
19	4.38	3.52	3.13	2.90	2.74	2.63	2.48	2.38
20	4.35	3.49	3.10	2.87	2.71	2.60	2.45	2.35
21	4.32	3.47	3.07	2.84	2.68	2.57	2.42	2.32
22	4.30	3.44	3.05	2.82	2.66	2.55	2.40	2.30
23	4.28	3.42	3.03	2.80	2.64	2.53	2.37	2.27
24	4.26	3.40	3.01	2.78	2.62	2.51	2.36	2.25
25	4.24	3.39	2.99	2.76	2.60	2.49	2.34	2.24
26	4.23	3.37	2.98	2.74	2.59	2.47	2.32	2.22
27	4.21	3.35	2.96	2.73	2.57	2.46	2.31	2.20
28	4.20	3.34	2.95	2.71	2.56	2.45	2.29	2.19
29	4.18	3.33	2.93	2.70	2.55	2.43	2.28	2.18
30	4.17	3.32	2.92	2.69	2.53	2.42	2.27	2.16
40	4.08	3.23	2.84	2.61	2.45	2.34	2.18	2.08
60	4.00	3.15	2.76	2.53	2.37	2.25	2.10	1.99
80	3.96	3.11	2.72	2.48	2.33	2.21	2.05	1.95
120	3.92	3.07	2.68	2.45	2.29	2.17	2.02	1.91
∞	3.84	3.00	2.60	2.37	2.21	2.10	1.94	1.83

Degrees of Freedom for the Denominator

*Adapted from Table V of R. A. Fisher and F. Yates, *Statistical Tables for Biological, Agricultural and Medical Research*, 1948 edition (Edinburgh and London: Oliver & Boyd, Limited) by permission of the authors and publishers.

TABLE D *F*-TABLE FOR THE .05 LEVEL OF SIGNIFICANCE *(Continued)*

Degrees of Freedom for the Denominator

df_2 \ df_1	*Numerator Degrees of Freedom*							
	12	*15*	*20*	*30*	*40*	*60*	*120*	*∞*
1	243.9	245.9	248.0	250.1	251.1	252.2	253.3	254.3
2	19.41	19.43	19.45	19.46	19.47	19.48	19.49	19.50
3	8.74	8.70	8.66	8.62	8.59	8.57	8.55	8.53
4	5.91	5.86	5.80	5.75	5.72	5.69	5.66	5.63
5	4.68	4.62	4.56	4.50	4.46	4.43	4.40	4.36
6	4.00	3.94	3.87	3.81	3.77	3.74	3.70	3.67
7	3.57	3.51	3.44	3.38	3.34	3.30	3.27	3.23
8	3.28	3.22	3.15	3.08	3.04	3.01	2.97	2.93
9	3.07	3.01	2.94	2.86	2.83	2.79	2.75	2.71
10	2.91	2.85	2.77	2.70	2.66	2.62	2.58	2.54
11	2.79	2.72	2.65	2.57	2.53	2.49	2.45	2.40
12	2.69	2.62	2.54	2.47	2.43	2.38	2.34	2.30
13	2.60	2.53	2.46	2.38	2.34	2.30	2.25	2.21
14	2.53	2.46	2.39	2.31	2.27	2.22	2.18	2.13
15	2.48	2.40	2.33	2.25	2.20	2.16	2.11	2.07
16	2.42	2.35	2.28	2.19	2.15	2.11	2.06	2.01
17	2.38	2.31	2.23	2.15	2.10	2.06	2.01	1.96
18	2.34	2.27	2.19	2.11	2.06	2.02	1.97	1.92
19	2.31	2.23	2.16	2.07	2.03	1.98	1.93	1.88
20	2.28	2.20	2.12	2.04	1.99	1.95	1.90	1.84
21	2.25	2.18	2.10	2.01	1.96	1.92	1.87	1.81
22	2.23	2.15	2.07	1.98	1.94	1.89	1.84	1.78
23	2.20	2.13	2.05	1.96	1.91	1.86	1.81	1.76
24	2.18	2.11	2.03	1.94	1.89	1.84	1.79	1.73
25	2.16	2.09	2.01	1.92	1.87	1.82	1.77	1.71
26	2.15	2.07	1.99	1.90	1.85	1.80	1.75	1.69
27	2.13	2.06	1.97	1.88	1.84	1.79	1.73	1.67
28	2.12	2.04	1.96	1.87	1.82	1.77	1.71	1.65
29	2.10	2.03	1.94	1.85	1.81	1.75	1.70	1.64
30	2.09	2.01	1.93	1.84	1.79	1.74	1.68	1.62
40	2.00	1.92	1.84	1.74	1.69	1.64	1.58	1.51
60	1.92	1.84	1.75	1.65	1.59	1.53	1.47	1.39
80	1.88	1.80	1.70	1.60	1.54	1.49	1.41	1.32
120	1.83	1.75	1.66	1.55	1.50	1.43	1.35	1.25
∞	1.75	1.67	1.57	1.46	1.39	1.32	1.22	1.00

TABLE D *F*-TABLE FOR THE .01 LEVEL OF SIGNIFICANCE

Degrees of Freedom for the Denominator

df_1 df_2	Numerator Degrees of Freedom							
	1	2	3	4	5	6	8	10
1	4052	4999.5	5403	5625	5764	5859	5982	6056
2	98.50	99.00	99.17	99.25	99.30	99.33	99.37	99.40
3	34.12	30.82	29.46	28.71	28.24	27.91	27.49	27.23
4	21.20	18.00	16.69	15.98	15.52	15.21	14.80	14.55
5	16.26	13.27	12.06	11.39	10.97	10.67	10.29	10.05
6	13.75	10.92	9.78	9.15	8.75	8.47	8.10	7.87
7	12.25	9.55	8.45	7.85	7.46	7.19	6.84	6.62
8	11.26	8.65	7.59	7.01	6.63	6.37	6.03	5.81
9	10.56	8.02	6.99	6.42	6.06	5.80	5.47	5.26
10	10.04	7.56	6.55	5.99	5.64	5.39	5.06	4.85
11	9.65	7.21	6.22	5.67	5.32	5.07	4.74	4.54
12	9.33	6.93	5.95	5.41	5.06	4.82	4.50	4.30
13	9.07	6.70	5.74	5.21	4.86	4.62	4.30	4.10
14	8.86	6.51	5.56	5.04	4.69	4.46	4.14	3.94
15	8.68	6.36	5.42	4.89	4.56	4.32	4.00	3.80
16	8.53	6.23	5.29	4.77	4.44	4.20	3.89	3.69
17	8.40	6.11	5.18	4.67	4.34	4.10	3.79	3.59
18	8.29	6.01	5.09	4.58	4.25	4.01	3.71	3.51
19	8.18	5.93	5.01	4.50	4.17	3.94	3.63	3.43
20	8.10	5.85	4.94	4.43	4.10	3.87	3.56	3.37
21	8.02	5.78	4.87	4.37	4.04	3.81	3.51	3.31
22	7.95	5.72	4.82	4.31	3.99	3.76	3.45	3.26
23	7.88	5.66	4.76	4.26	3.94	3.71	3.41	3.21
24	7.82	5.61	4.72	4.22	3.90	3.67	3.36	3.17
25	7.77	5.57	4.68	4.18	3.85	3.63	3.32	3.13
26	7.72	5.53	4.64	4.14	3.82	3.59	3.29	3.09
27	7.68	5.49	4.60	4.11	3.78	3.56	3.26	3.06
28	7.64	5.45	4.57	4.07	3.75	3.53	3.23	3.03
29	7.60	5.42	4.54	4.04	3.73	3.50	3.20	3.00
30	7.56	5.39	4.51	4.02	3.70	3.47	3.17	2.98
40	7.31	5.18	4.31	3.83	3.51	3.29	2.99	2.80
60	7.08	4.98	4.13	3.65	3.34	3.12	2.82	2.63
80	6.96	4.88	4.04	3.56	3.25	3.04	2.74	2.55
120	6.85	4.79	3.95	3.48	3.17	2.96	2.66	2.47
∞	6.63	4.61	3.78	3.32	3.02	2.80	2.51	2.32

TABLE D *F*-ᴛᴀʙʟᴇ ꜰᴏʀ ᴛʜᴇ .01 Lᴇᴠᴇʟ ᴏꜰ Sɪɢɴɪꜰɪᴄᴀɴᴄᴇ (*Continued*)

Numerator Degrees of Freedom

df_1 / df_2	12	15	20	30	40	60	120	∞
1	6106	6157	6209	6261	6287	6313	6339	6366
2	99.42	99.43	99.45	99.47	99.47	99.48	99.49	99.50
3	27.05	26.87	26.69	26.50	26.41	26.32	26.22	26.13
4	14.37	14.20	14.02	13.84	13.75	13.65	13.56	13.46
5	9.89	9.72	9.55	9.38	9.29	9.20	9.11	9.02
6	7.72	7.56	7.40	7.23	7.14	7.06	6.97	6.88
7	6.47	6.31	6.16	5.99	5.91	5.82	5.74	5.65
8	5.67	5.52	5.36	5.20	5.12	5.03	4.95	4.86
9	5.11	4.96	4.81	4.65	4.57	4.48	4.40	4.31
10	4.71	4.56	4.41	4.25	4.17	4.08	4.00	3.91
11	4.40	4.25	4.10	3.94	3.86	3.78	3.69	3.60
12	4.16	4.01	3.86	3.70	3.62	3.54	3.45	3.36
13	3.96	3.82	3.66	3.51	3.43	3.34	3.25	3.17
14	3.80	3.66	3.51	3.35	3.27	3.18	3.09	3.00
15	3.67	3.52	3.37	3.21	3.13	3.05	2.96	2.87
16	3.55	3.41	3.26	3.10	3.02	2.93	2.84	2.75
17	3.46	3.31	3.16	3.00	2.92	2.83	2.75	2.65
18	3.37	3.23	3.08	2.92	2.84	2.75	2.66	2.57
19	3.30	3.15	3.00	2.84	2.76	2.67	2.58	2.49
20	3.23	3.09	2.94	2.78	2.69	2.61	2.52	2.42
21	3.17	3.03	2.88	2.72	2.64	2.55	2.46	2.36
22	3.12	2.98	2.83	2.67	2.58	2.50	2.40	2.31
23	3.07	2.93	2.78	2.62	2.54	2.45	2.35	2.26
24	3.03	2.89	2.74	2.58	2.49	2.40	2.31	2.21
25	2.99	2.85	2.70	2.54	2.45	2.36	2.27	2.17
26	2.96	2.81	2.66	2.50	2.42	2.33	2.23	2.13
27	2.93	2.78	2.63	2.47	2.38	2.29	2.20	2.10
28	2.90	2.75	2.60	2.44	2.35	2.26	2.17	2.06
29	2.87	2.73	2.57	2.41	2.33	2.23	2.14	2.03
30	2.84	2.70	2.55	2.39	2.30	2.21	2.11	2.01
40	2.66	2.52	2.37	2.20	2.11	2.02	1.92	1.80
60	2.50	2.35	2.20	2.03	1.94	1.84	1.73	1.60
80	2.41	2.28	2.11	1.94	1.84	1.75	1.63	1.49
120	2.34	2.19	2.03	1.86	1.76	1.66	1.53	1.38
∞	2.18	2.04	1.88	1.70	1.59	1.47	1.32	1.00

Degrees of Freedom for the Denominator

TABLE E TRANSFORMATION OF r TO Z'*

r	.000	.001	.002	.003	.004	.005	.006	.007	.008	.009
.000	.0000	.0010	.0020	.0030	.0040	.0050	.0060	.0070	.0080	.0090
.010	.0100	.0110	.0120	.0130	.0140	.0150	.0160	.0170	.0180	.0190
.020	.0200	.0210	.0220	.0230	.0240	.0250	.0260	.0270	.0280	.0290
.030	.0300	.0310	.0320	.0330	.0340	.0350	.0360	.0370	.0380	.0390
.040	.0400	.0410	.0420	.0430	.0440	.0450	.0460	.0470	.0480	.0490
.050	.0501	.0511	.0521	.0531	.0541	.0551	.0561	.0571	.0581	.0591
.060	.0601	.0611	.0621	.0631	.0641	.0651	.0661	.0671	.0681	.0691
.070	.0701	.0711	.0721	.0731	.0741	.0751	.0761	.0771	.0782	.0792
.080	.0802	.0812	.0822	.0832	.0842	.0852	.0862	.0872	.0882	.0892
.090	.0902	.0912	.0922	.0933	.0943	.0953	.0963	.0973	.0983	.0993
.100	.1003	.1013	.1024	.1034	.1044	.1054	.1064	.1074	.1084	.1094
.110	.1105	.1115	.1125	.1135	.1145	.1155	.1165	.1175	.1185	.1195
.120	.1206	.1216	.1226	.1236	.1246	.1257	.1267	.1277	.1287	.1297
.130	.1308	.1318	.1328	.1338	.1348	.1358	.1368	.1379	.1389	.1399
.140	.1409	.1419	.1430	.1440	.1450	.1460	.1470	.1481	.1491	.1501
.150	.1511	.1522	.1532	.1542	.1552	.1563	.1573	.1583	.1593	.1604
.160	.1614	.1624	.1634	.1644	.1655	.1665	.1676	.1686	.1696	.1706
.170	.1717	.1727	.1737	.1748	.1758	.1768	.1779	.1789	.1799	.1810
.180	.1820	.1830	.1841	.1851	.1861	.1872	.1882	.1892	.1903	.1913
.190	.1923	.1934	.1944	.1954	.1965	.1975	.1986	.1996	.2007	.2017
.200	.2027	.2038	.2048	.2059	.2069	.2079	.2090	.2100	.2111	.2121
.210	.2132	.2142	.2153	.2163	.2174	.2184	.2194	.2205	.2215	.2226
.220	.2237	.2247	.2258	.2268	.2279	.2289	.2300	.2310	.2321	.2331
.230	.2342	.2353	.2363	.2374	.2384	.2395	.2405	.2416	.2427	.2437
.240	.2448	.2458	.2469	.2480	.2490	.2501	.2511	.2522	.2533	.2543
.250	.2554	.2565	.2575	.2586	.2597	.2608	.2618	.2629	.2640	.2650
.260	.2661	.2672	.2682	.2693	.2704	.2715	.2726	.2736	.2747	.2758
.270	.2769	.2779	.2790	.2801	.2812	.2823	.2833	.2844	.2855	.2866
.280	.2877	.2888	.2898	.2909	.2920	.2931	.2942	.2953	.2964	.2975
.290	.2986	.2997	.3008	.3019	.3029	.3040	.3051	.3062	.3073	.3084
.300	.3095	.3106	.3117	.3128	.3139	.3150	.3136	.3172	.3183	.3195
.310	.3206	.3217	.3228	.3239	.3250	.3261	.3272	.3283	.3294	.3305
.320	.3317	.3328	.3339	.3350	.3361	.3372	.3384	.3395	.3406	.3417
.330	.3428	.3439	.3451	.3462	.3473	.3484	.3496	.3507	.3518	.3530
.340	.3541	.3552	.3564	.3575	.3586	.3597	.3609	.3620	.3632	.3643
.350	.3654	.3666	.3677	.3689	.3700	.3712	.3723	.3734	.3746	.3757
.360	.3769	.3780	.3792	.3803	.3815	.3826	.3838	.3850	.3861	.3873
.370	.3884	.3896	.3907	.3919	.3931	.3942	.3954	.3966	.3977	.3989
.380	.4001	.4012	.4024	.4036	.4047	.4059	.4071	.4083	.4094	.4106
.390	.4118	.4130	.4142	.4153	.4165	.4177	.4189	.4201	.4213	.4225
.400	.4236	.4248	.4260	.4272	.4284	.4296	.4308	.4320	.4332	.4344
.410	.4356	.4368	.4380	.4392	.4404	.4416	.4429	.4441	.4453	.4465
.420	.4477	.4489	.4501	.4513	.4526	.4538	.4550	.4562	.4574	.4587
.430	.4599	.4611	.4623	.4636	.4648	.4660	.4673	.4685	.4697	.4710
.440	.4722	.4735	.4747	.4760	.4772	.4784	.4797	.4809	.4822	.4835
.450	.4847	.4860	.4872	.4885	.4897	.4910	.4923	.4935	.4948	.4961
.460	.4973	.4986	.4999	.5011	.5024	.5037	.5049	.5062	.5075	.5088
.470	.5101	.5114	.5126	.5139	.5152	.5165	.5178	.5191	.5204	.5217
.480	.5230	.5243	.5256	.5279	.5282	.5295	.5308	.5321	.5334	.5347
.490	.5361	.5374	.5387	.5400	.5413	.5427	.5440	.5453	.5466	.5480

*From Albert E. Waugh, *Statistical Tables and Problems* (New York: McGraw-Hill Book Company, 1952, table A11, pp. 40–41, by permission of the author and publisher.

TABLE E TRANSFORMATION OF r TO Z' (Continued)

r	.000	.001	.002	.003	.004	.005	.006	.007	.008	.009
.500	.5493	.5506	.5520	.5533	.5547	.5560	.5573	.5587	.5600	.5614
.510	.5627	.5641	.5654	.5668	.5681	.5695	.5709	.5722	.5736	.5750
.520	.5763	.5777	.5791	.5805	.5818	.5832	.5846	.5860	.5874	.5888
.530	.5901	.5915	.5929	.5943	.5957	.5971	.5985	.5999	.6013	.6027
.540	.6042	.6056	.6070	.6084	.6098	.6112	.6127	.6141	.6155	.6170
.550	.6184	.6198	.6213	.6227	.6241	.6256	.6270	.6285	.6299	.6314
.560	.6328	.6343	.6358	.6372	.6387	.6401	.6416	.6431	.6446	.6460
.570	.6475	.6490	.6505	.6520	.6535	.6550	.6565	.6579	.6594	.6610
.580	.6625	.6640	.6655	.6670	.6685	.6700	.6715	.6731	.6746	.6761
.590	.6777	.6792	.6807	.6823	.6838	.6854	.6869	.6885	.6900	.6916
.600	.6931	.6947	.6963	.6978	.6994	.7010	.7026	.7042	.7057	.7073
.610	.7089	.7105	.7121	.7137	.7153	.7169	.7185	.7201	.7218	.7234
.620	.7250	.7266	.7283	.7299	.7315	.7332	.7348	.7364	.7381	.7398
.630	.7414	.7431	.7447	.7464	.7481	.7497	.7514	.7531	.7548	.7565
.640	.7582	.7599	.7616	.7633	.7650	.7667	.7684	.7701	.7718	.7736
.650	.7753	.7770	.7788	.7805	.7823	.7840	.7858	.7875	.7893	.7910
.660	.7928	.7946	.7964	.7981	.7999	.8017	.8035	.8053	.8071	.8089
.670	.8107	.8126	.8144	.8162	.8180	.8199	.8217	.8236	.8254	.8273
.680	.8291	.8310	.8328	.8347	.8366	.8385	.8404	.8423	.8442	.8461
.690	.8480	.8499	.8518	.8537	.8556	.8576	.8595	.8614	.8634	.8653
.700	.8673	.8693	.8712	.8732	.8752	.8772	.8792	.8812	.8832	.8852
.710	.8872	.8892	.8912	.8933	.8953	.8973	.8994	.9014	.9035	.9056
.720	.9076	.9097	.9118	.9139	.9160	.9181	.9202	.9223	.9245	.9266
.730	.9287	.9309	.9330	.9352	.9373	.9395	.9417	.9439	.9461	.9483
.740	.9505	.9527	.9549	.9571	.9594	.9616	.9639	.9661	.9684	.9707
.750	.9730	.9752	.9775	.9799	.9822	.9845	.9868	.9892	.9915	.9939
.760	.9962	.9986	1.0010	1.0034	1.0058	1.0082	1.0106	1.0130	1.0154	1.0179
.770	1.0203	1.0228	1.0253	1.0277	1.0302	1.0327	1.0352	1.0378	1.0403	1.0428
.780	1.0454	1.0479	1.0505	1.0531	1.0557	1.0583	1.0609	1.0635	1.0661	1.0688
.790	1.0714	1.0741	1.0768	1.0795	1.0822	1.0849	1.0876	1.0903	1.0931	1.0958
.800	1.0986	1.1014	1.1041	1.1070	1.1098	1.1127	1.1155	1.1184	1.1212	1.1241
.810	1.1270	1.1299	1.1329	1.1358	1.1388	1.1417	1.1447	1.1477	1.1507	1.1538
.820	1.1568	1.1599	1.1630	1.1660	1.1692	1.1723	1.1754	1.1786	1.1817	1.1849
.830	1.1870	1.1913	1.1946	1.1979	1.2011	1.2044	1.2077	1.2111	1.2144	1.2178
840	1.2212	1.2246	1.2280	1.2315	1.2349	1.2384	1.2419	1.2454	1.2490	1.2526
.850	1.2561	1.2598	1.2634	1.2670	1.2708	1.2744	1.2782	1.2819	1.2857	1.2895
.860	1.2934	1.2972	1.3011	1.3050	1.3089	1.3129	1.3168	1.3209	1.3249	1.3290
.870	1.3331	1.3372	1.3414	1.3456	1.3498	1.3540	1.3583	1.3626	1.3670	1.3714
.880	1.3758	1.3802	1.3847	1.3892	1.3938	1.3984	1.4030	1.4077	1.4124	1.4171
.890	1.4219	1.4268	1.4316	1.4366	1.4415	1.4465	1.4516	1.4566	1.4618	1.4670
.900	1.4722	1.4775	1.4828	1.4883	1.4937	1.4992	1.5047	1.5103	1.5160	1.5217
.910	1.5275	1.5334	1.5393	1.5453	1.5513	1.5574	1.5636	1.5698	1.5762	1.5825
.920	1.5890	1.5956	1.6022	1.6089	1.6157	1.6226	1.6296	1.6366	1.6438	1.6510
.930	1.6584	1.6659	1.6734	1.6811	1.6888	1.6967	1.7047	1.7129	1.7211	1.7295
.940	1.7380	1.7467	1.7555	1.7645	1.7736	1.7828	1.7923	1.8019	1.8117	1.8216
.950	1.8318	1.8421	1.8527	1.8635	1.8745	1.8857	1.8972	1.9090	1.9210	1.9333
.960	1.9459	1.9588	1.9721	1.9857	1.9996	2.0140	2.0287	2.0439	2.0595	2.0756
.970	2.0923	2.1095	2.1273	2.1457	2.1649	2.1847	2.2054	2.2269	2.2494	2.2729
.980	2.2976	2.3223	2.3507	2.3796	2.4101	2.4426	2.4774	2.5147	2.5550	2.5988
.990	2.6467	2.6996	2.7587	2.8257	2.9031	2.9945	3.1063	3.2504	3.4534	3.8002

Index

logarithm (*cont'd*)
  use, 213, 597
logit analysis, 329n
log-linear models, 597
Lohnes, Paul R., 354, 357
longitudinal design, 285
loop-back boxes, 4
Lopreato, J., 341, 357
Los Angeles Times, 13, 407, 462
Lowenthal, Marjorie Fiske, 396,
  407

Magid, Lawrence, J., 9n, 13
Mantel, Nathan, 585, 607
mantissa (logarithms), 100
marginal totals (table), 167
Maris, Ronald, W., 93, 94, 106
Markoff, A., 6
Marsh, John F. Jr., 251, 252,
  265
Marsh, Robert M., 28, 37
Martin, Clyde E., 401, 407
Massey, Frank J. Jr., 557n, 563
matched (related) samples, 554
mathematical descriptions of
  form, 151
mathematics, approach, xii, 8,
  9, 328n
matrix, correlation, 251
McAllister, Ronald, 313, 314,
  324
McCarthy, 406
McDill, Edward L., 302n, 324
McKeown, Thomas, 480, 506
McLemore, S. Dale, 36
McMaster, Kirby, xv
mean
  arithmetic, 125
  confidence limits for, 458, 461
  geometric, 129
  harmonic, 129
  mean proportion, 425, 453
  moving average, 131
  overall (grand) for eta, 248
  sampling distribution of, 427
  standard error of, 433
  unbiased estimators of, 428,
    458
  weighted, 129
  within categories for eta, 248
meanings of perfect association,
  188, 190
means test
  k-sample, 519, 566
  one-sample, 508
    assumptions, 512
    computations, 511, 564
    normal approximation of t-
      distribution, 512
    selecting a sampling
      distribution, 449
  two-sample, 512
    assumptions, 518
    computations, 565

direct difference test, 554
  example, 516
  expected value of mean
    difference, 513, 513n
  randomization test, 551
  sampling distribution of
    differences, 513
  standard error of the
    difference, 513 (for
    correlated samples, *see*
    514n, 551, 554)
measurement
  definition, 14
  in sociology, 31
  levels of, 17
  non-response, 17, 48
  precision, 17
  to last unit, 27
  to nearest unit, 24
  unit, 48
measures of association, 184,
  201
  cautions in interpretation, 255
  equivalences in 2 by 2 table,
    189n
  for interval by nominal
    variables, 248
  for interval/ratio variables,
    230
  for nominal variables, 205
  for ordinal variables, 219
  organization of, 228, 256, 257
  PRE measures, 204
  selecting, 253
measuring variables, 14
median, 71, 123
Meecker, Marchia, 519, 563
Mendelsohn, Harold, 39, 67
Merton, Robert K., 258n, 324
Messer, Mark, 277, 278, 282,
  322, 324, 587, 591, 592,
  607
methodological problems with
  inferential leaps, 401
Meyers, Edmund D., xv, 302n,
  324
MIDCARS, 32n
midpoint, 25, 27, 49
Miller, Delbert, 31n, 36
minimizing decision errors, 499
MINITAB, 10, 13
mobility, 54, 262, 314
mode, 119, 122
models
  log-linear, 586, 597
  multiple variable effects, 587
  no association, 179, 188n
  no-effects (hierarchical), 588
  perfect association, 184, 188
  proportionate reduction in
    error (PRE) association
    measures, 206
  saturated (hierarchical), 588

testing, 586
Mohr, Lawrence, B., 505
moment system, 146
Moore, Davis-Moore model of
  stratification, 339, 340,
  343
Morrison, Denton E., 620, 626
Morse, Nancy C., 451, 456, 474
mortality ratio, 102
Mosteller, Frederick, 361n, 378,
  401, 407
moving average, 131
mu, population mean, 125n,
  384n
Mueller, John H., 36, 141n, 154
multiple partial correlation
  coefficient, 335n
multiple regression, 328
  assumptions, 329, 330, 330n
  curvilinear, 351
  dummy variable, 348
  equation, 332
  illustration, 336, 344, 347, 350
  intercept, 346
  standardized, 332
  stepwise, 350
  unstandardized, 343, 347
  use in path analysis, 338
multiple correlation coefficient,
  328, 334
  computation of corrected
    coefficient, 335
  corrected, 335
  discussion, 328, 334
  illustration, 336
  significance test for, 604
multiple-variable effects, 587
multiplication operation,
  symbol for, 590
multiplication rule for
  probabilities, 369, 370
multivariate
  descriptive statistics, 276
  inferential statistics, 586, 604
  sampling distributions, 416n
mutually exclusive, 17
  events, 364
  special addition rule, 367

N, sample size, 46, 48, 388, 496
Naperian (natural) logarithms,
  100, 597
National Opinion Research
  Center (NORC), 64, 347
Natrella, Mary G., 502, 503, 506
natural logarithms, 100, 597
nature of association, 176, 180
  versus strength, 254
nearest whole unit, 24
necessary condition, 191
Nett, Roger, 402, 406
Neuman, W. Lawrence, 199,
  200

unit (*cont'd*)
  of measurement, 19, 24, 25,
      27, 48, 50
United Nations, 541n
  Agriculture Organization, 480
  Children's Fund, 480, 506
  live births, 541
univariate distributions, 118
univariate statistics
  central tendency, 121
  form, 119., 146
  variation, 133
universal set, 362
universe, 4
  conceptual, 4
  general, 4, 401
  of indicators, 33
unstandardized multiple
      regression, 343, 347

validity, 15, 34
Van de Geer, John P., 349, 357
Vargo, Louis G., 22, 37
variability
  of sampling distribution, 416
  of univariate distributions,
      133
variable
  creating, 32, 33
  definition, 3, 14
  level of measurement, 17
  measuring, 14
  role in research, 29
variance, 137, 430
  assumptions for
      one-sample test, 540
      two-sample test, 544
  computational formulas, 433
  correction factors for, 432
  explained (regression), 239
  homoscedasticity assumption
      in regression, 247
  limits of, 139
  and standard deviation, 137
  of the estimate, 236
  total (regression), 239
  unexplained (regression), 239
variances, significance tests for
  one-sample test, 537
      assumptions for, 540
      example, 539

two-sample test, 540
  assumptions for, 544
  within-groups and between-
      groups in ANOVA, 523
variation measures (*see also*
      Variance)
  computing, 158
  measures of, 133
      average absolute deviation,
          136
      crude range, 135
      H_{rel}, 215n
      index of dispersion, 141
      index of qualitative
          variation, 141n
      interquartile range, 135
      relative uncertainty, 141n
      selecting, 144
      standard deviation, 137
      variance, 137
      note on ordinal, 134n
variation (correlation,
      regression)
  explained, 239
  percent explained, 242
  total, 239
  unexplained, 239
Venn diagram (*see* Euler
      diagrams)
Verwaller, Darrel J., 512, 563
voluntary samples, 400

Waldo, Gordon P., 60n, 67
Walker, Helen M., 474
Wallis, W. Allen, 13, 130n, 154,
      403, 407
Wampler, Roy H., 328n, 357
Wang, Deming, xv
Warner, W. Lloyd, 519, 563
Webb, Eugene J., 31, 37
Weber, Max, 81
weighted mean, 129
Weiss, Robert S., 199, 451, 456,
      474
Weitzman, Murry S., 79, 107
Whitt, Hugh P., 574, 607
width, interval (class), 49
Winkler, Robert L., 448
Winsborough, Hal H., 352

within-groups sum of squares,
      522
within-groups variance, 523
Wonnacott, Ronald J., 357
Wonnacott, Thomas H., 357
Wood, James, xv
working hypothesis, 477
World Health Organization, 153,
      154
Wray, Joe D., 101, 102, 107
Wright, Jack M., 31n, 37

X, score or class midpoint, 25,
      46
X-axis, 75

Yates, F., 6
Y-axis, 75
Y-intercept (regression
      constant)
  computing, 239
  discussion, 233
Yoels, William C., 266
Youtz, Robert, 597, 598, 606
Yu, Eui-Young, 548, 549, 550,
      563
Yule, G. Udny, 6
Yule's Q coefficient, 192, 255,
      256, 287n

Zeisel, Hans, 62, 67, 68, 199,
      285n, 323
Zeitlin, Maurice W., 199, 200
zero (no) association, 179, 188n
zero-order association, 179, 188n
zero-point, defined on a scale, 19
Zeller, Richard A., 36
Zetterberg, Hans, 403
Z-prime (Z'), conversion of r to
      Z', 601
  Table for conversion,
      Appendix, Table E, 639
z-score (standard score)
  definition, 145
  form of correlation
      coefficient, 244
  sum of, 145n
  uses, 243, 331n, 340